HIGH-VELOCITY CLOUDS

ASTROPHYSICS AND SPACE SCIENCE LIBRARY

VOLUME 312

HIGH-VELOCITY CLOUDS

Edited by

HUGO VAN WOERDEN
Kapteyn Institute, Groningen, The Netherlands

BART P. WAKKER
University of Wisconsin-Madison, U.S.A.

ULRICH J. SCHWARZ
Kapteyn Institute, Groningen, The Netherlands
and
Universiteit Nijmegen, The Netherlands

and

KLAAS S. DE BOER
Sternwarte, Universität Bonn, Germany

KLUWER ACADEMIC PUBLISHERS
DORDRECHT / BOSTON / LONDON

A C.I.P. Catalogue record for this book is available from the Library of Congress.

ISBN 978-90-481-6685-5 (PB)
ISBN 978-1-4020-2579-2 (e-book)

Published by Kluwer Academic Publishers,
P.O. Box 17, 3300 AA Dordrecht, The Netherlands.

Sold and distributed in North, Central and South America
by Kluwer Academic Publishers,
101 Philip Drive, Norwell, MA 02061, U.S.A.

In all other countries, sold and distributed
by Kluwer Academic Publishers,
P.O. Box 322, 3300 AH Dordrecht, The Netherlands.

Printed on acid-free paper

Contents

PREFACE

On the occasion of the retirement of Ulrich Schwarz, a symposium was held in Groningen in May of 1996, celebrating his contributions to the study of the interstellar medium, including his work on the high-velocity clouds. The coming together of many specialists in the latter field prompted the idea of compiling a book containing their contributions, and summarizing the status of our understanding of the high-velocity cloud phenomenon.

This seemed especially worthwhile at the time, since many exciting developments were taking place. After the discovery of some H I clouds with high velocities, about 40 years ago, the subject had been dominated by 21-cm observations of H I emission. Starting in the mid-1980s much progress was being made because of the availability of new instruments, such as large ground-based optical telescopes and UV observatories in space. The connections between the work on high-velocity clouds and other studies of the properties of the (hot) interstellar medium also became clearer.

Progress in the study of high-velocity clouds has been especially marked in the years between 1998 and 2003, during which time the Leiden-Dwingeloo Survey and *HIPASS* H I surveys became available, many extragalactic high-velocity cloud analogues were found, Hα measurements finally became practical, molecular hydrogen was discovered in IVCs, the first distances and metallicities were obtained, many highly-ionized high-velocity clouds were discovered by observing the O VI absorption lines with *FUSE*, and the model proposing an origin of the clouds in the Local Group was put forward, vigorously discussed, and analyzed in detail. With these new results, the high-velocity clouds moved from the category of peculiar phenomena to a place of significance in the study of the structure and evolution of the Milky Way and of other galaxies.

All authors were quick in their response to our request to write a chapter for this book, but some took a considerable time to complete their contribution (including some of the editors). For this and many other reasons, putting together this book took much longer than anticipated. Meanwhile, many important developments took place, and the editors had to request a complete rewrite of some chapters. We would like to thank the authors of those chapters (8, 9, 11, and 15) for their forbearance, and appreciate their extra effort. In one case (chapter 14) a rewrite turned out to be impractical, and the chapter is still in the form in which it was originally delivered in January 1999.

All authors were asked to (and did) update their texts in late 2003 and 2004, to incorporate the most recent developments in their respective subfields. In the later stages, authors had access to the other chapters, allowing a fair amount of homogenization. Yet, each chapter reflects the author's views and conflicting thoughts on the HVC phenomenon have not been polished away. This makes the present book, the editors hope, an up-to-date account of the understanding of the phenomenon of high-velocity clouds in the Milky Way Halo and elsewhere, and therefore a much more valuable overview than it would have been if it had been finished a few years back, as was originally planned.

The editors like to express their gratitude to all of the authors contributing to this volume. It is their confidence in the final product which made it all work. We also thank the publishers, notably Harry Blom, for their patience and encouragement. A book is easily conceived but the making of it requires foremost endurance.

April 2004

1. HISTORY OF HVC RESEARCH – AN OVERVIEW

BART P. WAKKER
Department of Astronomy, University of Wisconsin-Madison, USA; wakker@astro.wisc.edu
KLAAS S. DE BOER
Sternwarte, Universität Bonn, Germany
deboer@astro.uni-bonn.de
AND
HUGO VAN WOERDEN
Kapteyn Institute, Groningen, The Netherlands
hugo@astro.rug.nl

Abstract. The discovery of gas clouds at high velocity outside the disk of the Milky Way dates from the middle of the 20th century. Since their discovery, numerous new techniques and new instruments have allowed great strides in the understanding of the high-velocity cloud phenomenon. This chapter presents a review of these developments, organized by period (five sections each covering about one decade), and within each period summarizing the progress in each of several subfields, such as radio surveys, UV observations, and theories.

1. Discovery; research until 1963

Studies in the late 1930s had revealed multiple interstellar absorption components in spectra of early-type stars (Beals 1938). In his survey of interstellar absorption lines in 300 stars, Adams (1949) found eight stars at $b>+20°$ with absorption at velocities $|v_{\rm LSR}|>20$ km s^{-1} relative to the Local Standard of Rest (LSR). Then Guido Münch started a program to systematically observe distant stars at high galactic latitude, using the Mt. Wilson 100-inch Coudé spectrograph. First results for the high-latitude star HD 93521 (Münch 1952) revealed four separate clouds, with velocities up to -50 km s^{-1}. The full project data appeared much later (Münch & Zirin 1961), showing many absorption components at velocities substantially different from the 0 km s^{-1} expected from differential galactic ro-

H. van Woerden et al. (eds.), High-Velocity Clouds, 1–24

Figure 1. Lyman Spitzer, Jr. (1914-1997) in 1989.

Figure 2. J.H. Oort (1900-1992) in 1970.

tation for nearby galactic gas. These absorptions were called *high-velocity clouds* (HVCs), which at the time referred to interstellar gas with velocities differing by more than 20 km s^{-1} from the LSR (e.g. Schlüter et al. 1953). A circumstellar origin was suspected for many such lines in the spectra of O and B stars.

In those years, Lyman Spitzer regularly visited Mt. Wilson, and he was generally well informed about the work going on there. As related by Münch (see de Boer 1989), when Spitzer learned of Münch's findings, he surmised that if clouds of apparently neutral gas were present away from the Milky Way disk, a gaseous medium outside the disk should exist to pressure-confine these clouds. However, the confining gas had to be highly ionized, or it would have shown up in metal absorption lines too. Assuming hydrostatic equilibrium, Spitzer concluded that there should be gas with a temperature up to 10^6 K several kpc above the disk. Since very hot gases were at that time known only from the coronas of stars, Spitzer dubbed this gas the *Galactic Corona*, and he described the essence of this model in a very influential paper (Spitzer 1956). In it, he speculated that with further development of satellite techniques (Sputnik would be launched in 1957) high-resolution spectra in the ultraviolet might reveal the presence of absorption lines of Si IV, C IV, N V, and O VI. The physics of the Galactic Corona and questions about the required observations to study those gases showed up in Spitzer's class exercises for many years thereafter (Savage, priv. comm.). Exactly 40 years after the first paper on the Galactic Corona, Spitzer (1996) summarized his views of all the developments up to then.

2. Early radio research: 1963–1968

The first survey of Galactic H I was done from 1952 to 1955 with the 7.5-meter Kootwijk telescope in the Netherlands, complemented by data for the southern sky from a 36-ft aerial in Sydney, Australia. The results clearly showed the presence of differential galactic rotation and spiral arms (Oort et al. 1958).

Following the prediction by Spitzer (1956) that there should be a Galactic Corona, Oort suggested to van Woerden that such a corona might contain neutral hydrogen with velocities up to 100 km s^{-1}, and that such hydrogen might replenish the gas seen to be expanding away from the Galactic Center (van Woerden et al. 1957). Early searches were started by Raimond in 1958, using the then just 2-year old 25-m *Dwingeloo* telescope (shown in Fig. 3). These were unsuccessful, because of insufficient sensitivity (a brightness temperature detection limit of ~2 K). A few years later, however, new receiver technology led to several detections of H I with velocities

Figure 3. The 25-m Dwingeloo telescope in 1973. Photo: ASTRON.

of -110 to -175 km s^{-1} (Muller et al. 1963). These HVCs are now known as parts of core C I, and cores A IV and B. The paper was published in a relatively obscure journal (the Comptes rendus de l'Académie des Sciences de Paris), because of its very short publication time (a few weeks). A reproduction of this discovery paper is shown in Fig. 4.

C. R. Acad. Sc. Paris, t. 257, p. 1661-1662 (2 septembre 1963). Groupe 3.

RADIOASTRONOMIE. — *Hydrogène neutre dans la couronne galactique ?* Note de MM. CHRISTIAAN ALEXANDER MULLER, JAN HENDRIK OORT et ERNST RAIMOND.

En utilisant un amplificateur paramétrique quasi-dégénéré en tête de notre récepteur à huit canaux et du radiotélescope de 25 m de Dwingeloo ([1]), Pays-Bas, nous avons observé des profils de la raie 21 cm s'étendant de — 175 km/s jusqu'à + 175 km/s de vitesse radiale. Ces profils ont été relevés tous les 5 degrés en longitude et aux latitudes de — 40 et + 40°. La bande passante utilisée est de 20 kc/s (4,2 km/s).

En quatre endroits, des nuages d'hydrogène neutre ayant des vitesses supérieures à 100 km/s ont été détectés. On trouvera dans le tableau ci-dessous les coordonnées, la vitesse radiale moyenne, la demi-largeur et la température de brillance des maximums observés. Chaque maximum a été déterminé par deux mesures indépendantes. La vitesse radiale est mesurée par rapport au système local de référence, qui est défini par le mouvement moyen du gaz interstellaire dans le voisinage du Soleil. L'unité de température de brillance est approximativement 1°K. La dernière colonne du tableau donne le nombre d'atomes d'hydrogène présents dans une colonne de 1 cm² de section.

l^{II} (degrés).	b^{II} (degrés).	l^{I} (degrés).	b^{I} (degrés).	V_r (km/s).	W (km/s).	T_b (unités).	N (cm⁻²).
83,42	+40,68	50	+40	—116	27	1	5.10^{19}
88,47	+40,56	55	+40	—116	23	1	4 »
153,20	+38,99	120	+40	—174	25	4	20 »
167,95	+38,74	135	+40	—152	25	1,5	7 »

Le diamètre à demi-intensité du nuage à $l^{I} = 120°$ est de 2°. Les dimensions angulaires des autres objets n'ont pu encore être déterminées. Pour l'instant, on peut avancer que le rayonnement observé à $l^{I} = 50°$ et $l^{I} = 55°$ est probablement émis par un même nuage. L'objet observé à $l^{I} = 135°$ n'a pu être détecté aux longitudes 130 et 140. Par conséquent, sa dimension angulaire est limitée à moins de 10° en longitude.

Dans quelques semaines, des observations permettant de déterminer les dimensions seront disponibles. Un programme d'observation pour la recherche d'objets à grande vitesse dans d'autres parties du ciel est également entrepris.

Il n'y a actuellement pas de possibilité de mesurer les distances. Si les nuages étaient des objets proches et de courte vie situés près du plan galactique, leur nombre serait étonnamment élevé. Par conséquent, nous tendons à conclure que les objets observés sont situés dans la couronne de notre galaxie. Toutefois, la grande dispersion de vitesse observée dans les nuages est un phénomène intriguant.

(2)

En plus des résultats donnés ci-dessus, les profils montrent plusieurs maximums et de longues ailes indiquant des vitesses allant jusqu'à ± 70 km/s. Certains de ceux-ci s'étendent en longitude sur des dizaines de degrés, si bien qu'ils sont probablement peu distants. Une étude de ces particularités est entreprise par les observatoires de Groningue et Leyde.

(1) H. VAN WOERDEN, K. TAKAKUBO et L. L. E. BRAES, *B. A. N.*, 16, 1962, p. 323.

(Observatoires de Dwingeloo et de Leyde, Pays-Bas.)

Figure 4. Reproduction of the paper announcing the discovery of high-velocity H I (Muller et al. 1963). Note that the table shows positions in both old (l^I,b^I) and new (l^{II},b^{II}) galactic coordinates. The survey was done at five-degree steps in longitude at old-style latitude b^I=40°. Print supplied by E. Raimond.

Not long thereafter, Smith (1963) found a cloud with v=+90 km s^{-1} at (l,b)=(40°, −15°), currently also known as complex GCP (see Wakker & van Woerden (1991) for current HVC nomenclature), while Dieter (1964) found high-velocity gas near the South Galactic Pole, using the *Hat Creek* telescope. In the years after the discovery, astronomers from Leiden and Groningen used the *Dwingeloo* telescope to make a systematic search for high-velocity H I on a 10°×10° grid at latitudes >+20°. During the course of this survey, gas at velocities between −50 and −100 km s^{-1} was also found in many directions.

A division of labor was decided upon. Leiden astronomers would analyze the high-velocity gas ($v_{\rm LSR}$<−80 km s^{-1}), while those at Groningen would concentrate on the "intermediate-velocity clouds" (IVCs). This work led to a classic series of papers in the BAN (Blaauw & Tolbert 1966, Hulsbosch & Raimond 1966, Oort 1966), with Oort discussing possible origins of the HVCs: (a) nearby supernova remnants (he considered this unlikely); (b) condensations formed in a gaseous corona at high temperature; (c) clouds ejected from the Galactic Nucleus; (d) clouds ejected as cool clouds from the disk; (e) intergalactic gas accreted by the Galaxy; (f) small satellites of the Galaxy or independent galaxies in the Local Group. With the addition of (g) material tidally extracted from the Magellanic Clouds, this is still a good summary of the possible origins of HVCs, forty years later. In 1966, high-velocity gas found a place in the program of IAU Symposium 31, "Radio Astronomy and the Galactic System" (van Woerden 1967).

3. Developments 1968–1980

Following the discovery and early surveys, the decade from 1968 to 1980 saw much effort devoted to radio observations of the high-velocity H I. Little effort was put into the IVCs; Wesselius (1973) and Wesselius & Fejes (1973) completed the analysis of a *Dwingeloo* survey of the northern sky made by Tolbert (1971), but no further progress was made until the mid-1990s. On the other hand, surveys for the HVCs became ever more complete and many individual clouds were mapped in detail, while several papers put forward arguments for variants of Oort's suggested origins.

A small but steady stream of papers on HVCs kept appearing in these years (on average 6 a year), including a paper in the Annual Reviews (Verschuur 1975). In 1973 and 1978 there were special sessions on HVCs during IAU Symposia – No. 60 on Galactic Radio Astronomy (Kerr & Simonson 1974) and No. 84 on The Large-Scale Characteristics of the Galaxy (Burton 1979). In general, however, HVCs were only studied by a small number of astronomers, and they were considered something of a curiosity.

3.1. H I SURVEYS

Following the early surveys, sky coverage was quickly improved to a $5° \times 5°$ grid for declinations $>-35°$, leading to the first thesis on HVCs (Hulsbosch 1973). Over the next few years, many new clouds were discovered. Using the *Parkes* 60-ft, Mathewson et al. (1974) showed that the gas found near the South Galactic Pole by Dieter (1964) was part of a chain of clouds starting from the Magellanic Clouds, henceforth called the Magellanic Stream. Mathewson (1967) found a cloud at l=165°, b=+60°, which was named cloud "M" after him. Van Kuilenburg (1972a, b), exploring the southern galactic hemisphere, discovered gas with velocities <-150 km s^{-1} near the Anti-Center. Wannier et al. (1972) discovered widespread, faint positive-velocity gas in the region l=250° to 320°, b=+10° to +30°. By then the number of known HVCs had grown to the point that a scheme was adopted in which each cloud was given a name based on its galactic longitude, latitude and LSR velocity, e.g. HVC 165+55−150. Thus, by 1973 a fairly complete, but rough picture of the high-velocity sky existed, with varying sensitivity and grid in different parts of the sky.

Using a much improved receiver system at the *Dwingeloo* telescope, Hulsbosch (1978) mapped part of the Anti-Center region in detail, on a $1° \times 1°$ grid, with a 36′ beam. The difference with earlier surveys was that a much larger velocity range was searched (−500 to +500 km s^{-1}), although at a velocity resolution of just 16 km s^{-1}, and that 15-minute integrations were used, giving a 5-σ detection limit of 0.05 K, or $\sim 2 \times 10^{18}$ cm^{-2} for a typical 20 km s^{-1} wide line. This was a factor ~5 better sensitivity than any previous survey, a factor 2 to 3 broader velocity coverage, and a factor 4 greater sky coverage. Many new HVCs were discovered, including some with very high negative velocities (<-300 km s^{-1}), which were named the

VHVCs. At that time it was decided to extend this survey to almost the whole sky observable from Dwingeloo (declinations $>-17\overset{\circ}{.}2$). Because of limitations in computing power and memory (the data were reduced on a PDP 11/70 with 32 kbyte of RAM and 1 Mbyte disks), only the velocity and peak brightness temperature of each identified high-velocity profile component were saved. By 1980 observations were progressing rapidly.

In the same period, an unbiased survey of most of the northern sky (declinations -18 to $+55°$) was carried out by Giovanelli (1980, 1981), using the 300-ft *Green Bank* telescope ($9\overset{\prime}{.}7$ beam) on a $1°\times2°$ grid in right ascension and declination, with a velocity range of -900 to $+900$ km s^{-1}. This survey established that a) HVCs are only found at LSR velocities between -450 and $+350$ km s^{-1}, b) there are two populations of HVCs (the large complexes and the VHVCs, which stand out in a longitude-velocity plot), and c) a pronounced north-south asymmetry is present in the distribution of the HVCs. Giovanelli (1981) also argued that the angular and velocity distribution of the VHVCs is incompatible with that of Local Group galaxies, as well as incompatible with a location in the Galactic Disk. He concluded that they are likely either shreds of the Magellanic Stream or other objects orbiting the Galaxy.

3.2. DETAILED MAPPING

The period 1968 to 1980 also saw many papers with maps and analyses of individual clouds, at angular and velocity resolutions much higher than those of the surveys. One of the more important of these was the study by Giovanelli et al. (1973), who used the *Green Bank* 300-ft to map large parts of clouds A, C, M, and HVC 131+1−200 (currently also known as complex H) at intermediate ($10'$) angular resolution, showing that much structure is present, and discovering multiple cores, which they named A I, A II, A III, A IV, A V, C IA, C IB, C IC, C IIIA, C IIIB, C IIIC, M I, M II, M III. [Core C II is at lower velocities (~-80 km s^{-1}), and at the present time is considered part of the IV Arch, specifically core IV 8.]

From these and other maps with $10'$ beams and 1 km s^{-1} velocity resolution, covering degree-scales, several authors concluded that the brightest HVCs show a core-halo structure (Davies et al. 1976; Giovanelli & Haynes 1977). This was interpreted as evidence for the presence of the standard two phases of the ISM – a cold phase, with temperature $\sim$100 K and density $\sim$1 cm^{-3}, and a warm phase, with temperature $\sim10^4$ K and density $\sim$0.01 cm^{-3} (Field et al. 1969). Both of these would then be confined by the external pressure of the hot Galactic Corona proposed by Spitzer.

High-resolution maps of the cores made with interferometers started appearing in the mid-1970s (Greisen & Cram 1976, Schwarz et al. 1976;

Schwarz & Oort 1981). Especially the maps made with the new Westerbork Synthesis Radio Telescope (*WSRT*) showed the presence of much small-scale structure, down to the $1'$ resolution.

3.3. DISTANCES AND ABUNDANCES

Hypotheses for the origin of HVCs proliferated in this decade (see Sect. 3.5). Most of the proposals provided reasonable fits to some parts of the clouds' sky and velocity distributions, even though these models were very different. Predicted distances ranged from a few kpc to several hundreds of kpc. Discriminating between the models clearly required HVC distances and metallicities, which were completely lacking. That distances and metallicities held the key to understanding HVCs had been clear since their discovery. It had also been clear that deriving distances required high-resolution absorption-line spectra of faint background stellar targets with known distances. An early attempt was published just a few years after the discovery of HVCs (Prata & Wallerstein 1967). This study concentrated on the Ca II K line, which together with Na I D is the only useful line in the optical. Unfortunately, both these interstellar lines tend to be weak in HVCs.

The early failure to detect absorption from HVCs, together with their possible intergalactic origin, led some authors to suggest that they are primordial (Z=0) gas. Due to a lack of known good background probes, the absorption-line studies languished for many years, with no positive results (detections), nor significant negative results (non-detections giving limits).

3.4. UV AND X-RAY DATA; THE "GALACTIC FOUNTAIN" CONCEPT

After the 1956 paper, Spitzer pushed to get access to the UV part of the spectrum in order to observe the highly-ionized gas in the predicted Galactic Corona. This led to the development of the *Copernicus* satellite. Launched in 1973, it gave access to the 900 to 3000 Å spectral range, which includes the O VI, N V and C IV absorption lines. However, the tracking capability of that satellite was limited to stars brighter than $V \simeq 6.5$ mag, so that spectra could be obtained essentially only for nearby galactic stars. These showed the presence of O VI with $v_{\rm LSR} \sim 0$ km s^{-1} in all directions (Jenkins & Meloy 1974). One of the more distant stars observed, HD 93521, indicated the presence of O VI in the halo, at $v_{\rm LSR} = -50$ km s^{-1} (Jenkins 1978).

The 1970s also saw important developments in X-ray astronomy. First scans with rocket-borne X-ray detectors showed the presence of considerable diffuse X-ray emission (Williamson et al. 1974).

Together, the widespread O VI and diffuse X-rays led to the need to include the hot (10^6 K) phase in the models of the ISM in the Disk. This hot phase was thought to be produced by supernova explosions, leading to outflows from the disk. Once in the halo the hot gas could cool, condense, and might fall back to the Milky Way disk as HVCs, establishing a "Galactic Fountain" (Shapiro & Field 1976). In a first theoretical exploration of the kinematics, the estimated back-flow velocities were compared with observed HVC velocities and appeared to be largely in agreement (Bregman 1980).

The 1978 launch of the International Ultraviolet Explorer (*IUE*) satellite with its echelle spectrographs enabled measurements of the absorption lines of the high ions C IV and N V over long sight lines for large numbers of stars. Fe II, Mg II, etc. absorption was detected in an HVC at $+120$ km s^{-1} against background stars in the LMC. Interestingly, this HVC also showed C IV absorption (Savage & de Boer 1979). Spitzer et al. (1980) concluded that this was the first sign of the highly-ionized halo gas he had proposed in 1956. A first determination of metal column densities of the HVCs on the LMC sight line showed the metal content of the gas to be relatively high, but the weakness of the 21-cm emission prevented giving hard values (Savage & de Boer 1981). The z-extent of the C IV gas was found to be several kpc. Further, the Magellanic Clouds also appeared to possess an envelope of coronal gas (de Boer & Savage 1980; de Boer 1984).

3.5. HYPOTHESES FOR THE ORIGIN OF HVCS

After Oort's (1966) original summary of possible origins for the HVCs (see Sect. 2), many new ideas were put forward. These tended to concentrate on explaining complexes A and C. Wakker & van Woerden (1997) summarized and criticized each of these proposals, counting eighteen different models. The most well received of these were the following.

(a) The idea of Oort (1970) that the HVCs represent the last stage of condensations in a hot Galactic Halo, left over from the formation of the Milky Way, and now at distances $\sim$1 kpc above the plane. This was worked out in more detail in the paper by Oort & Hulsbosch (1978), in which they predicted a z-height of 1 kpc and a metallicity of 0.7 solar for HVC complex A.

(b) The proposals by Verschuur (1973) and Davies (1974) that some HVCs are part of the outer galactic spiral structure; this related complexes A, C and the Outer Arm. This model was worked out in detail in the review paper on HVCs by Verschuur (1975).

(c) The analysis by Bregman (1980), which views the HVCs as the return flow of a Galactic Fountain as proposed by Shapiro & Field (1976), at distances of a few kpc above the plane; this was considered very attractive because it provided a possible physical mechanism for generating HVCs.

(d) The proposal by Mirabel (1982) that the high-negative-velocity clouds in the general direction of the Galactic Center and Anti-Center represent infall to the Milky Way.

(e) The Magellanic Stream was explained as a tidal bridge and tail of gas pulled out from the Magellanic Clouds, although there were many competing ideas about the details of this process. Around 1980, the debate was about whether the Stream is a leading or a trailing feature; the most notable papers presenting tidal models being Fujimoto & Sofue (1976), Murai & Fujimoto (1980) and Lin & Lynden-Bell (1982).

4. Developments 1981–1991

This decade saw the culmination of the earlier radio efforts, which would pave the way for the work in the decade thereafter. Specifically, the completion of Hulsbosch's survey of the northern sky, and the high-angular-resolution observations of Schwarz & Oort (1981) and Wakker & Schwarz (1991). Optical and UV absorption-line studies of background targets of HVCs and of the hot Galactic Corona started to bear their first fruits. Most of the proposed models remained alive, though some were oversold and others remained underappreciated. The number of refereed papers concerning models or observations of the HVCs stayed about constant at an average of 7 per year. Four HVC-specific sessions were included in conferences – at the 1983 IAU Symposium 106 (The Milky Way Galaxy) in Groningen (van Woerden et al. 1985), in the 1985 NRAO workshop The Gaseous Galactic Halo, held in Green Bank (Bregman & Lockman 1986), at the 1989 IAU Colloquium 120, Structure and Dynamics of the Interstellar Medium, held in Granada (Tenorio-Tagle et al. 1989) and in IAU Symposium 144, The Interstellar Disk-Halo Connection in Galaxies, held at Leiden in 1990 (Bloemen 1991). The proceedings of the last of these gives a useful summary of the state of HVC research in the late eighties.

4.1. SURVEYS

The observations for Hulsbosch's *Dwingeloo* HVC survey (see Sect. 3.1) were finished in 1982, but several more years of mopping up passed, and Hulsbosch left astronomy. Wakker then prepared this survey for publication, resulting in the Hulsbosch & Wakker (1988) catalog. As the *Dwingeloo* survey was nearing completion, Argentinian astronomers used the 30-m dish at Villa Elisa for the first systematic HVC survey of the southern sky, with parameters chosen to match the northern survey, albeit on a $2°\times2°$ grid (Bajaja et al. 1985). They followed this by more detailed mapping of individual clouds (Bajaja et al. 1989). Combining the northern and southern data, Wakker & van Woerden (1991) produced the first all-sky HVC catalog. In this catalog they decided to name the HVC complexes and number the clouds, rather than using the l,b,v notation, since for many clouds and complexes it is not clear whether one should use the brightest core, the average for the cores, a position-weighted average or an intensity-weighted average to determine the l,b,v name. Thus, HVC complexes D, G, H, L, WA, WB, WC, WD, GCN and GCP came into existence. "D" and "L" were named after the constellations Draco and Libra, in which they lie. "H" was named after Hulsbosch, "G" is near the galactic plane and close to "H", "GCN" and "GCP" are near the Galactic Center. The "WA" through "WD" clouds are the positive-velocity complexes discovered by Wannier, Wrixon & Wilson (1972) – the author names of the discovery and naming papers all start with "W". Each cloud was also given a "WW" number in the catalog by Wakker & van Woerden (1991).

The *Dwingeloo* survey also led Wakker (1991) to propose a new definition for the HVCs, based on the "deviation velocity". This is the difference (negative or positive) between the observed velocity and the maximum radial velocity that can be understood from a simple model of galactic rotation. Defining HVCs in this manner allows one to eliminate clouds with $|v_{\rm LSR}|$>100 km s^{-1} near the galactic plane, while at the same time including clouds with $|v_{\rm LSR}|$~90 km s^{-1} at high galactic latitude.

4.2. DETAILED MAPPING

The results that Schwarz & Oort (1981) obtained by mapping core A I at 1′ and 1 km s^{-1} resolution using the *WSRT* led to the hope that such maps contained the key to understanding the internal physics of the clouds. As a result, a number of cores and small clouds were mapped in detail, which became the basis for Wakker's (1990) Ph.D. thesis. However, deriving physical parameters proved difficult, especially since the estimates depend on the assumed distance. Nevertheless, it was shown that the brighter cores have a hierarchical structure, and that column density variations of up to a factor a few exist on arcminute scales. Wakker & Schwarz (1991) also considered that the velocity field for two small VHVCs mimicked the rotation of a self-gravitating cloud. They rejected this as an explanation because it implied distances of several 100 kpc, and an average H I volume density of 10^{-3} cm^{-3}, small enough to expect that the clouds would be mostly ionized, in which case the calculations are no longer valid.

One important finding from the high-angular-resolution work was that HVC cores show structure down to (and below) 1 arcminute scales, implying that accurate determinations of the H I column density in the direction of background probes require a high-resolution H I map.

4.3. X-RAY OBSERVATIONS OF THE GALACTIC HALO

Following the earlier discovery of diffuse X-ray emission, further X-ray measurements by the Wisconsin group resulted in the proposal that the Sun was embedded in an X-ray-emitting cavity, the Local Hot Bubble. Others maintained that the emission also contained a component associated with the hot fountain-flow gas in the Halo. Thus the quest was on to find shadows of HVCs against the X-ray-emitting hot background gas. Such shadows were indeed found in data from the X-ray satellite *ROSAT* launched in 1990. The group in Bonn found the shadow of the Draco Cloud (Burrows & Mendenhall 1991; Snowden et al. 1991) and of a cloud in Hercules (Lilienthal et al. 1992). However, both were shadows of lower-velocity clouds closer to the disk.

Two competing models for the distribution of the X-ray-emitting gas were debated: 1) a Local Hot Bubble plus an extragalactic background was favored by the Wisconsin group (see review by McCammon & Sanders 1990); 2) a multi-phase model of hot and cool pockets on all lines of sight, with the halo being predominantly hot and radiating, was favored by others (see Hirth et al. 1991). The competition for data and data rights of the *ROSAT* All Sky Survey, with which both models could be tested, was fierce.

4.4. DEVELOPMENTS IN THE ULTRAVIOLET

With the *IUE*, launched in 1978, numerous spectra of distant stars were obtained, showing absorption by C II, Mg II, and Fe II, etc. on halo sight lines (e.g. Savage & de Boer 1979, 1981; Pettini & West 1982). Savage & Massa (1987) detected substantial amounts of N V on halo sight lines in the direction of the Milky Way center. First attempts to measure extragalactic sources showed absorption at halo velocities (3C 273; York et al. 1983). The explosion of SN 1987A in the LMC provided the light to make in-depth studies of the gas in front of the LMC (de Boer et al. 1987). In particular the abundance of the metals (Blades et al. 1988) and the optical-depth profiles, spatial distribution, and nature of the Al III, Si IV, and C IV-containing gas (Savage et al. 1989) were investigated.

4.5. DISTANCES AND ABUNDANCES

In 1981 Ca II absorption was finally detected in an HVC – in the Magellanic Stream, in the direction of the AGN Fairall 9 (Songaila & York 1981; Songaila 1981). Songaila et al. (1985, 1986) claimed detections of complex A and C, which were later shown to be in error (Lilienthal et al. 1990).

By 1990 the Ca II K line had been detected in a few directions: toward PKS 0837−12 in complex WB (Robertson et al. 1991), and toward

NGC 3783 in HVC 287+22+240 (or WW 187) (West et al. 1985). However, HVC distances and metallicities still remained basically unknown. By this time it was clear that progress required observations of stars and AGNs fainter than about V=15 at 15 km s^{-1} resolution or better, which was just beyond the reach of most telescopes.

Without knowing distances, it is difficult to properly understand the distribution of HVC radial velocities. However, by using the statistics some progress can be made. An observational indication that halo gas rotates slower than the Milky Way disk came from the velocity of absorption lines found toward halo globular clusters (de Boer & Savage 1983, 1984). Combining this with the Center vs Anti-Center velocity asymmetry in Giovanelli's (1980) data, Kaelble et al. (1985) found that inflow (both vertical and radial) as well as rotation slightly slower than that in the disk are characteristics of the kinematics of most HVCs.

Progress was also made in understanding the Galactic Corona. Sembach & Savage (1992) analyzed the distribution of C IV and N V absorption, based on *IUE* data. Danly (1989) and Danly et al. (1992) studied the intermediate-velocity clouds with *IUE*, finding that the velocity range of the absorption appeared to increase with increasing distance from the plane, but only in the northern sky. However, these studies required bright, relatively nearby stars, often not projected onto H I HVCs, so they did not provide direct data on HVC distances.

4.6. HYPOTHESES FOR THE ORIGIN OF HVCS

Little progress was made in the theoretical understanding of HVCs in this decade, although the Galactic Fountain model started gaining favor. New models were proposed (e.g. HVCs as a polar ring – Haud 1988), but none of these has stood the test of time (see Ch. 13 for the current status of the polar ring idea).

5. Developments 1992–1999

During the nineties, HVC research came of age. At the beginning of the decade the old situation still prevailed, in which a small number of afficionados made slow progress on various aspects of the phenomenon. By 1999 the HVCs had become a hot topic that commanded widespread interest. The number of refereed HVC papers went up to an average of 10 a year, and by the end of the decade, the first separate HVC conference had taken place, in Canberra, Australia (Gibson & Putman 1999). The first distances and metallicities had been measured, making it possible to link individual clouds with some of the models proposed in the previous three decades, and indicating that examples might be found for several of these origins. HVC analogues were starting to be discovered in other galaxies, while Galactic HVCs were detected in Hα emission. Finally, Blitz et al. (1996) proposed that most (if not all) HVCs contained 10 times more dark matter than H I, and that they were the remnants of the formation of the Local Group, thus enlarging the context in which HVCs were to be understood.

5.1. SURVEYS

In the nineties, there were few 21-cm studies specifically aimed at observing HVCs. Nevertheless, four major developments took place. First, Kuntz & Danly (1996) used the Bell Labs Survey (Stark et al. 1992; declinations $>-40°$, 3° angular and 5 km s^{-1} velocity resolution) to create the first IVC catalog, after 20 years finally making progress on IVCs.

Second, in this timespan a new survey was done using a yet-again upgraded *Dwingeloo* telescope. By 1997 this produced what is now known as the Leiden-Dwingeloo Survey (*LDS*; Hartmann & Burton 1997), covering the whole sky visible from Dwingeloo (declination $>-30°$) on a $0\overset{\circ}{.}5\times0\overset{\circ}{.}5$ grid, with 1 km s^{-1} velocity resolution and 0.07 K rms. An important improvement over previous surveys is that all the *LDS* data have been corrected for stray radiation. This survey contains most (~90%) of the detections in the earlier Hulsbosch & Wakker (1988) data, but for the first time also comparable data for the IVCs. A proper analysis of the HVC and IVC information in this dataset is still lacking.

Third, a southern complement to the *LDS* was done in Argentina, using the *Villa Elisa* telescope (Arnal et al. 2000); the HVCs in this survey were listed by Morras et al. (2000). A global analysis of the new surveys is given in Ch. 2. The current status of research on IVCs is presented in Ch. 4.

Fourth, the *HIPASS* (H I Parkes All Sky Survey) was completed (Barnes et al. 2001). This survey aimed at detecting galaxies out to several tens of Mpc, but it also allowed the mapping of Galactic HVCs at declinations $<0°$ at 17′ angular resolution (and full angular coverage), though only at 26 km s^{-1} velocity resolution. Putman et al. (2002) used these data to produce the most detailed map yet of the Magellanic Stream, including the first clear H I map of the Leading Arm (Putman et al. 1998). Chapter 5 presents these data, and discusses the new results concerning the Stream.

Finally, from studies of the small-scale structure of the HVCs (see Ch. 7) some authors concluded that it is possible to use the morphology observed at high angular resolution to uncover evidence for interactions of the HVCs with their environment. This work is summarized in Ch. 12.

5.2. EXTRAGALACTIC HVCS AT 21 CM

By the end of the eighties, H I interferometers had finally become sensitive enough that searches for HVCs in external galaxies were possible. This search was pushed hard by Sancisi in Groningen. A first check of old data showed the possibilities (Wakker et al. 1989). The first bona-fide extragalactic HVCs were discovered in M 101 by van der Hulst & Sancisi (1988). These objects have velocities of 80 to 150 $\mathrm{km\,s^{-1}}$ relative to the underlying disk, and appear more massive than HVCs in the Milky Way. Several other galaxies were observed in great detail in the ensuing years, and H I with deviating velocities was found to be common. In a series of papers, Schulman et al. (1994, 1996, 1997) observed several galaxies with *Arecibo* and the Very Large Array (*VLA*), concluding that the high-velocity gas seems to be related to the underlying star-formation activity. High-sensitivity interferometer observations also showed many clouds unrelated to disk gas. Chapter 6 presents a review of this work, as well as a discussion of the post-1999 results and the implications for understanding the Galactic HVCs.

5.3. Hα OBSERVATIONS

Songaila et al. (1985) claimed to have detected Hα emission from complex C. However, the detection was marginal, and the paper had little impact (when Tufte et al. (1998) reobserved this direction, they found that the Hα emission is much brighter). Reynolds (1987) reported several non-detections, but Kutyrev & Reynolds (1989) detected faint emission associated with HVC 168−43−280. The next detection of high-velocity Hα was made by Weiner & Williams (1996), for a direction toward the Magellanic Stream. Then, Tufte et al. (1998) found emission from complexes A, C and M, and Bland-Hawthorn et al. (1998) detected complex GCP. These data appeared to hold the promise of determining HVC distances, by assuming that the Hα emission was the result of recombination after photoionization, an idea first proposed for HVCs by Ferrara & Field (1994). However, it has turned out to be more complicated than that, and more observations and theory will be necessary to properly interpret the Hα emission. A review of the current status of this work is given in Ch. 8, while the implications of the Hα detections are discussed in Ch. 16.

5.4. DISTANCES AND ABUNDANCES

The years 1991 to 1998 saw an explosion of absorption-line studies of HVCs, mostly based on data from space observatories: *IUE*, and the Hubble Space Telescope (*HST*). Danly et al. (1993) derived the first upper distance limit (4.5 kpc to cloud M III). Then de Boer et al. (1994) set a lower limit of 2 kpc to the distance of complex C, Wakker et al. (1996) found a lower limit of 4 kpc for complex A, and Wakker et al. (1998) used the *IUE* archive to derive a lower limit of 5 kpc for complex H. Work on IVC distances proceeded mostly with *IUE* (Kuntz & Danly 1996). Lu et al. (1994, 1998) combined data from *HST* and the Australia Telescope Compact Array (*ATCA*) to derive an abundance of 0.25 times solar for a bright high-positive-velocity HVC toward NGC 3783. Together with a new model for the Magellanic Stream (Gardiner & Noguchi 1996), this result allowed to exclude the proposal by Moore & Davis (1994) that the Magellanic Stream is caused by ram-pressure stripping rather than tidal forces.

Ground-based work did not languish either. Many results for IVCs were obtained by the group in Belfast, led by Keenan, using observations of blue halo and globular cluster stars with distances up to a few kpc. Van Woerden, Schwarz, Wakker & Peletier had many runs in the late 1980s and the 1990s with the Isaac Newton and William Herschel telescopes at La Palma, taking spectra of Blue Horizontal Branch (BHB) and RR Lyrae stars. Although a large fraction of the allocated nights were lost to bad weather (since complexes A and C are high in the sky in early spring), in 1997 complex A was detected in absorption toward the RR Lyrae star AD UMa, establishing the first HVC distance bracket (van Woerden et al. 1999). During the same series of observations, the first detections of Ca II in complex C were made against the Seyferts Mrk 290 and PG 1351+640 (Wakker et al. 1996). This result was used to obtain *HST* time, which showed that complex C has a metallicity ~0.1 times solar, firmly establishing that it does not have a Galactic origin (Wakker et al. 1999). Chapter 10 describes these and post-1999 results on distance and abundance determinations in detail.

5.5. PROGRESS IN THE ULTRAVIOLET

The sensitivity of *IUE* was insufficient to observe extragalactic targets at high resolution, except for the bright supernova SN 1993J in M 81. In its spectrum de Boer et al. (1993) found an HVC containing both low-ionization-stage metals (C II, Mg II, etc.) and C IV, at velocities clearly indicating infall onto M 81.

Savage et al. (1997) analyzed high-resolution *HST* observations of C IV and N V toward extragalactic objects, and estimated the scaleheight of the hot gas, deriving ~4 kpc for C IV, and ~2 kpc for N V. This difference shows a complication with the observational interpretation of the highly-ionized atoms in terms of Spitzer's hot halo – some of these ions may be photoionized. The presence of hot gas in the lower halo now is firmly established, however, vindicating Spitzer's idea that neutral halo clouds are confined by a source of external pressure.

The long absence of observing facilities in the far-UV (912 to 1200 Å) after the termination of *Copernicus* was ended in the middle of the decade, when the Space Shuttle experiment named "Orbiting and Retrievable Far and Extreme Ultraviolet Spectrometers" (*ORFEUS*) flew twice, containing the Berkeley medium-, the Tübingen-Heidelberg high-, and the Princeton very-high-resolution spectrographs. Spectra of a fair number of halo stars suggested a scale height for O VI gas on the order of 5 kpc (Widmann et al. 1998); Savage et al. (2003) later derived 2.3 kpc. H_2 was also discovered in the HVC projected against the LMC (Richter et al. 1999).

6. New developments since 1999

6.1. THE LOCAL GROUP HYPOTHESIS

Broad interest in HVCs was generated by the proposal by Blitz et al. (1996, 1999) that HVCs represent the missing population of dark-matter halos in the Local Group (see Ch. 14 for a full description). Although few people familiar with the details of the HVC phenomenon accepted this idea, it generated a lot of work in trying to understand HVC metallicities, mapping HVC Hα emission and comparing the gas content of the Local Group with that of other galaxy groups (see Chaps. 2, 6, 8, 15). The original proposal by Blitz et al. (1999) seems untenable, but a variant was proposed by Braun & Burton (1999), who singled out the small HVCs as possible Local Group objects, calling these objects "Compact HVCs" (CHVCs). An extensive study of these CHVCs formed the basis of another HVC thesis, published as a series of papers (de Heij et al. 2002a, b, c) (see also Ch. 15). Note, however, that the larger HVCs may just be nearby examples of the same population, as Blitz et al. (1999) proposed.

6.2. RE-OPENING THE FAR-ULTRAVIOLET

The launch of the Far Ultraviolet Spectroscopic Explorer (*FUSE*) occurred in June of 1999. With it, Spitzer's hot Corona is finally being studied in detail. *FUSE* (shown in Fig. 5) also has revolutionized the study of HVCs, providing many measurements of metallicities, and showing the ubiquity of the highly-ionized HVCs, which were previously only known from a few directions (Sembach et al. 1999). Highlights of the discoveries made with *FUSE* are summarized in Chaps. 9, 10 and 11. The papers based on *FUSE* and *HST* data by Murphy et al. (2000), Gibson et al. (2001), Richter et al. (2001), Collins et al. (2003) and Tripp et al. (2003) list measurements of many ions for complex C. *FUSE* also found H_2 in some HVCs and many IVCs (Richter et al. 2003; Ch. 9).

The O VI absorption associated with the highly-ionized HVCs was an-

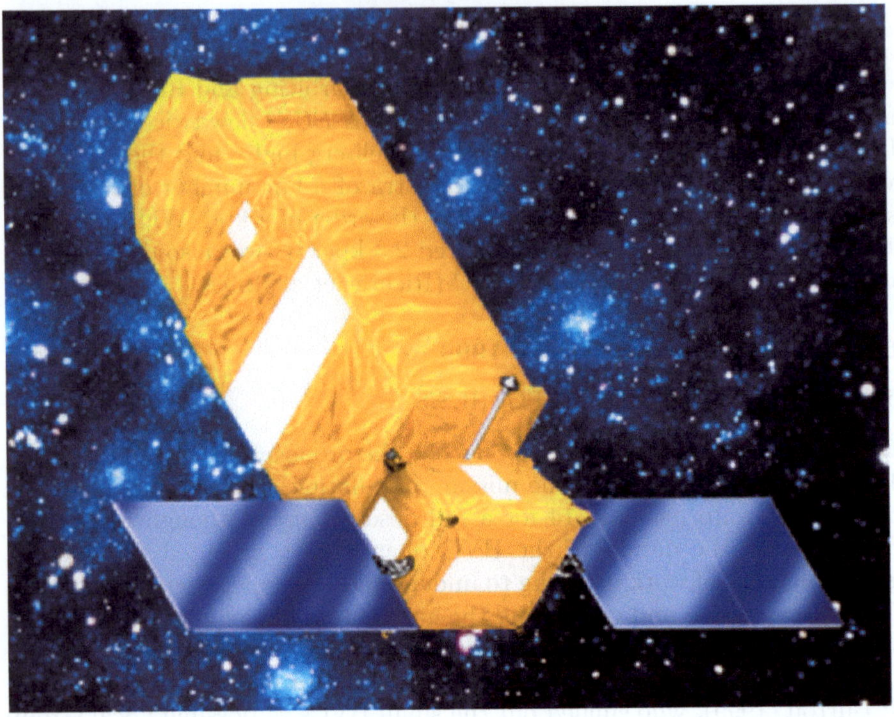

Figure 5. Artist impression of *FUSE* in orbit. Available on the *FUSE* web site http://fuse.pha.jhu.edu/Photos/pub_quality/bestof_fuse.html. Graphic by Orbital Sciences Corp.

alyzed by Wakker et al. (2003), Sembach et al. (2003) and Savage et al. (2003) in a special issue of the ApJ Supplement (see summary in Ch. 11). The properties of the O VI indicate the presence of hot gas surrounding complex C (Fox et al. 2004) and the Magellanic Stream, suggesting a new kind of Galactic Corona, one much larger (out to at least 50 kpc) than envisioned by Spitzer. In many directions high-velocity O VI absorption without associated high-velocity H I is found, which might represent hot gas in the Local Group. More study of this new kind of HVC is underway.

Many questions about the origin and properties of the HVCs remain open. Chapters 13 and 16 discuss the evidence for some of the proposed origins. Chapter 17 presents directions for future work.

Acknowledgements

B.P. Wakker would like to acknowledge financial support from NASA LTSA grant NAG5-9179.

References

Adams, W.S. 1949, ApJ, 109, 354
Arnal, E.M., Bajaja, E., Larrarte, J.J., Morras, R., Pöppel, W.G.L. 2000, A&AS, 142, 35
Bajaja, E., Cappa de Nicolau, C.E., Cersosimo, J.C., Martin, M.C., Loiseau, N., Morras, R., Olano, C.A., Pöppel, W.G.L. 1985, ApJS, 58, 143
Bajaja, E., Cappa de Nicolau, C.E., Martin, M.C., Morras, R., Olano, C.A. 1989, A&AS, 78, 345
Barnes, D.G., Staveley-Smith, L., de Blok, W.J.G., et 36 al. 2001, MNRAS, 322, 486
Beals, C.S. 1938, ApJ, 87, 568
Blaauw, A., Tolbert, C.R. 1966, BAN, 18, 405
Blades, J.C., Wheatley, J.M., Panagia, N., Grewing, M., Pettini, M., Wamsteker, W. 1988, ApJ, 334, 308
Bland-Hawthorn, J., Veilleux, S., Cecil, G.N., Putman, M.E., Gibson, B.K., Maloney, P.R. 1998, MNRAS, 299, 611
Blitz, L., Spergel, D.N,, Teuben, P.J., Hartmann, D., Burton, W.B. 1996, BAAS, 189, 6101
Blitz, L., Spergel, D.N., Teuben, P.J., Hartmann, D., Burton, W.B. 1999, ApJ, 514, 818
Bloemen, H., ed. 1991, IAU Symp. 144, The Interstellar Disk-Halo Connection (Dordrecht: Reidel)
Braun, R., Burton, W.B. 1999, A&A, 341, 437
Bregman, J.N. 1980, ApJ, 236, 577
Bregman, J.N., Lockman, F.J., eds. 1986, Gaseous Halos of Galaxies, Proc. Green Bank workshop No. 12 (Green Bank: NRAO)
Burrows, D.N., Mendenhall, J.A. 1991, Nature, 351, 629
Burton, W.B., ed. 1979, IAU Symp. 84, The Large-Scale Characteristics of the Galaxy (Dordrecht: Reidel)
Collins, J.A., Shull, J.M., Giroux, M.L. 2003, ApJ, 585, 336
Danly, L. 1989, ApJ, 342, 785
Danly, L., Lockman, F.J., Meade, M.R., Savage, B.D. 1992, ApJS, 81, 125
Danly, L., Albert, C.E., Kuntz, K.D. 1993, ApJ, 416, L29
Davies, R.D. 1974, in IAU Symp. 60, Galactic Radio Astronomy, eds. F.J. Kerr, S.C. Simonson (Dordrecht: Reidel), 599
Davies, R.D., Buhl, D., Jafalla, J. 1976, A&AS, 23, 181
de Boer, K.S. 1984, in IAU Symp. 108, Structure and Evolution of the Magellanic Clouds, eds. S. van den Bergh, K.S. de Boer (Dordrecht: Reidel), 375
de Boer, K.S. 1989, in IAU Coll. 120, Structure and Dynamics of the Interstellar Medium, eds. G. Tenorio-Tagle, M. Moles, J. Melnick (Berlin: Springer), 432
de Boer, K.S., Savage, B.D. 1980, ApJ, 238, 86
de Boer, K.S., Savage, B.D. 1983, ApJ, 265, 210
de Boer, K.S., Savage, B.D. 1984, A&A, 136, L7
de Boer, K.S., Grewing, M., Richtler, T., Wamsteker, W., Gry, C., Panagia, N. 1987, A&A, 177, L37
de Boer, K.S., Rodriguez Pascual, P., Wamsteker, W., Sonneborn, G., Fransson, C., Bomans, D.J., Kirshner, R.P. 1993, A&A, 280, L15
de Boer, K.S., Altan, A., Bomans, D., Lilienthal, D., Moehler, S., van Woerden, H., Wakker, B.P., Bregman, J.N. 1994, A&A, 286, 925
de Heij, V., Braun, R., Burton, W.B. 2002a, A&A, 391, 67
de Heij, V., Braun, R., Burton, W.B. 2002b, A&A, 391, 159
de Heij, V., Braun, R., Burton, W.B. 2002c, A&A, 391, 417
Dieter, N.H. 1964, AJ, 69, 288
Ferrara, A., Field, G.B. 1994, ApJ, 423, 665
Field, G.B., Goldsmith, D.W., Habing, H.J. 1969, ApJ, 155, 149
Fox, A.J., Savage, B.D., Wakker, B.P., Richter, P., Sembach, K.R., Tripp, T.M. 2004, ApJ, 602, 738

Fujimoto, M., Sofue, Y. 1976, A&A, 47, 263
Gardiner, L.T., Noguchi, M. 1996, MNRAS, 278, 191
Gibson, B.K., Giroux, M.L., Penton, S.V., Stocke, J.T., Shull, J.M., Tumlinson, J. 2001, AJ, 122, 3280
Gibson, B.K., Putman, M.E., eds. 1999, ASP Conf. Ser. 166, Stromlo Workshop on High-Velocity Clouds
Giovanelli, R. 1980, AJ, 85, 1155
Giovanelli, R. 1981, AJ, 87, 1468
Giovanelli, R., Haynes, M.P. 1977, A&A, 54, 909
Giovanelli, R., Verschuur, G.L., Cram, T.R. 1973, A&AS, 12, 209
Greisen, E.W., Cram, T.R. 1976, ApJ, 203, L119
Hartmann, D., Burton, W.B. 1997, Atlas of Galactic Neutral Hydrogen (Cambridge: Cambridge University Press)
Haud, U. 1988, A&A, 198, 125
Hirth, W., Mebold, U., Dahlem, M., Müller, P. 1991, ApSpSci, 186, 211
Hulsbosch, A.N.M. 1973, Ph.D. thesis, Rijks Universiteit Leiden
Hulsbosch, A.N.M. 1978, A&A, 66, L5
Hulsbosch, A.N.M., Raimond, R. 1966, BAN, 18, 413
Hulsbosch, A.N.M., Wakker, B.P. 1988, A&AS, 75, 191
Jenkins, E.B. 1978, ApJ, 219, 845
Jenkins, E.B., Meloy, D.A. 1974, ApJ, 193, L121
Kaelble, A., de Boer, K.S., Grewing, M. 1985, A&A, 143, 408
Kerr, F.J., Simonson S.C., eds. 1973, IAU Symp. 60, Galactic Radio Astronomy (Dordrecht: Reidel)
Kuntz, K.D., Danly, L. 1996, ApJ, 457, 703
Kutyrev, A.S., Reynolds, R.J. 1989, ApJ, 344, L9
Lilienthal, D., Meyerdierks, H., de Boer, K.S. 1990, A&A, 240, 487
Lilienthal, D., Hirth, W., Mebold, U., de Boer, K.S. 1992, A&A, 255, 323
Lin, D.N.C., Lynden-Bell, D. 1982, MNRAS, 198, 707
Lu, L., Savage, B.D., Sembach, K.R. 1994, ApJ, 426, 563
Lu, L., Savage, B.D., Sembach, K.R., Wakker, B.P., Sargent, W.L.W., Oosterloo, T.A. 1998, AJ, 115, 162
Mathewson, D.S. 1967, PASAu, 1, 21
Mathewson, D.S., Cleary, M.N., Murray, J.D. 1974, ApJ, 190, 291
McCammon, D., Sanders, W.T. 1990, ARA&A, 28, 657
Mirabel, I.F. 1982, ApJ, 256, 1112
Moore, B., Davis, M. 1994, MNRAS, 270, 209
Morras, R., Bajaja, E., Arnal, E.M., Pöppel, W.G.L. 2000, A&AS, 142, 25
Muller, C.A., Oort, J.H., Raimond, E. 1963, C.R. Acad. Sci. Paris, 257, 1661
Münch, G. 1952, PASP, 64, 312
Münch, G., Zirin, H. 1961, ApJ, 133, 11
Murai, T., Fujimoto, M. 1980, PASJ, 32, 581
Murphy, E.M., Sembach, K.R., Gibson, B.K., Shull, J.M., Savage, B.D., Roth, K.C., Moos, H.W., Green, J.C., York, D.G., Wakker, B.P. 2000, ApJ, 538, L35
Oort, J.H. 1966, BAN, 18, 421
Oort, J.H. 1970, A&A, 7, 381
Oort, J.H., Hulsbosch, A.N.M. 1978, in Astronomical Papers Dedicated to Bengt Strömgren, eds. A. Reiz, T. Anderson (Copenhagen: Copenhagen University Observatory), 478
Oort, J.H., Kerr, F.J., Westerhout, G. 1958, MNRAS, 118, 379
Pettini, M., West, K.A. 1982, ApJ, 260, 561
Prata, S.W., Wallerstein, G. 1967, PASP, 79, 202
Putman, M.E., Gibson, B.K, Staveley-Smith, L., et 23 al. 1998, Nature, 394, 752
Putman, M.E., de Heij, V., Staveley-Smith, L., et 29 al. 2002, AJ, 123, 873
Reynolds, R.J. 1987, ApJ, 323, 553

Richter, P., de Boer, K.S., Widmann, H., Kappelmann, N., Gringel, W., Grewing, M., Barnstedt, J. 1999, Nature, 402, 386
Richter, P., Sembach, K.R., Wakker, B.P., Savage, B.D., Tripp, T.M., Murphy, E.M., Kalberla, P.M.W., Jenkins, E.B. 2001, ApJ, 559, 318
Richter, P., Wakker, B.P., Savage, B.D., Sembach, K.R. 2003, ApJ, 586, 230
Robertson, J.G., Schwarz, U.J., van Woerden, H., Murray, J.D., Morton, D.C., Hulsbosch, A.N.M. 1991, MNRAS, 248, 508
Savage, B.D., de Boer, K.S. 1979, ApJ, 230, L77
Savage, B.D., de Boer, K.S. 1981, ApJ, 243, 460
Savage, B.D., Massa, D. 1987, ApJ, 314, 380
Savage, B.D., Jenkins, E.B., Joseph, C.L., de Boer, K.S. 1989, ApJ, 345, 393
Savage, B.D., Sembach, K.R., Lu, L. 1997, AJ, 113, 2158
Savage, B.D., Sembach, K.R., Wakker, B.P., Richter, P., Meade, M., Jenkins, E.B., Shull, J.M., Moos, H.W., Sonneborn, G. 2003, ApJS, 146, 125
Schlüter, A., Schmidt, H., Stumpff, P. 1953, Zeitschrift für Astroph., 33, 194
Schulman, E., Bregman, J.N., Roberts, M.S. 1994, ApJ, 423, 180
Schulman, E., Bregman, J.N., Brinks, E., Roberts, M.S. 1996, AJ, 112, 960
Schulman, E., Brinks, E., Bregman, J.N., Roberts, M.S. 1997, AJ, 113, 1559
Schwarz, U.J., Oort, J.H. 1981, A&A, 101, 305
Schwarz, U.J., Sullivan III, W.T., Hulsbosch, A.N.M. 1976, A&A, 52, 133
Sembach, K.R., Savage, B.D. 1992, ApJS, 83, 147
Sembach, K.R., Savage, B.D., Lu, L., Murphy, E.M. 1999, ApJ, 515, 108
Sembach, K.R., Wakker, B.P., Savage, B.D., Richter, P., Meade, M., Shull, J.M., Jenkins, E.B., Sonneborn, G., Moos, H.W. 2003, ApJS, 146, 165
Shapiro, P.R., Field, G.B. 1976, ApJ, 205, 762
Smith, G.P. 1963, BAN, 17, 203
Snowden, S.L., Mebold, U., Hirth, W., Herbstmeier, U., Schmitt, J.H. 1991, Science, 252, 1529
Songaila, A. 1981, ApJ, 243, L19
Songaila, A., York, D.G. 1981, ApJ, 242, 976
Songaila, A., York, D.G., Cowie, L.L., Blades, J.C. 1985, ApJ, 293, L15
Songaila, A., Cowie, L.L., Weaver, H.F. 1986, ApJ, 329, 580
Spitzer, L. 1956, ApJ, 124, 20
Spitzer, L. 1996, ApJ, 458, L29
Spitzer, L., de Boer, K.S., Savage, B.D. 1980, Sky & Tel., 60, 189
Stark, A.A., Gammie, C.F., Wilson, R.W., Bally, J., Linke, R.A., Heiles, C., Hurwitz, M. 1992, ApJS, 79, 77
Tenorio-Tagle, G., Moles, M., Melnick, J., eds. 1989, Structure and Dynamics of the Interstellar Medium (Berlin: Springer)
Tolbert, C.A. 1971, A&AS, 3, 349
Tripp, T.M., Wakker, B.P., Jenkins, E.B., Bowers, C.W., Danks, A.C., Green, R.F., Heap, S.R., Joseph, C.L., Kaiser, M.E., Linsky, J.L., Woodgate, B.E. 2003, AJ, 125, 3122
Tufte, S.L., Reynolds, R.J., Haffner, L.M. 1998, ApJ, 504, 773
van der Hulst T., Sancisi R. 1988, AJ, 95, 1354
van Kuilenburg, J. 1972a, A&A, 16, 276
van Kuilenburg, J. 1972b, A&AS, 5, 1
van Woerden, H., ed. 1967, IAU Symp. 31, Radio Astronomy and the Galactic System (London: Academic Press)
van Woerden, H., Rougoor, G.W., Oort, J.H. 1957, C.R. Acad. Sci. Paris, 244, 1691
van Woerden, H., Allen, R.J., Burton, W.B., eds. 1985, IAU Symp. 106, The Milky Way Galaxy (Dordrecht: Reidel)
van Woerden, H., Schwarz, U.J., Peletier, R.F., Wakker, B.P., Kalberla, P.M.W. 1999, Nature, 400, 138
Verschuur, G.L. 1973, A&A, 22, 139
Verschuur, G.L. 1975, ARA&A, 13, 257

Wakker, B.P. 1990, Ph.D. thesis, Rijks Universiteit Groningen

Wakker, B.P. 1991, A&A, 250, 499

Wakker, B.P., Schwarz, U.J. 1991, A&A, 250, 484

Wakker, B.P., van Woerden, H. 1991, A&A, 250, 509

Wakker, B.P., van Woerden, H. 1997, ARA&A, 35, 217

Wakker, B.P., Broeils, A.H., Tilanus, R.P.J., Sancisi, R. 1989, A&A, 226, 57

Wakker, B.P., van Woerden, H., Schwarz, U.J., Peletier, R.F., Douglas, N.G. 1996, A&A, 306, L25

Wakker, B.P., Howk, J.C., Schwarz, U.J., van Woerden, H., Beers, T.C., Wilhelm, R., Kalberla, P.M.W., Danly, L. 1996, ApJ, 473, 834

Wakker, B.P., van Woerden, H., de Boer, K.S., Kalberla, P.M.W. 1998, ApJ, 493, 762

Wakker, B.P., Howk, J.C., Savage, B.D., van Woerden, H., Tufte, S.L., Schwarz, U.J., Benjamin, R., Reynolds, R.J., Peletier, R.F., Kalberla, P.M.W. 1999, Nature, 402, 388

Wakker, B.P., Savage, B.D., Sembach, K.R., et 18 al. 2003, ApJS, 146, 1

Wannier, P., Wrixon, G.T., Wilson, R.W. 1972, A&A, 18, 224

Weiner, B.J., Williams, T.B. 1996, AJ, 111, 1156

Wesselius, P.R. 1973, A&A, 24, 35

Wesselius, P.R., Fejes, I. 1973, A&A, 24, 15

West, K.A., Pettini, M., Penston, M., Blades, J.C., Morton, D.C. 1985, MNRAS, 215, 481

Widmann, H., de Boer, K.S., Richter, P., Krämer, G., Appenzeller, I., Barnstedt, J., Gölz, M., Grewing, M., Gringel, W., Mandel, H., Werner, K. 1998, A&A, 338, L1

Williamson, F.O., Sanders, W.T., Kraushaar, W.L., McCammon, D., Borken, R., Bunner, A.N. 1974, ApJ, 193, L133

York, D.G., Wu, C.C., Blades, J.C., Ratcliff, S., Cowie, L.L., Morton, D.C. 1983, ApJ, 274, 136

2. HVC/IVC MAPS AND HVC DISTRIBUTION FUNCTIONS

BART P. WAKKER
Department of Astronomy, University of Wisconsin-Madison, USA; wakker@astro.wisc.edu

Abstract. This chapter presents an overview of the basic sky, velocity, column density, volume density, area, and flux distributions of the high-velocity clouds (HVCs). All-sky maps are shown of their deviation and GSR velocities, and their column densities. A comparison is made between the HVCs defined by using a limit in the deviation velocity (logically preferred) versus using a limit in the LSR velocity (historically done). A comparison is also made between the HVC maps obtained using the Leiden-Dwingeloo Survey (Hartmann & Burton 1997) versus the maps obtained with the combined Hulsbosch & Wakker (1988) and Morras et al. (2000) surveys. The maps of the negative-velocity HVCs and IVCs in the northern galactic hemisphere are compared and it is shown that they are probably unrelated.

A discussion is presented of the column density distribution of high-velocity gas. Previous studies had concluded that the HVCs have a typical column density of log$N$(H I)=19.2 (Giovanelli 1980) and that there is little HVC gas with N(H I)<7×10^{17} cm^{-2} (Murphy et al. 1995). Here it is shown that these conclusions were the result of not properly taking into account the effects of binning log$N$(H I), instead of binning N(H I). In fact, the HVCs have no typical column density (the most common column density in each survey is at the detection limit), and there probably is much high-velocity gas with N(H I)<7×10^{17} cm^{-2}. A corollary is that a calculated "average column density" always scales with the minimum column density used to calculate it.

Finally, some of the implications of the proposal by Blitz et al. (1999) that HVCs are massive, virially stable objects spread throughout the Local Group, are analyzed. It is shown that there would be 10 to 100 times more clouds of a given H I mass than are observed in the form of dwarf galaxies. Only if the ratio, f, of neutral hydrogen mass to total cloud mass is <0.01 is there no discrepancy between data and model. This, however, implies that the HVCs must lie within about 200 kpc from the Milky Way.

H. van Woerden et al. (eds.), High-Velocity Clouds, 25–54

1. Introduction

The studies presented in this chapter follow previous studies by Wakker (1991) and Wakker & van Woerden (1991). More recent data allow a better insight into the southern sky, and into the intermediate-velocity clouds (IVCs) and their relation to the HVCs. Also, some distributions that were not included in the analysis of Wakker (1991) have new-found relevance.

Early surveys for HVCs concentrated on high galactic latitudes, where normal disk gas has low velocities. Velocities were reported relative to the Local Standard of Rest (LSR), using a limit of $|v_{\rm LSR}|>90$ or 100 $\rm km\,s^{-1}$ to define the HVCs. Objects with $40<|v_{\rm LSR}|<90$ $\rm km\,s^{-1}$ are then called intermediate-velocity clouds (IVCs). However, at low latitudes differential galactic rotation may result in high LSR velocities. The range of radial velocities that can be considered "normal" can be found from a model of the spatial structure of the gaseous Galactic Disk, an assumed galactic rotation curve, and the coordinates of the line of sight. Wakker (1991) thus defined the "deviation velocity", $v_{\rm DEV}$, as the difference between $v_{\rm LSR}$ and the extreme value of the allowed velocity range. This quantity is described in Sect. 2.1. The lower velocity limit for defining IVCs is rather diffuse. Some authors (e.g. Albert & Danly, Ch. 4) use $\sim$20 $\rm km\,s^{-1}$. However, velocities below 40 $\rm km\,s^{-1}$ are often compatible with differential galactic rotation. Some authors have argued that the division between HVCs and IVCs based on LSR velocity may be artificial. However, Wakker (2001) finds that IVCs tend to lie about 1 kpc from the galactic plane, while HVCs with known distance limits lie at $|z|>3$ kpc (see Ch. 10). Thus, a division based on LSR or deviation velocity may be physical after all.

Recent H I surveys are summarized in Sect. 2.2, and their relevance to HVC studies is discussed. A series of all-sky maps is presented in Sect. 2.3, showing both the HVCs and the IVCs. Section 3 summarizes the properties of the HVC complexes. A discussion of the distribution of H I column densities (Sect. 4) reaches a very different conclusion from those presented in previous papers (Giovanelli 1980, Murphy et al. 1995). Next, HVC volume densities are discussed in Sect. 5. Finally, in Sect. 6, the distributions of HVC area and total flux are presented. The model of Blitz et al. (1999) is applied to derive virial distances and virial masses for the HVCs, and the resulting H I mass function is compared to the observations of Zwaan et al. (1997), using an analysis resembling the one done by Zwaan & Briggs (2000).

2. Basic HVC and IVC data

2.1. DEFINITION OF THE DEVIATION VELOCITY

Since much of this chapter will use the deviation velocity, $v_{\rm DEV}$, this section describes how to calculate it. Basically: for each direction (l,b) find the maximum and minimum value of $v_{\rm LSR}$ compatible with a simple model of galactic structure and rotation ($v_{\rm max}$ and $v_{\rm min}$). Then:

$$v_{\rm DEV} = v_{\rm LSR} - v_{\rm min} \text{ if } v_{\rm LSR} < 0; \quad v_{\rm DEV} = v_{\rm LSR} - v_{\rm max} \text{ if } v_{\rm LSR} > 0.$$

The run of $v_{\rm LSR}$ with distance in the line of sight is found from:

$$v_{\rm LSR} \;=\; \left(\frac{R_0}{R}\, v(R) \;-\; v(R_0)\right)\, \sin l\, \cos b,$$

where R is the galactocentric radius of a point in the sightline, and $v(R)$ gives the rotation velocity at that galactocentric radius. R_0=8.5 kpc is the distance of the Sun to the Galactic Center. Rotation is assumed to be solid-body for galactocentric radii less than R_s=0.5 kpc, and to have constant velocity $v(R)$=220 km s^{-1} further out. For a given distance d in the line of sight:

$$R \;=\; R_0 \sqrt{\cos^2 b \left(\frac{d}{R_0}\right)^2 - 2\,\cos b \cos l \left(\frac{d}{R_0}\right) + 1}; \quad \text{also}: z \;=\; d\, \sin b.$$

Then calculate $v_{\rm LSR}$ for all distances between 0 and $d_{\rm max}$, with $d_{\rm max}$ defined as the distance where the sightline leaves the disk:

$$R > R_{\rm max} \;\text{ or } |z| > z_{\rm max}$$

$$z_{\rm max} = z_1 \text{ if } R < R_0; \quad z_{\rm max} = z_1 + (z2 - z1) \times \frac{(R/R_0 - 1)^2}{4} \text{ if } R > R_0.$$

The disk has a diameter of $R_{\rm max}$=26 kpc, and a thickness of $2{\times}z_1$=2 kpc for R<R_0, increasing to $2{\times}z_2$=6 kpc at R=$3R_0$.

For an HVC lying *inside* the gaseous Galactic Disk, its velocity relative to the disk gas near it ($v_{\rm pec}$) is always greater than $v_{\rm DEV}$; at longitudes $-90°<l<+90°$, $v_{\rm pec}$ may even exceed $v_{\rm LSR}$; at $90°<l<270°$, $v_{\rm pec}$ will be intermediate between $v_{\rm DEV}$ and $v_{\rm LSR}$. On the other hand, at low galactic latitudes, some objects classified as HVCs on the basis of $v_{\rm LSR}$ may have modest peculiar velocities. In this chapter we use $v_{\rm DEV}$ to select HVCs whenever feasible. Note that de Heij et al. (2002) present a slightly different way of calculating deviation velocities – they allow for a Galactic Warp.

2.2. OVERVIEW OF RELEVANT H I SURVEYS

Wakker & van Woerden (1997) reviewed HVC surveys published before 1988. The HVC cloud catalog of Wakker & van Woerden (1991) was based on the data of Hulsbosch & Wakker (1988) and Bajaja et al. (1985). The former covered the northern sky (declinations $>-18°$) on a $1°\times1°$ grid, with a $0\overset{\circ}{.}5$ beam, 16 $\mathrm{km\,s^{-1}}$ velocity resolution, and 5-σ detection limit 0.05 K. The southern sky (declinations $<-18°$) was covered on a $2°\times2°$ grid, with detection limit 0.08 K. In addition, Bajaja et al. (1988) presented column density maps of selected regions on a $0\overset{\circ}{.}5\times0\overset{\circ}{.}5$ grid. For a typical line width of 25 $\mathrm{km\,s^{-1}}$, the detection limits correspond to column densities of $\sim$2.5 and 3.5×10^{18} $\mathrm{cm^{-2}}$. Both datasets were published as a list of HVC profile components, giving longitude, latitude, LSR velocity and peak brightness temperature. Information about line widths and shapes was mostly lost.

Until recently, H I surveys containing useful data for IVCs were rare. The earliest H I surveys only covered $|v_{\mathrm{LSR}}|\lesssim\pm80$ $\mathrm{km\,s^{-1}}$. Tolbert (1971) mapped the northern sky with a $36'$ beam on a fairly sparse grid (6% of the sky north of b=+15° was covered) at 3.4 $\mathrm{km\,s^{-1}}$ velocity resolution. This was used by Wesselius & Fejes (1973) to analyze the IVCs. Stark et al. (1992) presented a survey with 5.3 $\mathrm{km\,s^{-1}}$ resolution and 3° beam, which Kuntz & Danly (1996) used for a detailed analysis of the northern IVCs.

These surveys have now been superseded by the "Leiden-Dwingeloo Survey" or *LDS* (Hartmann & Burton 1997), which covers the sky north of declination $-30°$ on a $0\overset{\circ}{.}5\times0\overset{\circ}{.}5$ grid, with 1.03 $\mathrm{km\,s^{-1}}$ velocity channels, an rms noise of 0.07 K, and is corrected for stray-radiation effects. After integrating over 16 $\mathrm{km\,s^{-1}}$ (15 channels) and 4 gridpoints, the 5-σ detection limit is $\sim$0.05 K, similar to that of the earlier HVC surveys. With the *LDS* it will be possible to study line shapes and line widths for HVCs and IVCs. However, the faint edges of the large HVC complexes are not as clearly seen (see Sect. 2.3). The *LDS* data are especially useful to map the IVCs, and study their connection to high- and low-velocity gas.

For the southern sky (declinations $<-23°$) the *Villa Elisa* or *IAR* (Instituto Argentino de Radioastronomia) survey (Arnal et al. 2000) is comparable to the *LDS*, although the data are not yet freely available. Morras et al. (2000) presented a list of HVC components extracted from this survey, with 5-σ detection limit 0.07 K, or $\sim3\times10^{18}$ $\mathrm{cm^{-2}}$. In combination with the Hulsbosch & Wakker (1988) list, HVC components have thus been listed for almost the whole sky (except between declinations $-23°$ and $-18°$) on a grid of $1°\times1°$ or finer, with a detection limit of 2 to 3×10^{18} $\mathrm{cm^{-2}}$.

Two other recent southern surveys include HVCs. The "H I Parkes All-Sky Survey" (*HIPASS*; Barnes et al. 2001) covers the sky south of declination 0° on a $15'\times15'$ grid, although only with 26 $\mathrm{km\,s^{-1}}$ velocity resolution.

The standard calibration of this dataset filters out structures larger than 2° and has unreliable baselines near Milky Way velocities; these spectra are thus not useful for HVC and IVC studies. Putman et al. (2002) recalibrated the part of the data containing the Milky Way, recovering the emission unless structures completely fill an 8° scan. In addition to the *HIPASS* data, the "Parkes Narrow Band Survey" (Brüns et al. 2004) covers the Magellanic Stream with 15′×15′ angular and 1 km s^{-1} velocity resolution. A catalog of southern HVCs based on these data is now available (Putman et al. 2002).

2.3. MAPS OF HVCS

A number of all-sky and polar-projection maps of the H I sky are presented in Figs. 1 to 4. Detailed maps of many HVC and IVC complexes can be found in Wakker (2001). Figure 1a shows the deviation velocities, $v_{\rm DEV}$, of the HVCs. Clouds in grey have $|v_{\rm LSR}|>90$ km s^{-1}, but $|v_{\rm DEV}|<90$ km s^{-1}, rendering their classification as HVCs uncertain; they should probably be reclassified as IVCs. A major feature of Fig. 1a is that at $0°<l<180°$ most HVCs have negative deviation velocities, while at $180°<l<360°$ most values of $v_{\rm DEV}$ are positive. This is even more pronounced when $v_{\rm LSR}$ is used (Wakker 1991, Fig. 3a; Wakker & van Woerden 1991, Fig. 2b). This velocity asymmetry is clearly caused by the motion of the LSR at 220 km s^{-1} toward l=90°, b=0°.

Outside the gaseous Galactic Disk, ordered rotation may be lacking, and the Local Standard of Rest is undefined. Also, at distances much greater than the Sun's galactocentric radius, the line-of-sight component of galactic rotation would be small. In that situation, the preferred coordinate system is the "Galactic Standard of Rest" (GSR): a system which is at rest relative to the Galactic Center. To correct $v_{\rm LSR}$ for the effects of this rotation reduces the velocities to the GSR:

$$v_{\rm GSR} = v_{\rm LSR} + 220 \sin l \cos b.$$

Note that this is *not* a velocity relative to the Galactic Center, as is often mistakenly stated. It is *the velocity relative to the Sun* in a reference frame in which the Milky Way rotates.

Figure 1b shows an all-sky map of the $v_{\rm GSR}$ velocities, selecting HVCs with $|v_{\rm LSR}|>90$ km s^{-1}. It is clear that the run of $v_{\rm GSR}$ with galactic longitude is much flatter than that of $v_{\rm LSR}$ or $v_{\rm DEV}$ (see Wakker 1991, Fig. 3a; Wakker & van Woerden 1991, Fig. 2c), although the spread in $v_{\rm GSR}$ at each longitude is quite large. These facts suggest that, on average, the HVCs do not follow the rotation of the Galaxy, and that they possess a large velocity dispersion (these findings suggest, but do not prove, that most HVCs are at great distances).

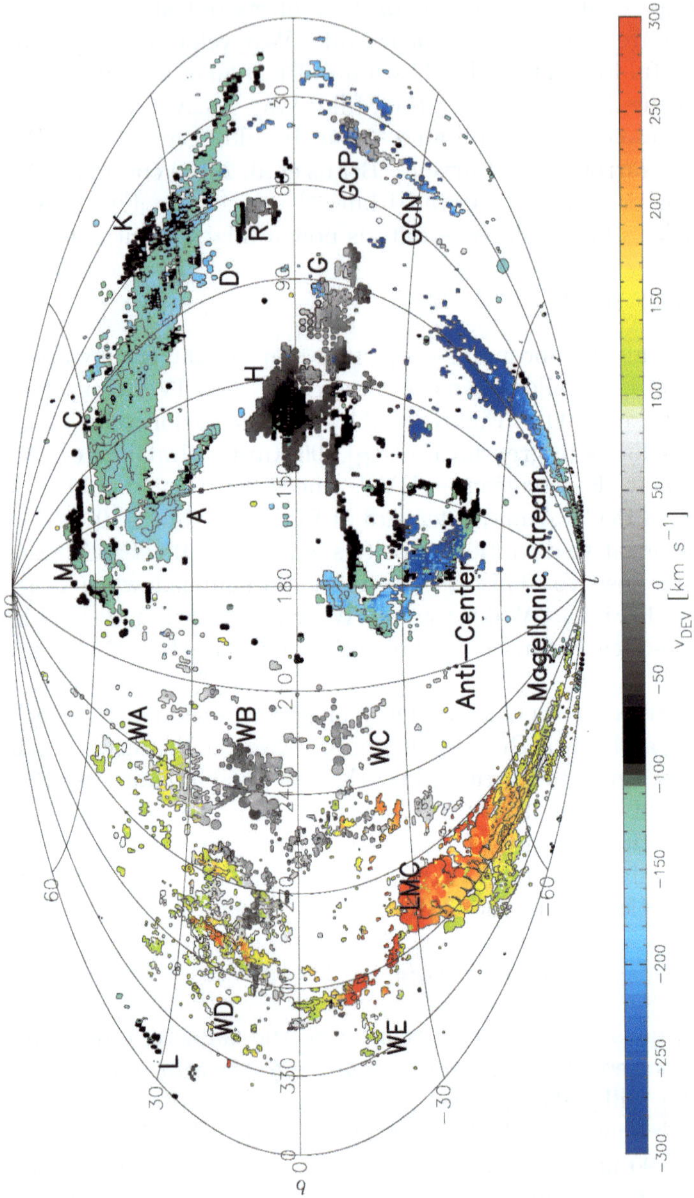

Figure 1a. Aitoff projection all-sky map of the HVCs, in galactic coordinates, with the Anti-Center in the middle. Based on the Hulsbosch & Wakker (1988) and Morras et al. (2000) datasets. Contour levels at 0.05 and 0.5 K brightness temperature. Colors code deviation velocities (Sect. 2.1), with the scale given by the wedge at the bottom. HVCs were selected using $|v_{\rm LSR}|$>90 km s^{-1}. Grey clouds have $|v_{\rm LSR}|$>90 km s^{-1}, but $|v_{\rm DEV}|$<90 km s^{-1}. Figure composed for this article.

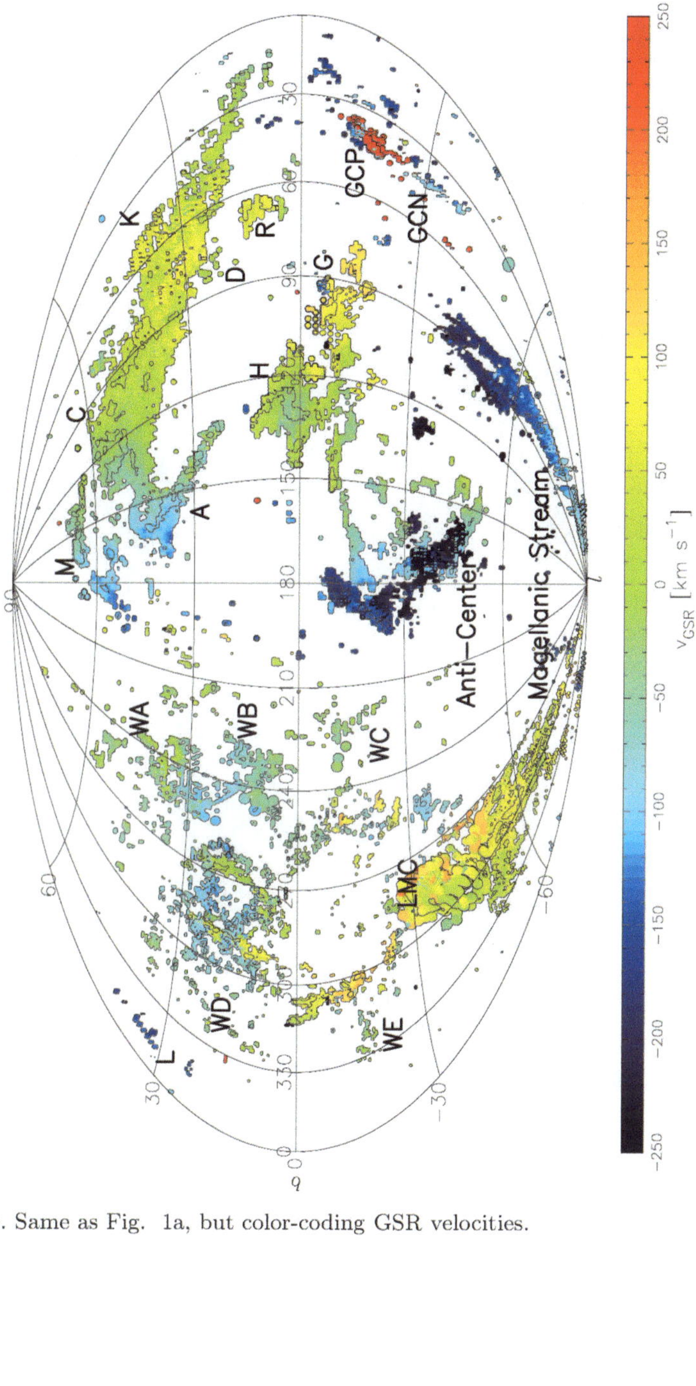

Figure 1b. Same as Fig. 1a, but color-coding GSR velocities.

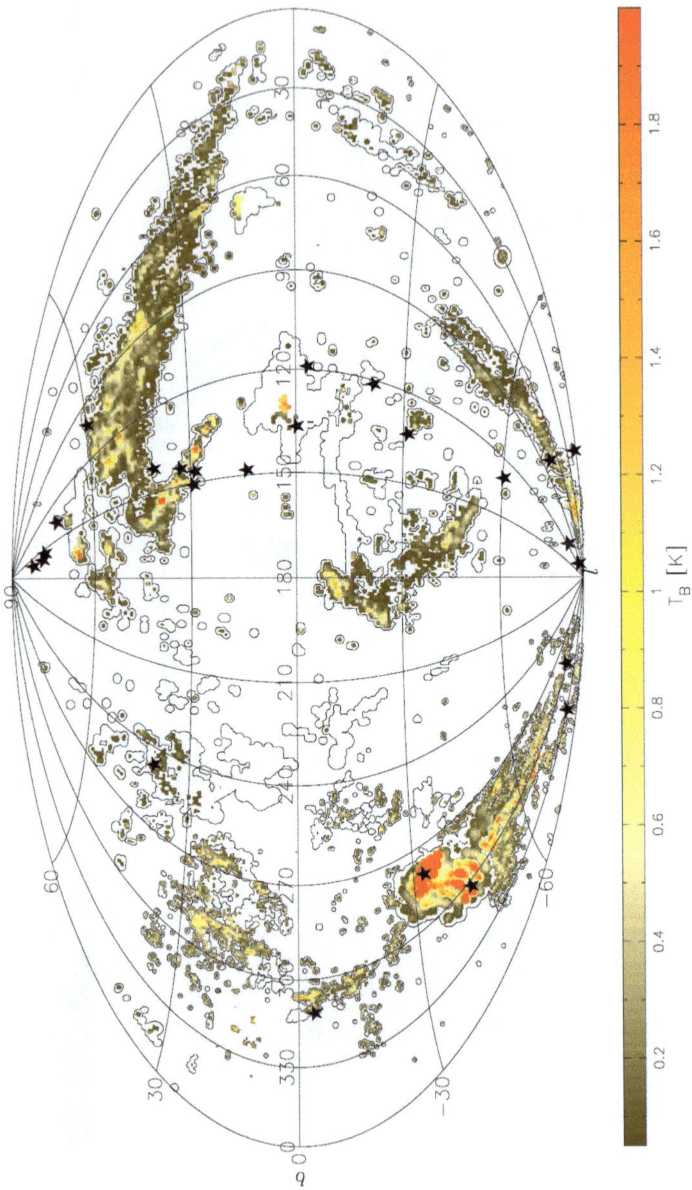

Figure 2a. Peak brightness temperature (T_B) of H I with $|v_{\rm LSR}|$>90 km s^{-1}, based on the Hulsbosch & Wakker (1988) and Morras et al. (2000) datasets. For HVCs with $|v_{\rm DEV}|$>90 km s^{-1} T_B is color-coded, while HVCs with $|v_{\rm DEV}|$<90 km s^{-1} are shown by the outer 0.05 K contour. Aitoff projection map in galactic coordinates, with the Anti-Center in the middle. Stars show the H I-bright galaxies with 90<$|v_{\rm DEV}|$<350 km s^{-1}. Figure composed for this article.

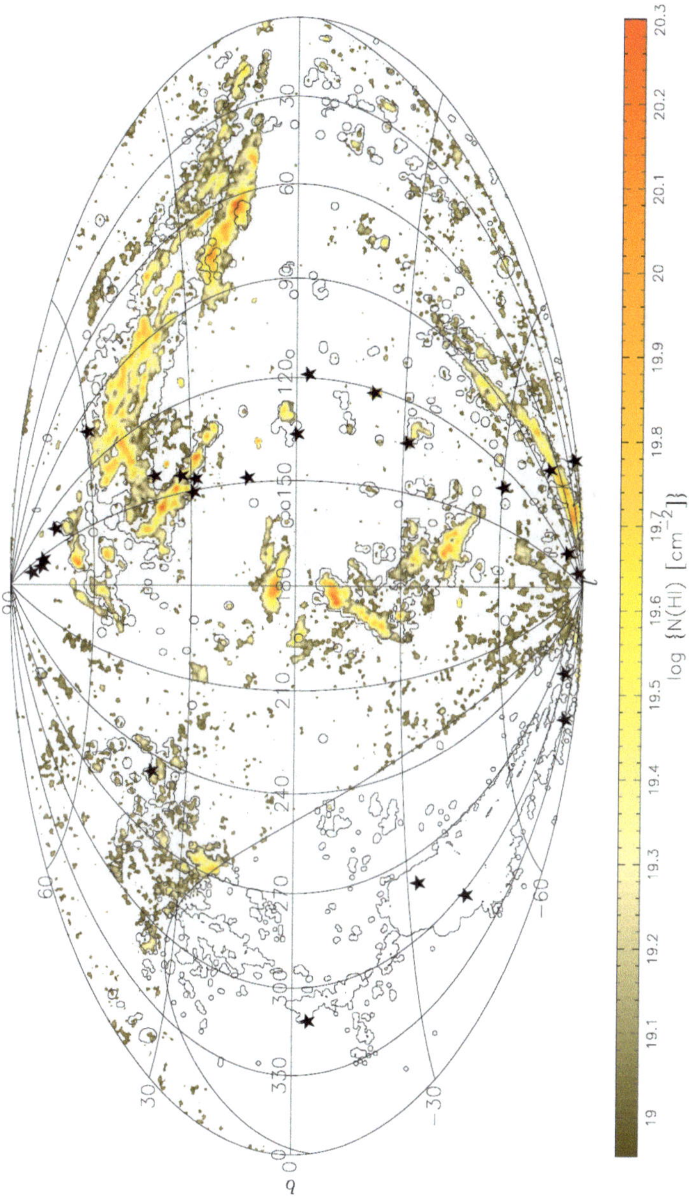

Figure 2b. Color-coded column density of H I with $|v_{\rm DEV}|>90$ km s^{-1}, based on the *LDS*, with an outer contour corresponding to 0.05 K in the Hulsbosch & Wakker (1988) and Morras et al. (2000) surveys. Aitoff projection map in galactic coordinates, with the Anti-Center in the middle. No *LDS* data are available below declination −30°, shown by the curved line. Stars show the H I-bright galaxies with $90<|v_{\rm DEV}|<350$ km s^{-1}. Figure composed for this article.

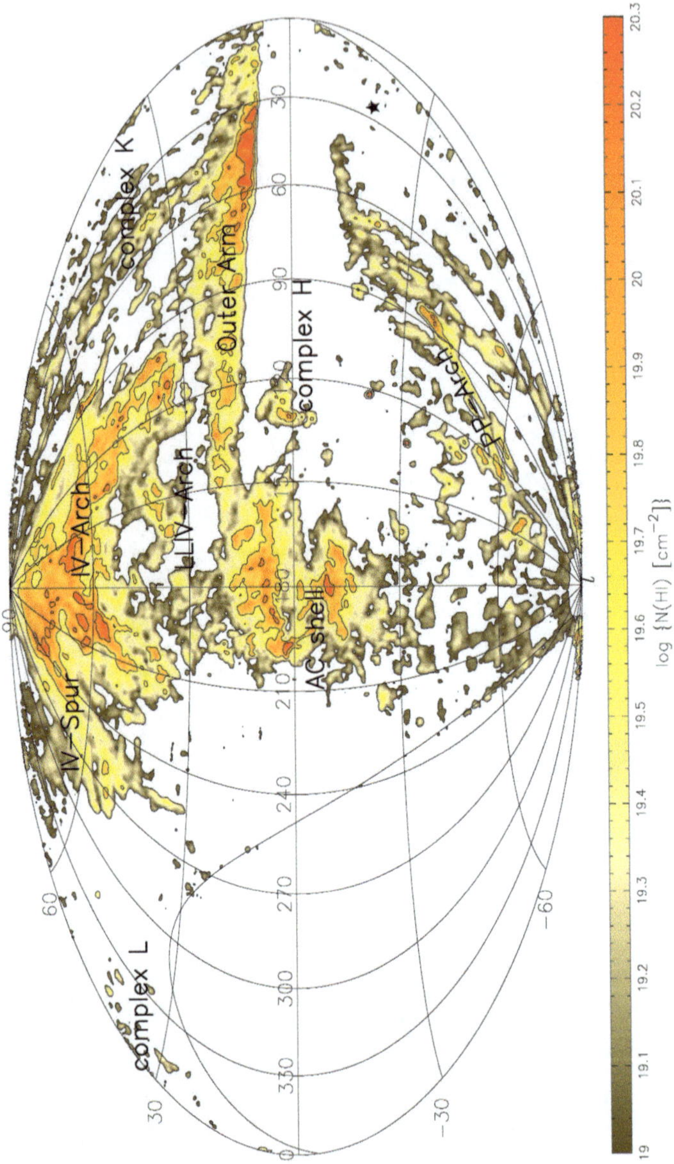

Figure 3a. Colors show the column density (aitoff projection) of the intermediate-velocity gas with deviation velocity between −90 and −35 km s^{-1}, based on the *LDS* (Hartmann & Burton 1997). For clarity the resolution was degraded to 2°. Contour levels show column densities of 10, 50 and 120×10^{18} cm^{-2}. Figure composed for this article.

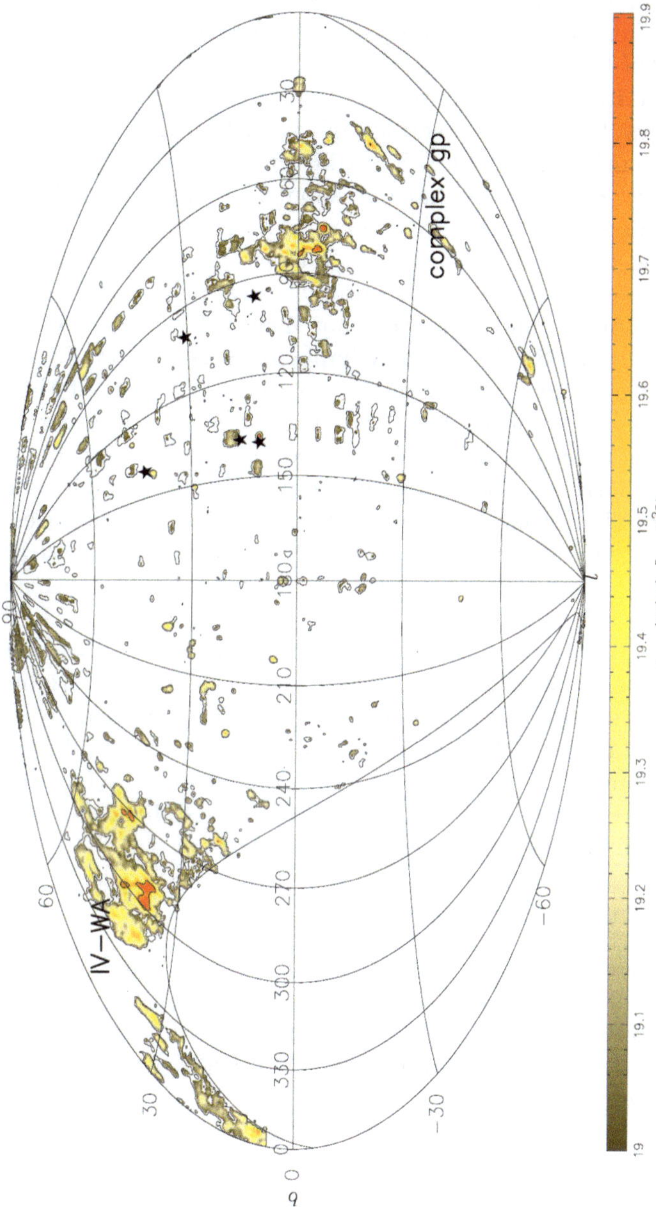

Figure 3b. As 3a, but for positive-velocity gas with $v_{\rm DEV}$ between +35 and +90 km s^{-1}. Some of the blocky features at $b>+50°$ are artifacts related to low-level baseline problems.

Figure 2 presents intensity maps of the HVC sky, based on the Hulsbosch & Wakker (1988) and Morras et al. (2000) data (Fig. 2a), as well as on the *LDS* (Fig. 2b). For both maps $|v_{\rm DEV}|>90$ km s^{-1} was used as the selection criterion. The contour in Fig. 2a shows the HVCs defined by $|v_{\rm LSR}|>90$ km s^{-1}. The 4-σ detection limit is $\sim$2 to 3×10^{18} cm^{-2} for Fig. 2a, but $\sim8\times10^{18}$ cm^{-2} for the *LDS*. The difference can be seen from the outer contour in Fig. 2b, showing that one should be careful in using the *LDS* to map the full extent of the HVC complexes. That requires smoothing to a 1° beam and careful selection of the velocity range. Another part of the difference can be attributed to the profile wings of H I clouds which have $v_{\rm LSR,peak}\lesssim90$ km s^{-1}, and thus were not included in the survey by Hulsbosch & Wakker (1988). This clearly happens for the Outer Arm (at l=30° to 90°, $b\sim$+20°), and also around (l,b)=(180°,+5°) and (200°,0°). Some clouds (especially complexes A, C, and WA) cover a much larger area when the lower column density data are included. The shapes of complex A and the Magellanic Stream also are qualitatively different: at high column densities only the line of cores is seen, but at low column densities a wider edge shows up at high latitudes. Braun & Thilker (2004) find that for a column density limit of 2×10^{17} cm^{-2} the Stream has an even larger extent (l=90° to 115°, up to b=−10°). The very faint, scattered clouds near (l,b)=(280°,+70°) were apparently missed by Hulsbosch & Wakker (1988).

In Fig. 2 stars are placed at the positions of nearby H I-bright galaxies with velocities in the selected ranges. These were found from the RC3 (de Vaucouleurs et al. 1991). Five galaxies have $v_{\rm DEV}<-90$ km s^{-1} (M 31, M 33, WLM, IC 10, and IC 1613), one $v_{\rm DEV}$=−35 to −90 km s^{-1} (NGC 6822), five $v_{\rm DEV}$=+35 to +90 km s^{-1} (NGC 1560, IC 2574, NGC 6503, NGC 6946, and UGCA 86), and twenty $v_{\rm DEV}$= +90 to +350 km s^{-1}.

2.4. MAPS OF IVCS

Figure 3 shows column density maps for the IVCs ($35<|v_{\rm DEV}|<90$ km s^{-1}). The well-known north-south dichotomy for the negative-velocity IVCs at longitudes l<180° is clearly visible. A number of well-defined structures are visible in these maps, as labeled. In the northern sky Kuntz & Danly (1996) defined the "Intermediate-Velocity Arch" (IV Arch), the "Intermediate-Velocity Spur" (IV Spur), and the "Low-Latitude Intermediate-Velocity Arch" (LLIV Arch). Wakker (2001) gave names to three more groupings of intermediate-velocity gas: "complex K", the "Pegasus-Pisces Arch" (PP Arch), and "complex gp" (Sect. 3 below gives more details).

In combination with Fig. 2b, Fig. 3b shows that positive-velocity gas is widespread in the region l=210° to 330°, b=+10° to +60°. The structure of this gas is extremely complicated, both spatially and in velocity. The

TABLE 1. Summary of properties of HVC and IVC complexes

#	Name	l range [°]	b range [°]	$v_{\rm LSR}$ range [km s^{-1}]	$\overline{v}^a_{\rm DEV}$ [km s^{-1}]	Ω^b [°^{2}]	D^c [kpc]	Mass^d [10^6M$_\odot$]	$\dot{\rm M}^e$ [M$_\odot$ yr^{-1}]	Z^f [$Z_\odot$]
1	A	131:174	+22:+51	−209:−91	−141	288	8−10	1.0	−0.018:−0.057	0.1?
2	C	39:147	+18:+61	−223:−91	−122	1546	>6	[5.0]	[−0.085:−0.200]	0.14
3	H	110:152	−16:+13	−221:−91	−48	479	>5	[2.4]	[−0.002:−0.140]	?
4	MS	11:358	−86:−21	−433:+412	+155	1525	50	280	0.28	
5	EP	243:332	−34:+28	+103:+354	+99	395	50	32	0.25	
6	ACHV	140:197	−54:−6	−170:−91	−100	397	?	[1.0]	[−0.010:−0.052]	?
7	ACVHV	156:191	−54:−7	−337:−134	−221	338	?	[1.0]	[−0.021:−0.096]	?
8	L	340:348	+23:+41	−188:−91	−72	21	?	[0.04]	?	
9	GCN	1:50	−40:+10	−341:−169	−243	130	>10?	[0.22]	[−0.003:−0.021]	?
10	D	72:84	+17:+27	−206:−151	−178	17	?	[0.02]	[−0.000:−0.002]	?
11	EN	28:287	−84:+41	−465:−132	−233	46	50	[1.7]	?	
12	P	107:132	−39:−30	−429:−325	−360	30	?	[0.06]	?	
13	GCP	35:65	−42:−9	+91:+126	+25	58	?	[0.70]	[0.001:0.013]	?
14	WA	232:269	+27:+43	+100:+193	+126	102	?	[0.13]	[0.002:0.006]	?
15	WB	225:268	+1:+45	+91:+168	+52	289	?	[0.70]	[0.002:0.014]	1
16	WC	209:260	−34:+3	+92:+220	+10	116	?	[0.29]	[0.001:0.011]	?
17	WD	259:324	+9:+38	+94:+278	+106	253	?	[1.1]	[0.007:0.061]	0.1
18	WE	292:332	−28:−8	+94:+195	+106	51	<12.8	<0.12	0.001:0.008	?
19	OA	54:185	−5:+31	−165:−91	+35	2374	10?	[44]	1?	
20	M	132:194	+46:+69	−141:−91	−95	174	<4?	<0.15	−0.006:−0.008	1.0?
21	R	63:75	+6:+15	−149:−108	−84	50	?	[0.36]	?	
22	G	77:121	−19:−1	−199:−92	+31	214	>1.3	[2.1]	?	
23	IV Arch	80:220	+30:+85	−30:−94	−40	1300	0.8−1.8	0.13	∼−0.005	1.0
24	IV Spur	210:290	+45:+90	−30:−80	−35	100	0.3−2.1	0.18	∼−0.006	1
25	LLIV Arch	110:185	+27:+37	−30:−70	−20	400	0.9	0.15	∼−0.004	1.0
26	K	30:70	+25:+68	−60:−95	−80	35	<6.8	<0.75		1
27	PP Arch	83:170	−78:−33	−85:−45	−50	375	<1.1	<0.05		0.5
28	gp	34:68	−44:−23	+55:+90	+35	21	0.8−4.3	0.16		1
29	IV−WA	270:320	+30:+62	+45:+90	+40	900	?	0.03		?

Notes: (a) $\overline{v}_{\rm DEV}$ is the column density-weighted average of $v_{\rm DEV}$ for each beam on the cloud. (b) Ω is the cloud area. (c) The measured or most-likely theoretical distance. (d) H I mass of complex at the given distance; numbers in square brackets indicate that 10 kpc was assumed. (e) Mass infall rate, calculated as the ratio of mass to time-to-impact, with the latter given by D/v_z, where v_z is calculated as either $v_{\rm DEV} \times \sin b$ or $v_{\rm DEV}/\sin b$ (projecting $v_{\rm DEV}$ on the vertical or assuming that the space velocity is vertical). (f) Measured metallicity of the complex.

most notable as-yet-unnamed grouping is the positive-velocity IVC in the region around (l,b)=(290°,+40°). Neither detailed maps nor absorption-line studies have been made of this object. As can be seen by comparing Figs. 2 and 3, this IVC complex connects closely to HVC WA; I therefore propose the name "IV-WA" for it. Finally, the high-velocity gas straddling declination −30° that forms complex WD may have low-latitude IV gas associated with it that lies below the declination limit of the *LDS*.

In Fig. 3b, many squarish ("blocky") positive-velocity clouds appear to be visible in the region l=60° to 150°, b~45°, as well as around the North Galactic Pole. These are artifacts – in the *LDS* spectra there is often an extended positive-velocity wing, which does not show up in spectra taken in the same directions with other telescopes. Most likely, this wing is a low-level (<0.1 K) baseline problem, present in this part of the sky because of the extensive high- and intermediate-velocity gas.

3. Features of the H I sky

3.1. OVERVIEW OF HVC AND IVC COMPLEXES

Wakker & van Woerden (1991), Kuntz & Danly (1996), and Wakker (2001) defined a total of 28 different HVC and IVC complexes. A 29th was defined in the previous section. Table 1 summarizes the main parameters of all of these complexes. A complete discussion of the derived distances and metallicities is given in Ch. 10. Values for the infall rate represented by each cloud are only given when it makes sense to do so. Many small, previously unstudied HVCs and IVCs can also be recognized in Figs. 2 and 3. Most of these were included in the catalog of Wakker & van Woerden (1991), but especially the smaller IVCs have not been cataloged or studied.

3.2. COMPARISON OF HVC AND IVC MORPHOLOGY

Figure 4 shows an overlay of the HVCs (from Hulsbosch & Wakker 1988) on the IVCs (from the *LDS*) in the northern sky, using the deviation velocity to define them. As can be seen from Fig. 4, parts of complex A overlap the LLIV Arch, and the southern edges of both clouds (from (l,b)= (140°,+30°) to (170°,+45°)) seem to align. However, this is probably a chance coincidence, as a) the main line of complex A cores extends past the LLIV Arch, b) the main line of LLIV Arch cores lies perpendicular to the main line of complex A cores, c) the distance of the LLIV Arch is ~1 kpc, while that of complex A is ~10 kpc (Wakker 2001).

Complex C and the IV Arch overlap over a significant fraction of their extent. However, there are many differences in detail. For instance, the main cores of complex C lie on a line from (l,b)= (70°,+40°) to (135°,+55°), while the IV Arch lies along (l,b)= (105°,+35°) to (170°,+75°). Also, much of complex C lies at longitudes <100°, where there is no IV Arch gas, and conversely, much of the IV Arch and IV Spur lie at l>150°, where complex C is absent. Some complex C cores align with an IV Arch core (e.g. C III and IV 19 near (l,b)= (120°,+55°), but others lie away from IV gas (e.g. C I near

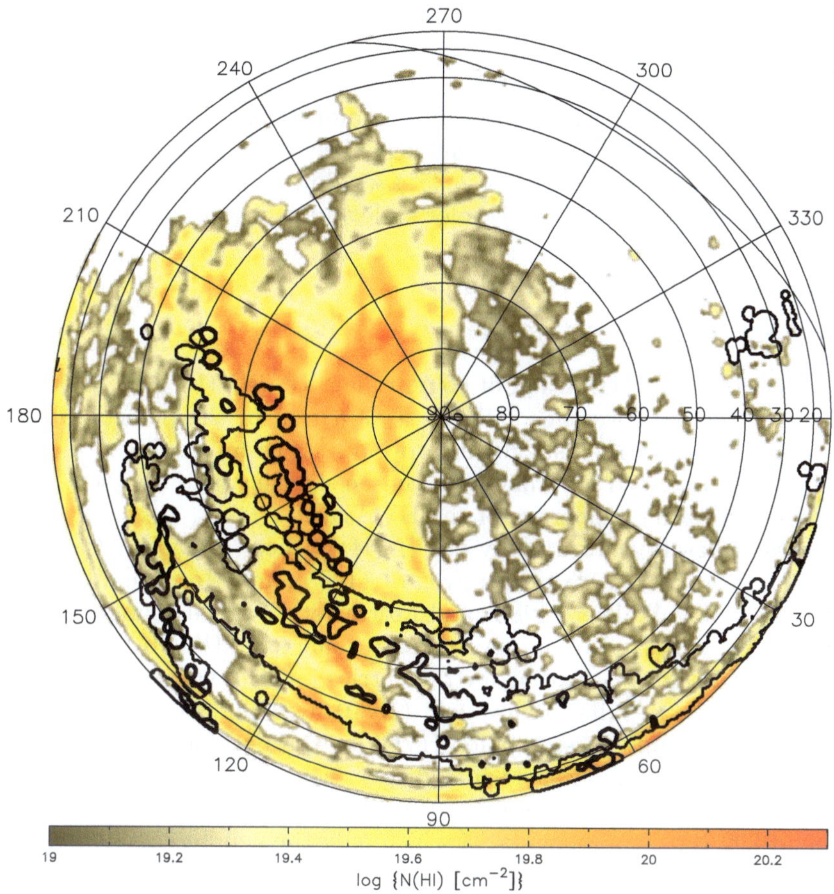

Figure 4. Polar view of the negative-velocity HVCs ($v_{\rm DEV}$<−90 km s^{-1}; thick contours at 0.05 and 0.5 K; from Hulsbosch & Wakker (1988) and Morras et al. (2000)), overlaid on a color-coded map of the column density of the negative-velocity IVCs ($v_{\rm DEV}$=−90 to −35 km s^{-1}; from the *LDS*; Hartmann & Burton 1997). Galactic longitude is indicated along the outer circle, galactic latitude on the x-axis. The line shows the −30° declination limit of the *LDS*. Figure composed for this article.

(l,b)= (90°,+45°)). The southern edge of complex C shows no morphological relation with that of the IV Arch. Finally, the distance of the IV Arch is ~1 kpc, while complex C lies at z>4 kpc (Wakker 2001). Thus, even though the northern sky is filled with large, bright HVC complexes (A and C) as well as large, bright IVC complexes (IV Arch, IV Spur, LLIV Arch), and even though such large, bright complexes are less common elsewhere, it appears that these HVCs and IVCs are not directly related.

Cloud M I appears to be part of the IV Arch. This is apparent in Fig. 4, where the thick contour around (l,b)= (160°,+65°) that outlines cloud M I ($v_{\rm DEV}$~-100 km s^{-1}) lies on top of a bright spot in the map of the intermediate-velocity gas. Additionally, the column density in cloud M I is slightly above 8×10^{19} cm^{-2}, very similar to the values in the IV Arch near M I. The situation is less clear-cut for clouds M II at (l,b)= (175°,+55°) and M III at (180°,+55°), but they are also likely to be part of the IV Arch.

Complex L (Fig. 4, (l,b) ~345°,+35°) straddles $v_{\rm DEV}$~-90 km s^{-1}, and therefore partly shows in the IVC color scale and partly as an HVC contour.

4. Column density distribution

4.1. RELEVANT OBSERVATIONS

In this section the distribution of column density, $P[N(\text{H I})]$, is discussed. This gives the sky coverage of HVCs as function of the detection limit, helps to predict how often strong absorption lines (from e.g. Mg II, C II and O I) should be detected in a randomly chosen sightline, and says something about the ionization properties of the HVC gas.

Several datasets allow a study of $P[N(\text{H I})]$: those of Giovanelli (1980), Hulsbosch & Wakker (1988), Murphy et al. (1995), and Morras et al. (2000). The Hulsbosch & Wakker (1988) and Morras et al. (2000) papers list T_B, which can be converted to an approximate value for N(H I) by assuming a line width of 25 km s^{-1} (except for the entries (~10%) where the lists give a measured (larger) value). Giovanelli (1980) used the 300-ft *Green Bank* telescope (9.′7 beam) to search for HVCs in the declination range $-18°$ to $+55°$, on a grid that was 1° in right ascension and 2.°5 in declination. The nominal 5-σ detection limit of this survey was about 3×10^{18} cm^{-2}. He found 799 HVC components, of which only the good components, with "quality-factor" 1 or 2, are used here. Finally, Murphy et al. (1995) used the 140-ft *Green Bank* telescope (21′ beam) to obtain spectra for 171 directions toward and near 118 AGNs observed with the Faint Object Spectrograph on the Hubble Space Telescope. Their 5-σ detection limit was 7×10^{17} cm^{-2}. Savage et al. (2000) list the detections.

4.2. PREVIOUS ANALYSES

The easiest way to compare the N(H I) distributions derived from the different surveys is to use the percentage of sky covered by HVCs in the area covered by the survey. This is the sum over all detections of the quantity [cell area, in square degrees]/[square degrees of sky covered by the survey]. The cell area is 2.5 square degrees for the Giovanelli (1980) data, 0.25 square degrees for the Morras et al. (2000) list, and $\Delta l \Delta b \cos b$ square degrees for the Hulsbosch & Wakker (1988) list, where Δb=1° and Δl is (1°, 2°, 2.°5) for $|b|$(<45°, 45° to 60°, >60°). For the Murphy et al. (1995) list, the quantity "percentage of sample" was used. The column density distributions were derived using a velocity limit of $|v_{\rm DEV}|>90$ km s^{-1}. However, there are only marginal differences with the distributions obtained if $|v_{\rm LSR}|>100$ km s^{-1} were to be used, as was done by Giovanelli (1980) and Murphy et al. (1995).

Figure 5 shows the resulting distributions of N(H I) in several ways, differing in how the binning is done and on whether or not logarithmic scales are used. This matters because these differences in presentation lead Giovanelli (1980), Murphy et al. (1995) and this article to different conclusions.

The top row of Fig. 5 shows the distributions that are usually shown: P[log(N(H I))] and log P[log(N(H I))]. That is, the percentage of sky covered in constant 0.1 dex intervals of log(N(H I)). The dotted lines show the distribution presented by Giovanelli (1980). This peaks near log(N(H I)) $\sim$19.2, a factor $\sim$3 above the 5-σ detection limit. From this, Giovanelli (1980) concluded that a) incompleteness at low column densities was not a problem and b) HVCs have a typical column density of log(N(H I))=19.2. For log(N(H I))>19.5 there clearly is a very good correspondence between the distributions of Giovanelli (1980; dotted line) and of Hulsbosch & Wakker (1988; thin solid line). At lower column densities, Giovanelli's (1980) survey seems to be incomplete. This is most easily explained as a result of the large grid/beam ratio for this survey (2.5 square degrees over $(10')^2$, or $\sim$100), while that of the Hulsbosch & Wakker (1988) data is $\sim$4. Thus, faint and small clouds were more easily missed in Giovanelli's (1980) survey.

The distribution presented by Murphy et al. (1995) is given by the thick solid line in Fig. 5a. This matches up well with the Hulsbosch & Wakker (1988) data. Murphy et al. (1995) noted the downturn at N(H I)<10^{18} cm^{-2}, somewhat above their 5-σ completeness limit of 7×10^{17} cm^{-2}. From this they surmised that there is little neutral gas below their sensitivity limit.

The N(H I) distributions for the northern sky (declination >−18°; Hulsbosch & Wakker 1988) and the southern sky (declination <−23°; Morras et al. 2000; not shown) have identical shapes. To compare to Giovanelli (1980) and Murphy et al. (1995) only the northern data will be used, as a) each of these surveys only covers the northern sky and b) the detection limit of the southern survey is slightly worse than that of the northern survey.

4.3. CURRENT RESULTS

Based on Fig. 5a, the conclusions of Giovanelli (1980) and Murphy et al. (1995) seem reasonable. However, for the Hulsbosch & Wakker (1988) *Dwingeloo* survey, the modus lies at log(N(H I)) = 18.7, rather than at log(N(H I)) = 19.2 as in the Giovanelli (1980) dataset, or log(N(H I)) = 18.2 as in the Murphy et al. (1995) data. Thus, in each case the "typical" value lies about 0.5 dex above the 5-σ detection limit! A little thought

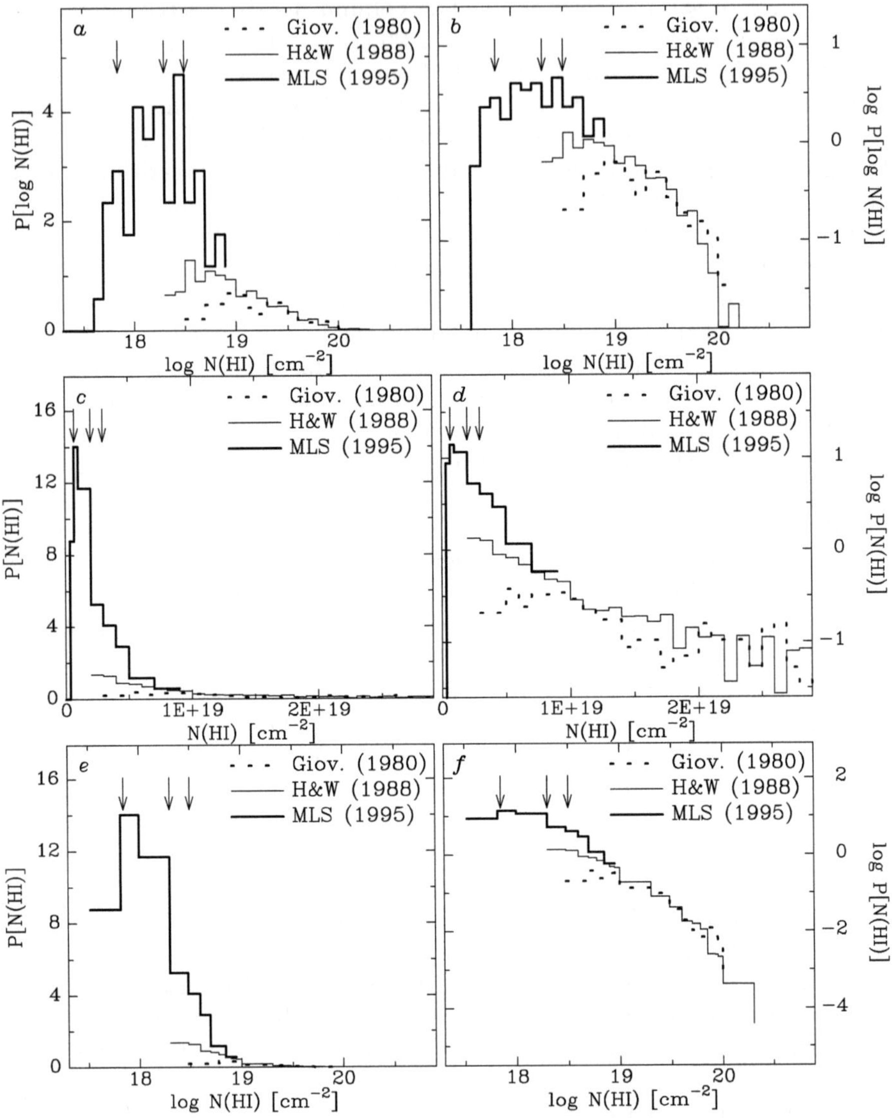

Figure 5. Distributions of $N(\text{H I})$ for H I with v_{DEV}>90 km s^{-1}. *Left:* percentage of sky covered. *Right:* logarithm of percentage of sky covered. *Top row:* standard logarithmic binning of log($N(\text{H I})$) in intervals of 0.1 dex. *Middle row:* linear binning of $N(\text{H I})$ in intervals of 10^{18} cm^{-2}, with linear $N(\text{H I})$ scale on the x-axis. *Bottom row:* linear binning of $N(\text{H I})$ in varying intervals with logarithmic $N(\text{H I})$ scale (see text). Dotted histogram: data from Giovanelli (1980) with quality index Q<=2. Thin solid histogram: data from Hulsbosch & Wakker (1988). Thick solid histogram: data from Murphy et al. (1995). Arrows point to the nominal detection limits for these surveys. Figure composed for this article.

shows that binning data in the manner shown in Figs. 5a, b easily lends itself to misinterpretation. The reason is that $P[\log(N(\mathrm{H\,I}))]$ d $\log(N(\mathrm{H\,I}))$ $= P[N(\mathrm{H\,I})]$ d$N(\mathrm{H\,I})$, so that $P[\log(N(\mathrm{H\,I}))] = N(\mathrm{H\,I})\ P[N(\mathrm{H\,I})]$. Thus, if at low $N(\mathrm{H\,I})$ the slope of $P[N(\mathrm{H\,I})]$ as function of $N(\mathrm{H\,I})$ is >-1, the modus of the distribution of $\log(N(\mathrm{H\,I}))$ does not correspond to the most common value of $N(\mathrm{H\,I})$ and does not indicate a "typical" column density.

To determine whether the sky coverage as function of $N(\mathrm{H\,I})$ increases or decreases at low $N(\mathrm{H\,I})$, linear intervals should be used. However, $N(\mathrm{H\,I})$ has a range of a factor >100 and binning in small intervals results in many empty bins at high column densities. One possible solution to this dilemma is shown in Figs. 5c and d, where $P[N(\mathrm{H\,I})]$ is binned in 10^{18} cm^{-2} intervals, and only a limited $N(\mathrm{H\,I})$ range is shown. This makes it clear that there is no "typical" value of $N(\mathrm{H\,I})$. In fact, for the Murphy et al. (1995) and the Hulsbosch & Wakker (1988) datasets, the most common value is the lowest value, while the Giovanelli (1980) data seem incomplete below $\sim 7\times10^{18}$ cm^{-2}. Another solution would be to present cumulative histograms. However, these do not allow to determine whether or not there is a "typical" column density based on a simple visual impression.

Figures 5e and f present a different solution, achieved by binning $N(\mathrm{H\,I})$ in intervals of 10^{18} cm^{-2} below 10^{19} cm^{-2}, in intervals of 10^{19} cm^{-2} between 10^{19} and 10^{20} cm^{-2} (with the result scaled down by a factor 10), and in intervals of 10^{20} cm^{-2} between 10^{20} and 10^{21} cm^{-2} (with the result scaled down by a factor 100). The final sky-coverage percentages are equivalent to binning $N(\mathrm{H\,I})$ in intervals of 10^{18} cm^{-2} for all values between 10^{18} and 10^{21} cm^{-2}, but avoiding the problem that most high column density bins would be empty.

Figures 5e and f gives the most honest representation of the data, if the purpose of the plot is to compare the sky coverage at different column densities and to estimate incompleteness. It is clear that the distribution obtained by Murphy et al. (1995) connects rather smoothly to that derived from Hulsbosch & Wakker (1988) in the overlap region between 5×10^{18} and 10^{19} cm^{-2}. This comparison suggests that the latter survey is incomplete by a factor 2 for $N(\mathrm{H\,I})\sim 3$ to 5×10^{18} cm^{-2} ($T_B\sim 0.07$ to 0.10 K) and a factor 3 for $N(\mathrm{H\,I})\sim 2$ to 3×10^{18} cm^{-2} ($T_B\sim 0.05$ to 0.07 K). Since the Murphy et al. (1995) data extend the distribution down to $N(\mathrm{H\,I})\sim 7\times10^{17}$ cm^{-2}, but no strong turnover is seen in the distribution $P[N(\mathrm{H\,I})]$, it is likely that gas with lower column densities may be fairly common.

The fact that the relative number of detections is largest at the smallest column densities shows that one can *not* find a typical column density for the high-velocity gas, as has previously been claimed by many authors. Therefore, *it is not possible to derive a meaningful average or median column density.* A sample average or median will be proportional to the low-end cutoff, and will be independent of the intrinsic cloud properties. Also, the most common value for the column density can not be found from $P[\log(N(\mathrm{H\,I}))]$, as it will be the lowest measurable column density.

4.4. SKY COVERAGE

High-velocity gas with $N(\mathrm{H\,I})>7\times10^{17}$ cm^{-2} was seen in 48% (82 of 171) of the directions observed by Murphy et al. (1995), while 27% (46 of 171) have $N(\mathrm{H\,I})>2\times10^{18}$ cm^{-2}. The latter can be compared to the 15% sky coverage in the Hulsbosch & Wakker (1988) HVC survey (excluding the Outer Arm; see Wakker 1991). Most (25 out of 36) sightlines with $8\times10^{17}<N(\mathrm{H\,I})<2\times10^{18}$ cm^{-2} are located near or toward the large HVCs (the Magellanic Stream, complexes A, C, M). At first sight, these numbers suggest that the large clouds have extended halos, and that the sky coverage of gas with $N(\mathrm{H\,I})=7\times10^{17}$ to 2×10^{18} cm^{-2} is 21%.

However, none of these 25 detections lie more than 2° from the cloud edges defined by the 2×10^{18} cm^{-2} contour. Further, for complex C three of the seven low column density detections are secondary components falling within the 2×10^{18} cm^{-2} contour of the main component, thus they do not really increase the sky coverage. Eight low column density detections are in the tip of the Magellanic Stream, and lie within the 2×10^{18} cm^{-2} contour seen with *Dwingeloo*, also not increasing the sky coverage; small-scale structure can easily result in lower column densities when observing with the smaller *Green Bank* 140-ft beam. Eliminating these components reduces the sky coverage between 7×10^{17} and 2×10^{18} cm^{-2} to 15% (25 of 171). Thus, for a randomly chosen direction on the sky, the probability is 30% that high-velocity gas with $N(\mathrm{H\,I})>7\times10^{17}$ cm^{-2} will be found, and 15% to find high-velocity gas with $N(\mathrm{H\,I})>2\times10^{18}$ cm^{-2}.

4.5. HYDROGEN IONIZATION

Interferometer observations of the H I disks of galaxies appear to show a column density cutoff at the outer edges. One of the deepest observations is that of NGC 3198 by van Gorkom et al. (1993 – unpublished, but quoted in Maloney 1993), who find that, with a detection limit of 4×10^{18} cm^{-2}, the H I disk stops where $N(\mathrm{H\,I})<4\times10^{19}$ cm^{-2}. The explanation put forward by Maloney (1993) is that the extragalactic ionizing background ionizes the hydrogen when the column density drops below this level. The precise degree of ionization depends on the details of the volume density structure and on the strength of the extragalactic radiation field. How can this be reconciled with the fact that the Galactic HVCs become more common at low column densities, with >95% of the HVC gas seen in H I emission having $N(\mathrm{H\,I})<4\times10^{19}$ cm^{-2}?

In the example given by Maloney (1993), the amount of H^+ is on the order of 5×10^{19} cm^{-2} for total hydrogen column densities above 5×10^{19} cm^{-2}, so that the fraction N(H I)/N(H)(tot) decreases from ~0.66 at N(H I)= 10^{20} cm^{-2} to ~0.1 at N(H I)=5×10^{18} cm^{-2}. At N(H I)= 2×10^{18} cm^{-2} the neutral fraction is only 0.04. Thus, if the HVCs are objects in the Galactic Halo, and if they are bathed in the extragalactic radiation field, most of the HVC gas should be in the form of H^+, with the H I being the tip of the iceberg. This is especially true for the many faint and small HVCs with N(H I) in the range 2 to 10×10^{18} cm^{-2}.

Two studies of sightlines toward extragalactic sources shine light on this possibility. In the direction toward Mrk 290, Wakker et al. (1999) found N(H I)=9×10^{19} cm^{-2}, and $N(H^+)$=1 to 2×10^{19} cm^{-2}. This is a higher N(H I)/N(H)(tot) ratio than in Maloney's (1993) example, but not incompatible with that model. On the other hand, toward Mrk 876 Murphy et al. (2000) need H^+/H I>3 to reconcile the apparent Fe II/H I ratio of 0.5 times solar for complex C with the intrinsic abundance of ~0.14 solar implied for this HVC by other sightlines. Since N(H I)~2×10^{19} cm^{-2}, N(H) must be >8×10^{19} cm^{-2}. Although the model provided by Maloney (1993) predicts N(H)=7×10^{19} cm^{-2} for this N(H I), the implied column densities are not inconsistent with the idea that lower values of N(H I) imply higher H^+/H I ratios. A problem, however, is that no Hα emission associated with complex C is observed in the direction of Mrk 876, which is only possible if the H^+ is thinly spread over a many kpc (>5) thick layer.

To determine whether most of the HVC gas is indeed ionized, two kinds of observations are needed: deep Hα exposures for clouds with known distances, and more observations of interstellar absorption lines of atoms in different, adjacent ionization stages. This can then be compared to photoionization models, such as those of Ferrara & Field (1994) and Bland-Hawthorn & Maloney (1999).

5. Volume density distribution

The mean volume density of an H I cloud can be calculated if its mass and volume are known. The catalog of Wakker & van Woerden (1991) lists the H I mass, $M(\text{H I}) = 0.236\, S\,(D/1\ \text{kpc})^2\ \text{M}_\odot$ (with the flux, S, in Jy km s^{-1}). Here, an update to this catalog is used, now including the components in the list of Morras et al. (2000); about 60 new (mostly southern) clouds were added. Further, about 20 previously cataloged clouds were split, combined, or deemed to be non-existent. A volume estimate can be found from the cloud's radius: R=αD, where α is the cloud's angular size. The radius can be estimated from the cloud's surface area: Ω=$\pi\alpha^2$. The average volume density then is: $n = 0.0101\,(0.236S)\,\Omega^{-3/2}\,D^{-1}$ cm^{-3} (expressing S in

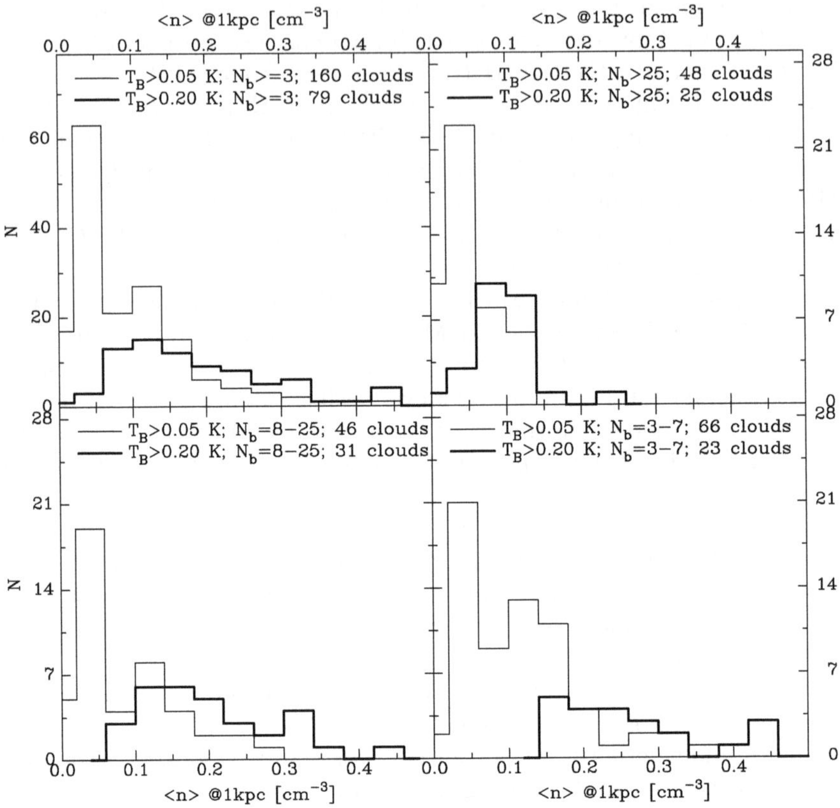

Figure 6. Distribution of volume density at assumed distance of 1 kpc for the clouds in the updated catalog of Wakker & van Woerden (1991) (see Sect. 5). Panels b–d give the distributions for clouds detected on >25, 8–25 and 3–7 beams. Panel a shows a combined histogram. The thin-lined histogram uses a detection limit of 0.05 K, the thick-lined histogram is the result that would have been obtained if the detection limit of the Hulsbosch & Wakker (1988) survey had been 0.20 K. Figure composed for this article.

Jy km s^{-1}, Ω in square degrees, D in kpc, and assuming spherical clouds). Usually one does not know the distance to the cloud and only the angular radius can be measured. Thus, the estimated volume density remains uncertain up to a factor inversely proportional to the distance.

There are some problems with this calculation. First, the clouds are probably not spherical. For a disk-shaped cloud that is $5^\circ \times 1^\circ$ on the sky, the volume will be underestimated by a factor 2.3, while for a cigar-shaped cloud it will be overestimated by a factor 2.1. These factors are smaller than the width of the distribution of volume densities (see Fig. 6), so that a statistical conclusion is still possible. Second, chances are that the density is not constant in the cloud, so that the relevance of the *average* density remains somewhat unclear. Third, the radius estimate depends on the brightness temperature cutoff of the data, and for small clouds also on the beam size. If the brightness cutoff is set too high only the cloud cores are included, whereas most of the cloud area is in the fainter outer parts. This effect is illustrated in Fig. 6.

Figure 6 shows the resulting distribution of average densities. To better understand the selection effects, separate distributions are shown for small, intermediate-size and large clouds and for two different brightness temperature cutoffs. Clouds that were only detected on one or two gridpoints have been omitted, as for these the radius calculation is rather uncertain.

Using all data (T_B>0.05 K), the most common average density is about $0.05\,(D/1\ \mathrm{kpc})^{-1}\ \mathrm{cm}^{-3}$, independent of cloud size. However, the effect of a brightness temperature cutoff can clearly be seen by comparing the thin and thick lines. The latter give the distributions using only detections with T_B>0.20 K. For the larger clouds this slightly increases the modus of the distribution to $\sim$$0.1\,(D/1\ \mathrm{kpc})^{-1}\ \mathrm{cm}^{-3}$, while for the smaller clouds the modus increases substantially to $0.25\,(D/1\ \mathrm{kpc})^{-1}\ \mathrm{cm}^{-3}$. Therefore, within the uncertainties introduced by selection effects and intrinsic problems with the volume-density estimation, the average density of the HVCs is on the order of 0.05 to $0.2\,(D/1\ \mathrm{kpc})^{-1}\ \mathrm{cm}^{-3}$. Such average densities are compatible with the directly measured value of $\sim$0.08 cm^{-3} found for complex C from the combination of H I, Hα and [S II] emission and S II absorption (Wakker et al. 1999).

6. Distributions of cloud parameters

6.1. DISTRIBUTIONS OF CLOUD AREA AND FLUX

Blitz et al. (1999; Ch. 14) proposed that most HVCs are intra-Local-Group clouds, with a median distance of 1 Mpc, a median radius of 14 kpc, and a median H I mass of 1.9×10^7 M$_\odot$. They based their median distance on the assumption that the clouds randomly fill the Local Group (i.e. they assumed that the median distance is 1 Mpc) and then derived the other parameters from the Wakker & van Woerden (1991) catalog. Several of the distributions relevant to this model are shown in Fig. 7. The left row uses the standard binning of constant intervals in log$x$, while on the right, x is binned linearly, with variable bin size, in the manner described in Sect. 4.3 (here x stands for cloud size, flux, virial distance and virial mass).

Figures 7a and b show the distribution of cloud areas, using the updated Wakker & van Woerden (1991) catalog (see Sect. 5). Binning the cloud areas in linear bins (Fig. 7b) yields a power-law distribution over the full range of values. There is a turnover at $\sim$1 square degree, but that is due to incomplete sky coverage and the fact that the smallest clouds (0.25

to 0.5 square degrees) can only be found in the southern survey. Taking this into account, there is no real turnover at the lowest areas, and no median cloud size can be determined. This is even apparent when binning areas logarithmically, as the slope of the distribution is $\sim$−1, so that the logarithmic binning does not introduce an artificial turnover (see Sect. 4.3).

Figures 7c and d show the distribution of cloud fluxes, calculated as the integral of $T_B W$ over the cloud area (with appropriate unit conversion factors; W is the line width). This distribution is a power law over most of its range. The logarithmically-binned function does appear to show a turnover at the lowest fluxes, but this is an artifact because the slope of the function is not steep enough to compensate for the effects of binning.

6.2. VIRIAL DISTANCES AND MASSES

The crux of the model of Blitz et al. (1999) is that each HVC is supposed to contain a sufficient amount of dark matter to make it self-gravitating, so that the virial theorem can be applied:

$$\sigma_{\rm 3D}^2 = \frac{GM}{2R}.$$

The 3-dimensional dispersion $\sigma_{\rm 3D}$ is $\sqrt{3}$ times the observed dispersion, σ. The cloud radius, R, can be approximated from the cloud area, Ω, as $\sqrt{\Omega/\pi}D$. The mass M is the total mass: M(H I+H^++H_2+He+dark). Observable is the H I mass, which can be corrected for M(He) to arrive at the "luminous mass", calculated as M(luminous) = M(H I+He) = $1.39 \times 0.236\, S\, D^2$ $M_\odot$, with the flux, S, expressed in Jy km s^{-1}, and the distance, D, in kpc. The total mass can then be expressed as M=M(H I+He)/f, where f parametrizes the unknown amounts of H^+, H_2 and dark matter. Blitz et al. (1999) prefer f=0.1. Then, the cloud distance can be derived from the observables as:

$$D_{\rm vir} = \frac{6\,\sqrt{\Omega/\pi}\,\sigma^2\,f}{1.39 \times 0.236\,G\,S}.$$

When expressing D in kpc, Ω in steradian, σ in kpc Myr^{-1}, and S in Jy km s^{-1}, the value of G is 4.53×10^{-12} kpc^3 Myr^{-2} $M_\odot{}^{-1}$.

Figures 7e and f show the distribution of $D_{\rm vir}$ calculated in this manner, for two values of f: f=0.1 (thin solid line) and f=0.01 (thick solid line). The large HVC complexes (MS, C, A, M, G) are probably nearby and not virially stable, and for clouds seen on less than 3 grid points the calculation of σ is less reliable. Therefore, a separate histogram is shown with these clouds excluded (dotted line). Clearly, the shape of the distribution does not change qualitatively, although only one fifth of the HVCs are included. With f=0.1 the virial stability distances in Fig. 7f show a more or less flat distribution between $D_{\rm vir}$=100 kpc and 2 Mpc.

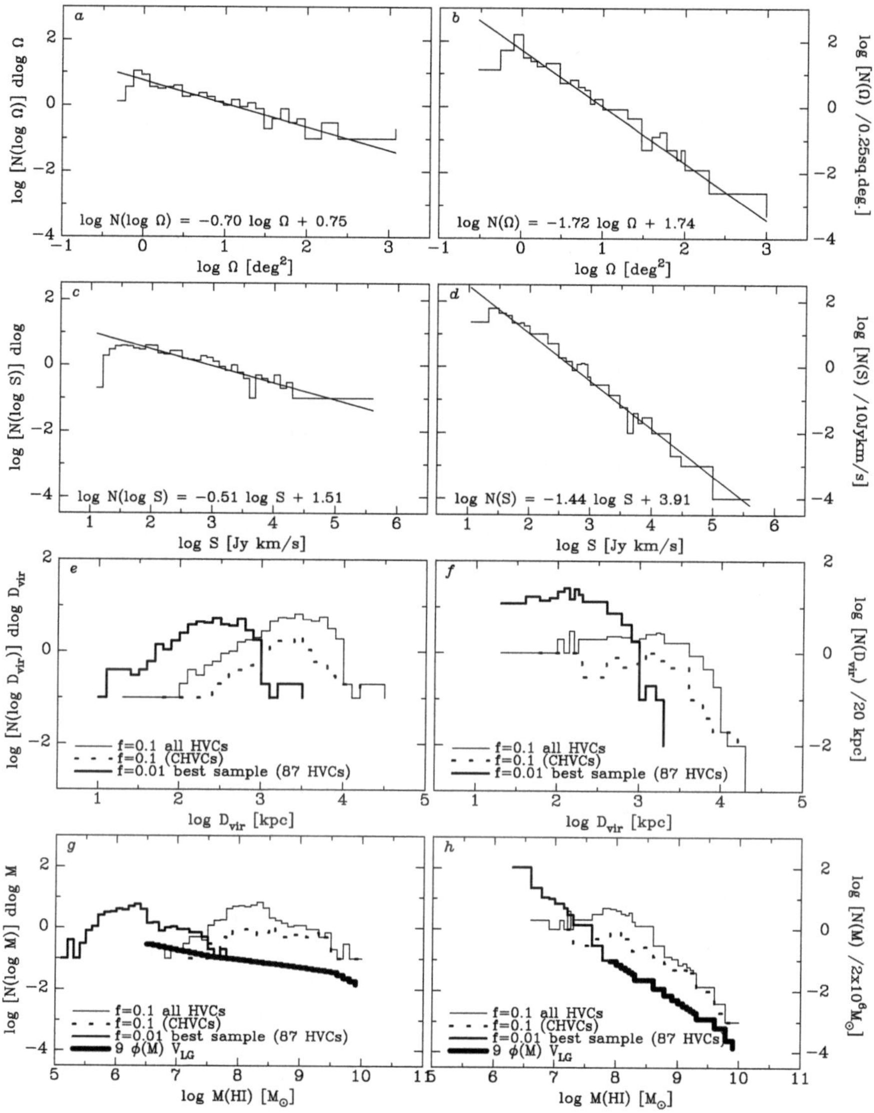

Figure 7. Distribution of cloud parameters, based on the (update of the) Wakker & van Woerden (1991) catalog. *Left:* the logarithm of the x-axis parameter (area, flux, virial distance, virial H I mass) was binned in intervals of 0.1 dex. *Right:* the x-axis parameter was linearly binned in bins of 0.25 square deg, 10 Jy km s^{-1}, 20 kpc, and 2×10^6 M$_\odot$, using the method described in Sect. 4.3. For the area and flux distributions the parameters of a least-squares fit are shown. Three histograms are shown for the virial distance and mass, with criteria summarized in the label. The very-thick solid histogram is the Zwaan et al. (1997) H I mass function, $\phi(M)$, scaled by a factor 9 to account for the Local Group overdensity and a factor V_{LG} to account for the Local Group volume. Figure composed for this article.

If the cloud distance and flux are known, the cloud's luminous mass can be calculated. The resulting distributions in Fig. 7g and h are shown for a) full HVC sample with f=0.1 (thin lines), f=0.01 (thick lines), and excluding complexes MS, C, A, M, and G, and clouds seen on less than 3 grid points (dotted lines). The latter is a power-law for M(H I)>10^8 $M_\odot$, but flatter at lower masses (Fig. 7h).

6.3. CONSTRAINTS ON THE MODEL FOR LOCAL GROUP HVCS

To determine whether one expects the Local Group to contain many massive H I clouds, one can estimate the expected number of clouds in the Local Group as function of H I mass. This estimate can be obtained by observing other small groups, or by combining the H I mass function of field galaxies and an estimate of the overdensity of the Local Group.

The first method was applied by Zwaan (2001), who used the *Arecibo* telescope to observe five galaxy groups similar to the Local Group, on five 3′ strips lying about 1° apart, with a 5-σ detection limit of ~8×10^6 $M_\odot$. He found no star-less H I clouds and concluded that the total H I mass of intra-group clouds must be less than 10% of the total H I mass of the group.

A constraint on the value of f can be derived from the second method, in a manner similar to that presented by Zwaan & Briggs (2000). Zwaan et al. (1997) observed a 3′×300° strip out to ~100 Mpc (less for galaxies with M(H I)<10^9 $M_\odot$) using the *Arecibo* telescope, and used this to derive the H I mass function – the number of galaxies per Mpc^3 per unit mass interval. This requires assuming a value for the Hubble constant to convert redshift to distance; 75 $km\,s^{-1}\,Mpc^{-1}$ is used here. The overdensity of the Local Group can be estimated by combining the H I masses of the Local Group galaxies with Local Group volume. H I masses were taken from van den Bergh (1999; M 31 and the Milky Way), Huchtmeier (1978; M 33), Kim et al. (1998; LMC), Stanimirovic et al. (1999; SMC) and Mateo (1998; the dwarfs). The Local Group volume can be estimated by noting that the zero-velocity surface of the Local Group lies 1180 kpc from its barycenter (van den Bergh 1999), thus enclosing a volume of about 7 Mpc^3. Binning the H I masses in intervals of 0.5 dex, and comparing the number of galaxies in each bin to the H I mass function shows that there are 9 times more galaxies with M(H I)>3×10^7 $M_\odot$ in the Local Group than in the field.

The resulting H I mass distribution for the Local Group is shown as the very-thick-lined histogram in Fig. 7g and h. Clearly, if HVCs are virially stable and have f=0.1, the number of massive HVCs in the Local Group (thin solid and dotted lines in Fig. 7h) would be 10 to 100 times larger than the number of galaxies of the same mass in the Local Group (thick solid line in Fig. 7h). Also, the number of HVCs per unit volume would be

100 to 1000 times larger than the number of objects detected at *Arecibo* in a comparable volume. Even if only the best sample (no large nearby complexes and only clouds with well-determined dispersions) is used, these ratios are still 2 to 20 and 20 to 200, respectively.

There are four ways to reconcile this discrepancy. First, the Local Group may be unusual and indeed have many more massive starless H I clouds than the general field. Second, Zwaan et al. (1997) may have underestimated the H I mass function, although a factor ~10 is required. This would require that HVC-like objects are small and are being missed due to beam dilution. However, an "average" HVC with a radius of 3° at a typical distance of 500 kpc would still have a radius of 9′ at a distance of 10 Mpc, and thus it would easily be seen in the *Arecibo* data. Third, the value of f could be lower. As the Zwaan et al. (1997) H I mass function is unreliable for $\log M<7.5$, the thick-lined histogram in Fig. 7h shows that f=0.01 appears to be sufficient to get rid of the incompatibility. However, this would imply that most virial distances are in the range 20 to 300 kpc, and hence the HVC ensemble would not fill the Local Group, but rather be concentrated around the Milky Way. Lowering f can be done by increasing the relative amount of dark matter, but also by having a large ionized fraction, e.g. 90%. As discussed in Sect. 4.5 this can not be ruled out, and in fact seems rather likely. Fourth, only a small fraction of the HVC sample may be Local Group objects. For f=0.1, 5 to 10 such objects could exist without causing a discrepancy. For lower f proportionally more are allowed to exist.

Thus, the HVC model as originally proposed by Blitz et al. (1999) is not compatible with the Zwaan et al. (1997) H I mass function. A population of dark-matter-containing HVCs at distances up to 300 kpc is not ruled out, however, if the H I represents a small fraction of the total mass (<1%), which is possible if the ionized fraction is high. To reconcile such a population with the H I mass function requires that the clouds concentrate around the Milky Way, rather than fill the Local Group as a whole.

7. Conclusions and outlook

The results presented in this chapter lead to the following conclusions.

1) Several new surveys containing HVC information have provided a wealth of new information. The Morras et al. (2000) list of southern profile components provides a much needed complement to the list of northern profile components (Hulsbosch & Wakker 1988). The Leiden-Dwingeloo Survey (*LDS*) of Hartmann & Burton (1997) is extremely useful, especially for the study of the intermediate-velocity clouds. However, its sensitivity for faint HVCs is less than that of the other surveys, because of low-level baseline problems (especially for positive velocities in the northern sky).

2) From the *LDS* it will be possible to create a catalog of the IVCs. A systematic approach to component fitting will be most useful for this purpose. A catalog for the southern IVCs is also within reach, using the data of Arnal et al. (2000); however, these are not yet publicly available.

3) Using the *LDS* the issue of angular correlations between HVCs and IVCs can be studied in much more detail. So far, the results are inconclusive (see Sect. 3.2) and more work is needed. However, taking into account the different distances indicated for HVCs and IVCs (see Ch. 10), it seems likely that the two are not generally interacting.

4) The distribution of HVC column densities does not have a modal value, contrary to earlier claims by Giovanelli (1980) and Murphy et al. (1995). These claims were artifacts based on the improper binning applied by these authors. In fact, the distribution increases monotonically all the way down to the detection limit. At the lowest H I column density levels (below about 3×10^{19} cm^{-2}), hydrogen ionization becomes important.

5) The sky coverage fraction of HVCs is about 30% for column densities $>7\times10^{17}$ cm^{-2}. This number should be compared with statistics from absorption-line data, which are more sensitive to low values of N(H I).

6) The mean volume density of HVCs is about 0.05 to 0.25 $(D/1\ \mathrm{kpc})^{-1}$ cm^{-3}. Using methods such as the one applied by Wakker et al. (1999) or from absorption-line studies of excited ions (such as C^{+*}), it may be possible to directly measure more volume densities. If these values come out to be on the order of 0.1 cm^{-3}, as was the case for complex C, it may become possible to determine an ensemble-averaged distance for the HVCs.

7) There is no modal cloud area, but instead the distribution of areas is a power law over the full observable range from 1 to 2000 square degrees.

8) The distribution of cloud virial masses (derived in the manner proposed by Blitz et al. 1999) was compared to the distribution expected from the number density of Local Group galaxies and the field H I mass function (Zwaan & Briggs 1997). This shows that the proposed population of virially-stable clouds in which M(H I)=0.1 M(total) is incompatible with the data, unless the Local Group is a special environment. An H I mass fraction of 0.01 is possible, but this then implies that 99% of the cloud consists of dark matter or 90% of the gas is ionized *and* that the typical cloud distance is 20 to 300 kpc, i.e. the clouds concentrate around the Milky Way. In fact, only up to about ten free-floating Local Group HVCs are allowed to exist within the statistical constraints.

9) The origin of the general HVC population still remains unknown. Measuring metallicities and distances is still the best way to determine their origin. Statistically, an origin in or near the Galaxy is the most likely.

Acknowledgements

The author acknowledges financial support from NASA grant NAG5-9179.

References

Arnal, E.M., Bajaja, E., Larrarte, J.J., Morras, R., Pöppel, W.G.L. 2000, A&AS, 142, 35
Bajaja, E., Cappa de Nicolau, C.E., Cersosimo, J.C., Martin, M.C., Loiseau, N., Morras, R., Olano, C.A., Pöppel, W.G.L. 1985, ApJS, 58, 143
Bajaja, E., Morras, R., Pöppel, W.G.L. 1987, Pub. Astr. Inst. Czech. Ac. Sci., 69, 237
Bajaja, E., Cappa de Nicolau, C.E., Martin, M.C., Morras, R., Olana, C.A. 1988, A&AS, 78, 345
Barnes, D.G., Staveley-Smith, L., de Blok, W.J.G., et 36 al. 2001, MNRAS, 322, 486
Bland-Hawthorn, J., Maloney, P.R. 1999, ApJ, 510, L33
Blitz, L., Spergel, D., Teuben, P., Hartmann, D., Burton, W.B. 1999, ApJ, 514, 818
Braun, R., Thilker, D.A. 2004, A&A, 417, 421
Brüns, C., Kerp, J., Staveley-Smith, L., Mebold, U., Putman, M.E., Haynes, R.F., Kalberla, P.M.W., Muller, E., Filipovic, M.D. 2004, A&A, submitted
de Heij, V., Braun, B., Burton, W.B. 2002, A&A, 391, 159
de Vaucouleurs, G., de Vaucouleurs, A., Corwin Jr., H.G., Buta, R., Paturel, G., Fouque, P. 1991, Third Reference Catalog of Bright Galaxies (RC3) (Springer: New York)
Ferrara, A., Field, G.B. 1994, ApJ, 423, 665
Giovanelli, R. 1980, AJ, 85, 1155
Hartmann, D., Burton, W.B. 1997, Atlas of Galactic Neutral Hydrogen (Cambridge: Cambridge University Press)
Huchtmeier, W.K. 1978, in IAU Symp. 77, Structure and Properties of Nearby Galaxies, eds. E.M. Berkhuijsen, R. Wielebinski (Dordrecht: Reidel), 197
Hulsbosch, A.N.M., Wakker, B.P. 1988, A&AS, 75, 191
Kim, S., Staveley-Smith, L., Dopita, M.A., Freeman, K.C., Sault, R.J., Kesteven, M.J., McConnell, D. 1998, ApJ, 503, 674
Kuntz, K.D., Danly, L. 1996, ApJ, 457, 703
Maloney, P. 1993, ApJ, 414, 41
Mateo, M.L. 1998, ARA&A, 36, 435
Morras, R., Bajaja, E., Arnal, E.M., Pöppel, W.G.L. 2000, A&AS, 142, 25
Murphy, E.M., Lockman, F.J., Savage, B.D. 1995, ApJ, 451, 616
Murphy, E.M., Sembach, K.R., Gibson, B.K., Shull, J.M., Savage, B.D., Roth, K.C., Moos, H.W., Green, J.C., York, D.G., Wakker, B.P. 2000, ApJ, 538, L35
Putman, M.E., de Heij, V., Staveley-Smith, L., et 29 al. 2002, AJ, 123, 873
Savage, B.D., Wakker, B.P., Bahcall, J.N., et 12 al. 2000, ApJS, 129, 563
Stanimirovic, S., Staveley-Smith, L., Dickey, J.M., Sault, R.J., Snowden, S.L. 1999, MNRAS, 302, 417
Stark, A.A., Gammie, C.F., Wilson, R.W., Bally, J., Linke, R.A., Heiles, C., Hurwitz, M. 1992, ApJS, 79, 77
Tolbert, C.R. 1971, A&AS, 3, 349
van den Bergh, S. 1999, A&ARv, 9, 273
Wakker, B.P. 1991, A&A, 250, 499
Wakker, B.P. 2001, ApJS, 136, 463
Wakker, B.P. 2004, ApJ, submitted
Wakker, B.P., van Woerden, H. 1991, A&A, 250, 509
Wakker, B.P., van Woerden, H. 1997, ARA&A, 35, 217
Wakker, B.P., Howk, J.C., Savage, B.D., van Woerden, H., Tufte, S.L., Schwarz, U.J., Benjamin, R., Reynolds, R.J., Peletier, R.F., Kalberla, P.M.W. 1999, Nature, 402, 388
Wesselius, P.R., Fejes, I. 1973, A&A, 24, 15
Zwaan, M.A. 2001, MNRAS, 325, 1142

Zwaan, M.A., Briggs, F.H. 2000, ApJ, 530, L61
Zwaan, M.A., Briggs, F.H., Sprayberry, D., Sorar, E. 1997, ApJ, 490, 173

3. KINEMATICS OF HV AND IV GAS

ULRICH J. SCHWARZ
Kapteyn Institute, Groningen, The Netherlands, and Department of Astrophysics, Universiteit Nijmegen, The Netherlands; schwarz@astro.rug.nl

AND

KLAAS S. DE BOER
Sternwarte, Universität Bonn, Germany deboer@astro.uni-bonn.de

Abstract. Gas at high galactic latitudes is observed to have a large range of radial velocities. To understand its kinematics, various aspects have to be considered, including observational selection effects, as well as the absence of information about the velocity in the plane of the sky and about distances. The statistics of the velocities of both high- and intermediate-velocity (HV and IV) gas are presented. Simple models for their distribution and motion can help to interpret the observed velocities.

1. General considerations

Clouds of neutral gas detected at high galactic latitudes that have velocities deviating from gas rotating in the disk (HV and IV gas) show certain systematics in their velocities. Several ideas exist for the origin of those gases, but none of these can explain all detections. To get a feel for how the gas really moves, the simplest situations can be taken as a starting point. Before doing so, it is necessary to summarize the observed systematics in the cloud velocities.

Since only the radial velocity of the gas is observed, while the transverse velocity and (in almost all cases) the distance are unknown, there is initially little or no information about the space velocity of the gas clouds. Yet, using all-sky properties, such as the distributions of radial velocities and column density (N(H I)), clues to spatial motions can be obtained.

The most important aspect to note is that we see the gas clouds *from our vantage point*, a point which takes part in the rotation of the Milky

H. van Woerden et al. (eds.), High-Velocity Clouds, 55–72

Way Disk. All measured velocities reflect the motion of the Local Standard of Rest (LSR). One way to correct for this is to take out the rotation of the LSR, leading to velocities in the so-called Galactic Standard of Rest (GSR) system. Gas co-rotating with the LSR, but not moving with respect to it, has $v_{\rm LSR}$=0 km s^{-1}. Similarly, gas at rest with respect to the Galactic Center has $v_{\rm GSR}$=0 km s^{-1}. However, zero velocity in either system need not mean the gas has no space motion. Gas moving perpendicularly to the line of sight has a radial velocity $v_{\rm LSR}$=0 km s^{-1}. Note that gas at very large distances and rotating around the Milky Way has $v_{\rm GSR}$∼0 km s^{-1}. Unfortunately, both the LSR and the GSR systems have misled many a superficial reader to think that such velocities are the space velocities of the clouds, which they aren't. However, for extended objects the angular variation of the velocity can give additional information about the space velocity.

It must also be emphasized that, in the Milky Way Halo (and elsewhere), gas may exist exhibiting $v_{\rm LSR}\simeq$0 km s^{-1}. That gas is not easily detectable, since its emission or absorption blends with that of gas with $v_{\rm LSR}\simeq$0 km s^{-1} near the Sun in the Disk. One must conclude that distant gas with $v_{\rm LSR}\simeq$0 km s^{-1} is probably missing in most considerations.

There are several possible approaches for deriving the spatial distribution of the observed HVCs and IVCs. The kinematical method is to assume some model for origin, space distribution, and velocities, and then derive the observable quantities, either from simple arguments or in analytic form. A more sophisticated method is to make dynamical models and let them evolve in time; this must be done using a Monte Carlo-type of hydrodynamical or gravitational code. In this chapter we will give the general characteristics of the observed velocity patterns (Sects. 3, 4, and 5) and present the simplest approaches to help understand the kinematics of HVCs (Sects. 6 and 7).

2. Defining the HV and IV gas sample

Soon after the discovery of gas deviating from simple galactic rotation it became apparent that there is not only H I with large negative velocities ($v_{\rm LSR}<-100$ km s^{-1}), but also a lot of gas that has $v_{\rm LSR}>-100$ km s^{-1}, but still is distinguishable from the gas in the Galactic Disk. This gas was called intermediate-velocity gas (IV gas). Some parts of one of the large-scale features (the Outer Arm) fall just below the limit of $|v_{\rm LSR}|$=100 km s^{-1}, which was therefore softened to 90 km s^{-1} to allow it to be included as a continuous feature. In practice, for several decades these limits turned out to be very useful for high galactic latitudes.

From a theoretical point of view such a distinction is problematic, because gas with a given space velocity may, when seen in one direction, have a high radial velocity. In another direction it may have an intermediate or even low radial velocity (think of the effect of galactic latitude for a cloud with only a vertical velocity). Comparison with models to project the space velocities onto the observer's vantage point should take care of this. Furthermore, the distinction between HV and IV gas was not made exclusively on kinematical grounds, but also takes into account the density distribution on the sky. An important point to mention is that, astonishingly, as yet no systematic overview exists of the kinematics of the IV gas. In this chapter we present for the first time the velocity field of the IVCs.

For low latitudes Wakker (1991) introduced the "deviation velocity" concept, defining new limits to avoid confusion between the HVCs and galactic gas which, depending on its position, sometimes can have velocities beyond the LSR HVC velocity limits. The original definition of the deviation velocity, $v_{\rm DEV}$, is described in Ch. 2. Later, refinements were added (de Heij et al. 2002), e.g. taking into account the effects of the Galactic Warp. Figure 1 shows velocity contours of the most negative and most positive velocities ($v_{\rm min}$, $v_{\rm max}$) expected from galactic rotation. The difference between the observed LSR velocity and this extreme velocity is called the deviation velocity ($v_{\rm DEV}$). Gradually, new limits in terms of $v_{\rm DEV}$ came into use: for HVCs $|v_{\rm DEV}|>90$ km s^{-1} and for IVCs $35<|v_{\rm DEV}|<90$ km s^{-1}.

As can be seen from Fig. 1, it is not correct to use the same velocity limits at high and low latitudes. On the one hand, not all velocities at low (or even zero) latitude are kinematically hidden behind the Galactic Disk. Approximately one third of the longitude range is kinematically "transparent"; the location of these parts depends on the sign of the velocity. On the other hand, there are 4 regions with high $v_{\rm min}$ or $v_{\rm max}$. This makes it necessary to pay close attention to selection effects when doing a statistical analysis, even at latitudes up to $\simeq 35°$! At longitudes around 180° and latitudes above 40° the deviation velocity essentially equals $v_{\rm LSR}$.

Some objects traditionally considered HVCs turn out to lie close to or even below these new limits, therefore indicating that they may not be HVCs, but instead are IVCs or even ordinary galactic gas. This mostly concerns clouds at low latitude, belonging to the complexes H and (especially) G at negative velocities, and at positive velocities to complexes GCP, WA and WB (see Fig. 1a in Ch. 2).

For studies of the IV gas, the deviation velocity concept is even more important. The border between the IVCs and the low-velocity gas (LV gas) is strongly contaminated with gas becoming almost exponentially stronger with decreasing $|v_{\rm LSR}|$. For the first time Wakker (Ch. 2) shows column density maps based on the Leiden-Dwingeloo Survey (*LDS*), using the new limits in $v_{\rm DEV}$. In these maps IV gas, especially at negative velocities, is

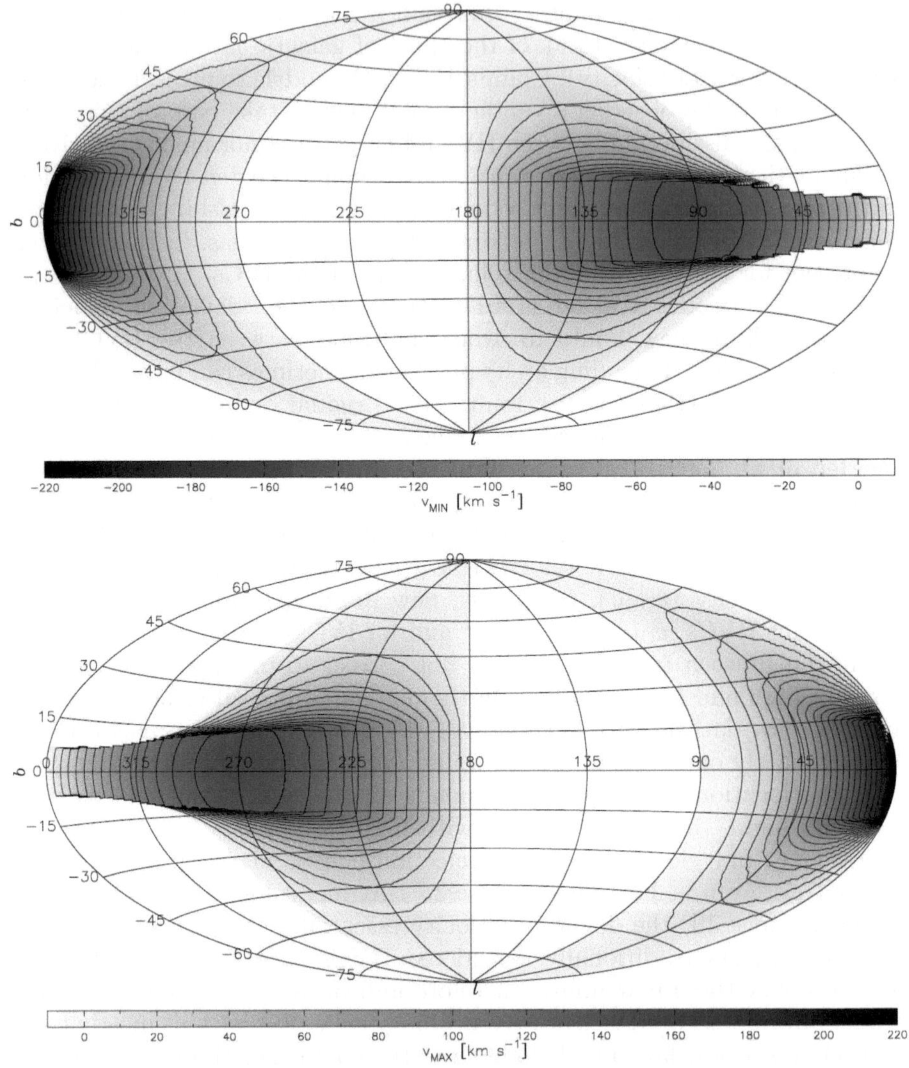

Figure 1. Velocity contours of the most negative ($v_{\rm min}$, top) and most positive ($v_{\rm max}$, bottom) LSR velocities expected from galactic rotation. The contour interval and the first contour are 10 $\rm km\,s^{-1}$. LSR velocities with respect to these limits are the "deviation velocities", $v_{\rm DEV}$. Figure composed for this article.

visible over a large part of the sky. Unfortunately, the southern equivalent to the *LDS* is not yet generally available, so that large parts of the sky are still inaccessible. The kinematics of the IV gas will be discussed in Sect. 4.

3. Systematic motions of high-velocity (HV) gas

3.1. RESULTS FROM EARLY SURVEYS AND ASPECTS OF DISPLAY

Looking at large samples of detections of high-velocity gas one readily sees in a plot of $v_{\rm LSR}$ versus galactic longitude a sine-wave-like distribution of data points. Giovanelli (1980) was the first to point out that this reflects the motion of the LSR. He concluded that the HV gas did not take part in differential galactic rotation and must be distant, either at the edge of the Galaxy or within the Galactic Halo. He also concluded that a few large-scale features (in particular complex C) were exceptions, being closer and showing evidence for taking part in galactic rotation.

The H I data come in the form of a data-cube of brightness with two sky positions and the velocity as coordinates; other observational parameters can sometimes be added as extra dimensions. For display purposes this always poses a problem; using averages in one of the dimensions or using color can often be helpful. Another difficulty for displaying the data is that clouds have at least two orders of magnitude variation in angular size and brightness. One can choose to display either all datapoints of the survey or to use the average parameters of the "clouds" or "complexes". In the latter case extended clouds are presented by just one point. Both possibilities have their merits.

3.2. DISCUSSION OF NEW RESULTS

All-sky maps of the data can be found in Ch. 2, in the two velocity systems $v_{\rm DEV}$ and $v_{\rm GSR}$ (see Fig. 1a and Fig. 1b there). Here we will give updated versions of the diagrams of Giovanelli (1980), using the currently-available whole-sky HVC data. Similar plots can be found in the recent literature.

We use the combination of the high-velocity profile components from the Hulsbosch & Wakker (1988) and Morras et al. (2000) surveys, and the clouds in the Wakker & van Woerden (1991) catalog, updated using the southern survey. Both updated datasets were made available by Wakker (priv. comm.).

Figure 2 shows the sample of all components, from which we removed co-rotating clouds, the Outer Arm, the Magellanic Stream, and complexes A, C, and M, as well as all components belonging to populations and complexes with $|\overline{v}_{\rm DEV}|$<90 km s^{-1}. The Magellanic Stream (MS) is excluded because it is not considered to be an HVC. Complexes A, C, and M belong to the HVC sample, but since they are so extended they overload diagrams like Fig. 2, where the entries of the survey are used. In Sect. 3.3 we give this plot separately for complex C. Figure 2 also shows diagrams based on the revised cloud catalog of Wakker & van Woerden.

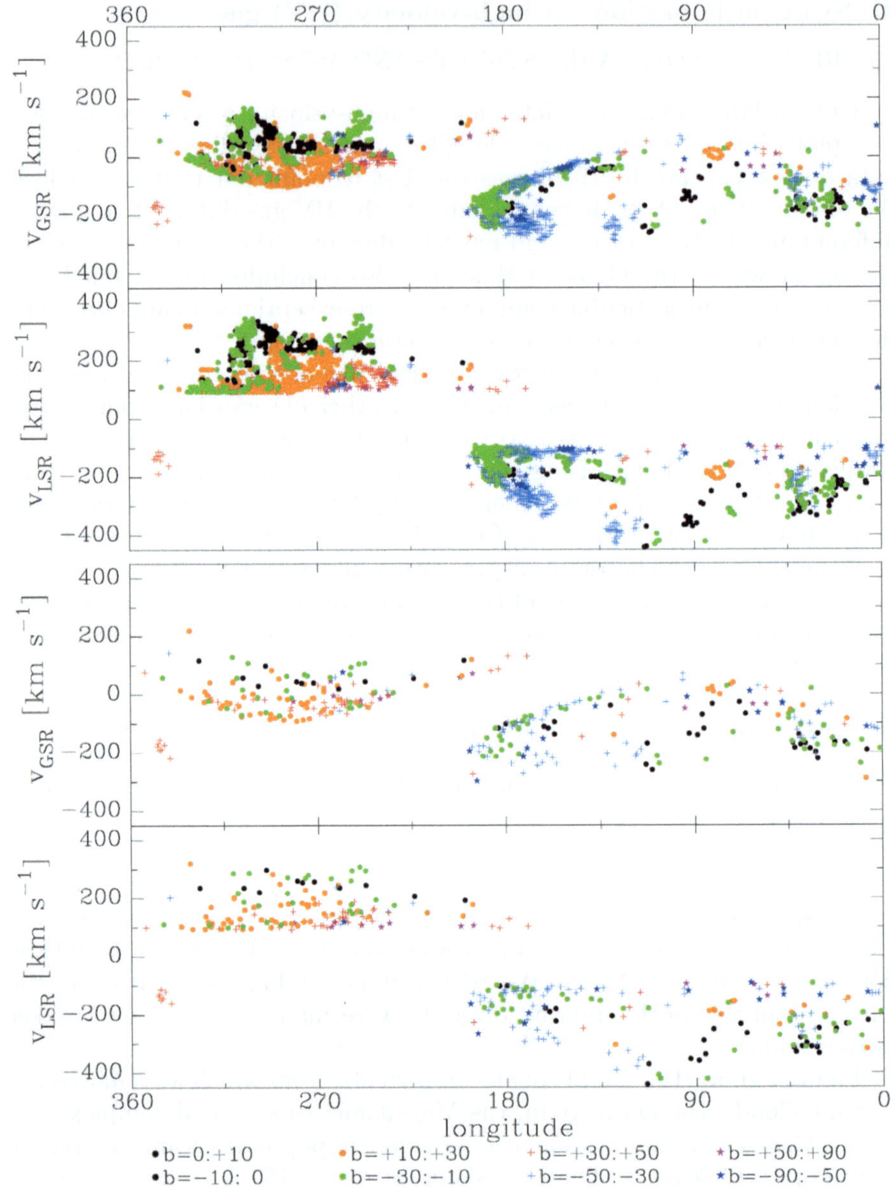

Figure 2. *Top two panels:* Scatter diagrams of all detections of the combined Hulsbosch & Wakker (1988) and Morras et al. (2000) data (see Sect. 3.2), in $v_{\rm LSR}$ (bottom) and $v_{\rm GSR}$ (top). "Co-rotating" clouds, the detections of the OA, MS, and complexes A, C, and M have been excluded, as are those with $|\overline{v}_{\rm DEV}|<90$ km s^{-1}. *Bottom two panels:* Same distributions, but for all clouds in the updated Wakker & van Woerden (1991) catalog that have $|\overline{v}_{\rm DEV}|>90$ km s^{-1} (see Sect. 3.2). Figure composed for this article.

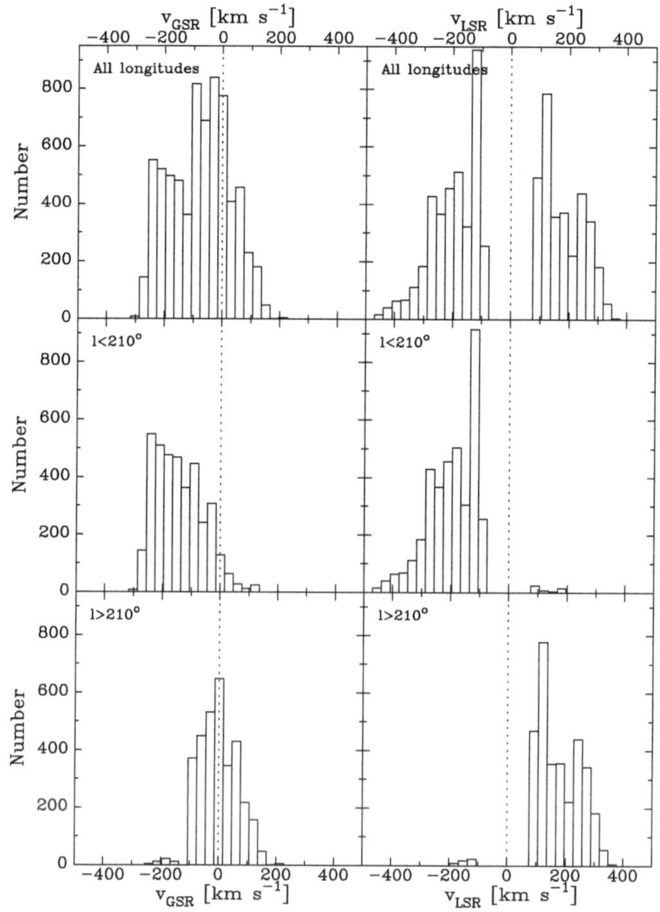

Figure 3. Histograms based on the Hulsbosch & Wakker (1988) and Morras et al. (2000) data sets, without the large-scale populations (as in Fig. 2), and with limits in v_{DEV} applied (see Fig. 1). The top histograms show all detections. The two middle histograms are for $l<210°$, the bottom ones for $l>210°$. The diagrams at the right are plotted as function of v_{LSR} (where the reflection of the solar motion is clearly visible), at the left as function of v_{GSR}. Figure composed for this article.

Although we left out several large clouds (populations OA, MS and complexes A, C, M), the scatter diagram is very patchy. The empty region due to the v_{LSR} cutoff of 100 $\mathrm{km\,s^{-1}}$ is obvious; but it becomes distorted in the v_{GSR} plot. Still, some gross features are apparent. In the bottom panels (v_{LSR} vs. longitude) a clear sine-wave is visible, but the wave crosses zero not at l=180° but rather at l~200°. Bajaja et al. (1987) proposed to simply explain this by a downward shift in velocity. This would be expected from a non-rotating shell around the Galaxy, falling in toward the Galactic Center, but also reflecting the solar motion (see also Sect. 7).

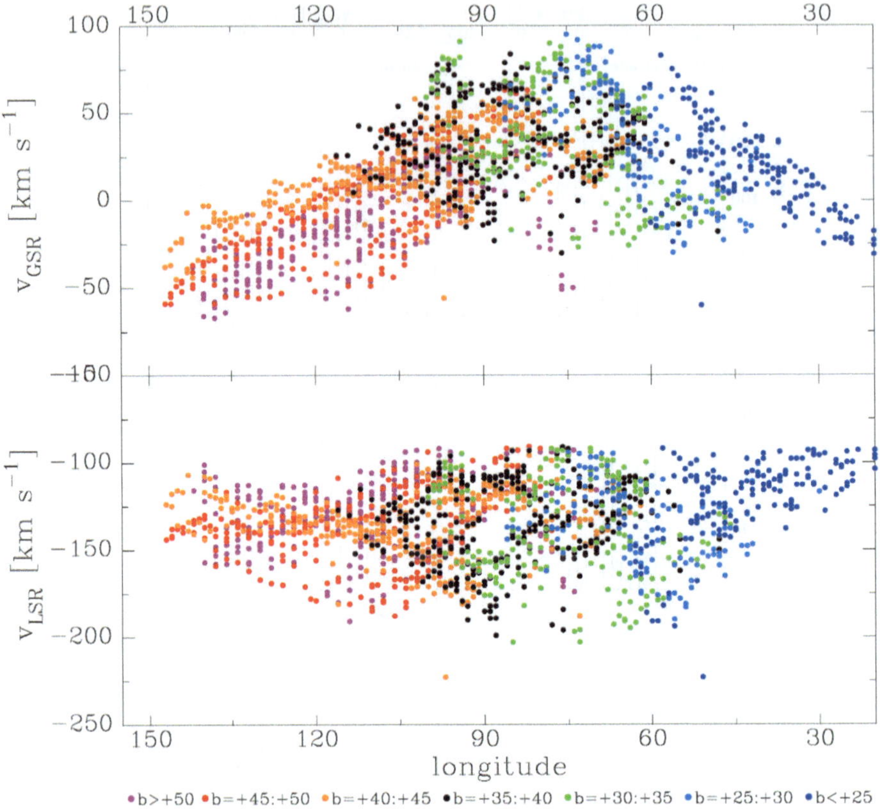

Figure 4. Velocity field of complex C plotted in v_{LSR} (bottom) and v_{GSR} (top). Complex C seems to have rotation and to be infalling (see Sect. 3.3). Figure composed for this article.

The figure for cataloged clouds (lower part of Fig. 2) is similar to that for components, but shows a more regular pattern, hence a clearer sine wave. Also interesting is that the most extreme velocities are predominantly at low latitudes, whereas the lower values are at high latitudes; one would expect that, because of projection effects on the radial velocities, it would be the other way round. As in Fig. 2, the non-zero velocity crossing at l=180° is present too, which will be discussed in Sect. 7.

A disadvantage of these plots is that all latitudes are mixed together. By splitting into positive and negative latitudes cleaner diagrams were obtained (Giovanelli 1980). Instead we used different colors and symbols respectively for 7 latitude ranges in Fig. 2. Some of the main features can also be seen in histograms (Fig. 3). In the frames at right, we plot histograms based on v_{LSR}, at left on v_{GSR}. The bottom histograms are split in longitude at l=210°; in the top panels all longitudes are included. The solar motion cannot naturally explain all features, as the histograms of $v_{\rm GSR}$ most clearly show: at l<210° the velocities are dominated by negative velocities, for l>210° the velocity distribution is symmetric. We will come back to this question in Sect. 7.

3.3. THE LONGITUDINAL VELOCITY DISTRIBUTION OF COMPLEX C

The longitudinal velocity distribution of complex C is shown separately (Fig. 4), in a diagram like that for the ensemble of the HVCs (Fig. 2). In contrast to what the $v_{\rm LSR}$ diagram of the HVC ensemble shows, the $v_{\rm LSR}$ velocity of complex C is roughly constant, but $v_{\rm GSR}$ varies strongly. If complex C is distant, the variation of $v_{\rm GSR}$ implies it is rotating while the bulk $v_{\rm GSR}$ indicates an infalling motion. If nearby, then surely it is infalling. However, the large extent of complex C and the unknown distance makes clear that detailed modeling will be necessary.

4. Systematic motions of intermediate-velocity (IV) gas

The study of the IV gas is not simple, because it is not nicely separated from disk gas in velocity (the situation with HV gas is much more clear-cut). The disk brightness is large, while its intensity falls off both with latitude (but not in a smooth manner) and with velocity (roughly exponentially). On top of this disk background intensity we find widespread IV gas, manifesting itself as secondary peaks in spectra, sometimes superposed on a strongly-inclined spectral "baseline".

In Ch. 4 polar-projection column density maps are shown for $|b|>10°$ and $40<|v_{\rm LSR}|<100$ km s^{-1}. In Ch. 2, Fig. 3 gives an all-sky map with $35<|v_{\rm DEV}|<90$ km s^{-1}. For both these figures the data are from the *LDS* survey (there is no publicly available counterpart in the southern hemisphere) and therefore cover only 3/4 of the sky. The large-scale distribution of negative-velocity gas can be seen. There is little gas at positive velocities.

To also get an overview of the velocities of the IV gas, one may display the data in so-called velocity-color-coded N(H I) diagrams (see Fig. 5). These were made by integrating the *LDS* spectra, weighting the intensities by three triangular functions, centered at equally-spaced velocities within the defined $v_{\rm LSR}$ range. In the integration the data were restricted to the range of 0.12 to 2 K in brightness temperature (T_B); data below the lower limit were not included, above the upper limit T_B was set to the upper limit. Each sum of the three triangles is displayed with colors blue, green and red. For low intensities, the brightness in the diagram is proportional to N(H I). The data selection used a limit in $v_{\rm DEV}$, but the triangles were set in the $v_{\rm LSR}$ system. This sounds confusing, but if one thinks of displaying clouds from a catalog, an object can be chosen according to the

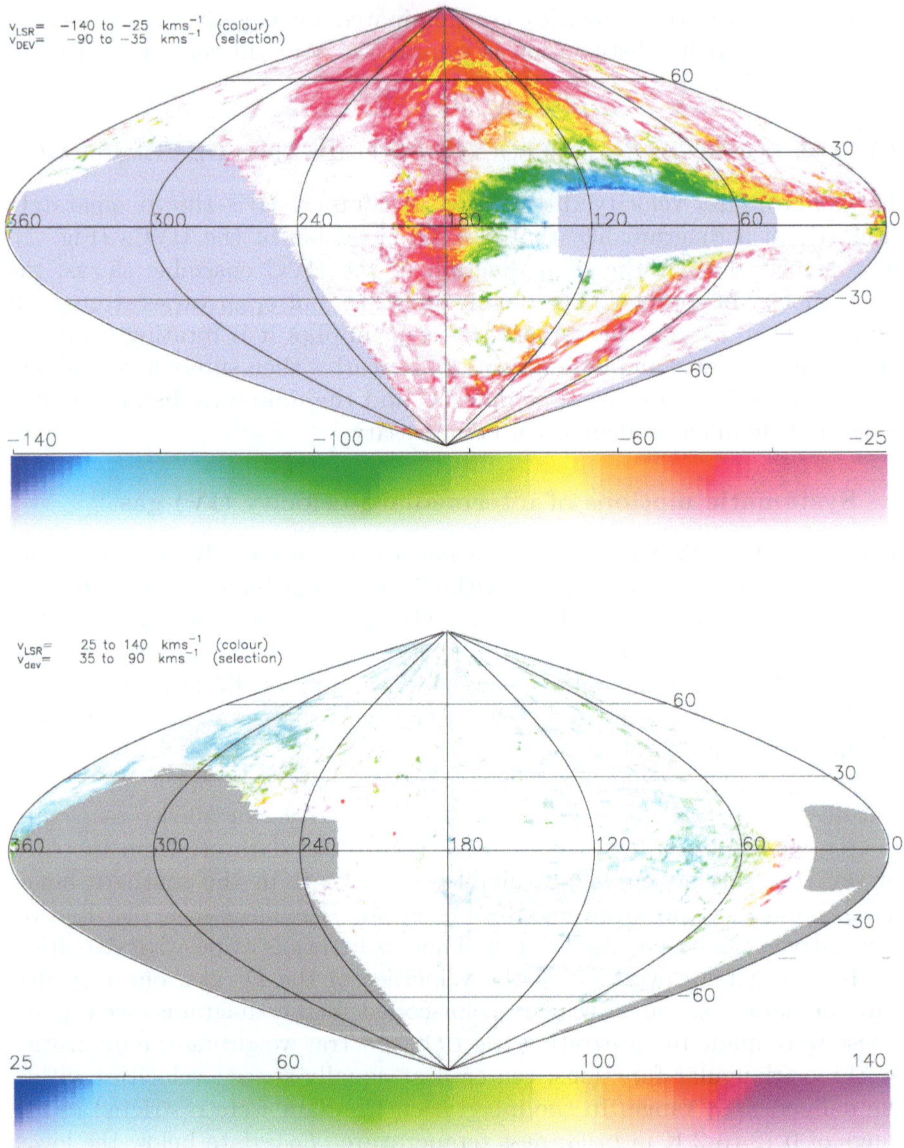

Figure 5. Velocity color-coded $N(\text{H I})$ diagrams of the IV gas at negative (top) and positive velocities (bottom). These plots are based on the *LDS* survey. Three colors are superimposed; each color is obtained by integrating the spectrum at each pixel, weighted by a triangle shifted to the corresponding velocity (see Sect. 4).. Figure composed for this article by Schwarz.

selection criterion and displayed in color in agreement with any parameter, e.g. the LSR velocity. If the $v_{\rm DEV}$ limits happen to fall completely outside the chosen $v_{\rm LSR}$ range, then the point at the given position is undefined.

Several aspects of the IV gas are discussed in two chapters in this book (Ch. 2, Ch. 4), but we restrict ourselves in this chapter to the kinematics. Looking at Fig. 5 it is immediately apparent that the IV-gas phenomenon manifests itself mainly at $v_{\rm LSR}<0$ km s^{-1}. At positive velocities there is less IV gas, and most of it is in the lower-velocity part of the selected range ($+35<v_{\rm DEV}<+90$ km s^{-1}), showing a light blue color; it could be the phenomenon called "disk froth" by Albert & Danly in Ch. 4. There are a few exceptions with higher velocities: at $l<50°$ lies the "Smith-cloud" (with $l=40°$, $b=-15°$, $v_{\rm LSR}=+100$ km s^{-1}, $v_{\rm DEV}=+20$ km s^{-1}), in the past classified as an HVC. There also are a few small clouds at $l>200°$ and low b. At negative velocities a large fraction of the sky is covered with IV gas. There are a few large filamentary structures visible at $b>0°$ and many thin filaments at $b<0°$. The former include the "IV Arch", the "LLIV Arch" (with $v_{\rm LSR}=-40$ to -90 km s^{-1}), the "IV Spur" (with $v_{\rm LSR}\simeq-45$ km s^{-1}) and the "Outer Arm" (see names labeled in Fig. 3a of Ch. 2). Albert & Danly (Ch. 4) give a summary of the many distance brackets for the above features obtained from optical absorption observations.

The Outer Arm was first systematically described by Habing (1966), and later again analyzed by Haud (1992). Other papers about this feature can be found in Lindblad (1967). As can be seen in Fig. 5a, $v_{\rm LSR}$ varies strongly, a signature of galactic differential rotation of gas at considerable distance. From the minimum velocity around $l=100°$ of -125 km s^{-1} a distance of $R/R_0\simeq2$ to 2.5 is expected. Distance measurements would also here be of key importance.

Most of the IV gas is extended as a foreground or background gas layer with $v_{\rm LSR}$ approximately -50 km s^{-1}. The brightest regions are near the North Galactic Pole and the AC shell. Since the LSR velocity is almost constant, this gas rotates with the Disk and also has an infall velocity of $\simeq50$ km s^{-1}. There also are numerous optical absorption observations with upper limits of 1 to 2 kpc (see Ch. 4). A sample of absorption observations in probes in the direction of HVCs illustrates nicely how absorption by the IV-gas is strong and present in spectra of almost all stars with distances above 1 to 2 kpc (Schwarz et al. 1995).

5. Systematic motions from optical and UV observations

Most observations of HV gas have been made at 21 cm, clearly not allowing the derivation of distances for the gas in the line of sight. Another large set of observations was made in the optical and UV using extragalactic

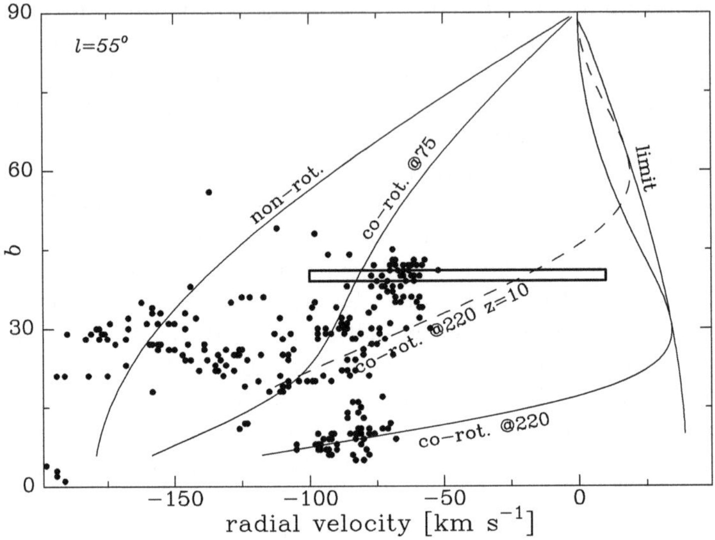

Figure 6. Velocities an observer would detect from a layer of gas parallel to the Galactic Disk having a z-height of 3 kpc, when looking near l=55° at various latitudes, As an example, three values for the velocity of the layer are indicated: corotating at 220 km s^{-1}, corotating at just 75 km s^{-1}, or non-rotation. The dashed line gives expected velocities for gas with 220 km s^{-1} co-rotation but at z=10 kpc. The line "limit" (at right) gives the positive-velocity limit for gas at any z with 220 km s^{-1} co-rotation. Data points are for the clouds detected at 50°<l<60° in the Hulsbosch & Wakker (1988) list. The horizontal bar indicates the velocities of absorption detected in gas in front of the globular cluster M 13 (at l=59°, b=41°, z=4 kpc). The halo gas in this sight line clearly rotates more slowly than the gas of the Milky Way Disk. Note that if the slab also has a vertical velocity v_z, it would shift each modeled curve by $v_z \sin b$. Figure updated from de Boer (1983).

background sources, also not yielding distance limits. A few absorption detections exist using halo objects, providing distance limits for the gas and thus more detailed kinematic information.

The absorption detected in front of the globular cluster M 13 showed that on this line of sight halo gas rotates slower than the disk gas (de Boer 1983; de Boer & Savage 1983). The cluster is at d=6.3 kpc and z=4 kpc. UV absorption is seen between approximately −80 and 0 km s^{-1} (see Fig. 6), while in that direction a simple halo-gas co-rotation model predicts velocities between −10 and +20 km s^{-1}. Thus, this sightline provides a clear case of halo gas with slow rotation.

A part of complex A with a z-height of $\simeq$6 kpc (van Woerden et al. 1999), together with the detected velocity of emission and absorption near −160 km s^{-1}, also yields information about systematic motions. Using the approach of Kaelble et al. (1985; see Sect. 6.1) it is clear that the gas in this direction has a considerable infall velocity. Note that at the longitudes of complex A (130° to 165°), the contribution of the rotational component of the velocity of halo gas is very small in the detected v; the velocities seen are dominated by the infall component.

Together with the non-zero LSR velocity sine-wave crossing discussed in Sect. 3.2, these two cases where gas distance limits are known provide badly needed insight into systematic motions of halo gas. Yet, the two cases above may still be special. Nevertheless, one may conclude that this Halo gas rotates slower than the Disk, falls toward the Disk, and has a velocity component directed toward the rotation axis of the Milky Way.

6. Simple kinematical considerations

A very useful approach to get insight into what one would see in the sky is to take a cloud or a simplified distribution of gas with an assumed space velocity. We discuss two such simple models, one with clouds at constant height z, and a second one with a shell of gas at constant distance R from the Galactic Center. These models are not meant to represent the actual space distribution and velocities of the gas responsible for the observed HV and IV gas, but are rather of the category of "Gedankenexperiment" or "toy-models". Let us then "play" with these models and see what we can learn from them.

6.1. A LAYER OF GAS AT SOME Z-DISTANCE

A very simple toy model assumes a localized layer of gas, parallel to the Milky Way disk, at a height z, rotating along with the Disk with speed Θ, using the orthogonal system $\Theta = v_{\rm rot}$, Φ, v_z, for the galactic velocity components. Note that the vector sum $\Phi + v_z$ represents $v_{\rm infall}$. One now can explore what an observer would detect (following Kaelble et al. 1985).

Consider a radio telescope looking up from low latitudes at constant galactic longitude and observing such an extended layer of gas. At low latitudes the detected $v_{\rm LSR}$ essentially equals that of the Disk gas. Toward higher b the detected $v_{\rm LSR}$ becomes smaller until at the Galactic Pole it becomes 0 $\rm km\,s^{-1}$. Changing the simple case by varying z or $v_{\rm rot}$ or any of the velocity components allows to see what the effect would be for the observer. If the aforementioned layer has a vertical velocity $v_z = -100$ $\rm km\,s^{-1}$ (thus toward the Disk), then at low latitudes little or nothing of that vertical speed is visible, but at the pole it shows its full size. In this way systematics in relevant parts of the sky can be explored. For small ranges of galactic longitude one can plot the velocities expected for a variety of layer parameters chosen (see Kaelble et al. 1985). As an example, Fig. 6 gives the results of such a consideration in the longitude range containing the globular cluster M 13.

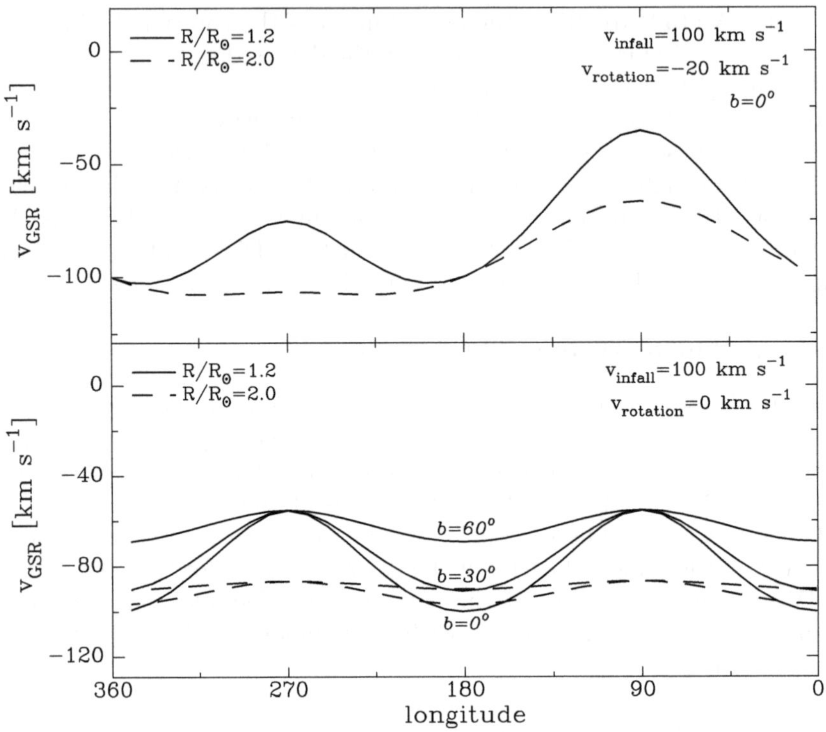

Figure 7. Velocities an observer would detect from an infalling gas shell that has $v_{\rm infall}$=100 km s^{-1}, radius R=1.2 or 2 R_0, and some rotation ($v_{\rm rot}$=−20 km s^{-1}, top) or no rotation ($v_{\rm rot}$=0 km s^{-1}, bottom). Figure composed for this article.

6.2. MODEL OF CONSTANT DISTANCE FROM THE GALACTIC CENTER

In a simple purely-kinematical model the existence of a spherical shell around the center of the Galaxy is assumed, having radius R and space velocities given by two parameters: (i) infall (or expansion) velocity ($v_{\rm infall}$) and (ii) rotation velocity ($v_{\rm rot}$) around the vertical galactic rotation axis. Radial velocities are computed with respect to the position of the LSR. The radial velocities are primarily in the $v_{\rm GSR}$ velocity system, but naturally the transformation to $v_{\rm LSR}$ can be made easily. No intensities are computed or any other sophistications made.

Figure 7 shows the results for two shells, with radii R given in units of R_0=8.5 kpc. We note that with the simple geometry that is assumed velocities are symmetric with respect to the galactic plane; therefore we give only results for $b \geq 0°$. Clearly, a small rotational component has a large influence on $v_{\rm GSR}$ for clouds near the Sun, but already at R=2 R_0, the range in $v_{\rm GSR}$ is only about 30 km s^{-1}. Similarly, an infall component only shows a significant longitude dependence for clouds with R<2 R_0.

7. N(H I) distribution and infall velocity

A statistical approach may give insight into another aspect of the observed velocities and cloud statistics. The histograms in Figs. 8d and e show the longitude and latitude distributions of the HVCs, using $|v_{DEV}|>90$ km s^{-1} to select HVCs. The longitude distribution (Fig. 8d) is more or less flat in the ranges 50° to 190° and 230° to 300°. There are two minima, near l=210° and l=330°. As we explain below, these minima probably are related to an average infall of the HVC population, which shifts the l-v distribution down by about 70 km s^{-1}, and thus changes the longitude range where $|v_{DEV}|<90$ km s^{-1}; such an infall was proposed by Bajaja et al. (1987). Similarly, the non-zero-velocity sine-wave crossing in Fig. 2 shows basically the same effect, and was interpreted by Kaelble et al. (1985) as a component of inflow aimed at the galactic rotation axis (see Sect. 6.1). The latitude distribution (Fig. 8e) is corrected for the smaller sky area at high latitudes, i.e. the integrated flux is multiplied by $\Delta b/\Delta \sin b$. With this correction, the distribution is almost flat from b=−85° to +60°, with a slight dip at low latitudes. The Magellanic Stream is responsible for most of the flux at latitudes below −40°, although part of the lowest-latitude gas may be misclassified and not really part of the Stream.

The constant-R model of Sect. 6.2 shows that, for gas only moderately more distant from the Galactic Center than the Sun, the v_{GSR} velocities become approximately constant, with small deviations. On the other hand, v_{LSR} shows the reflection of differential galactic rotation. In combination with the selection of HVCs based on the v_{DEV} velocity, this gives a strongly-varying probability to detect a cloud at given velocity, l and b. We now describe how this can be modeled, and how this explains many features of the longitude, latitude and velocity distributions of the HVCs.

Assuming a population of distant clouds at large R ($>2\,R_0$), but with galactocentric radial velocities between −140 and +0 km s^{-1}, the LSR velocities observed from our vantage point fall in a sinusoidal band in a v_{LSR}-longitude diagram, as illustrated in Figs. 8a and b. In these figures, the thick lines in the top two panels are for b=0° and b=30°, respectively. If there were no confusing galactic H I, the same number of clouds would be seen at all longitudes. However, the confusing foreground Galactic H I obscures some of these clouds, and we can only see the ones with high LSR velocities. The thin lines in Figs. 8a and b show v_{DEV} limits, which depend on longitude and latitude. Only model clouds falling outside these limits (i.e. having positive or negative v_{DEV}) can be detected. These detectable regions are shown as hatched areas in the longitude-velocity diagrams. At some longitudes almost the full velocity range of the 140 km s^{-1} wide band can be observed (near $l\sim$90°, $b\sim$30°). At some other longitudes no clouds have

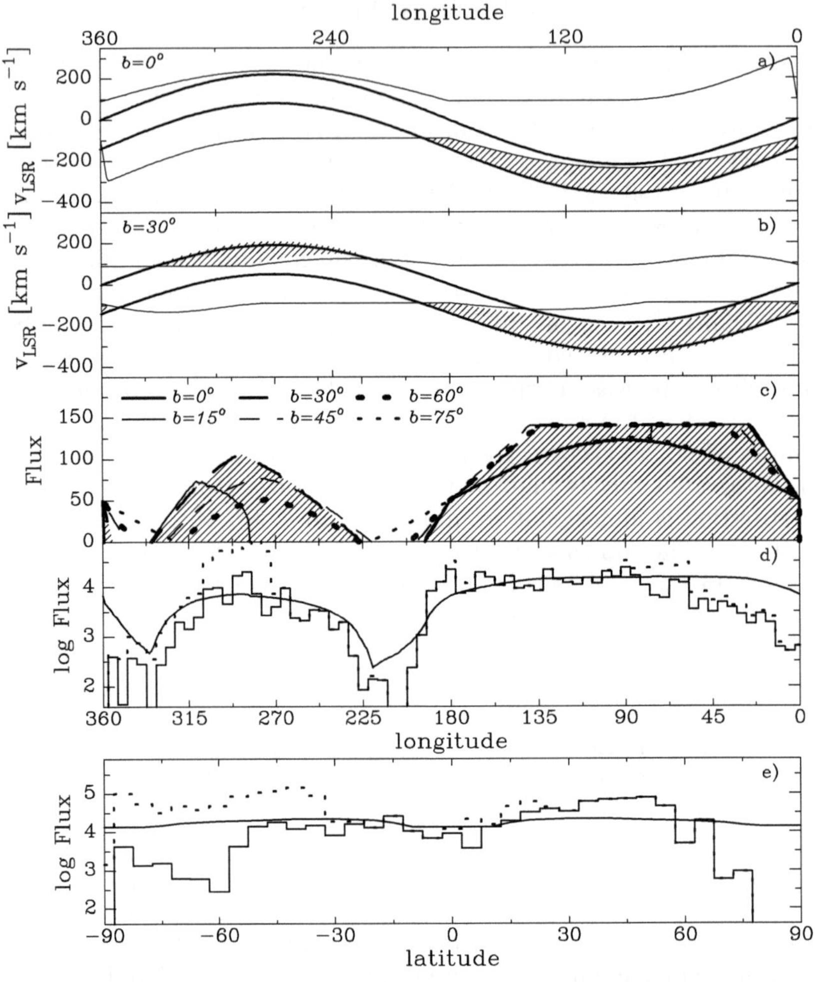

Figure 8. Schematic diagram to explain the longitude, latitude, and velocity distributions of HVCs. The upper two panels show the sinusoidal band of HVC velocities (thick lines), as well as the v_{DEV} limits (thin lines), for latitudes b=0° (top panel) and b=30° (second panel). This assumes a population of clouds with v_{infall}=70 km s^{-1}, and a band that is 140 km s^{-1} wide. The predicted number of clouds (for the calculation method see Sect. 7) is shown in the third panel for latitudes 0° to 75° in steps of 15°, as indicated. The shaded regions show where $|v_{LSR}|>|v_{DEV}|$ for b=30°; their contribution is represented with shading in the top panel too. In the bottom two panels the results for all latitudes/longitudes are added, taking into account the different sky areas in each 10° interval around the nine latitudes at which velocity ranges were calculated. Histograms of the observations are also shown. Dotted histograms include all HVCs with $|v_{DEV}|>90$ km s^{-1}, while for the solid histograms the Magellanic Stream and Outer Arm were removed. There is good qualitative agreement between the data and this very simple model. Figure composed for this article.

apparent deviating velocities (near $l{\sim}220°$, $b{\sim}30°$). The expected number of clouds (or flux) is proportional to the observable velocity range, which is shown by the hatched area in panel c for b=30°. For other latitudes the expected fluxes are illustrated by the other curves.

To compare the longitude distribution of HVCs expected from this simple model with the observations, we sum the expected number counts for the different latitudes, scaling with $(\sin(b+5°)-\sin(b-5°))$ to correct for the different amount of sky in each latitude interval. This sum is shown on a logarithmic scale in Fig. 8d, comparing it to the observed distribution. The gross features of the data are reproduced, namely the presence of two minima shifted away from l=0° and l=180°, and also the asymmetry in number between HVCs with l=240° to 300° vs HVCs with l<180°. The top two panels (Figs. 8a and b) also explain why most clouds with high velocities are found at latitudes of about 30° (see also Fig. 2). These results illustrate how seemingly mysterious features in the data can be strongly influenced by observational selection criteria.

8. Some results from further statistical analyses

The available full-sky 21-cm surveys allow deeper studies of the systematics of the kinematics of high-velocity gas. The many extragalactic sightlines sampled in the UV with *FUSE* add valuable information, too. Mean velocities and velocity dispersions can be calculated from high-velocity gas samples and survey data. This may provide insight into large-scale motion patterns and perhaps lead to clues about the distance of gas ensembles.

A general warning is in order for doing statistical analyses of such data. The completeness of the data is, in general, a function of the limitations of the observations (see Wakker 2004). At 21 cm these include the detection limit in brightness temperature, the resolution in velocity, and also the velocity limits chosen to define the sample (these limits may vary across the sky, e.g. depending on galactic coordinates). For the absorption-line data the sightlines are not evenly distributed over the sky, since the sampling depends on the sources available as well as on the brightness compared to the sensitivity of the detectors used.

Based on these considerations, mean velocities and velocity dispersions given from an analysis of such surveys are mostly of very limited value (and depend on the particular survey limits chosen). Wakker (2004) elaborated on this for mean velocities and velocity dispersions of 21-cm surveys. He showed that the often-given result σ(LSR)>σ(GSR) is mostly based on sample statistics in combination with the effects of the coordinate transformation. Similar effects are present in the sample of *FUSE* O VI data.

A careful analysis of the compact HVC (CHVC) sample (i.e. accounting properly for sensitivity and kinematic selection) show that they can be understood as clouds distributed widely around the Milky Way (and around M 31) having random motions, but with a net infall velocity of $\simeq$50 km s^{-1} (de Heij et al. 2002; Wakker 2004).

Wakker (2004) presents calculations for various statistical models, and the results are then compared with the available survey data. From such and similar investigations it becomes clear that HVCs have different origins and thus that their overall kinematics shows diverse features.

9. Conclusions

The kinematics of HVCs and IVCs are still ill defined, although the signs for true infall are becoming more substantial. As soon as badly needed HVC and IVC distances are determined, new analyses may yield well-documented results for the overall kinematics.

A final thought: high-velocity gas will, in a relatively short time, move to another location and will be seen in a different direction. It will therefore, for a later observer, appear to have kinematics different from that suspected now. And at another time, the number, distribution and velocities of the features may be quite different from the ones detected today.

Acknowledgements

The authors appreciate the generous help of Bart Wakker in preparing the figures and providing us with the updated HVC survey and catalog data.

References

Bajaja, E., Morras, R., Pöppel, W.G.L. 1987, Pub. Astr. Inst. Czech. Ac. Sci., 69, 237
de Boer, K.S. 1983, in Highlights of Astronomy, 6, 657
de Boer, K.S., Savage, B.D. 1983, ApJ, 265, 210
de Heij, V. Braun, R., Burton, W.B. 2002, A&A, 392, 417
Giovanelli, R. 1981, AJ, 85, 1155
Habing, H.J. 1966, BAN, 18, 323
Haud, U. 1992, MNRAS, 257, 707
Hulsbosch, A.N.M., Wakker, B.P. 1988, A&AS, 75, 191
Kaelble, A., de Boer, K.S., Grewing, M. 1985, A&A, 143, 408
Lindblad, P.O. 1967, in IAU Symp. 31, Radio Astronomy and the Galactic System, ed. H. van Woerden (London: Academic Press), 143
Morras, R., Bajaja, E., Arnal, E.M., Pöppel, W.G.L. 2000, A&AS, 142, 25
Schwarz, U.J., Wakker, B.P., van Woerden, H. 1995, A&A, 302, 364
van Woerden, H., Schwarz, U.J., Peletier, R.F., Wakker, B.P., Kalberla, P.M.W. 1999, Nature, 400, 138
Wakker, B.P. 1991, A&A, 250, 499
Wakker, B.P. 2004, ApJ, submitted
Wakker, B.P., van Woerden, H. 1991, A&A, 250, 509

4. INTERMEDIATE-VELOCITY CLOUDS

C. ELISE ALBERT
Physics Department, United States Naval Academy, Annapolis, USA; albert@usna.edu

AND

LAURA DANLY
Department of Physics and Astronomy, University of Denver, USA; ldanly@du.edu

Abstract. Interstellar gas is widely detected at intermediate velocities, both in H I emission at 21 cm and in optical and ultraviolet observations of trace elements seen in absorption toward distant stars. In the northern hemisphere, the distribution of intermediate-velocity clouds is dominated by several large complexes, and two weaker complexes have been identified recently in the southern hemisphere. Intermediate-velocity clouds are located in the lower Galactic Halo and most of them appear to be falling toward the galactic plane. They display a variety of physical conditions, but abundances consistently indicate grain disruption and a gas composition that is nearly solar. A range of models has been proposed to explain these basic characteristics but, at this time, no single scenario offers a compelling description of the intermediate-velocity cloud observations. Overall, evidence appears to be mounting in favor of identifying at least some of the intermediate-velocity clouds with the return flow of a Galactic Fountain. However, in individual cases, several different processes could also be at work and the relationship, if any, between high- and intermediate-velocity clouds is unclear. Intermediate-velocity clouds are a widespread phenomenon which may well play an important role in the dynamics and evolution of the gaseous component of our Galaxy.

1. Introduction: what are intermediate-velocity clouds?

A meaningful definition of intermediate-velocity clouds (IVCs) is hard to establish. Generally speaking, IVCs refer to "dynamically significant" gas with velocities outside the range that can be ascribed to normal galac-

H. van Woerden et al. (eds.), High-Velocity Clouds, 73–100

tic rotation plus "normal" galactic velocity dispersion, but not so extreme in velocity as to be considered high-velocity clouds. Although various authors, over the years, have defined "intermediate velocity" by slightly different ranges of velocity with respect to the local standard of rest (LSR), "IVCs" generally refer to interstellar gas with absolute LSR velocities from 20 to 100 $\mathrm{km\,s^{-1}}$. There are many caveats to this simple definition, however. For example, in many directions, absolute LSR velocities greater than 20 $\mathrm{km\,s^{-1}}$ are expected from galactic rotation. Similarly, gas with a velocity that is dynamically unusual or significant in its local frame may appear near 0 $\mathrm{km\,s^{-1}}$ with respect to the LSR, so that the emission and absorption it produces is blended with that of local zero-velocity gas. Although it is clear that galactic rotation plays a role along most lines of sight, it is not always easy to decipher its specific effects. As a result, a complete cataloging or description of dynamically significant intermediate-velocity gas is essentially impossible. The standard definition of IVCs based on velocity limits is a convenient starting point, but it is an oversimplification of interstellar kinematics that must be used with care.

The interpretation of IVCs is further complicated by gas motions associated with stellar phenomena (such as winds, bubbles and shocks) which can easily give rise to velocities in the IVC range and beyond. This review will focus on bulk motions of galactic gas and overall galactic interstellar dynamics, rather than on specific features that arise from individual stellar environments. At some level, this may be an arbitrary distinction: many of the observed intermediate-velocity features may trace their recent histories to gas accelerated by star formation and supernovae. In fact, the transition from individual events to overall gas dynamics may well lie at the heart of at least part of the IVC phenomenon.

Explanations for the origins of IVCs range from extragalactic infall to fountain material to supershells, mirroring the explanations offered for HVCs. There are many reasons to consider HVCs and IVCs together. Drawing arbitrary boundaries may only serve to obscure their true nature. Indeed, IVCs as a class derive their name from comparison to the HVC phenomenon. As part of a volume devoted to high-velocity clouds, this review will emphasize efforts to study the intermediate-velocity clouds as a global galactic phenomenon in their own right and to explore their relationship to the HVCs. Several questions immediately jump to the fore:

- Are the IVCs indeed a separate phenomenological class or do they merely reside on the outskirts of both the HVC and low-velocity disk cloud populations? If they are a phenomenological class, are they a dynamically significant one?
- Are there separate subclasses of IVCs? Are there differences between disk IVCs and halo IVCs?

• What is the source of their unusual motions? If there are subclasses of IVCs, do different subclasses have different origins?
• Are IVCs related to HVCs? If so, how? In what ways are they different?
• What do IVCs tell us about the evolutionary history of the Milky Way?

There are as yet no definitive answers to these questions, perhaps because a clear observational picture of the phenomenon has yet to be established. It is inherently difficult to define and pinpoint significantly different types of IVCs. Our picture is also obscured by the way the data are taken, analyzed and published. Unlike the growing body of work on HVCs, there are few IVC studies dedicated to solving these puzzles. Instead, the data are scattered throughout the literature, and differences in technique between emission- and absorption-line studies can make it difficult to weave together a complete picture.

The study of IVCs has progressed through two complementary techniques: direct radio detection of intermediate-velocity neutral hydrogen gas in emission at 21 cm, and optical and ultraviolet observations of trace elements seen in absorption at these velocities toward distant stars. H I data on intermediate-velocity gas are usually found as part of comprehensive surveys at all velocities, and conclusions drawn from examining channel maps at intermediate velocities can be biased toward large, coherent structures that are easily recognizable by eye. In addition, the lack of distance information, profile blending, and the positional superposition of emission features can result in confusing or even misleading interpretations.

IVC absorption-line data, in contrast, are usually found nested in detailed analyses of individual sight lines which have multiple velocity components, only a few of which may be at intermediate velocity. The lines of sight observed are necessarily limited by the availability of suitable background probes, but have the advantage that they include interstellar environments missing in H I studies and are not biased by emission-feature selection effects. However, the accumulation of data can be painstakingly slow; for example, the Hubble Space Telescope typically devotes observing time for detailed studies of only a very few stars per year. The extra effort is worth it since, unlike their HVC counterparts, absorption-line data on IVCs are more readily available, providing additional insights into their natures.

Finally, the concept of an interstellar "cloud" suggests an individual velocity component of gas in the cool, or perhaps the warm, phase of the interstellar medium and does not really address inter-cloud gas or the hot, highly-ionized medium. Sensitive ultraviolet absorption-line measures have shown that gas with a range of ionization is distributed ubiquitously throughout the interstellar medium on nearly all sight lines that extend beyond the local solar neighborhood. For the higher-ionization species in particular, the ultraviolet absorption is usually not identified with individual velocity components, but is most often characterized by its broad velocity limits.

2. The distribution of IVCs in space and in velocity

As early as 1949, Adams published an extensive study of optical absorption by singly ionized calcium toward 300 OB stars and noted individual cloud components whose velocity exceeded 50 $\mathrm{km\,s^{-1}}$ in eight stars. In the Galactic Halo, Münch & Zirin (1961) identified 10 clouds with velocities greater than 20 $\mathrm{km\,s^{-1}}$. However, even though the study of intermediate-velocity clouds dates back to such early investigations of the interstellar medium, it was not until the first all-sky H I surveys that large-scale flows of gas with non-standard velocities were recognized as a dynamically significant phenomenon in the Galaxy.

Even the earliest H I surveys recognized that galactic gas could not be described fully by a simple rotational model. In a paper titled "Intermediate Velocity Features in the Local Hydrogen Layer", Blaauw & Tolbert (1966) reported a preponderance of negative intermediate-velocity features, a near absence of low-velocity emission in the second quadrant of the Galaxy, and a sinusoidal ridge in the longitude-velocity diagrams that suggested large-scale flows in that direction. Maps from a follow-up survey and its analysis (Tolbert 1971; Wesselius & Fejes 1973) showed that the IV gas has predominantly negative velocities and can be found in a few large complexes. Confirming the absence of local low-velocity disk gas in the second quadrant of the Galaxy and asserting a physical anti-correlation between the IV and LV gas, Wesselius & Fejes equated the mass of the "missing" gas to the mass in the infalling gas to derive a distance to this infalling gas of about 70 pc. They interpreted their data as representing either a nearby supernova remnant or as the result of the impact of a cloud hitting the Galactic Disk.

Three decades of subsequent work have resulted in several large-scale surveys of galactic H I, mostly in the northern hemisphere, which include (but are not limited to) the Hat Creek Survey (Heiles & Habing 1974), the Bell Labs Survey (Stark et al. 1992), the Leiden-Dwingeloo Survey (*LDS*, Hartmann & Burton 1997), and the compilation by Wakker et al. (2001). In the southern hemisphere, the picture is less complete. Early southern surveys were certainly conducted, although published results tended to focus on specific phenomena such as the Magellanic Stream (Mathewson et al. 1974), or a compilation of HVCs for comparison to northern surveys (Bajaja et al. 1985). New surveys of the southern sky include the *Villa Elisa* survey (Arnal et al. 2000, Morras et al. 2000), the *HIPASS* survey (Barnes et al. 2000) and, for the Magellanic Stream, the Parkes Narrow

Band Survey (Brüns et al. 2000). A consistent full-sky picture of IVCs will be easier to develop when data from the *Villa Elisa* survey become freely available.

In summary, current data show that, to first order, evidence of galactic rotation dominates disk kinematics (i.e. positive velocities in the first and third quadrants, negative velocities in the second and fourth quadrants). Beyond the rotational signature, however, the H I data reveal a highly-disturbed interstellar medium. At low velocities, within 20 $km\,s^{-1}$ of the LSR, the whole sky is filled with features that appear elongated and filamentary, probably arising from relatively nearby disk disturbances such as superbubbles and shells. Even at these low values, the velocity distribution of the gas is not symmetric about 0 $km\,s^{-1}$; instead, there is a clear excess of gas with negative velocities over positive velocities.

On the velocity outskirts of the low-velocity gas, between +20 and +40 $km\,s^{-1}$ from zero LSR velocity, evidence for this "disk froth" continues. In all quadrants, the angular extent of the froth decreases as velocity increases, as one would expect if we were simply seeing the same phenomenon at the greater distances implied by galactic rotation. And again, over the entire sky, negative-velocity gas is favored over positive-velocity gas. A tremendous amount of observational and theoretical effort has gone into attempting to understand the nature of interstellar superbubbles and froth, both as a whole and along numerous individual lines of sight (Heiles 1984). They are widely believed to be created by star formation and supernovae.

At velocities beyond 40 $km\,s^{-1}$ with respect to the LSR, a clear distinction is evident both in the hemispheric distribution and the velocity distribution of the IV gas.kinematics The trend continues whereby the angular extent of the disk froth diminishes as velocity (and therefore distance) becomes more extreme. But at galactic latitudes greater than 20° the gas in both hemispheres is almost exclusively at negative velocities. In the northern hemisphere, the distribution is dominated by several IV complexes reaching to high latitude, which are shown in Fig. 1. Kuntz & Danly (1996) describe three large complexes: the Intermediate-Velocity Arch (IV Arch) (up to b=+70°), the Intermediate-Velocity Spur (IV Spur) (reaching nearly to the North Galactic Pole), and the Low-Latitude IV Arch (LLIV Arch; at about b=+40°). All three features are composed of smaller concentrations, or clumps, spanning the range in velocity from disk velocities to high velocities. There are large-scale velocity structures within the IV Arch, and Kuntz & Danly found that there is some evidence that the infall velocity is highest at the positions of the highest column density concentrations. Distance estimates, as discussed in the next section, place them well above the disk gas, in the lower halo. An additional IV complex, complex K, was defined by Wakker (2001), in a comprehensive study of

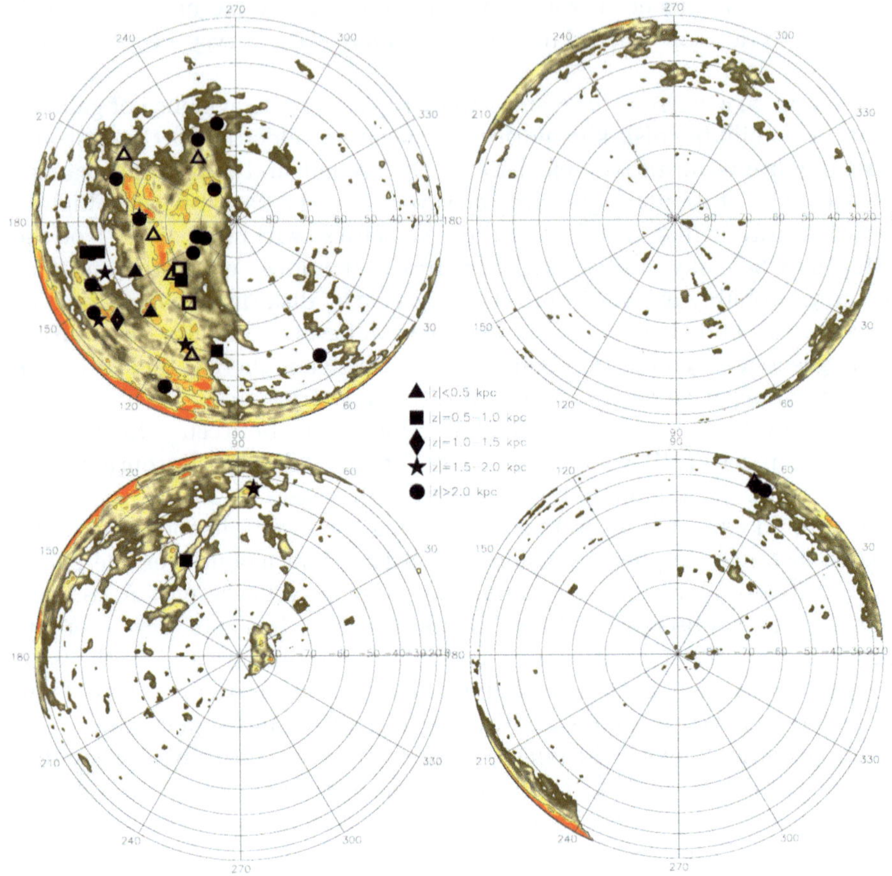

Figure 1. Plots of 21-cm emission seen toward the NGP (top) and SGP (bottom) over the LSR velocity ranges -100 to -40 km s^{-1} (left) and $+40$ to $+100$ km s^{-1} (right). The H I data are from the Leiden-Dwingeloo Survey (Hartmann & Burton 1997). Contours represent H I column densities of 2.5 and 10×10^{19} cm^{-2}. Symbols indicate the positions of the lines of sight listed in Tables 1 and 2. Filled symbols are lines of sight along which intermediate-velocity gas has been detected in absorption; open symbols denote the lack of a detection. Symbol shapes correspond to the absolute value of the z-height of the source (see legend). Figure composed for this article.

high- and intermediate-velocity clouds based on recent H I surveys. Complex K is centered on l=55°, b=+35°, and includes the line of sight toward M 13. The southern hemisphere also has several negative-velocity clouds, but their distribution appears less coherent than the structures in the north. Wakker (2001) defined two new southern complexes: the Pegasus-Pisces Arch (l=90°, b=−40° to l=130°, b=−60°), which is much weaker than the IV Arch but is its closest southern counterpart, and complex gp, composed of positive-velocity IVCs near l=50°, b=−25° and which includes the M 15 sightline.

The presence of IV gas in the disk and halo of the Galaxy is also revealed by absorption-line data. Pioneering optical surveys described above (Adams 1949; Münch & Zirin 1961) discovered numerous IV components along selected lines of sight. Ultraviolet absorption lines, produced by resonant transitions of abundant elements, are sensitive to much lower column densities of gas than optical lines or 21-cm surveys. Ultraviolet studies from the *Copernicus* satellite (Cohn & York 1977; Cowie & York 1978) showed that roughly half of the stars surveyed with distances beyond 1 kpc exhibit absorption from gas with absolute LSR velocities greater than 40 $km\,s^{-1}$. The *Copernicus* data also showed that gas with a range of ionization conditions shares similar velocity ranges; that is, the IV gas contains both neutral and ionized components. Similarly, subsequent data from the International Ultraviolet Explorer (*IUE*) satellite toward halo stars revealed that gas at 40 $km\,s^{-1}$ or greater is common beyond z=1 kpc, even in directions where no known IV emission complexes are evident (Danly 1989). Interestingly, the strong hemispheric asymmetry between negative- and positive-velocity emission in the north was also confirmed in these absorption-line data, suggesting that negative-velocity gas is favored over positive velocities for low column density diffuse gas as well as for the higher column density complexes.

3. Distances of intermediate-velocity clouds

3.1. TECHNIQUES AND LIMITATIONS

Of all the characteristics of intermediate-velocity clouds which we wish to understand, their distances are perhaps the most difficult and yet the most crucial to determine. The size, mass, density, and energy of the clouds depend critically upon distance as, therefore, do theories of their origin and of the role that IVCs play in the evolution of our Galaxy. The most direct method of locating an interstellar cloud in space is through its absorption, in the lines of trace elements such as Ca II, Na I, Si II, C II and O I, of radiation from stars behind the cloud. In principle, if absorption from a cloud is detected along the line of sight toward a distant star, but not toward a nearly aligned foreground star, the cloud is assumed to be located between the two stars. In practice, this "standard" approach is complicated by difficulties in technique (accurate stellar distances), measurement (precise absorption-line detection limits toward foreground stars) and interpretation (patchiness of the interstellar medium).

The reliability of stellar distances is limited by the spectral type vs absolute magnitude relation adopted for Population I stars and by the model atmospheres calculated for Population II stars. These issues are addressed by Jaschek & Gomez (1998) and Ryans et al. (1996a), respectively. Extinction is often ignored in the calculation of distances to halo stars, in particular, although there is some evidence that dust may be associated with IVCs (de Heij 1998; Knude & Hog 1999). Stellar distances are generally considered reliable to within 20%, although experience with data from the Hipparcos catalog has shown that errors can be significantly larger.

Upper limits on cloud distances are easy to set by detecting absorption toward a distant star, but it is more tricky to claim that not detecting absorption toward a foreground star establishes a meaningful lower distance limit. Several authors, such as Danly (1987), Schwarz et al. (1995), Ryans et al. (1997a) and Gladders et al. (1998, 1999), have treated this problem with great care by estimating the absorption that would be produced toward the foreground star, given the H I emission along its line of sight and the ratio of column densities of the observed ion and H I seen toward the background star. If one assumes that the abundance of the absorbing element is constant throughout the cloud, if the cloud is homogeneous, and if the spectrum of the foreground star is sensitive enough to detect the predicted absorption, a non-detection does indeed provide a lower limit to the cloud's distance.

However, the inhomogeneity of the interstellar medium poses a fundamental problem in interpreting any interstellar data. Although a line of sight may include closely-aligned target stars at well-defined distances and a valid non-detection, toward the foreground star, of an interstellar cloud seen clearly toward the background star, these observations could be explained equally well by a cloud which is nearby but unfortunately patchy. There have been many excellent recent studies of fine-scale structure in the interstellar medium. Frail et al. (1994) reported patchiness over dimensions from 5 to 100 AU in the 21-cm H I absorption toward seven pulsars. Heiles (1997) reviewed radio observations indicating small-scale structure on the order of 15 to 50 AU related to the edge-on orientation of interstellar sheets and filaments. In a classic application of optical absorption-line observations, Meyer & Blades (1996) found variations in very high resolution Na I and Ca II absorption toward the two members of the binary star μ Crucis, indicating structure on scales less than their projected separation of 6600 AU. Observations of the binary system HD 32039/40 revealed structure on scales less than 4800 AU (Watson & Meyer 1996), and time variations in the Na I line profiles toward HD 32040 over three years implied structure on scales of 15 to 21 AU, consistent with the star's proper motion (Lauroesch et al. 2000). Similarly, several studies of globular clusters have found variations in interstellar absorption over sub-parsec scales: M 13 (Bates et al. 1995; Shaw et al. 1996), M 15 (Pilachowski et al. 1998; Kennedy et al. 1998; Meyer & Lauroesch 1999; Lehner et al. 1999a), M 22

(Bates et al. 1990) and M 92 (Pilachowski et al. 1998). It appears that one can indeed conclude that an hierarchical pattern of fine-scale structure is a general property of interstellar material even if, as has been seen in many cases, gas at a given velocity is spread widely, but inhomogeneously, over a large area (Albert et al. 1993; Danly et al. 1993; Kennedy et al. 1998).

This conclusion is somewhat discouraging from the viewpoint of using background detections and foreground non-detections of interstellar absorption to bracket the distances of clouds. Even if the H I emission profile is identical toward the two stars, absorption lines probe much narrower lines of sight through the non-uniform cloud than does the larger 21-cm beamwidth. Although the specter of patchiness can never be banished completely, there have been recent attempts (Wakker 2001) to analyze the problem in more detail.

Other, less direct, techniques have also been employed to infer distances to IVCs. In the plane, kinematic distances to IV loops and shells cataloged by Heiles (1984) were determined, assuming that the centers of shells participate in normal galactic rotation. In the halo this does not apply, because the degree of co-rotation between the halo and disk is unknown and because, at large b, the bulk of the observed motion is in the z direction. Other approaches to finding distances to halo clouds have relied on models of either physical state (e.g. Wolfire et al. 1995) or motion (Benjamin & Danly 1997) as a function of z. However, because of both the uncertainties in the model assumptions and the uncertainty inherent in using kinematic information from gas with peculiar velocities, none of these indirect methods can be considered reliable until further demonstration of their applicability is confirmed. At present the direct detection of clouds via interstellar absorption is the only method of determining distance ranges reliably.

3.2. RESULTS

Despite the uncertainties inherent in the process of obtaining and interpreting interstellar distances, the locations of various intermediate-velocity clouds and cloud complexes are being estimated with some success and are presented in Tables 1 and 2. Most of the distances listed for individual background objects are the recent values given by Wakker (2001). The largest intermediate-velocity feature, the IV Arch, has been placed at a height, z, above the galactic plane of between 0.8 and 1.5 kpc by Kuntz & Danly (1996). These distance limits were determined from observations of ultraviolet absorption lines toward 10 stars distributed along the Arch. Ryans et al. (1997a) refined the limits to several of the clumps along the Arch identified by Kuntz & Danly. For more discussion, see Wakker (2001). The Low-Latitude Intermediate-Velocity Arch is also discussed in detail by

Kuntz & Danly (1996) and by Wakker et al. (1996). A summary of results by Wakker (2001) gives a z distance for this feature of 0.6 to 1.2 kpc. The third major intermediate-velocity feature identified by Kuntz & Danly in 1996, the IV Spur, does not appear to have been studied more recently. Kuntz & Danly's data place the Spur at z between 0.23 and 2.1 kpc. Only upper distance limits are currently available for complex K and the Pegasus-Pisces Arch. Complex K has been detected in absorption toward several stars in M 15, placing the gas less than 4.5 kpc above the galactic plane. The northern part of the Pegasus-Pisces Arch is fairly well-determined, by detections toward the star HD 215733, to be within a z distance of 1.6 kpc, while observations of the southern part of the Arch, toward PG 0039+049, suggest that it may be within a z height of 1.1 kpc. The z distance to complex gp is better constrained, although only by the HD 203664/HD 203699 lines of sight, to lie between 0.3 and 2.0 kpc.

In addition to the large IVC complexes identified from H I surveys, individual IVCs are discovered frequently in interstellar studies along specific lines of sight (see, for example, Albert et al. 1993 and Wakker 2001). Several lines of sight include nearly-aligned foreground and background stars which, subject to the caveats discussed earlier, provide both upper and lower limits to the distance of the intervening cloud. In all cases, such individual clouds appear to be located in the lower Galactic Halo. These detections may indeed represent separate individual features, but it is also possible that many, if not most, of these clouds may be clumps within more extended IV-gas complexes. For example, Kennedy et al. (1998) pointed out that a $+70\ \mathrm{km\,s^{-1}}$ cloud at least 3.1° long (now identified by Wakker (2001) as part of complex gp) could cover both HD 203664 and the globular cluster M 15. Similarly, Shaw et al. (1996) suggested that, although the globular cluster M 13 is some 50° in longitude away from the highest column densities that define the IV Arch, the gas at $-70\ \mathrm{km\,s^{-1}}$ seen toward M 13 may be part of the boundary region of that larger complex. The more recent analysis by Wakker (2001) defined the gas seen toward M 13 as part of the intermediate-velocity complex K.

It is indeed worth the considerable effort involved to determine the precise distances to as many IVCs as possible. As it stands, we can confirm and extend with some confidence the early results of Albert (1983) that the bulk of the intermediate-velocity gas is located above the Galactic Disk and in the lower Halo. This conclusion holds for both the large, coherent intermediate-velocity structures (IV Arch, LLIV Arch, IV Spur, Pegasus-Pisces Arch, and complexes K and gp) and numerous individual lines of sight. The implications for theories of the origin of the IVCs and their possible relation to HVCs will be discussed in Sects. 4 and 6.

TABLE 1. Stars giving distance limits to IV complexes – Part I

Cloud	v_{LSR} [km s^{-1}]	Line of Sight	l [°]	b [°]	z [kpc]	Refs
IV Arch: Upper Limits						
IV 4	−101	BD +49 2137	134	+67	0.74	1
IV 6	−66	HD 93521	183	+62	1.7	2,3
	−73	BD +38 2182	182	+62	3.5	4
IV 7	−64	PG 0955+291	200	+52	3.9	4
IV 9	−70	HD 121800	113	+50	1.6	5
IV 17	−63,−76	BD +49 2137	134	+67	0.74	4,6
IV 19	−37,−57	HD 121800	113	+50	1.6	5,6
IV 21	−45	BD +63 985	134	+53	0.32	7
IV 26	−50	H.O. +41B	145	+75	3.0	4,6
	−41	PG 1212+369[a]	160	+78	2.6	4
off core	−45	PG 0955+291	200	+52	3.9	4
	−44,−56	BD +36 2268	153	+80	3.6	4
	−44,−58	BD +38 2182	182	+62	3.5	4
	−57,−66,−76,−96	PG 1212+369	160	+78	2.6	4
		BA 90700003	114	+30	2.1	6
	−83	BT Dra	99	+51	1.5	8
	−50	AG +53 783[b]	154	+57	0.4	4
IV Arch: Lower Limits						
IV 4,17		HD 105058	141	+66	0.17	6
		HD 106420	141	+69	0.51	6
IV 6,26		HD 98152	171	+66	0.37	6
IV 9,19		HD 127557	109	+47	0.21	6
IV 11		PG 1255+547	121	+63	0.71	6
IV 24		HD 87015	211	+51	0.38	1,4
LLIV Arch: Upper Limits						
	−57	PG 0832+676	148	+35	4.6	9
	−45,−67	PG 0833+699	145	+35	1.9	9
LLIV 1	−41,−51	PG 1008+689	141	+42	1.2	9
LLIV 3	−50	PG 0859+593	157	+40	2.6	10
LLIVext	−44	HDE 233622	168	+44	3.1	1,9
	−47	HDE 237844	159	+47	1.8	9
	−37	HDE 77770	169	+42	0.8	11
	−34	HD 83206	168	+47	0.66	2
LLIV Arch: Lower Limits						
LLIV 3		PG 0906+597	156	+41	0.46	10

TABLE 2. Distance limits to IV complexes – Part II

Cloud	v_{LSR} [km s^{-1}]	Line of Sight	l [°]	b [°]	z [kpc]	Refs
IV Spur: Upper Limits						
	−55,−64	HD 100340	259	+61	2.6	5,12,13
		Feige 40	245	+64	2.6	6
S 1		H.O. +23B[c]	236	+79	2.1	6
IV Spur: Lower Limits						
S 1		HD 100600	239	+69	0.28	5,6
Complex K: Upper Limits						
		M 13 B 29	59	+41	4.5	14,15
	−72	M 13 I −48				15
	−69	M 13 II −67				15
	−69	M 13 L 598				15
Pegasus-Pisces Arch: Upper Limits						
	−43,−56,−93	HD 215733	85	−36	−1.6	16,17
	−47	PG 0039+048	119	−58	−0.93	18
Complex gp: Upper Limits						
	+54 to +80	M 15	65	−27	−4.6	19,20,21 22,23,24
	+70,+80	HD 203664	62	−27	−2.0	20,5,25
Complex gp: Lower Limits						
		HD 203699	66	−25	−0.34	20

References: (1) Danly et al. 1992; (2) Lehner et al. 1999b; (3) Spitzer & Fitzpatrick 1993; (4) Ryans et al. 1997a; (5) Albert et al. 1993; (6) Kuntz & Danly 1996; (7) Benjamin et al. 1996; (8) Songaila et al. 1988; (9) Ryans et al. 1997b; (10) Wakker et al. 1996; (11) Welsh et al. 1996; (12) Savage & Sembach 1996a; (13) Ryans et al. 1999; (14) de Boer & Savage 1983; (15) Shaw et al. 1996. (16) Albert 1983; (17) Spitzer & Fitzpatrick 1997; (18) Centurion et al. 1994; (19) Songaila & York 1981; (20) Kennedy et al. 1998; (21) Lehner et al. 1999a; (22) Langer et al. 1990; (23) Morton & Blades 1986; (24) Meyer & Lauroesch 1999; (25) Ryans et al. 1996b. Notes (a) PG 1212+369 is HZ 22. (b) AG +53 783 may be HDE 233791 (Wakker 2001). (c) H.O. +23B is PG 1205+228.

4. IVCs and their relationship to low- and high-velocity gas

Much attention has been drawn to the fact that the major IV complexes can be found in roughly the same part of the sky as the major high-velocity cloud complexes C, A, and M. The question naturally arises whether the IV and HV clouds are somehow connected, either spatially or in their origins. In addition, an intriguing spatial coincidence can be found between several major high- and intermediate-velocity clouds (complexes A, C, M and the

IV Arch) and the position of the so-called "Lockman Window", the region with the lowest galactic H I column density. Other authors (Wesselius & Fejes 1973; Morras 1980) have also noted potential anti-correlations between IV gas and low-velocity disk gas. Perhaps these spatial and velocity distributions provide clues to the origins of all three phenomena.

Establishing a relationship between intermediate- and high-velocity gas is made difficult by the distance ambiguity that results from using emission-line data alone. However, in some directions the data suggest a relationship. For example, using higher-resolution data from *Effelsberg*, Pietz et al. (1996, and references therein) have identified features they call "velocity bridges" which show emission from a patchy, low column density envelope of H I apparently extending from the low- and intermediate-velocity gas to well-known high-velocity features in the same, or very nearby, directions. Kuntz & Danly (1996) also noted systematic velocity trends that suggest a connection between the IV Arch and complexes C and M. If a physical connection between the IV and HV cloud complexes can be established, it could be indicative of a common origin. The fact that the HV gas is systematically more distant than the IV gas suggests that perhaps the IVCs were, at one time, HVCs which have been decelerated as they approached the disk. A connection between high- and intermediate-velocity gas could be ruled out by comparing their metallicities, and it is commonly believed that the metallicities of HVCs are lower than those of IVCs. However, very few abundance measurements, have been made, and these show a range of values (Wakker 2001). More work is necessary, on a case by case basis, to explore the possibility of HVC/IVC connections.

With the established distance estimates for the IVCs (Sect. 3) and the HVCs (Ch. 10), the combined mass contained in complexes C, A, and M and the IV Arch is probably several times 10^6 solar masses. If the material in these clouds originated in the disk, the minimum energy required to lift this it out of the disk and accelerate it to the observed radial velocity is about 10^{53} erg. This value is a lower limit since we cannot measure the tangential velocity of the gas and because we do not know its history. The energy requirements could be significantly greater if the gas was accelerated to very large distances above the plane before it reversed course and fell back toward the disk. If the gas contained in these complexes were lifted into the halo as the result of supernova activity, both a large number of supernovae and a very efficient conversion of the supernovae energy to the kinetic energy of the clouds would be required. It is interesting to note in comparison that the energy contained in the gas of the 30 Doradus nebula is also on the order of 10^{53} erg (Chu & Kennicutt 1994). The energy is also consistent with estimates of that associated with the collision of a high-velocity cloud with the disk (Mirabel & Morras 1990; Rand & Stone 1996). Further, the supposition that the observed IV and HV gas in the second quadrant is displaced disk gas has drawn additional support from the observed lack of low-velocity H I in the same quadrant. Kuntz & Danly (1996) revisited the proposed anti-correlation between the intermediate- and low-velocity gas first suggested by Wesselius & Fejes (1973), and concluded that no detailed anti-correlation is observed in the angular distribution of the gas. In addition, the distance of the IV gas indicates that both its size and mass are many times greater than that which is "missing" from the disk. As tantalizing as the coincidence of these features may be, it is likely that the IV (and HV, if related to the IV) gas in the second quadrant is not the missing disk gas accelerated to high z by an energetic disk event that evacuated the gas from the Lockman Window.

A second line of evidence suggestive of an interaction between high-, intermediate- and low-velocity gas comes from the appearance of X-ray emission enhancements (e.g. Kerp et al. 1996, 1999 and references therein) and even γ-ray enhancements (Blom et al. 1997) in the directions of high-velocity clouds. Claiming an enhancement of X-ray emission that is associated with the velocity bridges in particular, Kerp and collaborators suggested that the X-rays are released in the collision of HVCs with the disk, and that the IVCs are the rapidly decelerated high-velocity material impinging on galactic disk gas. In this scenario, some of the soft X-ray background in the Galaxy originates from such HVC-disk collisions, and may make up one class of soft X-ray emitting objects. Interactions between HVCs and the disk are discussed in detail in Ch. 12.

5. Abundances and physical conditions

Savage & Sembach (1996b) presented an excellent review of the extensive work that had been done on interstellar abundances up to that date, including progress along integrated lines of sight through the Galactic Halo and on individual high-velocity clouds. More recently, comprehensive data for both high-and intermediate-velocity clouds were summarized by Wakker (2001), who tabulated absorption-line measurements along with new H I column densities and abundances. Unfortunately, there are far fewer specific studies of intermediate- than of high-velocity clouds and these clouds are often discovered and analyzed peripherally, as part of general interstellar surveys or as part of detailed studies of a particular line of sight. In any case, there are many difficulties involved in estimating and interpreting interstellar abundances. Calculation errors can be produced by inaccurate f values and by line saturation, while uncertainties in interpretation arise from necessary, but problematic, assumptions about the relative depletion and ionization effects between trace and dominant ions, the mix of H I and

H II gas, H II region abundances, and the degree to which helium is ionized in H II regions. Further, absolute abundance determinations, from the ratio of the column density of an ion observed in absorption to the column density of H I observed in the 21-cm emission line, introduce possible errors from differences between the absorption and emission pathlengths and from 21-cm beam dilution. The first effect is usually assumed to be small toward distant halo stars, where the column density of neutral hydrogen obtained from emission is generally comparable to that obtained from Lyman α absorption (Lockman et al. 1986; Danly et al. 1992). Beam dilution is a more serious problem. A compact cloud can be detected easily in absorption if it happens to be aligned with a distant star. But, if the cloud is significantly smaller in angular diameter than the 21-cm beam size, its H I emission will be effectively spread out, or diluted, perhaps to the point of non-detectability. For instance, this effect may be important for the components seen at -63 and $+76.5$ km s^{-1} toward HD 100340 (Ryans et al. 1999), which are detected in Ca II absorption but not in H I emission and, indeed, it may be a problem wherever there is significant small-scale structure along the line of sight (Albert et al. 1993; Sembach 1995).

5.1. GENERAL TRENDS

Only limited information on interstellar abundances is available from absorption-line studies at optical wavelengths. Although observations of Ti II, the dominant ionization stage of this element in H I regions, and the 21-cm emission line provide a good lower limit to the gas phase abundance of titanium, and although this phase is enhanced at higher velocities, nevertheless the line at 3384 Å is weak and particularly difficult to detect in low column density clouds. The stronger K line of Ca II is more readily observed and has been measured in many IVCs. It suffers from the ambiguity of uncertain ionization corrections, but follows the same depletion pattern as titanium, and provides valuable insight into the grain-processing history of interstellar clouds. Albert et al. (1993) determined the ratio $N(\text{Ca II})/N(\text{H I})$ along 17 lines of sight for 26 discrete IVCs with estimated distance ranges that place them in the lower Galactic Halo. The ratio $N(\text{Ca II})/N(\text{H I})$, which is a lower limit to the relative amount of singly-ionized calcium, varies from 2.5×10^{-9} to greater than 300×10^{-9} and is consistent with the average value of 30×10^{-9} found for IVCs by Albert (1983) in a smaller study. In all cases, the ratio is larger than the value $N(\text{Ca II})/N(\text{H I})=1.8\times10^{-9}$ expected for a "standard" (Spitzer 1978) diffuse cloud. This could indicate either enhanced gaseous calcium, due perhaps to grain disruption, or ionization of hydrogen in the clouds. Detections of Ti II absorption can resolve this question and have been made for 4 of the IVCs. The titanium gas-phase abundances in these cases are among the highest found: these are indeed neutral clouds with enhanced gaseous heavy element abundances, consistent with grain disruption in shocked material. Ryans et al. (1997a) reached a similar conclusion for 11 IVCs observed in Ca II absorption and H I emission toward 7 halo stars in the region of the IV Arch. They determined limits on the Ca II abundance along these sightlines which are within a factor of 5 of the solar value of 2.2×10^{-6} (Anders & Grevesse 1989).

The trend of enhanced gas-phase abundances consistent with grain disruption seen in optical absorption-line data is generally followed by species observed in the ultraviolet. Wakker (2001) presents a detailed discussion of abundance determinations currently available for large IVCs. The IV Arch appears to have near-solar abundance, based on observations of the S II/H I ratio toward three sources and the O I/H I ratio along a fourth line of sight (Richter et al. 2001b). The LLIV Arch may have a somewhat lower metallicity, averaging about 0.8 solar, from measurements of O I/H I and N I/H I toward three sources (de Boer et al. 1993; Richter et al. 2001a; Ryans et al. 1997b). Observations of the Pegasus-Pisces Arch, which Wakker (2001) calls the closest southern counterpart to the northern IV Arch, are even more scarce, but the S II/H I ratio in intermediate-velocity components toward one star averages 0.5 solar (Fitzpatrick & Spitzer 1997). The depletion patterns in these three large IVCs vary: while depletion in the LLIV Arch is similar to that of warm disk gas, enhanced gas phase abundances in the IV and Pegasus-Pisces Arches are typical of the halo.

At this time, physical conditions have been studied in detail in only a very small sample of IVCs, including both the larger complexes and individual clouds, and these "hallmark" lines of sight are described in the following subsection. Not surprisingly, they represent a variety of physical conditions; however, none of them are typical cold diffuse clouds. Grain destruction appears to have occurred in all cases. The gas-phase abundances of refractory elements such as Ca, Ti, Fe, Mg, and Mn are enhanced over their average values in either disk or low-velocity gas and can approach a significant fraction of the solar abundances. Fitzpatrick (1996) has argued that the elements Fe, Mn, Cr and probably Ti have apparently well-defined upper limits of gas-phase abundance on the order of one-third solar. He pointed out that two very different scenarios could be responsible for this effect: either interstellar dust grains contain essentially indestructible cores, or the grains are destroyed and the total interstellar abundances of these elements are subsolar. But, in either case, it appears that the gas in IVCs is not primordial and that, probably due to at least partial grain disruption, gas-phase abundances are enhanced over those typical of low-velocity gas.

Interstellar dust grains appear to be widespread in IVCs. The presence of grains is believed to be necessary for the efficient formation of diffuse molecular hydrogen (Shull & Beckwith 1982) and, indeed, H_2 absorption was first observed in an intermediate-velocity cloud by Gringel et al. (2000). In a recent large survey with the Far Ultraviolet Spectroscopic Explorer (*FUSE*), Richter et al. (2003) found a relatively high detection rate of H_2 in IVC gas in both the northern and the southern sky. They concluded that, if the H_2 remains in formation-dissociation equilibrium, it is the fortuitous result of the presence of dust and the absence, in the lower halo, of a strong far-ultraviolet radiation field. These very interesting observations reveal a ubiquitous cold neutral component, probably small (about 0.1 pc) dense (about 30 cm^{-3}) filaments, within the IVCs. In fact, observations by Howk & Savage (1999) of nearby edge-on spiral galaxies implied that complex webs of dense dusty material extending to z heights of several kiloparsecs are a common property of massive spiral galaxies.

Both photoionization and collisional ionization calculations have been used successfully to model IVCs along the six well-studied "hallmark" lines of sight discussed below. Absorption from highly-ionized species is detected in all of these directions but, in most cases, is observed over a broad velocity range within which it is generally difficult to identify specific IVC components. Along four of these sightlines, and also toward the halo star HD 167756, Savage & Sembach (1996a) noted that the derived temperatures, on the order of a few times 10^3 K, of the IVCs are characteristic of the warm neutral medium and are too low to support significant amounts of highly-ionized gas. They concluded that the highly-ionized gas can exist in regions near but not identical to the IVCs. Subsequent *FUSE* spectra (Savage et al. 2003) revealed substantial, widespread amounts of hot (3×10^5 K) gas in the halo. O VI features are distributed irregularly, moving both toward and away from the galactic plane.

5.2. ANALYSES OF "HALLMARK" LINES OF SIGHT

Comprehensive studies of interstellar abundances first became possible with high-resolution ultraviolet spectra from the Goddard High Resolution Spectrograph (*GHRS*) aboard the Hubble Space Telescope (*HST*) and have been continued with the Space Telescope Imaging Spectrograph and, at somewhat shorter wavelengths, with *FUSE*. Spitzer & Fitzpatrick were among the first to investigate abundances in the Galactic Halo, and their work toward HD 93521, HD 149881 and HD 215733 remains classic in the field. In addition, detailed analyses of IVCs along halo lines of sight have been performed toward HD 18100, HD 100340, and HD 203664. Results for all 6 of these "hallmark" sightlines are outlined below.

• HD 93521 (l=183°, b=+62°): The line of sight toward HD 93521 is a particularly interesting direction for interstellar studies, as it probes the IV

Arch and is generally regarded to define the abundance patterns of halo gas (Savage & Sembach 1996b). Optical observations show strong absorption by Ca II, mirrored by weaker Ti II absorption, at both low and intermediate velocities (Albert 1983). Spitzer & Fitzpatrick's (1993) analysis of ultraviolet observations along this line of sight identified 5, out of a total of 9, components at LSR velocities more negative than -20 km s^{-1}. The S II/H I ratio, averaged over the four most negative-velocity components, has an essentially solar value of 0.97$\pm$0.06. All of the other elements studied (Fe, Si, Mn, Mg, and Ti) show depletions, with respect to sulphur, that decrease with increasing absolute velocity. For iron, the average gas-phase abundance, D(Fe), is -1.20 dex for absolute velocities below 10 km s^{-1} and -0.62 dex for absolute velocities above 35 km s^{-1}. Electrons and H atoms are apparently mixed together at all velocities and the average electron density varies with velocity, from 0.11 cm^{-3} in components at absolute velocities less than 20 km s^{-1} to 0.04 cm^{-3} for absolute velocities greater than 35 km s^{-1}. From an analysis of the observed ratio N(S II)/N(S III) and measured electron densities, Spitzer & Fitzpatrick (1993) concluded that essentially all of the S II is associated with the H I; however, a different approach (Sembach et al. 2000) suggests that a substantial ionization correction is necessary. Sembach et al. (2000) found that ionization conditions vary between the clouds along this line of sight and that, overall, roughly 50% of the S II is in the warm neutral medium and 50% is in the warm ionized medium. Slavin et al. (2000) have studied photoionization of the warm ionized medium along this line of sight. They found that the S II to S III ratio, as well as the Hα intensity and C II column density, could be explained by models of the EUV/soft-X-ray flux from cooling hot gas. More observations are needed to determine the importance of the warm ionized medium in abundances derived toward HD 93521.

Interestingly, Spitzer & Fitzpatrick (1993) found that, with the exception of only one narrow low-velocity component, the 21-cm emission-line widths of all clouds in this path are consistent with a kinetic temperature of about 6000 K, typical of the warm neutral medium. Although this may be an upper limit, in view of the possible presence of turbulence, they concluded that this direction is unusual in that it may not intersect any cold diffuse clouds. However, more recent detections of H_2 absorption, with the Orbiting and Retrievable Far and Extreme Ultraviolet Spectrometer (Gringel et al. 2000) and with *FUSE* (Richter et al. 2003) at a velocity of -62 km s^{-1} revealed the additional presence of a cold neutral component within the intermediate-velocity cloud. Highly-ionized species, C IV and Si IV (Spitzer & Fitzpatrick 1992), show broad IVC absorption features centered, respectively, at -67 and -60 km s^{-1} and exceeding the extent of the negative-velocity low-ion absorption by 25 km s^{-1}. Although a detailed analysis of this line of sight is still somewhat uncertain, Spitzer & Fitzpatrick (1992) concluded that the data are qualitatively consistent with a fountain model if the slower gas has cooled and recombined more than the faster gas.

• HD 149881 (l=31°, b=+36°): In contrast to the multiple components readily seen over a wide velocity range along the HD 93521 line of sight, optical observations of the interstellar medium toward HD 149881 show a single broad blended feature, confined to low velocities, extending from −20 to +22 km s^{-1} in Ca II and from −15 to +17 km s^{-1} in Ti II (Albert 1983). Subsequently, Spitzer & Fitzpatrick (1995) found that ultraviolet observations reveal a much more complex distribution of material in 9 H I regions, at least one H II region, and hot gas. The H I regions include two IVCs: a feature at +22.6 km s^{-1}, prominent in absorption by Si II and Fe II and also seen in Zn II, Mn II, Cr II and, weakly, in Ca II, and another feature at +34.4 km s^{-1} detected only in Fe II. Spitzer & Fitzpatrick (1995) concluded that the depletion levels toward HD 149881 seem generally characteristic of gas with absolute LSR velocities greater than 20 km s^{-1} but, owing to complicated velocity-projection effects in this direction, they hesitated to make any more detailed conclusions about depletion as a function of velocity. They estimated maximum temperatures (from 21-cm H I emission-line widths and the assumption of no turbulent broadening) of 2700±400 K and 2500±800 K, and actual temperatures (from a comparison of 21-cm emission and Fe II absorption line widths) of 2000±400 K and 2200±900 K, respectively, for these clouds. In addition, broad, smooth absorption lines of Si IV and C IV were observed at intermediate resolution. Over all velocities, the integrated column densities yield a ratio of N(C IV)/N(Si IV) which is consistent with the mean value found by Sembach & Savage (1992) for 13 lines of sight through the Galactic Halo. Although the data did not allow separate analyses of intermediate-velocity components, they do confirm the presence of one or more regions of hot gas along this line of sight, in which the ionization is determined by the time-dependent balance between electron collisional ionization and radiative plus dielectronic recombination.

• HD 215733 (l=85°, b=−36°): One of the most intriguing interstellar sightlines is that toward HD 215733, where early observations of Ca II detected 8 velocity components (Albert 1983), and the high-resolution study by Fitzpatrick & Spitzer (1997) revealed 23 features in both Ca II and ultraviolet absorption. Remarkably, 12 of these components are at $v_{\rm LSR}$ less than −20 km s^{-1} and Wakker (2001) associates these components with the Pegasus-Pisces Arch. Fitzpatrick & Spitzer (1997) decided that these are H I regions and that the pattern of gas-phase abundances is very similar to that seen toward both HD 93521 and HD 149881. As in the gas toward HD 149881, kinetic temperatures cover a wide range, again in contrast to the mildly warm gas seen at all velocities toward HD 93521. Of the IV components along the HD 215733 line of sight, only the three at the most negative velocities have temperatures above 1000 K: -74.9 km s^{-1} (6000+5000/-3100 K), -50.6 km s^{-1} (1400±600 K), and -45.6 km s^{-1} (7000±4100 K). Electron densities could be measured for the latter two clouds, giving values of 0.02 and 0.05 cm^{-3}, respectively. IVCs were also detected in absorption from the highly-ionized element C IV, but not from Si IV, at -57.6 km s^{-1}, and from both C IV and Si IV at -30.5 km s^{-1}. The ratio of column densities of C IV and Si IV in the second component is 5.6±1.2, not far from the average of 3.5±1.3 found by Sembach & Savage (1992) for halo stars. Comparing the ratio of b values between these two ions to the theoretical ratio expected in kinetic equilibrium gave a kinetic temperature of $6.3 \pm 1.1 \times 10^5$ K for the -30.5 km s^{-1} gas. Although Fitzpatrick & Spitzer (1997) regarded this calculated temperature as tentative, they discussed the fact that it is different from the temperature of about 8×10^4 K which would be required to produce the observed column density ratio if the gas were in collisional equilibrium, constant with time. They felt that this difference provides some support for the generally accepted assumption that transient conditions are needed to explain the high ions in the halo.

• HD 18100 (l=218°, b=-63°): The line of sight toward HD 18100 probes the region of the South Galactic Pole, which is known for a relative absence of IVCs (Danly 1989). In fact, 21-cm H I emission toward this star is observed over -40 to $+30$ km s^{-1} and only one weak intermediate-velocity component is detected in optical Ca II absorption (Albert et al. 1993). Ryans et al. (1996b) obtained very-high-resolution spectra of Na I and Ca II along this sightline with the *UHRF*, and the cloud components they defined were used by Savage & Sembach (1996a) in a subsequent analysis of ultraviolet absorption observed with the *GHRS*. Savage & Sembach (1996a) identified a weak pair of components at -68 and -58 km s^{-1} in the Mg II λ2800 Å doublet, and a corresponding blended feature at these velocities in C II, Si II, Si III and Fe II. Even though they pointed out that uncertainties in the nature of the ionization prevent a direct study of relative elemental abundances in the intermediate-velocity gas, Savage and Sembach were able to describe its general properties. They found that the intermediate-velocity gas has a ratio of column densities of Si III to Si II, N(Si III)/N(Si II), of 1.8±0.7, while singly-ionized gas dominates the total column density at lower velocities. This is also different from the ratio of 0.09 determined for these ions by Spitzer & Fitzpatrick (1993) in the intermediate-velocity gas toward HD 93521. Detailed comparisons of all the observed ionic ratios to those predicted by several photoionization and collisional ionization models showed that the intermediate-velocity gas toward HD 18100 is most consistent with harder photoionization models characterized by nearly solar abundance ratios. Savage & Sembach (1996a) suggested that these clouds may have been initially heated and accelerated by shocks capable of partially disrupting the grains and that they have now cooled to the point that photoionization is the dominant ionization mechanism. Absorption by highly-ionized gas in this direction lies within the velocity range of absorption by the neutral and weakly-ionized gas at both low and intermediate velocities. Overall ionic ratios are mostly consistent with those expected from "Type I" hot gas (Savage & Sembach 1994), which is associated with absorption by lower stages of ionization in warm gas and may occur in the interfaces or mixing layers between the warm and hot interstellar medium.

• HD 100340 (l=259°, b=+61°): In addition to their analysis of the line of sight toward HD 18100, Savage & Sembach (1994, 1996a) also presented *GHRS* ultraviolet absorption-line observations toward HD 100340, a star near the North Galactic Pole that probes the IV Spur. This sightline displays interstellar H I emission from −70 to +70 $\mathrm{km\,s^{-1}}$ and a complex Ca II absorption profile including components at −64.2, −32.1 and +73.4 $\mathrm{km\,s^{-1}}$ (Albert et al. 1993; Ryans et al. 1999). Savage & Sembach (1996a) studied IVCs seen in absorption by Fe II, O I, Si II and N I at velocities of +31.0, +58.6 and +78.4 $\mathrm{km\,s^{-1}}$. These clouds, like the IV gas toward HD 18100, appear to be partially ionized, in contrast to the bulk of the lower-velocity gas seen in this direction. Al III absorption is detected in the +31 $\mathrm{km\,s^{-1}}$ component, and its observed ionic ratios in this component are reproduced well by a low-photoionization model in which $N(\mathrm{H\,I})/N(\mathrm{H})$ is about 0.18. Savage & Sembach (1996a) pointed out that this model produces optical emission-line fluxes consistent with those of the warm ionized medium and has a ratio $N(\mathrm{H\,II})/N(\mathrm{H\,I})$ of approximately 4, as expected for the WIM if its temperature is in the range of 5000 to 6000 K. The higher-velocity components are a bit more problematic, but may also be partially ionized by O- and B-star radiation. Absorption by highly-ionized gas toward HD 100340, like that toward HD 18100, occurs within the velocity range of the neutral and weakly-ionized absorption. But, in this direction, the high-ion absorption represents a mixture of both "Type I" hot gas, produced, as toward HD 18100, in the interfaces or mixing layers between warm and hot regions, and "Type II" hot gas, which may be tracing the cooling gas of supernova bubbles or a Galactic Fountain. Cold neutral material is detected by H_2 absorption in the component at −29 $\mathrm{km\,s^{-1}}$ (Richter et al. 2003). This is a complex line of sight which deserves more concentrated study.

• HD 203664 (l=62°, b=−27°): The +70 $\mathrm{km\,s^{-1}}$ cloud observed along the sightline toward the halo star HD 203664, which probes the IVC complex gp (Wakker 2001), is one of the most interesting and best studied examples of intermediate-velocity gas (Habing 1969; Keenan et al. 1983; Bates et al. 1983; Albert et al. 1993; Little et al. 1994; Sembach 1995; Ryans et al. 1996b). From high quality *UHRF* spectra, Ryans et al. (1996b) were able to resolve the cloud into three components in absorption by the K line of Ca II: a broad feature at +69.8 km s^{-1} and narrower components at +75.2 and +80.0 km s^{-1}. They also detected Na I absorption in the strong +80 km s^{-1} component and found a value of the ratio N(Ca II)/N(Na I) of 3.7, compatible with neutral or mildly-shocked gas. In the weaker components, this ratio is in excess of 14, implying grain destruction in shocked gas, and is consistent with the conclusions reached earlier by Albert et al. (1993) and Little et al. (1994) from the N(Ca II)/N(H I) ratio in less well-resolved spectra. At lower resolution, Sembach (1995) observed this IVC with *IUE* in 17 ultraviolet absorption lines of 11 elements in various stages of ionization. He also proposed that processing of interstellar gas through shocks could be the mechanism responsible for the detection of C IV and Si IV, in a ratio expected for collisionally ionized cloud interfaces at a temperature of about 10^5 K. His determination that element abundances with respect to sulphur are within a factor of 5 of solar values is also indicative of significant grain destruction. The relative abundances of low-ionization species imply a rather high electron density of 0.15 to 0.34 cm^{-3}, which could be explained by contributions from the ionized cloud boundaries. Sembach (1995) concluded, from these relative abundances and from the line widths measured by Little et al. (1994), that the cloud itself has a temperature of about 6000 K and is similar to neutral or partially ionized warm diffuse clouds typically seen along low-density sightlines. He suggested that this IVC could be a condensation or an accelerated cloud in a hot (10^6 K) cooling flow due to supernova explosions in the Galactic Disk that expel fountains of gas into the lower halo. These general conclusions have been supported more recently by higher-resolution *GHRS* data (Sembach, private communication), and the line of sight toward HD 203664 and the related IVC complex gp have been broadly studied by Smoker et al. (2001, 2002).

6. Speculations on the origin and significance of IVCs

The widespread nature of the IVCs and their generally common attributes lead us to conclude that the bulk of these clouds probably do, indeed, define a phenomenon in their own right. A successful model for their origin must explain the basic observations, presented in the previous sections, which appear to characterize the majority of these clouds:

- They are located in the lower Galactic Halo.
- They are widespread interstellar features, including both large coherent structures, predominantly in the northern galactic hemisphere, and smaller

clouds seen along numerous individual lines of sight.

- The large IVC complexes and most of the individual IVCs appear to be falling toward the galactic plane, at negative velocity with respect to the Local Standard of Rest.
- They display a variety of physical conditions, but abundances consistently indicate evidence of grain disruption and a gas composition that is not primordial.

As noted earlier, the range of mechanisms that may explain IVCs parallels those that are invoked to explain HVCs. One class of models is related to processes associated with star formation and evolution, and includes both localized phenomena such as stellar winds, bubbles, supernovae, and superbubbles, and more global phenomena such as Galactic Fountains. A second class of models includes infall scenarios where material is being accreted onto the Galaxy. A third set of models stems from the interaction of HVCs with IVCs. Here, a subtle distinction needs to be drawn between scenarios where the IVCs derive their motions directly from the same source as the HVCs, versus those where the IVCs are originally "normal"-velocity disk clouds which are accelerated by the HVCs as they impinge on the disk. This "secondary-phenomenon" scenario is in some senses independent of where the HVCs got their original energy. It is likely that our current compilation of observations contains examples of all of these processes.

Several examples of IVCs that probably arise from localized star formation related activity can be found in the literature. For instance, Penprase & Blades (1992) proposed that the striking complexity of the absorption toward HD 93721 (l=281°, b=+13°) represents the interaction of an HVC with the local interstellar medium in the vicinity of the Gum nebula. Intermediate-velocity gas along the line of sight to the star HD 119608, seen at both low and high ionization, may well be associated with the Loop IV supernova remnant and may indicate a flow of gas into the Galactic Halo (Roberts 1992; Sembach et al. 1997).

Galactic Fountain models have been shown to be successful in predicting many aspects of the observed properties of the IVCs (Bregman 1980; Ch. 16). Three-dimensional fountain simulations by de Avillez (1999) have also produced an array of cloudlets, clouds and cloud complexes at intermediate velocities. The location of the large, coherent complexes in the northern galactic hemisphere could be an asymmetry produced by intermittent fountain activity. Observations of the predominantly negative velocities of the IVCs and their near-solar metallicities offer general support for the fountain model. In individual cases, Wakker (2001) pointed out that it is an open question which clouds are part of the return flow of a Galactic Fountain and which are previously-hot halo gas compressed by infalling material. He cited the LLIV Arch, falling between the Local and Perseus spiral arms, with disk-like dust, a substantial H^+ fraction, and embedded in hot gas (Richter et al. 2001a), as one of the best examples of possible fountain material. In a wide-reaching study of the distribution and kinematics of O VI in the Galactic Halo, Savage et al. (2003) concluded that, if a fountain flow dominates, the observations require a mass-flow rate of cooling hot gas to each side of the galactic plane of about 1.4 solar masses per year, which is comparable to that estimated for IVCs.

The possibility of a relationship between high-velocity and intermediate-velocity cloud complexes is controversial. Wakker (2001) argued that available distance and metallicity estimates point, in general, to their being independent phenomena. However, he suggested that the intermediate-velocity complex gp may be related to the high-velocity complex GCP. If the IVC and HVC cloud complexes in the north are related, their large energies (on the order of 10^{53} erg) are difficult to explain if they are derived from star formation and evolution processes. Even a large cluster of correlated supernovae (i.e. on the order of 10^3 or so) would require a highly efficient conversion to mechanical energy in order to accelerate the clouds to their observed motions. If the cluster responsible for heating and accelerating this gas was formed recently enough so that the same gas it accelerated is now returning to the Galaxy in a nearly ballistic flow, it is likely that other evidence of such a cluster (on the scale of 30 Dor) would be easily observable. Infall models have been invoked to explain comparable energies in the Galactic Anti-Center, where there is little star formation occurring (Mirabel & Morras 1990), and in external galaxies (Rand & Stone 1996) and could be applicable here. The X-ray, γ-ray, and radio-continuum data in these directions further suggest that non-thermal mechanisms associated with magnetic fields may also play an important role.

The fact that the abundances are close to solar values may suggest that the IV gas originated in the disk. However, very little is known about the actual abundances of circumgalactic and Local Group gas, so this observation cannot provide a binding constraint.

Independent of the clouds' origins, Benjamin & Danly (1997) addressed the detailed dynamics of infalling clouds as they undergo a phase change, lose buoyancy and, falling though a viscous medium, reach terminal velocity. Their "Galactic Rain" scenario successfully predicts trends of infall velocity with column density and z-height. More work needs to be done, however, to understand the relationship between the physics of the phase change and the resulting physical characteristics such as ion ratios, electron densities, line widths, and depletion patterns.

Finally, regardless of how the IV clouds were formed, they are headed our way, bringing their metals and energy with them. This clearly has important implications for galactic evolution overall, and a more thorough understanding of the IVC phenomenon, its magnitude, its constancy over time, and its applicability to other galaxies is clearly warranted.

7. Directions for future work

Our ultimate picture of the dynamics and evolution of the Galaxy must include a better understanding of the intermediate-velocity clouds. There are a number of exciting directions for future study in this field.

First, a complete, unbiased all-sky survey of H I is needed to provide a comprehensive statistical analysis of intermediate-velocity gas. Happily, such a survey is under preparation at the time of this writing, combining the *LDS* (Hartmann & Burton 1997) and the Villa Elisa (Arnal et al. 2000) data, which have nearly identical observational parameters (i.e. angular resolution, velocity coverage and resolution, and rms sensitivity). The two surveys are being jointly corrected for stray radiation, and preliminary results suggest that the resulting all-sky survey will be of extremely high quality (Hartmann 2000).

In order to understand better the full range of physical conditions and abundances in the IV gas, detailed high-resolution studies of a significant number of individual lines of sight are required. Our picture of the origins and histories of IVCs depends upon accurate abundances, which are currently available for only a very few lines of sight. The origin of the highly-ionized gas and its relation to intermediate-velocity gas is an important piece in the puzzle of galactic dynamics. In particular, observations of the O VI ion with *FUSE* are beginning to elucidate the role that hot gas plays in the large-scale development of the interstellar medium.

Of course, distance measurements remain crucial to all studies, and there is a great need to identify a larger sample of background halo stars and to determine their distances so they can serve as probes of interstellar gas at both intermediate and high velocities. A coordinated effort toward this goal was launched during the Stromlo Workshop on High-Velocity Gas, held at Mt. Stromlo Observatory in August 1998, and is described in Ch. 10. The prospects for such a concerted attack are hopeful.

If a large enough sample of halo IVCs with well-known distances, velocities and column densities can be compiled, perhaps the terminal velocity model of Benjamin & Danly (1997) can be verified. If it holds, it may eventually be possible to infer the vertical density structure of the interstellar medium and to use the relations to derive the distances to clouds from their velocities.

The relationship between the IVCs and the HVCs is still unclear and, indeed, may encompass a variety of interactions. The apparent connection, in the sky, of low-, intermediate-, and high-velocity features is tantalizing, but more specific distances to the gas are needed in order to confirm or refute the possibility of such H I bridges. A possible connection, in time, could be made if IVCs are a later stage of decelerating HVCs, but better metallicity measurements for more high-velocity cloud complexes are necessary to evaluate this suggestion.

And, finally, it is essential to maintain detailed communication between workers in this field, to identify new IVC observations and to make them readily accessible to others. Bart Wakker has established a website for the purpose of coordinating observations of both HVCs and IVCs. This website is described in Ch. 10. Everyone who encounters a cloud at intermediate velocity, whether as part of a dedicated IVC study or as a footnote to another work, is invited to list it there. This compilation will include observations of both 21-cm H I emission and optical and ultraviolet absorption lines and, we hope, will serve as a springboard for further progress.

Acknowledgements

We would like to gratefully acknowledge Kip Kuntz for creating the original version of Fig. 1 and for helpful discussions, and Ken Sembach and, especially, Bart Wakker, for very useful comments and suggestions.

References

Adams, W.S. 1949, ApJ, 109, 354
Albert, C.E. 1983, ApJ, 272, 504
Albert, C.E., Blades, J.C., Morton, D.C., Lockman, F.J., Proulx, M., Ferrarese, L. 1993, ApJS, 88, 81
Anders, E., Grevesse, N. 1989, Geochim. Cosmochim. Acta, 53, 197
Arnal, E.M., Bajaja, E., Larrarte, J.J., Morras, R., Pöppel, W.G.L. 2000, A&AS, 142, 35
Bajaja, E., Cappa de Nicolau, C.E., Cersosimo, J.C., Martin, M.C., Loiseau, N., Morras, R., Olano, C.A., Pöppel, W.G.L. 1985, ApJS, 58, 143
Barnes, D.G., Staveley-Smith, L., de Blok, W.J.G., et 36 al. 2000, MNRAS, 322, 486
Bates, B., Brown-Kerr, W., Giaretta, D.L., Keenan, F.P. 1983, A&A, 122, 64
Bates, B., Catney, M.G., Keenan, F.P. 1990, MNRAS, 245, 238
Bates, B., Shaw, C.R., Kemp, S.N., Keenan, F.P., Davies, R.D. 1995, ApJ, 444, 672
Benjamin, R.A., Danly, L. 1997, ApJ, 481, 764
Benjamin, R.A., Venn, K.A., Hiltgen, D.D., Sneden, C. 1996, ApJ, 464, 836
Blaauw, A., Tolbert, C.R. 1966, BAN, 18, 405
Blom, J.J., Bloemen, H., Bykov, A.M., et 8 al. 1997, A&A, 321, 288
Bregman, J.N. 1980, ApJ, 236, 577
Brüns, C., Kerp, J., Staveley-Smith, L. 2000, in ASP Conf. Ser. 218, Mapping the Hidden Universe, eds. R.C. Kraan-Korteweg, P.A. Henning, H. Andernach (San Francisco: ASP), 349
Centurion, M., Vladilo, G., de Boer, K.S., Herbstmeier, U., Schwarz, U.J. 1994, A&A, 292, 261
Chu, Y.-H., Kennicutt, R.C. 1994, ApJ, 425, 720
Cohn, H., York, D.G. 1977, ApJ, 216, 408
Cowie, L.L., York, D.G. 1978, ApJ, 223, 876
Danly, L. 1987, Ph.D. thesis, University of Wisconsin-Madison

Danly, L. 1989, ApJ, 342, 785
Danly, L., Lockman, F.J., Meade, M.R., Savage, B.D. 1992, ApJS, 81, 125
Danly, L., Albert, C.E., Kuntz, K.D. 1993, ApJ, 416, L29
de Avillez, M.A. 1999, in ASP Conf. Ser. 166, Stromlo Workshop on High-Velocity Clouds, eds. B.K. Gibson, M.E. Putman (San Francisco: ASP), 103
de Boer, K.S., Savage, B.D. 1983, ApJ, 265, 210
de Boer, K.S., Rodriguez Pascual, P., Wamsteker, W., Sonneborn, G., Fransson, C., Bomans, D.J., Kirshner, R.P. 1993, A&A, 280, L15
de Heij, V. 1998, in The Physics and Chemistry of the Interstellar Medium, Abstract Book of the 3rd Cologne-Zermatt Symposium, ed. V. Ossenkopf (Aachen: Shaker-Verlag)
Fitzpatrick, E.L. 1996, ApJ, 473, L55
Fitzpatrick, E.L., Spitzer, L. 1997, ApJ, 475, 623
Frail, D.A., Weisberg, J.M., Cordes, J.M., Mathers, C. 1994, ApJ, 436, 144
Gladders, M.D., Clarke, T.E., Burns, C.R., et 9 al. 1998, ApJ, 507, L161
Gladders, M.D., Clarke, T.E., Burns, C.R., et 14 al. 1999, in ASP Conf. Ser. 166, Stromlo Workshop on High-Velocity Clouds, eds. B.K. Gibson, M.E. Putman (San Francisco: ASP), 45
Gringel, W., Barnstedt, J., de Boer, K.S., Grewing, M., Kappelmann, N., Richter, P. 2000, A&A, 358, L37
Habing, H.J. 1969, BAN, 20, 177
Hartmann, D. 2000, in ASP Conf. Ser. 217, Imaging at Radio through Submillimeter Wavelengths, eds. J.G. Mangum, S.J.E. Radford (San Francisco: ASP), 58
Hartmann, D., Burton, W.B. 1997, Atlas of Galactic Neutral Hydrogen (Cambridge: Cambridge University Press)
Heiles, C. 1984, ApJS, 55, 585
Heiles, C. 1997, ApJ, 481, 193
Howk, J.C., Savage, B.D. 1999, AJ, 117, 2077
Jaschek, C., Gomez, A.E. 1996, A&A, 330, 619
Keenan, F.P., Dufton, P.L., McKeith, C.D., Blades, J.C. 1983, MNRAS, 203, 963
Kennedy, D.C., Bates, B., Keenan, F.P., Kemp, S.N., Ryans, R.S.I., Davies, R.D., Sembach, K.R. 1998, MNRAS, 297, 849
Kerp, J., Mack, K.-H., Egger, R., Pietz, J., Zimmer, F., Mebold, U., Burton, W.B., Hartmann, D. 1996, A&A, 312, 67
Kerp, J., Burton, W.B., Egger, R., Freyberg, M.J., Hartmann, D., Kalberla, P.M.W., Mebold, U., Pietz, J. 1999, A&A, 342, 213
Knude, J., Hog, E. 1999, A&A, 341, 451
Kuntz, K.D., Danly, L. 1996, ApJ, 457, 703
Langer, G.E., Prosser, C.F., Sneden, C. 1990, AJ, 100, 216
Lauroesch J.T., Meyer, D.M., Blades, J.C. 2000, ApJ, 543, L43
Lehner, N., Rolleston, W.R.J., Ryans, R.S.I., Keenan, F.P., Bates, B., Pollacco, D.L., Sembach, K.R. 1999a, A&AS, 134, 257
Lehner, N., Sembach, K.R., Lambert, D.L., Ryans, R.S.I., Keenan, F.P. 1999b, A&A, 352, 257
Little, J.E., Dufton, P.L., Keenan, F.P., Conlon, E.S., Davies, R.D. 1994, ApJ, 427, 267
Lockman, F.J., Hobbs, L.M., Shull, J.M. 1986, ApJ, 301, 380
Mathewson, D.S., Cleary, M.N., Murray, J.D. 1974, ApJ, 190, 291
Meyer, D.M., Blades, J.C. 1996, ApJ, 464, L179
Meyer, D.M., Lauroesch, J.T. 1999, ApJ, 520, L103
Mirabel, I.F., Morras, R. 1990, ApJ, 356, 130
Morras, R. 1980, A&A, 92, 315
Morras, R., Bajaja, E., Arnal, E.M., Pöppel, W.G.L. 2000, A&AS, 142, 25
Morton, D.C., Blades, J.C. 1986, MNRAS, 220, 927
Münch, G., Zirin, H. 1961, ApJ, 133, 11
Penprase, B.E., Blades, J.C. 1992, ApJ, 391, 276
Pietz, J., Kerp, J., Kalberla, P.M.W., Mebold, U., Burton, W.B., Hartmann, D. 1996,

A&A, 308, L37
Pilachowski, C., Langer, E., Sneden, C., Kraft, R. 1998, BAAS, 30, 870
Rand, R.J., Stone, J.M. 1996, AJ, 111, 190
Richter, P., Savage, B.D., Wakker, B.P., Sembach, K.R., Kalberla, P.M.W. 2001a, ApJ, 549, 281
Richter, P., Sembach, K.R., Wakker, B.P., Savage, B.D., Tripp, T.M., Murphy, E.M., Kalberla, P.M.W., Jenkins, E.B. 2001b, ApJ, 559, 318
Richter, P., Wakker, B.P., Savage, B.D., Sembach, K.R. 2003, ApJ, 586, 230
Roberts, B.R. 1992, Investigation of a Possible Galactic Chimney, USNA Trident Scholar Report No. 196, U.S. Naval Academy, Annapolis, MD
Ryans, R.S.I., Hambly, N.C., Dufton, P.L., Keenan, F.P. 1996a, MNRAS, 278, 132
Ryans, R.S.I., Sembach, K.R., Keenan, F.P. 1996b, A&A, 314, 609
Ryans, R.S.I., Keenan, F.P., Sembach, K.R., Davies, R.D. 1997a, MNRAS, 289, 83
Ryans, R.S.I., Keenan, F.P., Sembach, K.R., Davies, R.D. 1997b, MNRAS, 289, 986
Ryans, R.S.I., Keenan, F.P., Rolleston, W.R.J., Sembach, K.R., Davies, R.D. 1999, MNRAS, 304, 947
Savage, B.D., Sembach, K.R. 1994, ApJ, 434, 145
Savage, B.D., Sembach, K.R. 1996a, ApJ, 470, 893
Savage, B.D., Sembach, K.R. 1996b, ARA&A, 34, 279
Savage, B.D., Sembach, K.R., Wakker, B.P., Richter, P., Meade, M., Jenkins, E.B., Shull, J.M., Moos, H.W., Sonneborn, G. 2003, ApJS, 146, 125
Schwarz, U.J., Wakker, B.P., van Woerden, H. 1995, A&A, 302, 364
Sembach, K.R. 1995, ApJ, 445, 314
Sembach, K.R., Savage, B.D. 1992, ApJS, 83, 147
Sembach, K.R., Savage, B.D., Tripp, T.M. 1997, ApJ, 480, 216
Sembach, K.R., Howk, J.C., Ryans, R.S.I., Keenan, F.P. 2000, ApJ, 528, 310
Shaw, C.R., Bates, B., Kemp, S.N., Keenan, F.P. 1996, ApJ, 473, 849
Shull, J.M., Beckwith, S. 1982, ARA&A, 20, 163
Slavin, J.D., McKee, C.F., Hollenbach, D.J. 2000, ApJ, 541, 218
Smoker, J.V., Lehner, N., Keenan, F.P., Totten, E.J., Murphy, E., Sembach, K.R., Davies, R.D., Bates, B. 2001, MNRAS, 322, 13
Smoker, J.V., Haffner, L.M., Keenan, F.P., Davies, R.D., Pollacco, D. 2002, MNRAS, 337, 385
Songaila, A., York, D.G. 1981, ApJ, 242, 976
Songaila, A., Cowie, L.L., Weaver, H.F. 1988, ApJ, 329, 580
Spitzer, L. 1978, Physical Processes in the Interstellar Medium (New York: Wiley-Interscience), 163
Spitzer, L., Fitzpatrick, E.L. 1992, ApJ, 391, L41
Spitzer, L., Fitzpatrick, E.L. 1993, ApJ, 409, 299
Spitzer, L., Fitzpatrick, E.L. 1995, ApJ, 445, 196
Spitzer, L., Fitzpatrick, E.L. 1997, ApJ, 475, 623
Stark, A.A., Gammie, C.F., Wilson, R.W., Bally, J., Linke, R.A., Heiles, C., Hurwitz, M. 1992, ApJS, 79, 77
Tolbert, C.R. 1971, A&AS, 3, 349
Wakker, B.P. 2001, ApJS, 136, 463
Wakker, B.P., Howk, J.C., Schwarz, U.J., van Woerden, H., Beers, T.C., Wilhelm, R., Kalberla, P.M.W., Danly, L. 1996, ApJ, 473, 834
Wakker, B.P., Kalberla, P.M.W., van Woerden, H., de Boer, K.S., Putman, M.E. 2001, ApJS, 136, 537
Watson, J.K., Meyer, D.M. 1996, ApJ, 473, L127
Welsh, B.Y., Craig, N., Roberts, B. 1996, A&A, 308, 428
Wesselius, P.R., Fejes, I. 1973, A&A, 24, 15
Wolfire, M.G., McKee, C.F., Hollenbach, D., Tielens, A.G.G.M. 1995, ApJ, 453, 673

5. HVCS RELATED TO THE MAGELLANIC SYSTEM

MARY E. PUTMAN
Australia Telescope National Facility, CSIRO, Epping, Australia; currently at University of Michigan, Ann Arbor, USA; mputman@lsa.umich.edu

Abstract. This chapter provides a review of the observational and theoretical results on the HVCs known to have originated from the interaction of the Large and Small Magellanic Clouds with each other and the Milky Way. The neutral hydrogen observations of the Magellanic Bridge, Magellanic Stream, and Leading Arm are summarized and related to the optical and ultraviolet observations of these complexes. The theoretical models for their origin are largely based on either tidal or ram-pressure forces, with more sophisticated models using a combination of these forces. The Magellanic complexes account for the majority of the HVCs at positive velocities and serve as useful "standard candles" for the HVC population as a whole. They are also important current tracers of the galaxy formation process.

1. Introduction

The results of the interaction between the Milky Way and the Magellanic Clouds are revealed through several high-velocity complexes connected to the Clouds. The exact mechanism of their formation is under some debate, but they remain the only group of high-velocity clouds (HVCs) for which we have both an origin and (roughly) a distance. Given that, the Magellanic HVCs can be used as a calibrator for other HVCs, while also providing an opportunity to closely investigate the remnants of an interacting system. These HVCs may hold the key to the star formation history, kinematic structure, and present Hubble type of the Magellanic Clouds, and their proximity to the Milky Way allows us to estimate key Galactic parameters.

The HVCs of the Magellanic System can be classified into three major complexes: the Magellanic Bridge, an H I connection between the Clouds; the Magellanic Stream, which trails the clouds and is one of the largest H I features in the sky outside of our Galaxy; and the Leading Arm, a more diffuse H I filament which leads the Clouds. In terms of HVCs, these features

H. van Woerden et al. (eds.), High-Velocity Clouds, 101–124

have been studied rather extensively. In this review, I will first describe the observational results for each complex and subsequently discuss their origin and relationship to the overall HVC population. All of the H I data presented are from the H I Parkes All-Sky Survey (*HIPASS*) (Barnes et al. 2001; Putman et al. 2003a).

2. The Magellanic Bridge

2.1. H I STRUCTURE AND KINEMATICS

The Magellanic Bridge is a continuous filament of H I stretching from the body of the Small Magellanic Cloud (SMC) to an extended arm of the Large Magellanic Cloud (LMC) (see Fig. 1). The Bridge merges almost seamlessly with the SMC, but the boundary between the Bridge and SMC is usually defined at $(l,b)=(295°,-41.°5)$ and $v_{\rm LSR}=+125$ km s^{-1}, where the loop which extends from the SMC at approximately $(l,b)=(297°,-44°)$ rejoins the tail of the SMC (also known as Shapley's wing). This is also the boundary that was originally chosen based on stellar associations (Westerlund & Glaspey 1971); however, with the increasing numbers of stellar associations found in the Bridge, this boundary is also somewhat ambiguous (see Sect. 2.2). The Bridge emerges from the SMC's tail at the high column densities of 10^{21} cm^{-2} and remains clumpy, but gradually decreases in column density to 10^{20} cm^{-2} at $(l,b)=(287°,-34.°5)$. At this latter position, the Bridge joins with what appears to be an extended spiral arm of the LMC (Kim et al. 1998; Putman et al. 1999). South of the Bridge in galactic coordinates, especially on the SMC side, the chaotic beginnings of the Magellanic Stream are present. In general, the Bridge is a more orderly feature than the Stream, possibly representing the Bridge's shorter history or a more stable environment. The Bridge has an H I mass of approximately 5.5×10^7 M$_\odot$, but this value is highly dependent on whether extensions into the SMC, LMC and Stream are included.

The Bridge has a regular velocity gradient along its main filament, gradually increasing in velocity and decreasing in spatial width as it approaches the LMC. The final dense pockets of emission do not disappear until +350 km s^{-1} at $(l,b)=(283°,-42°)$. McGee & Newton (1986) report on line profiles which contain up to 5 components throughout the Bridge (with a velocity resolution of 4.1 km s^{-1}). They report systematic profile variations in the central Bridge region, but sporadic differences in the regions of the Bridge which extend into the Magellanic Stream, possibly indicating a more turbulent environment. High resolution Australia Telescope Compact Array (*ATCA*) and *Parkes* observations were recently presented by Muller et al. (2003b). On the larger scale they found two major velocity components to the Bridge, with the higher resolution revealing numerous

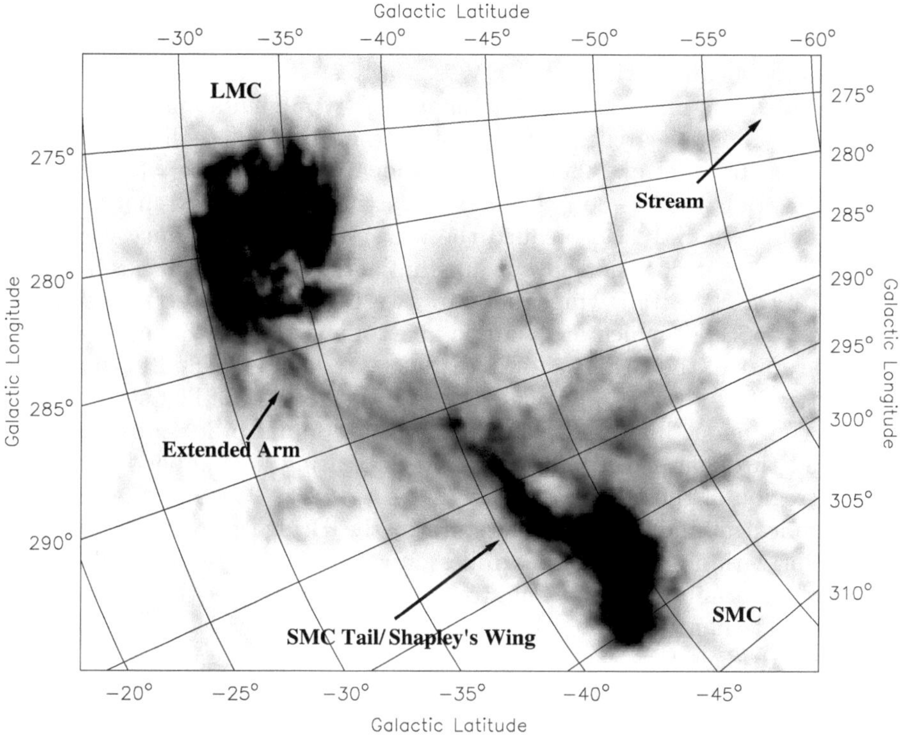

Figure 1. Neutral hydrogen column density map of the Magellanic Clouds and Bridge, with the main features discussed in the text labeled. The intensity scale is logarithmic, ranging from 10^{21} cm^{-2} (black) to 2×10^{18} cm^{-2}. Figure from Putman (2000).

shells and filaments. The shells have characteristics similar to those found in the SMC. H I absorption studies find that cool atomic-phase gas exists in the Bridge, indicating that the pressure in this region is surprisingly high and that stars may have formed from the Bridge material directly, rather than being drawn out from the SMC (Kobulnicky & Dickey 1999). Sensitive CO studies of the Bridge were recently completed by Muller et al. (2003a) and provide further evidence for star formation in this intergalactic environment. These observations also suggest lower metallicities for the Bridge than for the SMC and Stream.

2.2. STARS!

The Magellanic Bridge is the only HVC which has stars associated with it, and in this respect it may be inaccurate to call it an HVC. In addition, although the Bridge is an HVC in the galactic reference frame, it is not technically an HVC in the Magellanic reference frame, unlike the Stream and Leading Arm. The stars are very scarce and the gas-to-star ratio remains extremely high, so it is conceivable that future stellar searches may find stars associated with other HVCs. Early stellar searches in the SMC tail included the discovery of a number of B-type giants and dwarfs (e.g. Sanduleak 1969). Searches for blue stars then continued throughout the Bridge (e.g. Irwin et al. 1990), and were identified from $(l,b)=(296°,-41°)$ to at least $(l,b)=(287°,-36°)$. Demers & Battinelli (1998) find that the stars in the tail of the SMC (also called the wing) have little distance variation, indicating that it does not have a substantial depth. On the other hand, at the tip of the SMC tail/wing, there are two Bridge associations within 17′ (300 pc at 55 kpc) which are ~5 kpc apart along the line of sight. In general, the stars in the Bridge show a distance gradient expected for a feature linking the LMC (at 50 kpc) and the SMC (at 60 kpc). The stars do not form a continuous link as the H I does, but are found in loose associations scattered throughout the SMC tail and decreasing in number toward the central region of the Bridge.

Chemical abundances for the stars in the Bridge were thought to be consistent with an SMC origin (Rolleston et al. 1993; Hambly et al. 1994); however, determinations by Rolleston et al. (1999) suggest they are a factor of ~ 3 lower than the present-day metallicity of the SMC. The ages of the youngest Bridge stars range from 10 to 25 Myr, much younger than expected if they were torn from the SMC 200 Myr ago, as most tidal models predict. This (again) indicates that the Bridge is actually a star forming region. By considering all of the stars in the Bridge, Grondin et al. (1992) find that the Bridge's IMF is shallower than that of the Milky Way or the Clouds. This favors the formation of massive stars and may indicate that cloud-cloud collisions are the dominant star formation trigger (Scoville et al. 1986; Christodoulou et al. 1997). There has been no detection of a horizontal-branch star population in the Bridge, indicating that the halos of the two clouds do not meet (Grondin et al. 1992). Kunkel et al. (1997) have found an abundance of intermediate-age (several Gyr) carbon stars scattered throughout the Bridge region, with possible extensions into the beginning of the Stream. Diffuse Hα emission also appears to be prevalent in the Bridge region closest to the SMC (Johnson et al. 1982; Marcelin et al. 1985), as would be expected from the presence of hot young stars. However, there are also several non-detections in the central region of the Magellanic Bridge (Putman et al. 2003a). Finally, Far Ultraviolet Spectroscopic Explorer (*FUSE*) observations of two stars in the Magellanic Bridge show H_2, O VI, and numerous other atomic or ionic transitions in absorption, implying the presence of multiple gas phases (Lehner 2002).

3. The Magellanic Stream

3.1. H I STRUCTURE AND KINEMATICS

The Magellanic Stream, discovered 30 yrs ago (Wannier & Wrixon 1972; Mathewson et al. 1974), is a complex arc of neutral hydrogen starting at the Magellanic Clouds and trailing them for over 100°. The Stream contains $\sim 2\times 10^8$ $M_\odot$ of neutral hydrogen (at an average distance of 55 kpc) and has a velocity gradient of over 700 km s^{-1} from head to tip, 390 km s^{-1} greater than that due to galactic rotation alone. Recent *HIPASS* observations of the Stream provide an almost two-fold improvement in spatial resolution over previous survey data, and depict increasing complexity in the Stream's structure (see Fig. 2). In particular, the maps reveal multiple filaments at the Stream's head, a twisting ladder structure along the Stream's length, and small dense clouds which extend 20° from the Stream's main filament. A broad overview of the H I properties of the Magellanic Stream is presented below. See Putman et al. (2003a) for a full description.

The beginning of the Stream is rather chaotic, as it spews out from several locations north of the SMC and Bridge at $v_{\rm LSR}$=+90 to +240 km s^{-1} (see Figs. 2 and 3). There is a slight discontinuity in velocity as the H I enters the Stream from the Bridge. Figure 3 shows how the Stream becomes more negative in velocity as it extends away from the Clouds, and how there are multiple initial filaments which come to a clumpy end at $b\sim -60°$ and $v_{\rm LSR}\sim +85$ km s^{-1}. The main filament of the Stream continues toward the South Galactic Pole, where it reaches 0 km s^{-1}, making it difficult to disentangle from galactic emission. It then proceeds north to $(l,b)\sim(90°,-40°)$, $v_{\rm LSR}\sim -450$ km s^{-1}, and column densities of only a few $\times 10^{18}$ cm^{-2} (versus a few $\times 10^{19}$ cm^{-2} at the Stream's head). Relative to the Galactic Standard of Rest, the radial velocity of the Stream gradually becomes more negative from the head ($\sim$+50 km s^{-1}) to the tip (~ -200 km s^{-1}).

The main filament of the Stream is not as complex as the head, but it is also a complicated structure which appears to be made up of two distinct components. The splitting of the Stream into two filaments is evident throughout, but is most obvious beyond the multiple filaments at the Stream's head. The two filaments run parallel for the length of the Stream and begin to merge toward the tail (much as if one were looking down a long straight road). There are also several horse-shoe-shaped structures which join the two filaments at several positions. This helical structure may represent the orbit of the Magellanic Clouds about each other, with the two filaments representing material from the Bridge and SMC.

Small compact clouds are found throughout Figs. 2 and 3, surrounding the Stream's main filament in both position and velocity. Many of the small clouds, both in and about the Stream, show head-tail structures (i.e. a dense

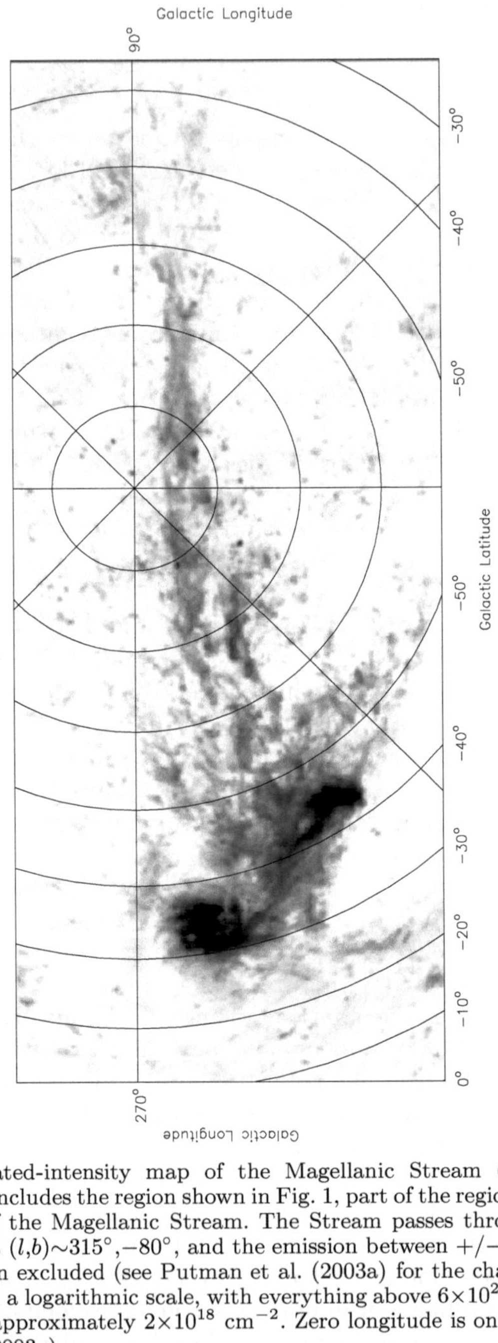

Figure 2. An integrated-intensity map of the Magellanic Stream ($v_{\rm LSR}= -400$ to $+400$ km s^{-1}), which includes the region shown in Fig. 1, part of the region shown in Fig. 4 and the full extent of the Magellanic Stream. The Stream passes through the velocity of galactic emission at $(l,b)\sim315°,-80°$, and the emission between $+/-20$ km s^{-1} (LSR) in this region has been excluded (see Putman et al. (2003a) for the channel maps). The intensity values are on a logarithmic scale, with everything above 6×10^{20} cm^{-2} black and the faintest levels at approximately 2×10^{18} cm^{-2}. Zero longitude is on the right. Figure from Putman et al. (2003a).

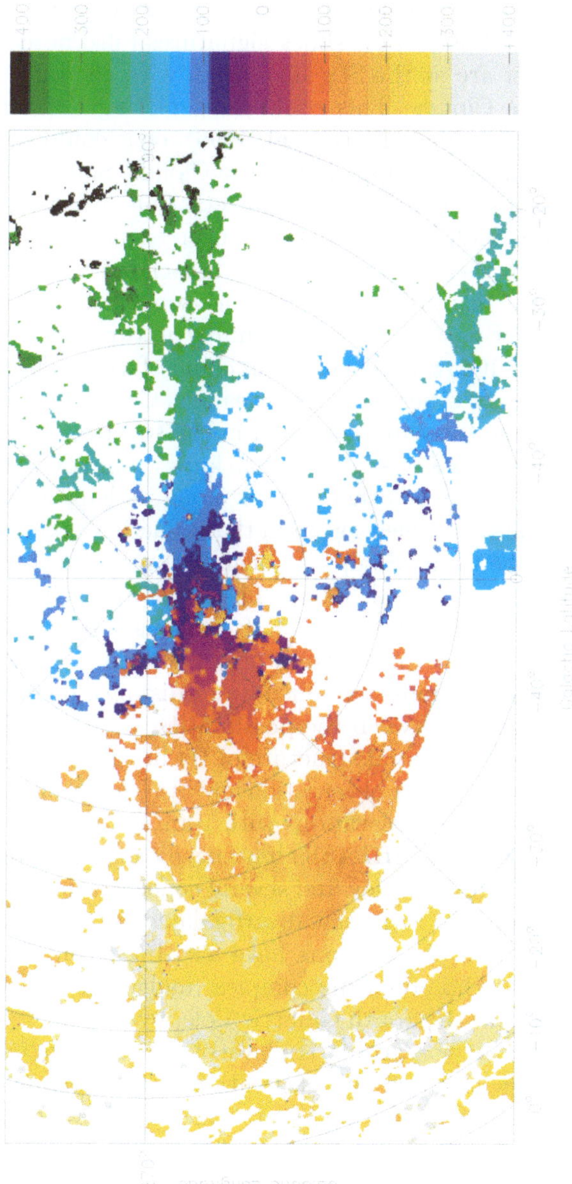

Figure 3. Velocity distribution (relative to the LSR) of the Magellanic Stream ranging from -450 km s^{-1} (black) to $+380$ km s^{-1} (white). The velocity gets progressively more negative as the Stream extends away from the Magellanic Clouds, and it passes through 0 km s^{-1} at approximately the South Galactic Pole. Zero longitude is on the right. Figure from Putman et al. (2003a).

core with a diffuse extension of approximately twice the diameter of the core (tadpoles)) and hollow bow-shock signatures (also noted by Mathewson et al. 1979). Most are at the Stream's head, with the tails generally pointing away from the Clouds. This could be depicting the Stream's interaction with the Galaxy's halo (Pietz et al. 1996), or it could simply represent the way the gas has been stripped from the Clouds. Some of the small clouds at positive-velocities near the South Galactic Pole in Figs. 2 and 3 are actually galaxies of the Sculptor Group. It has been argued that the abundant small clouds between these galaxies are not associated with the Stream, but are members of the Sculptor Group (Mathewson et al. 1975; Haynes & Roberts 1979). Considering the Stream's clumpy nature throughout this area, it would be difficult to make a confident claim of a cloud's association with the Sculptor Group or other dwarf galaxies (e.g. Carignan et al. 1998). However, it is curious how the clumps remain in the southern region of the Sculptor Group from velocities of -240 to $+240$ km s^{-1}, and do not follow the Stream as closely in velocity as other clumps along its length. Could these clumps be the remnants of an ejection from the Galactic Center, or intergalactic H I clouds along the Coma-Sculptor-Local Group supergalactic filament (Tully & Fisher 1987; Jerjen et al. 1998)? The observations indicate that a galactic phenomenon, such as the passage of previous satellites through the South Galactic Pole, is the most likely origin for these clouds (Putman et al. 2003a). Hα observations, metallicity and distance determinations will help further distinguish between the possibilities.

The small-scale spatial structure of H I in the Stream is currently being investigated with *ATCA* observations by Staveley-Smith, Brüns & Putman. *Parkes* narrow-band observations with 1 km s^{-1} velocity resolution (vs. the 26 km s^{-1} resolution of *HIPASS*) have been completed for the entire Stream by Brüns et al. (2004). They find several concentrations of gas in the Stream with the potential to form new dwarf galaxies in the Galactic Halo. Higher velocity resolution observations have also been completed in the past by Haynes (1979), who noted the complex, multi-profile nature of the Stream in the region near the South Galactic Pole, by Cohen (1982), who found the Stream to also have a strong transverse velocity gradient, and by Morras (1983, 1985), who noted the bifurcation of the Stream. The northern tip of the Stream was studied by Wayte (1989). He notes the continued bifurcation of the Stream and a complex velocity structure which may indicate that the tail of the Stream is breaking up into many individual clouds at different velocities. Stanimirovic et al. (2002) completed a detailed study of the tip of the Stream with *Arecibo* and found that a Galactic Halo density $>10^{-4}$ cm^{-3} is required to confine these clouds and prevent them from having already dispersed. The line profiles of the clouds at the tip show a core/envelope structure reminiscent of some non-Magellanic HVCs

(Wakker & van Woerden 1997). If these non-Magellanic HVCs are generally less distant than the majority of the Stream (see van Woerden et al. (1999) and Ch. 10 for some distances), this change in profile may indicate that the tip of the Stream is getting closer to the Galaxy. Tidal or ram-pressure forces may be responsible for stripping off the clouds' outer layers.

3.2. OPTICAL OBSERVATIONS

A new method of studying the Magellanic Stream has come with the discovery that the Stream can be detected in Hα emission (e.g. Weiner & Williams 1996). The detections vary tremendously in surface brightness (0.04 to 0.4 Rayleigh), and are usually in regions of high H I column density ($>10^{19}$ cm^{-2}; Putman et al. 2003b). There does not appear to be a correlation between the Hα emission measure and H I column density; however, the current lack of high-resolution H I data makes this difficult to test. The velocities of the H I line and the Hα line generally match within the resolution of the observations. It is possible that Hα emission will be detected beyond the H I contours of the Stream, as this has been observed for other complexes (Tufte et al. 1998) and may indicate the presence of an ionized sheath. Compact clouds (CHVCs) in the vicinity of the Magellanic Stream may be associated with the interactive debris of the Stream and have been detected in Hα emission (Tufte et al. 2002). [N II] has also now been detected, but [O III] has not (Putman et al. 2003b).

It is not clear how the emission-line results should be interpreted. Earlier suggestions that ram pressure is responsible for the Hα emission seem less secure in light of the strength of the emission, the line ratios, and the higher-resolution H I maps which show that the strong detections do not always correlate with the leading edges of H I condensations (Putman & Gibson 1999b). Bland-Hawthorn & Maloney (1999) conclude that shock ionization requires unrealistically high halo densities at $d{\sim}50$ kpc and suggest ionizing photons from the Galaxy are the main cause for the emission. On the other hand, the Hα emission measures in the Stream are generally ~2 times higher than for HVCs which have upper distance limits of ~10 kpc (Putman et al. 2003b). While it seems unlikely that the contribution of ionizing photons from the LMC can account for the difference, it remains to be seen if the presence of yet-unassociated young stars, pre-ionization due to interaction with a halo medium, or the effects of shadowing and nearby spiral arms can account for the elevated Hα emission. If the escape of ionizing photons from the Galaxy and the Magellanic Clouds can be accurately determined, the emission measures can be used to determine the distance to various points along the Stream (Bland-Hawthorn & Maloney 1999). A complete map of the Stream's ionized gas would be a very interesting complement to the H I data presented here.

There have been numerous searches for stars which are associated with the Stream, as they might be expected if the Stream were formed via a gravitational interaction. All of the searches for stars within the H I contours of the Stream have been negative, with most of the searches being based on the assumption that the Stream is young and should be populated by A to F stars. Brück & Hawkins (1983) claimed no stellar Stream counterpart based on star counts down to B magnitude 20.5 in the section of the Stream closest to the Clouds. Recillas-Cruz (1982) and Tanaka & Hamajima (1982) did a similar search of the tip of the Stream and found no excess of A-type stars. Guhathakurta & Reitzel (1998) recently used the Keck telescope to complete a deep stellar search in a $5' \times 7'$ region at $(l,b) \sim 60°, -68°$ (within MS IV) and claimed an upper limit on the Stream's star-to-gas ratio of 0.1 (5% that of the LMC). It is possible that these results are still not definitive, given the young population of stars searched for in the early searches and the limited area covered by the Keck search; but a more likely explanation is that the H I Stream does not contain stars. There is still the possibility of an offset stellar stream, as seen in many other interacting systems (Hibbard et al. 2000), or of a stellar stream that is significantly less extended than the Stream owing to the initial H I distribution of the Clouds being more extended (Yoshizawa 1998). A possible offset stellar tidal counterpart has been found by Majewski et al. (1999), who searched for giant stars about the Clouds and found interesting populations in a region north of the LMC.

3.3. METALLICITY AND DISTANCE DETERMINATIONS

Metallicity and abundance determinations for the Magellanic Stream are consistent with an origin in the Magellanic Clouds. The Stream's primary metallicity determination uses Fairall 9 as a probe and has been investigated by Gibson et al. (2000), Lu et al. (1994), and Songaila (1981), all of whom obtained consistent results. Gibson et al. (2000) used Goddard High-Resolution Spectrograph (*GHRS*) data and new H I observations to obtain a S/H ratio of 0.28 solar, extremely close to the metallicity of the SMC. They also detected Mg II near the tip of the Stream (toward III Zw 2=Mrk 1501), which indicates the Stream gas extends at least 15° from the H I shown in Fig. 2. Lu et al. (1994) found Si/H$\gtrsim$0.2 solar and S/H$\lesssim$0.3 solar along the Fairall 9 sightline. They find the implied Si/S ratio to be greater than or equal to 0.6 times the solar ratio, which indicates that dust depletion is not prevalent in the Stream (Si is easily depleted onto dust grains). This is consistent with the lack of extinction and infrared emission from the Stream (Fong et al. 1987). The extinction result is based primarily on galaxy counts (see also Mathewson et al. 1979) and, though inconclusive, the results suggest at most a very small level of extinction.

Sembach et al. (2003) have detected O VI associated with the Magellanic Stream, indicating that hot gas must be present. It is very difficult to produce O VI with photoionization and they suggest movement through a hot Galactic Halo medium may be responsible. Lu et al. (1994) also have a possible detection of C IV absorption at the position of Fairall 9. This suggests, along with the Hα detections discussed above, that the metallicity estimates are subject to an ionization correction. Another uncertainty in the metallicity determinations is the H I column density. The above determinations are based on fairly low spatial resolution H I data (15′ or 34′) and HVCs can vary by a factor of five in column density on scales of only 1′ (Wakker & van Woerden 1997; Ch. 7). The metallicity determinations remain clear in their indication of the Stream being made up of non-primordial gas and are consistent with the Stream originating from the Magellanic Clouds.

Distance estimates for the Stream are based largely on theoretical interaction models (see Sect. 5.2 for a full description). Watanabe (1981) made the assumption that the shape of the Stream clouds (i.e. elongation) is determined by the strength of the galactic tidal-disruption force and estimates that the Stream lies between 36 and 50 kpc. Hα observations also have the potential to provide distance information (see Sect. 3.2).

4. The Leading Arm

4.1. H I STRUCTURE AND KINEMATICS

The Leading Arm is made up of a string of clouds on the leading side of the Magellanic Clouds (to the galactic north) which have only recently been clarified as being connected to each other and the Magellanic System through *HIPASS* observations (Putman et al. 1998). The beginning of the Leading Arm protrudes from the Magellanic Bridge and LMC along several clumpy filaments (see Fig. 4). The multiple filaments give the appearance that the Leading Arm is associated with both of the Clouds. The Leading Arm is relatively thin ($\sim$1/4 the width of the trailing Stream), but roughly continuous until the galactic plane, where it abruptly shifts in galactic longitude from 307° to 290°. The Leading Arm is very clumpy, with diffuse filaments connecting the clumps. These filaments were missed in previous surveys owing to sparse spatial sampling (Mathewson & Ford 1984; Morras 1982), and it was thought that the clumps were isolated high-velocity clouds. There are also dense clouds about the main filament of the Leading Arm (primarily on the lower longitude side), similar to the small clouds which surround the Stream.

The Leading Arm's velocity distribution is somewhat confusing, and this may be due to the projection of the feature. It emanates from the

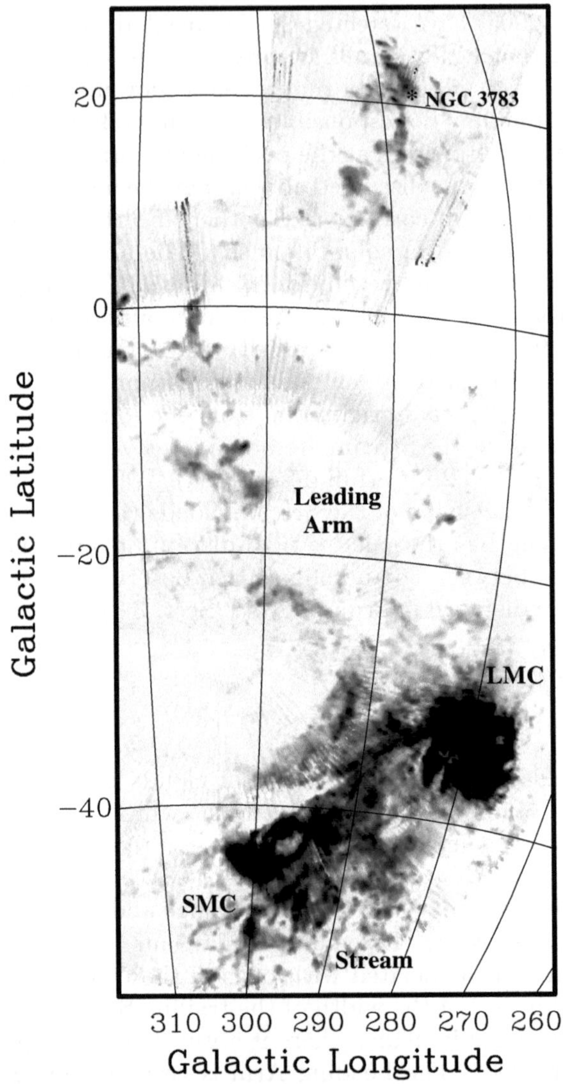

Figure 4. A *HIPASS* peak-intensity map which shows the full extent of the Leading Arm, as well as the Magellanic Clouds, the Bridge and the beginning of the Stream (as labeled). Note that this image uses only one scan of *HIPASS* data (Barnes et al. 2001), producing the stripe artifacts. The position of the background galaxy NGC 3783 is also noted (see Sect. 4.2). To avoid the emission from the galactic plane (which extends out to $+120$ km s^{-1} in this direction), only velocities between $+130$ and $+400$ km s^{-1} were used. (Thus the strange appearance of the SMC which begins at $\sim$+80 km s^{-1}.) Many features are intentionally saturated to bring out the low-level emission. The intensity scale is linear, ranging from approximately 0.1 to 2 K (black). Figure from Putman (2000).

Clouds at $v_{\rm LSR}\sim+180$ km s^{-1} and its velocity steadily increases until it reaches $+356$ km s^{-1} at $(l,b)=(302°,-17°)$. From this position it decreases in velocity to $\sim+200$ km s^{-1} as it moves toward the galactic plane (see Putman et al. (1998) for channel maps). When the Arm shifts in position by 15° in longitude at the plane, it also shifts in velocity, starting at $\sim+320$ km s^{-1} at latitude $+8°$ and extending to $+150$ km s^{-1} at latitude $+30°$. Relative to the Galactic Standard of Rest, the Leading Arm extends in velocity from $v_{\rm GSR}=-29$ to $+178$ km s^{-1}. The metallicity determination discussed below suggests that the feature at positive latitudes is a continuation of Magellanic material; however, it is difficult to reproduce the Leading Arm's initial $\sim$60° deflection angle from the great circle defined by the Stream, while also retaining the positive latitude clouds as tidal debris (Gardiner 1999). Verschuur (1975) suggested that the high-positive-velocity features which make up the Leading Arm are actually distant spiral features which form an intergalactic bridge between the Clouds and the Milky Way. This seems unlikely, since the *HIPASS* observations show the Leading Arm's velocity to be distinct from the velocity of the Galactic H I in this direction ($\sim+120$ km s^{-1}; Burton 1988).

It appears as if the data shown in Fig. 4 represent the full extent of the Leading Arm feature, as maps further north of $b=+30°$ do not show any obvious continuation of emission. It is curious that the filament abruptly ends at a relatively high column density; however, there could be more tenuous or fully ionized gas further along. The Leading Arm is not as ordered or massive as the Stream, possibly due to its leading position or age. The mass of the Leading Arm is approximately 2×10^7 $M_\odot$, an order of magnitude less massive than the Stream, assuming they are both at the distance of the Magellanic Clouds.

As in the case of the Magellanic Stream, high-resolution *ATCA* observations are in progress for many positions along the Leading Arm. The *Parkes* narrow-band-survey of Brüns et al. (2004) also includes the majority of the Leading Arm. They find the H I gas in the Magellanic Stream to be much more smoothly distributed than the H I gas in the Leading Arm, and note that the Leading Arm has two-component line profiles. Wakker et al. (2002) have already analyzed *ATCA* data for a position on the positive-latitude side of the plane and found velocity widths of 5 to 10 km s^{-1} and column density contrasts of a factor of 3 on arcminute scales. They also noted the two-component velocity structure of this cloud, similar to other non-Magellanic HVCs, and derived a pressure of 18000 $R^{-1}\,D_{\rm kpc}^{-1}$ K cm^{-3} (where R is the resolution in arcminutes and D is the distance to the cloud). Other observations of the Leading Arm include the work of Morras (1982), Morras & Bajaja (1983) and Bajaja et al. (1989), all of which are at a lower spatial resolution but higher velocity resolution than the data shown here.

4.2. METALLICITY DETERMINATION

Apart from the fact that the Leading Arm emanates from the Magellanic System, the strongest evidence that it is made of Magellanic material comes from the Lu et al. (1998) metallicity determination of HVC 287.5+22.5+240. Derived from *GHRS* spectra of the background galaxy NGC 3783 (see Fig. 4 for position), an S/H ratio of $\sim$0.25 $Z_\odot$ was found, consistent with the metallicity of the Magellanic Clouds. They also found Fe/H = 0.033 $Z_\odot$, with the subsolar Fe/S ratio indicating dust may be present. This filament lies spatially (and kinematically) in a region where tidal models predict gaseous tidal debris to reside, and the metallicity determination suggests that, despite the offset positioning of this filament, it is indeed part of the Magellanic Leading Arm. The Seyfert galaxy ESO 265−G23 is another possible background source which can be used to determine the Leading Arm's metallicity; however it appears to be just off the H I contours in the *HIPASS* map (see Putman & Gibson 1999a). This position may either have a very low column density or represent the ionized medium of the Arm.

5. Theoretical origin models

5.1. BRIDGE

It is generally agreed that the LMC and SMC are bound and that the Bridge was formed via a tidal encounter between the two Clouds (e.g. Gardiner & Noguchi 1996 (hereafter GN96); Moore & Davis 1994). The finding of stars in the Bridge region supports the tidal model, though the young stellar population may have been born in the Bridge (see Sect. 2.2). Few models can simultaneously reproduce both the Bridge and the Stream accurately. GN96 are relatively successful by refining the models of Lin & Lynden Bell (1982) and Murai & Fujimoto (1980). They find that the Magellanic Bridge was most likely pulled from the SMC 0.2 Gyr ago during a close encounter between the two Clouds (at 7 kpc separation). The GN96 model, in which the SMC is composed of both a disk and a halo, nicely explains the different bridge and tail H I components and the velocity distributions of the young (early-type) and old (carbon star) stellar populations.

In contrast to GN96, Kunkel et al. (1994) attempt to reproduce the properties of the SMC and Bridge by leaving the LMC and SMC unbound and ignoring the effect of the Galaxy (i.e. they do not reproduce the Magellanic Stream). They suggest that the carbon stars are part of the tidal bridge, separate from the H I and embedded in some type of ionized medium. Heller & Rohlfs (1994) agree with GN96 that the two Clouds are bound, but argue that they have remained in a stable binary system for the last 10^{10} years and that tidal forces from the Galaxy were not strong enough to pull out the Magellanic Stream. They suggest the Bridge or intercloud region was formed 0.5 Gyr ago when there was a close encounter between the LMC and SMC, and this also marks the beginning of the formation of the Magellanic Stream. The chaotic nature of the H I features north of the Bridge (at the head of the Stream) indicates that the Bridge and Stream were not formed in conjunction or that one is pulling material from the other. In the best model of Li (1999), the Clouds have only been gravitationally affecting each other for the past 2 Gyr, as he also finds that when the Clouds are a lifelong binary, the interaction between the two Clouds does not allow the Magellanic Stream to form.

All of the models assume that the Bridge is made up of material from the SMC, with the LMC ripping material from its less massive companion. The *HIPASS* data show an extension of the LMC which suggests that the LMC also contributes to the Bridge's mass (see Fig. 1). This feature may be reproduced when the potential of the LMC is modeled more realistically.

5.2. STREAM AND LEADING ARM

The Stream is the result of an interaction between the Galaxy and the Clouds; its link to the Magellanic System and spatial and kinematic continuity are the primary clues for this conclusion. The exact form of the interaction is not yet fully understood, but its striking appearance has attracted an abundance of theoretical attention. Many of the early models were created before the tangential velocity of the Clouds was known, or they were simply unable to reproduce the observed data. Proper motion measurements have now shown the Clouds to be leading the Stream with a total galactocentric transverse velocity of 215 ± 48 km s^{-1} (Jones et al. 1994). This section of the chapter summarizes the more recent developments in our theoretical understanding of the Magellanic HVC's formation and evolution. The models have generally been variations on two themes: gravitational tides from the Milky Way pulling the Stream from the Clouds, and ram-pressure stripping of the Stream gas as the Clouds interact with some form of galactic gas. The finding of the tidal Leading Arm feature (Putman et al. 1998) indicates that tidal forces are the dominant mechanism responsible for the formation of the Stream, but it is likely that other mechanisms also play a role in the Stream's evolution.

5.2.1. *Tidal models*

Tidal models have gradually become more complex, to match the increasing detail revealed in the observations. One of the most recent and advanced N-body tidal models is that of GN96, which simulates the SMC as a collection of self-gravitating particles and the LMC as a point mass. GN96 is an adaptation of early tidal models (e.g. Murai & Fujimoto 1980; Lin & Lynden-Bell 1982), where the Clouds are in a polar orbit leading the Stream and are presently close to perigalacticon (see also Gardiner et al. 1994; Lin et al. 1995). To achieve the high negative velocities at the Stream's tip, these models invoked a Galaxy with a massive halo ($\sim 10^{12}$ $M_\odot$) which extends out to $\sim$200 kpc, consistent with recent results (e.g. Kochanek 1996). GN96 (and other recent tidal models) predict that the Stream was pulled from the SMC 1.5 to 2 Gyr ago, when a tidal encounter between the two Clouds (at 14 kpc separation) coincided with their previous perigalactic passage. The Stream was drawn into its present position as the Clouds moved from apogalacticon ($\sim$0.9 Gyr ago) to their present position, just past perigalacticon. GN96 find that the Stream consists of two separate streams, a main filament along the observed position of the MS and a less densely populated secondary filament (this secondary stream is also a prediction of Tanaka (1981)). This splitting of the Stream is seen in the *HIPASS* data shown in Figs. 2 and 3; however, the separation of the two components is significantly larger in the model and the cause of the separation remains unclear. The dual filaments may represent multiple close encounters between the two Clouds, resulting in two major gas concentrations which were subsequently drawn into the Stream. The fact that the filaments are at approximately the same velocity argues for a similar origin. GN96 reproduce the velocity distribution of the Stream fairly accurately (see Fig. 5), but the variation in column density along the Stream requires further work.

An advancement on GN96 has been developed by Yoshizawa (1998), who incorporates gas dynamics (via a sticky-particle method) and star formation into the numerical code (see Fig. 6). The simulations find the beginning of the Stream to consist of multiple filaments, much as depicted in Figs. 2 and 3. The simulated Stream then becomes very narrow due to gas dissipation from cloud-cloud collisions, and the bifurcation found in previous models is lost. An important result of the Yoshizawa models is the demonstration that stars should *not* be drawn out along the Stream, but remain restricted to a $\sim$10 to 15 degree region surrounding the Clouds (appearing clump-like, or perhaps in several dispersed streams). The lack of stars in the Stream has been a major argument against the Stream having a tidal origin (e.g. Moore & Davis 1994). Yoshizawa (1998) preferentially disrupts the gas by having the initial gas distribution of the Clouds more extended than the stellar component (a common occurrence – e.g. Broeils & van Woerden 1994; Yun et al. 1994). As mentioned in Sect. 3.2, recent observational work indicates that there may be an excess of giants, at distances expected for tidal debris from the Magellanic Clouds, and distributed in patterns suggestive of the small stellar streams predicted by Yoshizawa's models (Majewski et al. 1999).

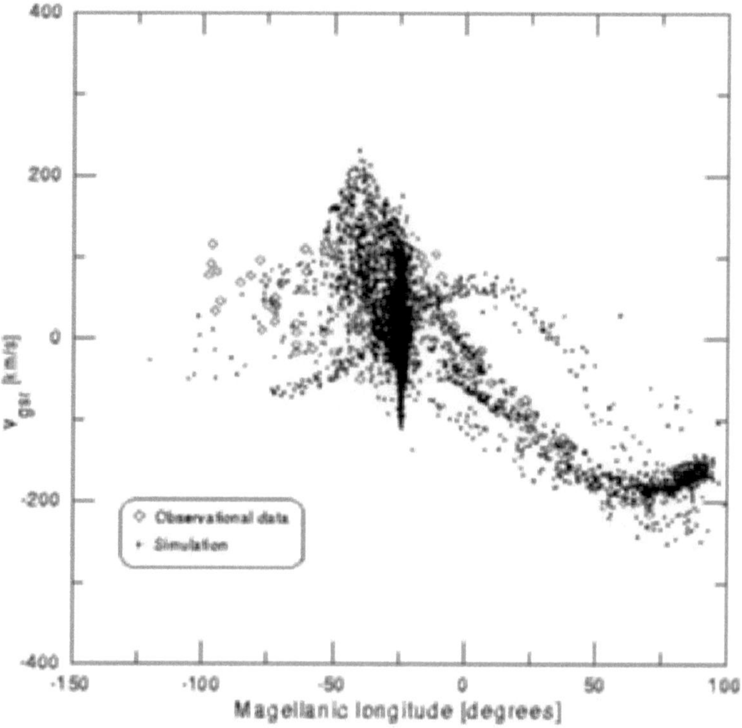

Figure 5. The velocity profile of the particles in the best tidal-plus-weak-drag model of Gardiner (1999), with the observational data of Mathewson et al. (1974) included for comparison. The velocities are shown in the Galactic Standard of Rest (GSR) reference frame, and the coordinates are Magellanic longitude as defined by Wannier & Wrixon (1972). The Stream extends from Magellanic longitude $-25°$ to $100°$ in this coordinate system. Figure from Gardiner (1999).

A natural result of the tidal model is a leading counterpart to the Stream, the Leading Arm. The original tidal models presumed that the interaction could be represented as a two-body problem between the Galaxy and the LMC, which resulted in symmetric leading and trailing streams of material. The more recent models of GN96 and Yoshizawa (1998) treat the interaction as a more realistic 3-body problem (Galaxy, LMC and SMC) and the perturbative nature of the LMC+SMC interaction leads to H I features which are clearly non-symmetric. The strong gravitational perturbation of the LMC pulls most of the material in the leading section back toward the Clouds, leaving a much weakened leading feature compared to the Stream. GN96 predicts a leading arm which, between the Magellanic Clouds and the galactic plane, has a mass $\sim$1/3 that of the entire trailing Stream, a relatively flat velocity gradient, and a deviation from the great circle defined by the trailing Stream of $\sim$30°. The newly-discovered Leading Arm has a mass $\sim$1/10 that of the Stream (assuming the Arm and the Stream

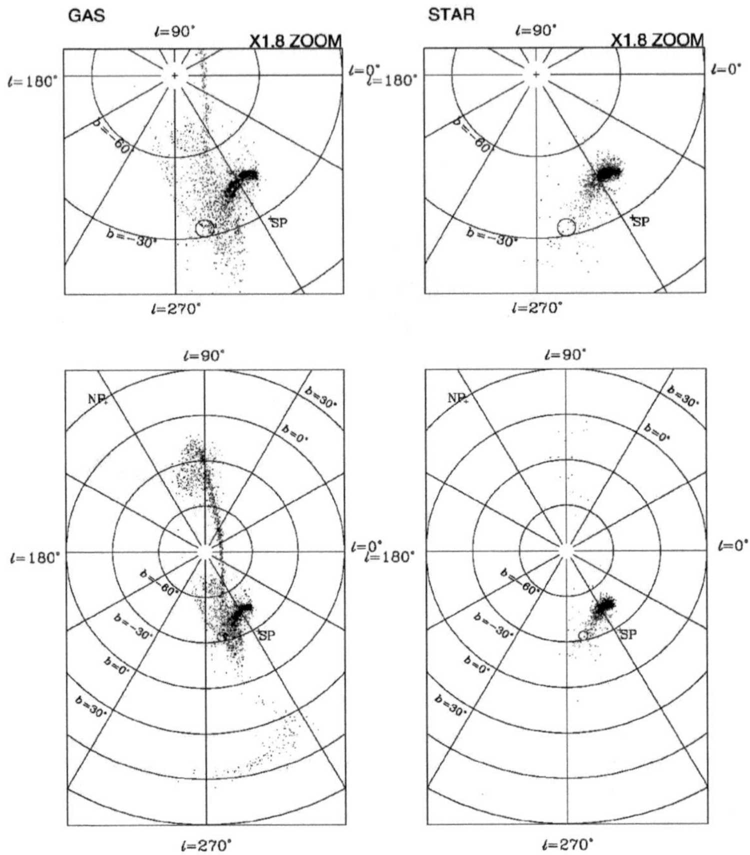

Figure 6. Yoshizawa's (1998) tidal simulation aimed at reproducing the Magellanic HVCs. The panel on the left shows the distribution of gas particles, while the one at the right shows the distribution of star particles. The SMC is at $(l,b)\sim303°,-44°$ and the LMC is at $(l,b)\sim280°,-33°$ (marked with the open circle). SP represents the position of the South Celestial Pole. Figure from Yoshizawa (1998).

are at the same distance) and a deviation angle closer to $\sim$60°. Though the Leading Arm does not match the predictions of the tidal models exactly, there are several additions to the current models which could change this situation. The differences in the mass and projected orientation could be due to the shape of the LMC's potential in tidal models (presently a rigid spheroid), a triaxial distribution of Galactic Halo mass (Lin et al. 1995), and/or a perturbation by another satellite of the Milky Way (e.g. the Sgr dwarf). The addition of a small amount of drag to the tidal model (Gardiner 1999) is able to reproduce the angle of deflection from the Stream's Great Circle and the velocity distribution of the Leading Arm (see Fig. 5), but it also introduces an extended anomalous component which wraps around to join the Stream and is not observed. The hydrodynamical models of Li (1999) indicate that a tidal interaction is not a tidy process and that multiple clumps of material would be drawn from the Clouds, along with the continuous streams. This could explain the rest of the debris seen in Figs. 2 and 3. They also find that the LMC has a substantial effect on the distribution of the leading gas, and that the stellar component of the Clouds remains largely confined.

5.2.2. *Ram-pressure models*

As noted above, combining aspects of the ram-pressure models with the tidal ones may be the key to reproducing all of the observational features of the Stream and Arm. Mathewson et al. (1977) were the first to suggest that the Stream was formed via thermal instabilities in the wake of the Clouds during their passage through the Galaxy's hot halo. These instabilities form cold clouds which lose their buoyancy and sink toward the Galactic Center. Variations on this model were subsequently developed and simulated. Liu (1992) proposed cold gas from the Clouds was dragged into their wake, and gravitational forces from the Milky Way accelerated the gas down the vortex. Meurer et al. (1985) simulated the tearing of cloudlets from the Bridge as the Magellanic Clouds passed through the hot gaseous halo of our Galaxy and they stretched out the Stream with tidal and drag forces. Meurer et al. (1985) are able to produce a reasonable model of the Stream (magnitude of the spatial and velocity extent within a factor of 2) with a broad range of parameters, but their best model puts the Stream clouds at an average distance of 38 kpc. Sofue (1994) also produces the Stream by passing the Clouds through Galactic Halo and Disk gases and elongates it with the Galaxy's potential. A leading stream is formed in the Sofue model when the Stream begins to rotate around the Galaxy at a higher angular velocity than the LMC and it wraps all the way around; but this is at the expense of the predicted extent and velocity profile of the Stream. In all of Sofue's simulations the Stream accretes onto the Galaxy within a few Gyr.

Moore & Davis (1994) have a similar, but more detailed model compared to the Meurer et al. (1985) and Sofue (1994) models. They pass the Clouds through an extended ionized Galactic Disk which strips off 20% of the Clouds' least bound H I into the Stream. The main interaction is thought to have taken place 0.5 Gyr ago at a distance of 65 kpc, and they propose that the material responsible for the stripping is an extension of the Galactic H I Disk which has column densities $<10^{19}$ cm^{-2} and is ionized by the extragalactic background radiation. Moore & Davis are able to explain the Stream's column density gradient and the high negative velocities at the Stream's tip, as the gas with the lowest column density loses the most orbital angular momentum and falls to a distance of ~20 kpc from the Galaxy with a velocity of $-380\ \mathrm{km\,s^{-1}}$ (v_{LSR}). Without the addition of a braking effect from an extended dilute halo of ionized gas, the stripped material actually begins to lead the Magellanic Clouds. This is because the gas clouds lose energy when initially stripped from the Clouds, which causes the apogalactic distance of their orbit to decrease. As their orbital period decreases they overtake the Clouds as projected on the sky.

A slightly different approach to the ram-pressure model was taken by Heller & Rohlfs (1994) and Mathewson et al. (1987). Heller & Rohlfs mark the beginning of the formation of the Stream at 0.5 Gyr ago, when the Magellanic Clouds had a close encounter with each other and much of the H I was disrupted from the core of the LMC and SMC into the Bridge region or some other extended configuration. The Stream was subsequently swept out by a strong wind, generated by the orbital motion of the Clouds through the Galactic Halo. Mathewson et al. (1987) proposed a discrete ram-pressure model in which the Stream's formation is due to the Clouds' interaction with high-velocity clouds presently found on the leading side of the Clouds. Detailed results are not available, but the fact that these leading clouds are shown here as the continuous Leading Arm casts doubt on this model.

Besides the further development and combination of the models currently in existence, there are several observational tests which can be carried out to distinguish between the origin scenarios. As previously discussed, continuing the search for a stellar counterpart to the H I Stream and Leading Arm is of importance. Abundance determinations will also be a crucial tool for confirming the origin of some of the more remote clouds which are proposed members of the Magellanic System. Searching for soft X-ray emission along the Stream and investigating the optical line ratios may provide insight into the mechanisms responsible for putting the Stream in its current state. Understanding the ionized component of the Stream is an important future goal. If it is determined that photoionization is the dominant ionizing mechanism, mapping the Stream in Hα emission has the potential to reveal the three-dimensional distribution of the Stream (Bland-Hawthorn & Maloney 1999; see also Sect. 3.2). Because ram-pressure models put the tip of the Stream at ~20 to 50 kpc (e.g. Moore & Davis 1994; Heller & Rohlfs 1994), and tidal models put the tip at ~70 to 100 kpc (e.g. Gardiner 1999), Hα observations of MS VI, in particular, might provide an elegant test for the competing models. It will be difficult to reproduce the detail revealed in the *HIPASS* Stream observations, but the gross properties should be matched before any interaction model is adopted. These properties include: a chaotic beginning consisting of multiple filaments, dual streams, narrowing of the main filament toward the tip and a broadening at the very end, compact clouds which surround the main filament, and a continuous velocity structure.

6. Relationships to other HVCs?

It has been suggested that the complexes described here are what is left of a polar ring of Magellanic debris (e.g. Mathewson et al. 1987). In fact, it is still possible that all HVCs originated as part of the Magellanic Cloud/Milky Way interaction (Mathewson et al. 1974), but the evidence points against it, considering the survival timescales, HVC abundance and distance measurements (Wakker 2001; Ch. 10), and the predicted position and velocity distributions of the Magellanic remnants (Wakker & Bregman 1990). A more likely possibility is that other high-velocity complexes are the remnants of previous non-Magellanic interactions and/or torn-apart Galactic satellites (e.g. Putman et al. 2004). When developing models to explain the global HVC population, it is important that the HVCs which are known to be part of the Magellanic System are excluded. For instance the identification of the Leading Arm eliminates many of the extreme-positive-velocity clouds, and the dense clumps about the Magellanic Stream are also likely to be interaction-related and should be excluded from analyses which attempt to match such clouds to a Local Group origin (e.g. Braun & Burton 1999). When the Magellanic HVCs are no longer included in the overall population of HVCs, the sky covering fraction goes down by at least 5% (Wakker & van Woerden 1997). It also leads to a serious deficiency of HVCs in the southern sky and at positive velocities. If most of the positive-velocity HVCs can be classified as Magellanic debris or galactic extensions, we may be able to reconsider various origin scenarios which are unable to produce high-positive-velocity gas.

In the quest to understand the global origin of HVCs, one of the most important roles of the Magellanic complexes is to provide a basis for observational comparisons. The major HVC questions concern their origin and environment, and are highly dependent on the clouds' distances. Since the Magellanic HVCs are known to have originated from the Clouds and lie at distances of several tens of kiloparsecs, they can be used as a calibrator to investigate the HVC phenomena. To begin, the overall spatial structure of the Magellanic complexes and other HVCs has similarities and differences on both large and small scales. The long filamentary structure of the Stream and Arm appears to be common throughout HVCs and may indicate that more clouds either have a tidal origin or are currently being tidally stretched. If assumptions are made about the density fall-off of the halo, the length of the head-tail and bow-shock structures may allow an estimate of the clouds' distances (Mebold, priv. comm.). The small-scale structure of HVCs supplies information about the physical conditions within the clouds, and an indication of the amount of or ordered structure. The future high-resolution H I observations of the Magellanic complexes should be compared to similar HVC observations (e.g. Wakker & Schwarz 1991). The Magellanic Bridge is useful, as it can be used to examine the effects of star formation on H I structure and to explore what has triggered this process. The kinematic structure of the H I may also provide clues to the formation and evolution of HVCs. While many of the Magellanic HVCs show a single-component line profile, other HVCs show two-component profiles, indicative of a core-envelope structure or a two-phase medium. The Magellanic Stream and Bridge also show a systematic variation in velocity and column density which is uncommon in high-velocity complexes. There are useful comparisons to be made at all wavelengths, including optical-emission and absorption-line observations to determine abundances and ionization conditions. These types of comparisons will clarify some of the current unknowns about the Magellanic complexes and their relation to the entire population of high-velocity clouds.

Acknowledgements

I would like to thank Ken Freeman, Lister Staveley-Smith, and Brad Gibson for providing useful feedback on the chapter. I would also like to thank Lance Gardiner and Akira Yoshizawa and the Astronomical Society of the Pacific Conference Series for permission to reproduce Figs. 5 and 6. The Australia Telescope National Facility is thanked for hosting me during much of this work.

References

Bajaja, E., Cappa de Nicolau, C.E., Martin, M.C., Morras, R., Olano, C.A., Pöppel, W.G.L. 1989, A&AS, 78, 345
Barnes, D.G., Staveley-Smith, L., de Blok, W.J.G., et 36 al. 2001, MNRAS, 322, 486
Bland-Hawthorn, J., Maloney, P.R. 1999, ApJ, 510, L33
Braun, R., Burton, W.B. 1999, A&A, 341, 437
Broeils, A.H., van Woerden, H. 1994, A&AS, 107, 129
Brück, M.T., Hawkins, M.R.S. 1983, A&A, 124, 216
Brüns, C., Kerp, J., Staveley-Smith, L., Mebold, U., Putman, M.E., Haynes, R.F., Kalberla, P.M.W., Muller, E., Filipovic, M.D. 2004, A&A, submitted
Burton, W.B. 1988, in Galactic and Extragalactic Radio Astronomy, 2nd edition, eds. G.L. Verschuur, K.I. Kellerman (Berlin: Springer), 295
Carignan, C., Beaulieu, S., Cote, S., Demers, S., Mateo, M.L. 1998, ApJ, 116, 1690
Christodoulou, D.M., Tohline, J.E., Keenan, F.P. 1997, ApJ, 486, 810
Cohen, R.J. 1982, MNRAS, 199, 281
Demers, S., Battinelli, P. 1998, AJ, 115, 154
Fong, R., Jones, L.R., Shanks, T., Stevenson, P.R.F., Strong, A.W. 1987, MNRAS, 224, 1059
Gardiner, L.T. 1999, in ASP Conf. Ser. 166, Stromlo Workshop on High-Velocity Clouds,

eds. B.K. Gibson, M.E. Putman (San Francisco: ASP), 292
Gardiner, L.T., Noguchi, M. 1996, MNRAS, 278, 191
Gardiner, L.T., Sawa, T., Fujimoto, M. 1994, MNRAS, 266, 567
Gibson, B.K., Giroux, M.L., Penton, S.V., Putman, M.E., Stocke, J.T., Shull, J.M. 2000, AJ, 120, 1830
Grondin, L., Demers, S., Kunkel, W.E. 1992, AJ, 103, 1234
Guhathakurta, P., Reitzel, D.B. 1998, in ASP Conf. Ser. 136, Galactic Halos: A UC Santa Cruz Workshop, ed. D. Zaritsky (San Francisco: ASP), 136
Hambly, N.C., Dufton, P.L., Keenan, F.P., Rolleston, W.R.J., Howarth, I.D, Irwin, M.J. 1994, A&A, 285, 716
Haynes, M.P. 1979, AJ, 84, 1173
Haynes, M.P., Roberts, M.S. 1979, ApJ, 227, 767
Heller, P., Rohlfs, K. 1994, A&A, 291, 743
Hibbard, J.E., Vacca, W.D., Yun, M.S. 2000, AJ, 119, 1130
Irwin, M.J., Demers, S., Kunkel, W.E. 1990, AJ, 99, 191
Jerjen, H., Freeman, K., Binggeli, B. 1998, AJ, 116, 2873
Johnson, P.G., Meaburn, J., Osman, A.M. 1982, MNRAS, 198, 985
Jones, B.F., Klemola, A.R., Lin, D.N.C. 1994, AJ, 107, 1333
Kim, S., Staveley-Smith, L., Dopita, M.A., Freeman, K.C., Sault, R.J., Kesteven, M.J., McConnell, D. 1998, ApJ, 503, 674
Kobulnicky, H.A., Dickey, J.M. 1999, AJ, 117, 908
Kochanek, C.S. 1996, ApJ, 457, 228
Kunkel, W.E., Demers, S., Irwin, M.J. 1994, in ESO Conf. Proc., The Local Group: Comparative and Global Properties, Third CTIO/ESO Workshop on The Local Group, eds. A. Layden, R.C. Smith, J. Storm (Garching: ESO), 200
Kunkel, W.E., Irwin, M.J., Demers, S. 1997, A&AS, 122, 463
Lehner, N. 2002, ApJ, 578, 126
Li, P.S. 1999, Ph.D. thesis, University of Wyoming
Lin, D.N.C., Lynden-Bell, D. 1982, MNRAS, 198, 707
Lin, D.N.C., Jones, B.F., Klemola, A.R. 1995, ApJ, 439, 652
Liu, Y. 1992, A&A, 257, 505
Lu, L., Savage, B.D., Sembach, K.R. 1994, ApJ, 437, L119
Lu, L., Savage, B.D., Sembach, K.R., Wakker, B.P., Sargent, W.L.W., Oosterloo, T.A. 1998, AJ, 115, 162
Majewski, S.R., Ostheimer, J.C., Kunkel, W.E., Johnston, K.V., Patterson, R.J., Palma, C. 1999, in IAU Symp. 190, New Views of the Magellanic Clouds, eds. Y.-H. Chu, N.B. Suntzeff, J.E. Hesser, D.A. Bohlender (San Francisco: ASP), 508
Marcelin, M., Boulesteix, J., Georgelin, Y. 1985, Nature, 316, 705
Mathewson, D.S., Ford, V.L. 1984, in IAU Symp. 108, Structure and Evolution of the Magellanic Clouds, eds. S. van den Bergh, K.S. de Boer (Dordrecht: Reidel), 125
Mathewson, D.S., Cleary, M.N., Murray, J.D. 1974, ApJ, 190, 291
Mathewson, D.S., Cleary, M.N., Murray, J.D. 1975, ApJ, 195, 97
Mathewson, D.S., Schwarz, M.P., Murray, J.D. 1977, PASAu, 3, 133
Mathewson, D.S., Ford, V.L., Schwarz, M.P., Murray, J.D. 1979, in IAU Symp. 84, The Large-scale Characteristics of the Galaxy (Dordrecht: Reidel), 547
Mathewson, D.S., Wayte, S.R., Ford, V.L., Ruan, K. 1987, PASAu, 7, 19
McGee, R.X., Newton, L.M. 1986, PASAu, 6, 471
Meurer, G.R., Bicknell, G.V., Gingold, R.A. 1985, PASAu, 6, 195
Moore, B., Davis, M. 1994, MNRAS, 270, 209
Morras, R. 1982, A&A, 115, 249
Morras, R. 1983, AJ, 88, 62
Morras, R. 1985, AJ, 90, 180
Morras, R., Bajaja, E. 1983, A&AS, 51, 131
Muller, E., Staveley-Smith, L., Zealey, W. 2003a, MNRAS, 338, 609
Muller, E., Staveley-Smith, L., Zealey, W., Stanimirovic, S. 2003b, MNRAS, 339, 105

Murai, T., Fujimoto, M. 1980, PASJ, 32, 581
Pietz, J., Kerp, J., Kalberla, P.M.W., Mebold, U., Burton, W.B., Hartmann, D. 1996, A&A, 308, L37
Putman, M.E. 2000, PASAu, 17, 1
Putman, M.E., Gibson B.K. 1999a, PASAu, 16, 70
Putman, M.E., Gibson B.K. 1999b, in ASP Conf. Ser. 166, Stromlo Workshop on High-Velocity Clouds, eds. B.K. Gibson, M.E. Putman (San Francisco: ASP), 276
Putman, M.E., Gibson, B.K, Staveley-Smith, L., et 23 al. 1998, Nature, 394, 752
Putman, M.E., Gibson, B.K., Staveley-Smith, L. 1999, in IAU Symp. 190, New Views of the Magellanic Clouds, eds. Y.-H. Chu, N.B. Suntzeff, J.E. Hesser, D.A. Bohlender (San Francisco: ASP), 51
Putman, M.E., Staveley-Smith, L., Freeman, K.C., Gibson, B.K., Barnes, D.G. 2003a, ApJ, 586, 170
Putman, M.E., Bland-Hawthorn, J., Veilleux, S., Gibson, B.K., Freeman, K.C., Maloney, P.R. 2003b, ApJ, 597, 948
Putman, M.E., Thom, C., Gibson, B.K., Staveley-Smith, L. 2004, ApJ, 603, L77
Recillas-Cruz, E. 1982, MNRAS, 201, 473
Rolleston, W.R.J., Dufton, P.L., Fitzsimmons, A., Howarth, I.D., Irwin, M.J. 1993, A&A, 277, 10
Rolleston, W.R.J., Dufton, P.L., McErlean, N.D., Venn, K.A. 1999, A&A, 348, 728
Sanduleak, N. 1969, AJ, 74, 877
Scoville, N.Z., Sanders, D.B., Clemens, D.P. 1986, ApJ, 310, 77
Sembach, K.R., Wakker, B.P., Savage, B.D., Richter, P., Meade, M., Shull, J.M., Jenkins, E.B., Sonneborn, G., Moos, H.W. 2003, ApJS, 146, 165
Sofue, Y. 1994, PASJ, 46, 431
Songaila, A. 1981, ApJ, 248, 945
Stanimirovic, S., Dickey, J.M., Krco, M., Brooks, A.M. 2002, ApJ, 576, 773
Tanaka, K.I. 1981, PASJ, 33, 247
Tanaka, K.I., Hamajima, K. 1982, PASJ, 34, 417
Tufte, S.L., Reynolds, R.J., Haffner, L.M. 1998, ApJ, 504, 773
Tufte, S.L., Wilson, J.D., Madsen, G.J., Haffner, L.M., Reynolds, R.J. 2002, ApJ, 572, L153
Tully, R.B., Fisher, J.R. 1987, Nearby Galaxy Atlas (Cambridge: Cambridge University Press)
van Woerden, H., Schwarz, U.J., Peletier, R.F., Wakker, B.P., Kalberla, P.M.W. 1999, Nature, 400, 138
Verschuur, G.L. 1975, ARA&A, 13, 257
Wakker, B.P. 2001, ApJS, 136, 463
Wakker, B.P., Bregman, J. 1990, in Ph.D. thesis B.P. Wakker, Rijks Universiteit Groningen, Chapter 5
Wakker, B.P., Schwarz, U.J. 1991, A&A, 250, 484
Wakker, B.P., van Woerden, H. 1997, ARA&A, 35, 217
Wakker, B.P., Oosterloo, T.A., Putman, M.E. 2002, AJ, 123, 1953
Wannier, P., Wrixon, G.T. 1972, ApJ, 173, L119
Watanabe, T. 1981, AJ, 86, 30
Wayte, S.R. 1989, PASAu, 8, 195
Weiner, B.J., Williams, T.B. 1996, AJ, 111, 1156
Westerlund, B.E., Glaspey, J. 1971, A&A, 10, 1
Yoshizawa, A. 1998, Ph.D. thesis, Tohoku University, Sendai, Japan
Yun, M.S., Ho, P.T.P., Lo, K.Y. 1994, Nature, 372, 530

6. HIGH-VELOCITY H I GAS IN EXTERNAL GALAXIES

TOM OOSTERLOO
Netherlands Foundation for Research in Astronomy, Dwingeloo, The Netherlands; oosterloo@astron.nl

Abstract. The great abundance of galactic high-velocity clouds indicates that they are important for the evolution of the Galaxy. It also suggests that other galaxies must contain neutral hydrogen clouds with kinematics very different from that of a settled disk. The occurrence and properties of such H I clouds with anomalous kinematics in external galaxies are reviewed.

In some cases, the high-velocity H I is likely to be associated with star formation in the disk and with Galactic Fountain type phenomena. Recent deep H I observations of a small number of nearby spirals have revealed the existence of extensive H I halos in these systems, containing gas 10 to 15 kpc above the plane. This halo gas appears to rotate slower than the gas in the disk, while also showing a small inward flow. A number of small clouds are observed that must have non-circular motions up to 200 $\mathrm{km\,s^{-1}}$. Outflows from the disk are also detected, often associated with sites of star formation in the disk.

Another origin for gas with anomalous kinematics is interactions, ranging from major interactions that lead to a transformation of the galaxy's structure, to small accretions that are connected with a slow evolution of the galaxy and the build-up of disks. In some cases, the accretions are only visible in neutral hydrogen. Such cases are relevant for understanding some of the larger galactic high-velocity gas complexes.

1. Introduction

High-velocity H I is defined, for our Galaxy, as H I with kinematics that significantly deviates from what is expected for gas in the thin, differentially rotating disk. A large fraction of the celestial sphere is covered by such gas. The nature and origin of these high-velocity clouds (HVCs) is a long-standing problem, although much progress has been made in recent years. A number of explanations have been suggested, and, to some extent, one reason why the nature of HVCs turns out to be such a complex problem is

H. van Woerden et al. (eds.), High-Velocity Clouds, 125–144

that many of the explanations proposed are probably relevant.

The large number of HVCs and the fact that several phenomena are involved indicate that they play a fundamental role in the evolution of our Galaxy. They also suggest that in other galaxies HVCs and related phenomena are equally important. Indeed, in several galaxies H I is found with a structure and kinematics that clearly indicate that it is not part of a regular, stable configuration such as a disk. Such gas could, following similar reasoning as for our Galaxy, be defined as high-velocity gas, in parallel to the galactic case, although a term that is used more commonly is *anomalous gas*. The scale of the phenomena can be quite different from what is seen in our Galaxy and a direct comparison is not always straightforward. However, the same processes are likely to be involved, and an overview of the properties of this anomalous-velocity gas in external galaxies may be instructive. More, in particular deep, H I observations of external galaxies, combined with detailed comparisons with models for the HVCs, are needed to fully understand the relation between the anomalous gas in external galaxies and the Galactic HVCs.

There are two processes that together are responsible for most of the anomalous gas observed in external galaxies and that I will discuss. First, *star formation* can drive significant amounts of H I into the halo of *normal* spiral galaxies, as shown by recent recent deep H I observations. These observations and the three-dimensional structure and kinematics of H I halos in normal spiral galaxies are described in Sect. 2. Second, a galaxy interacts with its environment. This is a very broad topic, and it is beyond the scope of this chapter to give a complete overview of galaxy interactions. Instead, in Sect. 3 we concentrate on some aspects that seem more relevant for a comparison with the Galactic HVCs, such as the slow infall of tidal material after a major merger and the possible accretion of small, purely gaseous objects by large spirals.

2. H I halos of spiral galaxies

2.1. OVERVIEW

It is well known that, although most interstellar matter is found in the disk of the Galaxy, a significant amount of gas is present in the halo. This halo gas is found in a broad range of physical states, ranging from cold and neutral gas to ionized gas at very high temperatures. The presence of the gas in the halo is thought to be related to star formation in the disk. This star formation drives, through supernova explosions and stellar winds, large-scale flows of gas from the disk into the halo. In the halo, this gas cools and falls back to the disk. Such a large-scale gas circulation is commonly referred to as a *Galactic Fountain*. The intermediate-velocity clouds, and some of the high-velocity clouds, are related to such a Galactic Fountain occurring in the Galaxy (see Ch. 16).

For a number of galaxies, optical observations have shown the presence of ionized gas in the halo, which is commonly referred to as the Diffuse Ionized Medium (Rand et al. 1990, 1992; Dettmar 1990; Rand 1996; Ferguson et al. 1996; Dahlem 1997; Hoopes et al. 1999; Howk & Savage 2000). In a few cases halo gas in external galaxies has also been seen in absorption (M 81: Vladilo et al. 1994; NGC 4319: Bowen et al. 1995). The information about H I in the halo is much more limited, because of the high sensitivities required to detect such gas in *normal* spiral galaxies. Recent deep H I observations of a small number of nearby normal spiral galaxies have allowed important progress in our understanding of the neutral gas in the halos of such galaxies, in particular with regard to its full three-dimensional distribution and kinematics. Such observations can now detect clouds that are equivalent to the larger of the HVCs seen in the Milky Way.

To obtain such a full observational view of the halo gas, very sensitive observations are needed for a number of galaxies of various inclinations, since projection effects allow to detect only certain components of the distribution and kinematics in a single galaxy. Edge-on galaxies provide information about the vertical structure and about the motions in the plane of the galaxy. Face-on galaxies can be used to study the vertical motions and the correlations of features in the halo with other features in the disk of the galaxy, such as sites of star formation. Observations of galaxies of intermediate inclination supply valuable additional information about the kinematics and the spatial distribution of the halo gas.

Although not many very deep H I observations on nearby spiral galaxies have been done, there is a general picture that seems to emerge. The observations indicate that, apart from the normal thin gas disk, there exists a layer of low-density H I several kpc in thickness in some galaxies, with filaments going out to 10 kpc. This layer can contain up to 10% of the H I. The thick H I layer has a lower rotation velocity than the gas in the thin disk. Superposed on this rotation, a small inward radial motion is present, on the order of 25 $\mathrm{km\,s^{-1}}$. Apart from this thick layer, small clouds with large non-circular motions are observed, with anomalous velocities up to 200 $\mathrm{km\,s^{-1}}$. The non-circular motions can be so large that gas is observed even in those regions of position-velocity diagrams where regularly rotating gas cannot appear (the so-called *forbidden quadrants* in such diagrams). Such clouds are often, but not always, associated with expanding shells of gas near sites of star formation. All these features suggest a complex circulation of gas between the disk and the halo of spiral galaxies.

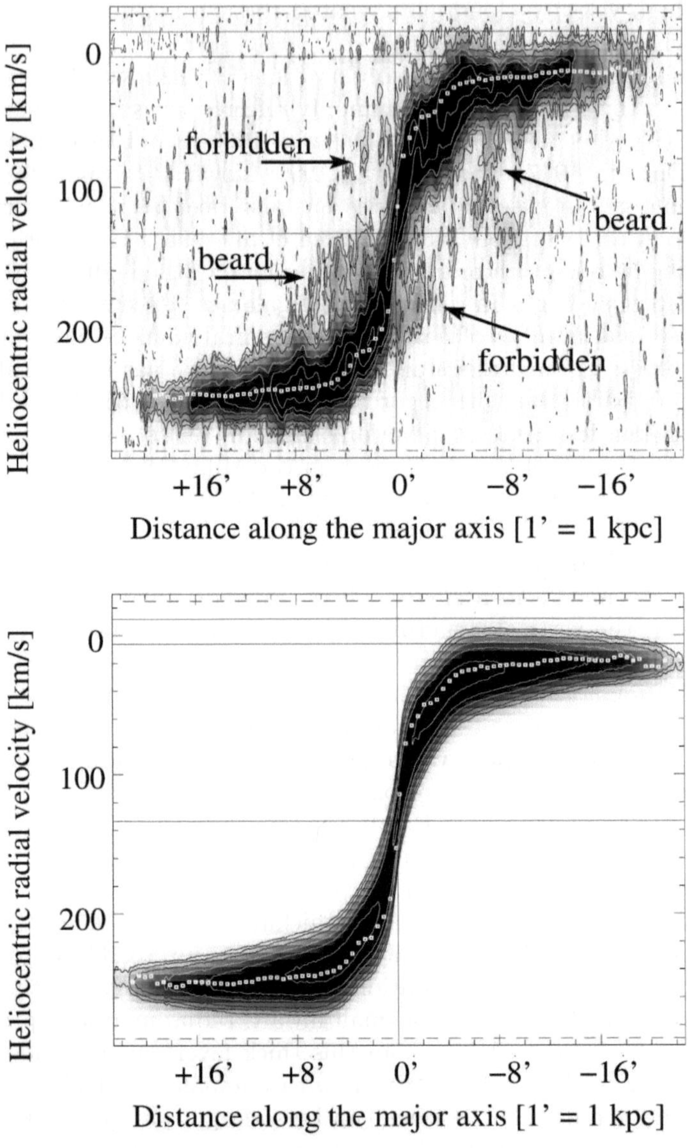

Figure 1. *Top:* Position-velocity diagram of the H I in NGC 2403, taken along the major axis of the galaxy. *Bottom:* the same diagram made using a model of an H I disk that has the same rotation properties as the thin H I disk in NGC 2403. The data show an extensive component of H I toward the systemic velocity (the "beard") that corresponds to a thick, slower rotating H I disk. Small clouds of H I are also observed in the forbidden quadrants. From Fraternali et al. (2001, 2002).

2.2. ANOMALOUS-VELOCITY GAS IN NGC 2403

Many of these features are well illustrated by the deep H I observations of NGC 2403 (Fraternali et al. 2001, 2002) that perhaps represent the most detailed study of H I in the halo of a spiral galaxy. Figure 1 shows the position-velocity diagram taken along the major axis of this galaxy, together with the same plot for a model dataset that contains only a thin disk having the same kinematics as the thin H I disk in NGC 2403. This figure clearly shows the existence of an extensive component of gas at velocities between the projected rotation velocity and the systemic velocity. Such gas at apparently lower rotation velocities is sometimes referred to as a "beard" and is present over the entire extent of the bright optical disk. Apart from this, more or less diffuse, component, small clouds exist with deviations from the local projected rotation velocity as large as 150 km s^{-1}. For example, at 8 kpc from the center (position $-8'$ in Fig. 1) there is a cloud observed at the systemic velocity instead of rotating with the disk, while in the central regions gas is even observed in the forbidden quadrants of the position-velocity diagram. The presence of such clouds underlines the existence of very large non-circular motions.

The total mass involved in these clouds with anomalous kinematics corresponds to about one tenth of the total H I mass of the galaxy. The masses of individual structures are typically $10^6 M_\odot$, while the largest complex has an H I mass of about 10^7 $M_\odot$ (see also Sect. 3). Anomalous-velocity gas is found throughout the bright optical disk of NGC 2403 and most of the halo gas appears to show, to a fair degree, a regular pattern of rotation. Models of the kinematics show that the rotation velocity of the diffuse halo gas is 25 to 50 km s^{-1} lower than that of the gas in the thin disk (Schaap et al. 2000, Fraternali et al. 2002). Such a lower rotation velocity of the halo gas is also observed in NGC 891 (Swaters et al. 1997; Fraternali et al. 2004) and in NGC 5775 (Lee et al. 2001). The ionized gas in the halo, as observed in e.g. NGC 891 and NGC 5775 (Rand 2000), also appears to rotate slower. Detailed modeling of the complex kinematics of the gas in the halo, combined with models for the total mass distribution, is required to interpret the slower rotation. If it is due to an asymmetric drift (i.e. reduced rotation due to large radial and vertical motions), it would indicate the presence of significant non-circular motions of the H I in the halo.

An interesting feature of the anomalous gas in NGC 2403 is that the kinematic minor and major axes are not perpendicular to each other. The interpretation that best fits the data is that, superposed on the rotation, a small radial *inward* motion of about 20 to 25 km s^{-1} is present. This is illustrated in Fig. 2, where a position-velocity diagram is shown that is taken along a line parallel to the minor axis with a central position slightly

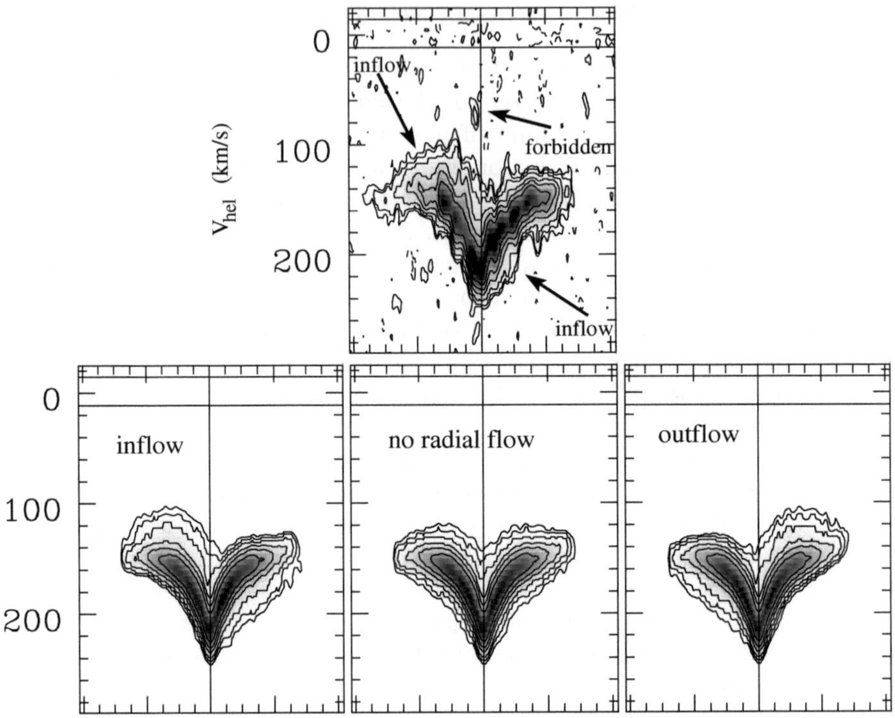

Figure 2. *Top:* Position-velocity diagram parallel to the minor axis of NGC 2403, centered somewhat offset from the center of the galaxy. The H I shows a distinct asymmetric pattern, indicated by the arrows. *Bottom:* Set of position-velocity diagrams, taken at the same position and orientation, of models for the thin+thick H I disk in NGC 2403. Each model assumes a slightly lower rotation for the thick disk, plus (as indicated) a radial motion of 20 km s^{-1} inward, no radial flow, or an outflow of 20 km s^{-1}. The data, except for the gas at forbidden velocities near the central position, are quite well reproduced by the inflow model. From Fraternali et al. (2001).

offset from the galaxy's center. In the same figure, similar position-velocity diagrams are shown for different models for the radial flow. The model with inward motion superposed on rotation describes the data quite well. Such a radial inflow is also observed in recent data taken with the Westerbork Synthesis Radio Telescope (*WSRT*) of NGC 4995 (Barbieri et al. 2004).

The gas near the center in the forbidden quadrants of position-velocity diagrams cannot be explained by this model of a thick H I layer with slower rotation plus a moderate radial inflow; an additional component has to exist. Standard models of Galactic Fountain type phenomena do not predict the existence of such "forbidden" motions, and the presence of these clouds in NGC 2403 indicates that some of the assumptions in these models, such as the conservation of angular momentum, may not be appropriate (Fraternali et al. 2001, 2002).

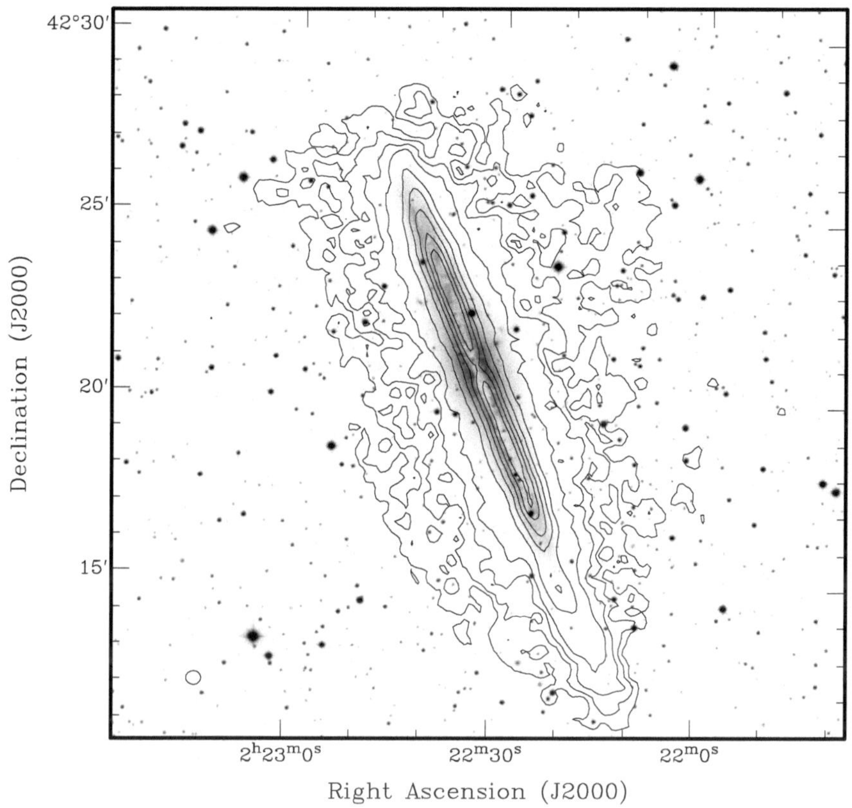

Figure 3. Total H I map (contours) of the edge-on spiral NGC 891 superposed on an optical image, clearly showing the large vertical extent of the H I layer. One arcminute corresponds to 2.8 kpc. Figure from Fraternali (2003).

Given the inclination of NGC 2403 (i=60°), the constraints on the thickness of the layers of the anomalous-velocity gas are not very strong, but the models suggest that the bulk of the gas is likely to be a few kpc from the midplane. This is probably characteristic of such H I layers in normal spirals. The vertical distribution of the thick H I layer is easier to study in edge-on galaxies, since it is directly observable. The best case is perhaps NGC 891, a well-known isolated, nearby edge-on galaxy. The total H I image of this galaxy (Swaters et al. 1997) shows H I throughout the halo, up to 5 kpc from the plane of the disk, while more recent very deep *WSRT* observations even show filaments out to 10 kpc from the disk (Fraternali et al. 2004; Fig. 3). Owing to projection effects, it is not straightforward to determine the nature of this H I and to distinguish between an H I halo, a flaring H I disk, or a warp. However, careful modeling of the full three-dimensional H I dataset shows that this gas does indeed correspond to an

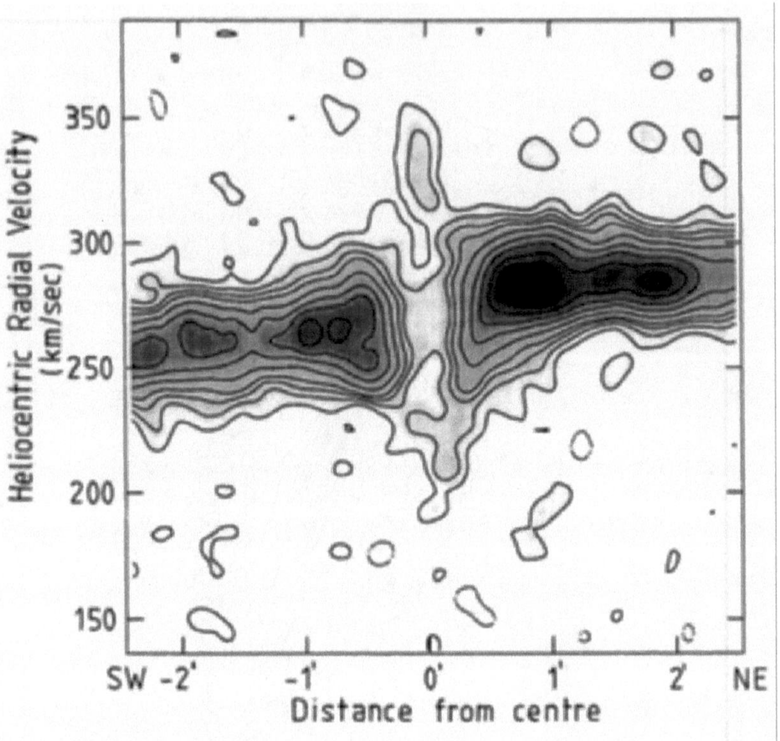

Figure 4. Position-velocity diagram of an expanding shell of H I in the galaxy M 101. From Kamphuis et al. (1991).

H I halo in the central regions of the galaxy. This modeling also shows that the rotation velocity of the gas in the halo is about 25 km s^{-1} lower than that of the H I in the disk, very similar to what is observed in NGC 2403.

2.3. ANOMALOUS-VELOCITY GAS IN OTHER GALAXIES

The thick H I layers are most likely related to star formation in the disk. Indeed, since several years observations of external galaxies, as well as of our own Galaxy, have indicated that star formation interacts with the disk of neutral hydrogen. Interaction of supernovae and stellar winds, associated with sites of star formation, with the gas in the disk is evidenced by holes in the gas disk and shells of expanding gas (e.g. Brinks & Bajaja 1986; Deul & den Hartog 1990; Puche et al. 1992; Thilker et al. 1998; Kamphuis et al. 1991; Kamphuis 1993). This is very well illustrated by the excellent H I images of the Magellanic Clouds (Stanimirovic et al. 1999; Kim et al. 1999), which show in great detail a very intricate structure of the H I distribution and clearly demonstrate the lively nature of the H I disk.

The vertical outflows are best observed in galaxies that are relatively

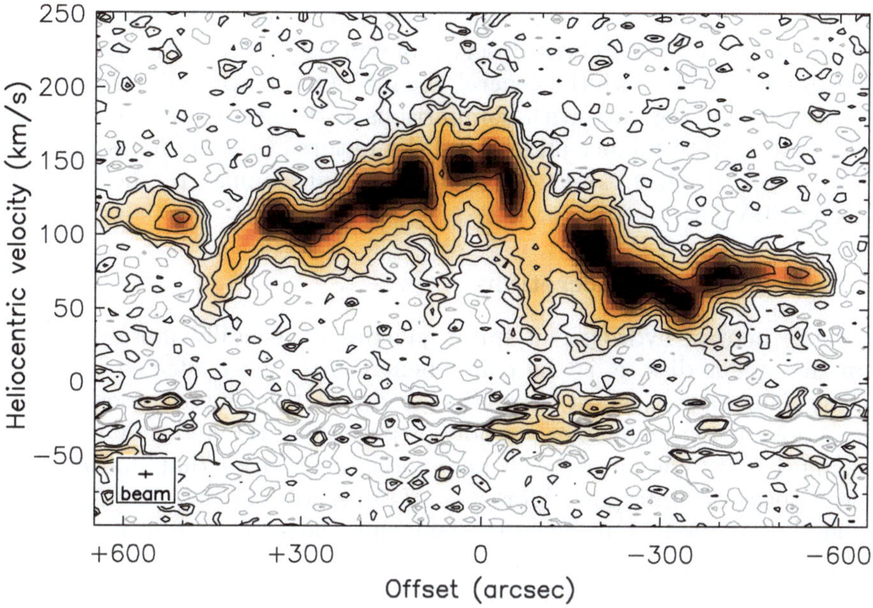

Figure 5. Position-velocity diagram taken through the disk of NGC 6946. The brightest emission corresponds to the thin H I disk in this galaxy. Two holes can be clearly distinguished, both associated with large vertical motions. A more diffuse component at lower apparent rotation can also be seen, and the data indicate the connection between the vertical outflows and the more general "beard" in this galaxy. The gas present with velocities between 0 and −50 km s^{-1} is galactic foreground H I emission. From Boomsma et al. (2002).

face-on. A textbook example of an expanding H I shell is found in M 101 (Fig. 4; from Kamphuis et al. 1991), where anomalous velocities exceeding 50 km s^{-1} are observed. The H I mass of the parts of the shell that are expanding outward is about 10^7 M$_\odot$ and contain a kinetic energy of about 2.5×10^{53} erg. This implies that on the order of a thousand supernovae are required to drive the expansion of this shell. The details of how these expanding shells feed the extensive H I halos, as well as how they connect to the entire large-scale gas flow, are not entirely understood and deep observations of the H I as well as the ionized gas in the halo of more nearby spiral galaxies are needed to study this problem. An example of the current state of the art is the deep *WSRT* observation of NGC 6946, a nearby, relatively face-on galaxy (Fig. 5; from Boomsma et al. 2002). This galaxy has an inclination of about 30° and previous observations suggested the existence of extensive high-velocity gas (Kamphuis et al. 1991). The new observations reveal this high-velocity gas in great detail. Many holes in the H I disk, as well as expanding shells, are detected, with anomalous velocities up to 200 km s^{-1} in some cases. Apart from the expanding shells, an extended anomalous H I component is now also detected, similar to that seen in e.g. NGC 2403. These observations begin to reveal the relation of the expanding shells with the general "beard". Observations of this kind for several galaxies, together with comparisons with models for the Galactic HVCs, will help to further understand the nature of the H I halos in spiral galaxies and to what extent their properties are similar to the Galactic IVCs and HVCs.

3. Interactions and accretions

Since the neutral hydrogen in spiral galaxies is generally more extended than the stellar disk, it is the component that is most affected by interactions and, in fact, a very sensitive tracer of tidal interactions. Often, interactions between galaxies are only observable in H I, and they may lead to extensive H I complexes with anomalous kinematics that do not have a clear optical counterpart. Moreover, H I observations do not only reveal ongoing interactions, but, given the long timescales involved at large radii, past interactions and mergers can also be recognized in the H I structure and kinematics. Many examples are known where H I features with a peculiar distribution and/or kinematics are observed around galaxies whose optical appearance is undisturbed, and such H I features are the only evidence that the galaxies are, or have been, interacting.

A distinction is often made between "major" and "minor" interactions. In reality, interactions occur on all scales and involve a continuum of mass ranges, and the classification of an interaction as major or minor is not always unambiguous. Nevertheless, the distinction between the two types of interaction is still meaningful because, depending on the relative sizes of the objects involved, the overall effects may be quite different.

The major interactions involve systems of comparable masses and usually produce large tidal effects. They lead to a *fast* transformation of the structure of the galaxies involved, and they often, but not always, lead to the *destruction of disks.* Such interactions generally lead to an elliptical galaxy as the end product. However, under certain conditions, the remnant may have a disk. During the interaction large H I complexes can be transported to large radius by tidal effects. This gas may remain bound to the galaxy and fall back to the remnant, where it will form a disk.

The minor interactions, or accretions, take place between a galaxy and one or more satellites or companions of relatively small mass (with a mass ratio less than, say, 0.1). During, or after, such interactions one often observes, sometimes small, H I complexes with anomalous kinematics around the main galaxy that are analogous to some of the larger HVC complexes in the Galaxy. The minor interactions, and the associated accretion of H I, play a fundamental role in the *slow* evolution of the main galaxy and of the *building up of disks*. In the course of such a minor interaction, the H I complexes may be either directly accreted, or displaced by the tidal forces and fall back later onto the interacting systems. This is a reservoir of fresh material which contributes to the process of disk building and star formation. An important issue is the nature of the small objects that are being accreted: are they small galaxies containing stars, or is there evidence for the accretion of small primordial gas clouds? It is important to note that for spiral galaxies a careful study of the structure and kinematics of the H I is necessary to distinguish between configurations that can be considered "normal", and configurations that are definitely "peculiar" and that point to recent interaction and infall. There are recognizable signatures in the H I that make this distinction possible, but it is not always easy to draw the line between effects due to the internal metabolism of the galaxy and those due to the environment. For example, it is interesting to consider asymmetries such as lop-sidedness, both kinematical and morphological, and warps, which affect spiral galaxies and seem to occur quite frequently. Are such asymmetries related to past interactions and accretion events? This is not at all obvious and there may be other explanations.

3.1. MAJOR INTERACTIONS

There are many systems known with two or more galaxies that show heavily perturbed H I images: in addition to the gas that is seen associated with the individual galaxies, there are cloud complexes that are clearly dislocated from the galaxies by tidal forces. These H I complexes have a wide range of morphology and kinematics, including one single or two long tails, bridges or ring-like structures, in regions near the galaxies, but also at large distances. The range of properties of the H I is very broad and I will only mention a few cases.

An impressive example of a complex interaction that is mainly visible in H I is the M 81 group (Yun et al. 1994), where large complexes of H I connect the various galaxies. Another well-known case is the NGC 4631 group (Rand & van der Hulst 1993). In the optical there are indications that interactions are occurring in this group, but they are visible much more dramatically in the form of several long tails of H I (Fig. 6). Seen from NGC 4631, there are at least four large streams of H I visible on the sky. Quite impressive are also the cases where two large spiral galaxies are involved, such as NGC 4038/9, NGC 4676 and NGC 7252 (Hibbard et al. 1994; Hibbard & van Gorkom 1996). The full spectrum of H I morphologies associated with galaxy interactions is given in the extensive collection of images of H I observations of galaxies by Hibbard et al. (2001; the "H I Rogues Gallery";

see http://www.nrao.edu/astrores/HIrogues/webGallery/index.html).

An important issue is the evolution of the H I complexes that are found at large radius. A number of optically-disturbed systems, characterized by bridges and long tails, were proposed by Toomre (1977) as a possible sequence of ongoing galaxy mergers. H I observations of some such systems (e.g. Hibbard & van Gorkom 1996) suggest trends along the merging sequence that are also supported by numerical work (e.g. Hibbard & Mihos 1995). In particular in compact groups of galaxies, where owing to the high space density of galaxies interactions are very common, the kinematics and distribution of the H I can be used to construct an evolutionary scenario of such systems (Barnes 1997; Verdes-Montenegro et al. 2001).

The general evolutionary trend is that in the early stages of the merger, most of the H I is still present within the galaxies. At a somewhat more advanced stage, long tidal tails develop. In interactions involving several galaxies, this H I often forms a diffuse envelope. At the same time, the large torques that occur during the interaction drive large amounts of gas toward the center of the merger. Owing to the resulting high central densities, a large fraction of the atomic hydrogen is converted into molecular form and star formation is occurring at a high rate in these central locations. In the final stages, this starburst dies out and the result is (often, but not always) a spheroidal system surrounded by clouds of H I.

One of the key questions is whether, depending on the kind of encounter, a disk may reform after the merger, by slow accretion of the H I that was not consumed by star formation. Numerical work (e.g. Barnes 1997; Hibbard & Mihos 1995) suggests that disks may form again under certain initial conditions. For example, the material that was tidally disrupted and transported to large radii might remain bound to the remnant. Since the tidal material at these large radii originates from the outer regions of the progenitors, it is likely to consist mainly, or even only, of gas. At a later stage, this material may slowly fall back toward the central remnant and may settle into a disk-like structure. This is well illustrated by the observations and modeling of NGC 7252 (Hibbard et al. 1994; Hibbard & Mihos 1995), where tidal material is seen to be falling back to the main galaxy.

There are more indications that remnants of a major merger that occurred in the recent past are surrounded by large H I complexes of low surface density and anomalous kinematics and that will evolve into a disk (e.g. van Gorkom & Schiminovich 1997). A good example is the galaxy NGC 520 (Hibbard & van Gorkom 1996), which is a merger remnant surrounded by a huge, slightly irregular, rotating H I disk. Another example of this later stage of evolution is NGC 5266 (Morganti et al. 1997), a dust-lane elliptical that has a large amount of H I (more than 10^{10} $M_\odot$, with a total extent of about 200 kpc) of which most is found in a large regular disk,

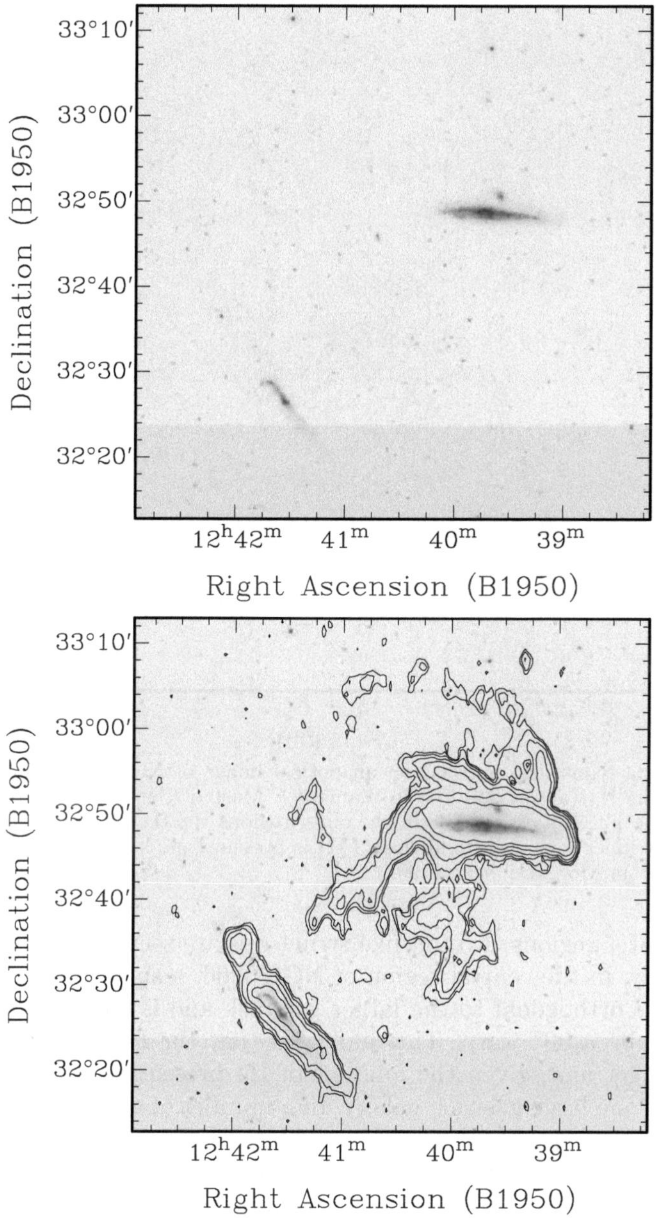

Figure 6. *Top:* Optical image of NGC 4631 (upper galaxy) and NGC 4656 (lower galaxy). Just above NGC 4631 lies NGC 4627, a small elliptical galaxy. *Bottom:* same optical image, but with the total H I distribution overlaid as contours. Several arms and other tidal distortions are visible in the H I. From Rand & van der Hulst (1993).

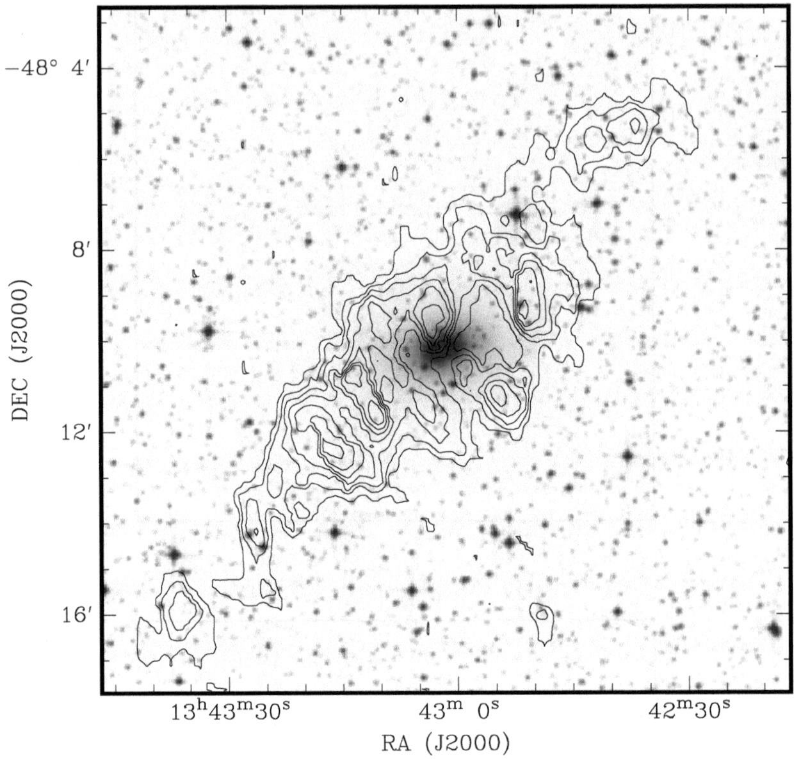

Figure 7. Total H I (contours) overlaid on an optical image of NGC 5266, showing the large extent of the H I (about 200 kpc in diameter). Most of the inner half of the H I shows regular rotation in a disk, while in the outer regions the H I has the shape of two arms. In the very inner region, a small ring of H I is present that is perpendicular to the larger H I disk. From Morganti et al. (1997).

while in the outer regions two tail-like structures are seen (see Fig. 7).

Interestingly, in the central region of NGC 5266, a small polar gas ring is found that is orthogonal to the larger gas disk and is connected to this outer gas disk by a 90° warp. This indicates that the H I structure is related to a merger and, given the amount of H I present, it must have involved at least one large gas-rich galaxy. In a number of gas-rich early-type galaxies such 90° warps are observed (e.g. MGC–5–7–1: van Gorkom et al. 1987; NGC 2685: Shane 1980; NGC 3718: Schwarz 1985, Verheijen & Sancisi 2001). The partial H I ring found at large radius around Centaurus A (Schiminovich et al. 1994) is probably also an example of such a strong warp. The origin of such extreme warps is likely to be similar to that of polar ring structures seen in many galaxies (see Ch. 13).

Merger remnants with a disk are not necessarily early-type galaxies. NGC 3310 is a peculiar (Arp 217) Sbc-type starburst galaxy. Mulder et al.

(1995) and Kregel & Sancisi (2001) have shown the presence of extended H I that has a well-developed two-tail structure. This must be an advanced merger of two galaxies, probably of similar masses, which has either preserved the old disk of one of the progenitors or, perhaps more likely, has led to the formation of a new disk.

3.2. MINOR INTERACTIONS AND ACCRETIONS

In contrast to the interactions where the structure of the galaxies is dramatically changed, many galaxies experience interactions with much smaller systems. Such minor interactions do not have an immediate large effect on the structure of the main galaxy, and the interactions do appear to be less spectacular. However, the cumulative effect of several such small interactions and accretions is large and they play a fundamental role in the evolution of galaxies. For spiral galaxies, the accretion of small companions constitutes a more or less continuous supply of gas from which the disk may be built up slowly.

Many of the minor interactions involve a small companion galaxy that is being captured and accreted by the main galaxy. This often involves a small, gas-rich companion that is also optically visible and that is tidally disrupted by the main galaxy. Many examples of such systems are given in the Rogue Gallery of H I observations (Hibbard et al. 2001). Neutral hydrogen found near ellipticals with dwarf companions, like NGC 4472 (McNamara et al. 1994) and NGC 3656 (Balcells & Sancisi 1996), indicates that the accretion phenomenon is probably playing an important role in all types of galaxies, including ellipticals.

In many galaxies, the small intruder can be easily identified and they probably represent the early stages of the interaction-accretion process. However, in other cases it is less straightforward to establish the details of the accretion as well as the nature of the object that is being accreted. Some accretions are only visible in neutral hydrogen and manifest themselves as pure H I complexes with anomalous kinematics. Such cases are relevant for understanding some of the Galactic HVCs.

A good example of such anomalous H I can be found in the nearby giant spiral M 101 (van der Hulst & Sancisi 1988; Kamphuis 1993). In this galaxy an H I complex of about 2×10^8 $M_\odot$, moving with velocities of up to 150 km s^{-1} with respect to the disk, has been found, as well as a large cavity in the disk near this filament. No optical counterpart is visible for this interaction.

A similar case is NGC 628 (Kamphuis & Briggs 1992). This galaxy has a clean, regular optical image. Only the H I betrays a possible recent accretion, as indicated by the presence of two high-velocity H I complexes in the

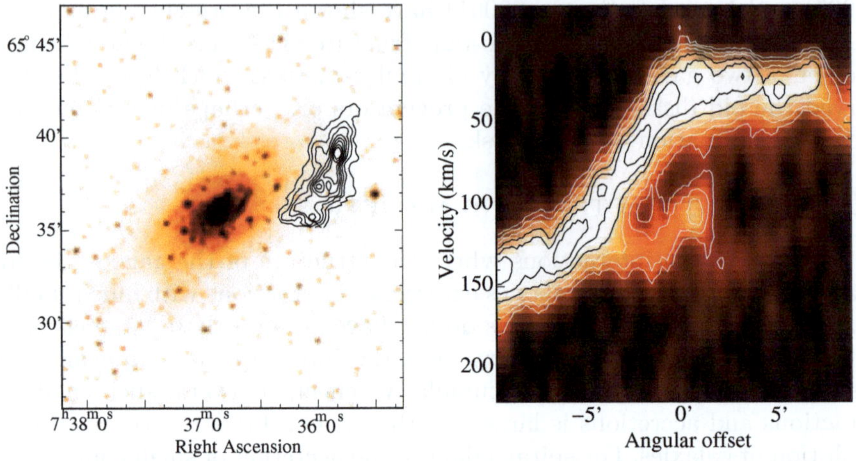

Figure 8. *Left:* optical image of NGC 2403, with the largest anomalous-velocity filament of H I superimposed (contours). The linear extent is about 8 kpc. *Right:* position-velocity diagram along this filament. The filament is observed at about the systemic velocity and the projected velocity difference with the normal H I disk is about 100 km s^{-1}. Data kindly provided by F. Fraternali.

outer regions, which are symmetrically placed with respect to the galaxy center, while also an extended tail of H I arcs over one third of the circumference of the galaxy. These complexes have sizes of tens of kiloparsecs, H I masses of 10^8 M$_\odot$, and maximum velocity excesses of 100 km s^{-1}. Another example of what appears to be a minor interaction with an H I cloud that has no obvious optical counterpart is NGC 925 (Pisano et al. 1998). In some galaxies H I is observed with very complex kinematics that indicate accretion of substantial amounts of H I. In IC 10 long, thin filaments of H I are observed that appear to be falling into the galaxy (Wilcots & Miller 1998). A similar situation is observed in NGC 4449 (Hunter et al. 1998).

On a slightly smaller scale, similar anomalous-gas features are observed in NGC 2403 (Fraternali et al. 2001, 2002). Several long filaments of H I are observed having properties similar to some of the larger HVC complexes in our Galaxy. The largest of these is 8 kpc in length and 10^7 M$_\odot$ in mass (see Fig. 8). In M 33 (Higdon & van der Hulst, priv. comm.) and in NGC 6946 (Boomsma et al. 2002) similar filaments are detected.

The nature of the objects that are being accreted is not clear. One possibility is that we are witnessing accretion at a very advanced stage of the process and all the optical evidence has disappeared. The features in M 101 could be the result of a collision with a dwarf companion which has gone through the H I layer of M 101, creating a large cavity; the high-velocity gas may eventually rain back down onto the M 101 disk (van der Hulst & Sancisi 1988; Kamphuis 1993). A similar scenario may apply to

NGC 628 (Kamphuis & Briggs 1992). A smaller companion may have been accreted in cases like NGC 2403, M 33 and NGC 6946.

On the other hand, the kinematics of the large filaments in NGC 2403 is similar to that observed for the more diffuse anomalous-velocity gas. This suggests that these filaments could just be bright features in the more diffuse halo gas and that they are connected to a galactic fountain. However, it is not clear how such long and relatively massive structures can be formed by a stochastic process such as a galactic fountain.

An interesting possibility is that we are witnessing the accretion of primordial material. The number of small companion objects predicted by models of galaxy formation is larger than the number of dwarf galaxies observed near larger galaxies (e.g. Moore et al. 1999). Perhaps the anomalous-velocity H I complexes discussed here are related to such a population.

Although statistical information on the presence of purely gaseous companions of external galaxies is limited, they are not likely to be present in large numbers for H I masses above 10^7 $M_\odot$. Even though they form a very heterogeneous database, one can use the many observations of nearby galaxies that have been imaged in H I to obtain some constraints. Only very few cases exist where a purely gaseous companion is detected that is not a possible tidal fragment. Often deep optical imaging is lacking and the nature of such gas clouds is still unclear. A possible candidate is NGC 925NGC 0925 (Wilcots & Miller 1998), where an H I streamer of about 10^7 $M_\odot$ is observed near the galaxy, although even in this case a tidal interaction with the neighbouring galaxy UGC 2023 cannot be entirely excluded.

A systematic search for small gas clouds in groups of galaxies was done by Zwaan (2001), who used the *Arecibo* telescope for both a blind survey and a targeted search. No pure H I clouds were detected and they concluded that there is not a large population of H I clouds surrounding larger galaxies. A similar conclusion was reached by de Blok et al. (2002), based on a small survey of the Sculptor and Centaurus A regions. However, few data are available on the occurrence of smaller ($<10^{6.5-7}$ $M_\odot$) H I objects and the situation is quite unclear. If the available results are interpreted in terms of the properties of the Galactic HVC population (or sub-populations), they imply that the bulk of the Galactic HVCs must be at distances from the Galaxy of at most a few hundred kpc. The latest version of the models for the Local Group nature of compact high-velocity clouds (CHVCs) (de Heij et al. 2002) places most of these objects at such distances and does indeed predict that such CHVCs are very hard to detect in galaxies outside the Local Group.

3.3. FREQUENCY OF INTERACTIONS

How often do interactions, within multiple systems or with small companions, take place? And how important are they for the formation of elliptical galaxies or for the building up of disks, the triggering of starbursts and for galaxy evolution in general? In the past years, a large number of galaxies have been imaged in H I. A first estimate made on the basis of about a hundred galaxies led to the conclusion (Sancisi 1992) that almost 50 percent of the galaxies show signs of either present or recent interactions. Obviously, the incompleteness and inhomogeneity of the sample examined make such an estimate rather uncertain.

A recent H I survey carried out by Verheijen & Sancisi (2001) of a magnitude- and volume-limited sample of galaxies from the Ursa Major Cluster allows a more solid estimate of the frequency of tidal interactions and of accretion phenomena. The Ursa Major Cluster differs in character from Virgo- or Coma-type clusters. It has a low velocity dispersion and a long crossing time, comparable to the Hubble time. It has no central concentration, nor X-ray emitting gas, and the cluster is dominated by late-type systems. It can therefore be considered as representative for a galaxy population in the field. Out of the 40 galaxies imaged in H I, about 10 show clear signs of interactions with small companions or have peculiar structures. About half of the sample galaxies show asymmetries in their kinematics or in the H I density distribution and at least 30 percent are warped.

The available data indicate that at least 25 percent of field galaxies are undergoing now, or have undergone in the recent past, some kind of interaction. If one is willing to accept the lopsided kinematics and structure as evidence of past interactions and mergers, as proposed in optical studies (Zaritsky & Rix 1997), then the conclusion would be that more than 50 percent of present day galaxies have been through one or more merger events in a recent past. If lumps of gas with H I masses of order 10^{8-9} $M_\odot$ (as indicated by the 21-cm observations) are accreted at a rate of 1 per 10^{8-9} yr, the typical accretion rate for the gas would be on the order of 1 $M_\odot$ yr^{-1}. If dark matter were also taken into consideration, the total mass accretion rate could easily be a factor 10 higher.

4. Concluding remarks

The above discussion illustrates that in many nearby galaxies H I complexes are found that have a location and/or kinematics that could be considered anomalous. Many of these anomalous H I clouds have similar characteristics as the Galactic HVCs. The data do show that the nature of anomalous gas in galaxies is very complex and that many processes are relevant. The data also show that further study of the properties of the anomalous H I in extragalactic systems will be illuminating for the case of our Galaxy. In at least two areas further progress can be made that will provide valuable input for the galactic case. Many of the properties of the H I in the halos of normal spirals are not understood. For example, how do the properties depend on the mass and mass distribution of the galaxy, and how do they depend on the star formation rate? Moreover, the gas at forbidden velocities, and the radial inflow, observed in galaxies like NGC 2403, are not easily explained by the current models. Deep H I observations of many nearby galaxies, combined with a detailed comparison with models for the Galactic HVCs, will help to better understand these issues. With regard to accretions and interactions, deep surveys are needed to study the statistics of the population of objects with H I masses well below 10^7 $M_\odot$, while the nature of the small gas clouds that are accreted should also be studied further. Such work would provide strong constraints on models for the accretion history of galaxies in general and on the properties of Galactic HVCs in particular.

Acknowledgements

I would like to thank Renzo Sancisi, Filippo Fraternali and Raffaella Morganti for their valuable input.

References

Balcells, M., Sancisi, R. 1996, AJ, 111, 1053
Barbieri, C., Fraternali, F., Oosterloo, T., Sancisi, R. 2004, in preparation
Barnes, J.E. 1997, in ASP Conf. Ser. 116, The Second Stromlo Symposium: The Nature of Elliptical Galaxies, eds. M. Arnaboldi, G.S. da Costa, P. Saha (San Francisco: ASP), 469
Boomsma, R., van der Hulst, J.M., Oosterloo, T.A., Sancisi, R. 2002, A&AS, 200, 4203
Bowen, D.V., Blades, J.C, Pettini, M. 1995, ApJ, 448, 662
Brinks, E., Bajaja, E. 1986, A&A, 169, 14
Dahlem, M. 1997, PASP, 109, 1298
de Blok, W.J.G., Zwaan, M.A., Dijkstra, M., Briggs, F.H., Freeman, K.C. 2002, A&A, 382, 43
de Heij, V., Braun, R., Burton, W.B. 2002, A&A, 392, 417
Dettmar, R.-J. 1990, A&A, 232, L15
Deul E., den Hartog, R. 1990, A&A, 229, 362
Ferguson, A.M.N., Wyse, R.F.G., Gallagher, J.S. 1996, AJ, 112, 2567
Fraternali, F. 2003, in IAU Symp. 217, Recycling Intergalactic and Interstellar Matter, eds. P.-A. Duc, J. Braine, E. Brinks (San Francisco: ASP), in press
Fraternali, F., Oosterloo, T.A., Sancisi, R., van Moorsel, G.A. 2001, ApJ, 562, L47
Fraternali, F., van Moorsel, G.A., Sancisi, R., Oosterloo, T.A. 2002, AJ, 123, 3124
Fraternali, F., Oosterloo, T.A., Sancisi, R., Swaters, R.A. 2004, in preparation
Hibbard, J.E., Mihos, J.C. 1995, AJ, 110, 140
Hibbard, J.E., van Gorkom, J.H. 1996, AJ, 111, 655
Hibbard, J.E., Guhathakurta, P., van Gorkom, J.H., Schweizer, F. 1994, AJ, 107, 67

Hibbard, J.E., van Gorkom, J.H., Rupen, M.P., Schiminovich, D.S., 2001, in ASP Conf. Ser. 240, Gas & Galaxy Evolution, eds. J.E. Hibbard, M.P. Rupen, J.H. van Gorkom (San Francisco: ASP), 659
(see also http://www.nrao.edu/astrores/HIrogues/webGallery/index.html)
Hoopes, C.G., Walterbos, R.A.M., Rand, R.J. 1999, ApJ, 522, 669
Howk, J.C., Savage, B.D. 2000, AJ, 119, 644
Hunter, D.A., Wilcots, E.M., van Woerden, H, Gallagher, J., Kohle, S. 1998, ApJ, 495, L47
Kamphuis, J. 1993, Ph.D. thesis, Rijks Universiteit Groningen
Kamphuis, J., Sancisi, R., van der Hulst, J.M. 1991, A&A, 244, L29
Kamphuis, J., Briggs, F. 1992, A&A, 253, 335
Kim, S., Dopita, M.A., Staveley-Smith, L., Bessell, M.S. 1999, AJ, 118, 2797
Kregel, M., Sancisi, R. 2001, A&A, 376, 59
Lee, S.-W., Irwin, J.A., Dettmar, R.-J., Cunningham, C.T., Golla, G., Wang, Q.D. 2001, A&A, 377, 759
McNamara, B.R., Sancisi, R., Henning, P.A., Junor, W. 1994, AJ, 108, 844
Moore, B., Ghigna, S., Governato, F., Lake, G., Quinn, T., Stadel, J., Tozzi, P. 1999, ApJ, 524, L19
Morganti, R., Sadler, E.M., Oosterloo, T., Pizzella, A., Bertola, F. 1997, AJ, 113, 937
Mulder, P.S., van Driel, W., Braine, J. 1995, A&A, 300, 687
Pisano, D.J., Wilcots, E.M., Elmegreen, B.G. 1998, AJ, 115, 975
Puche, D., Westpfahl, D., Brinks, E., Roy, J.-R. 1992, AJ, 103, 1841
Rand, R.J. 1996, ApJ, 462, 712
Rand, R.J. 2000, ApJ, 537, L13
Rand, R.J., van der Hulst, J.M. 1993, AJ, 108, 2098
Rand, R.J., Kulkarni, S.R., Hester, J.J. 1990, ApJ, 352, L1
Rand, R.J., Kulkarni, S.R., Hester, J.J. 1992, ApJ, 396, 97
Sancisi, R. 1992, in Physics of Nearby Galaxies: Nature or Nurture?, eds. T.X. Thuan, C. Balkowski, J.T.T. Van (Gif-sur-Yvette: Editions Frontieres), 31
Schaap, W.E., Sancisi, R., Swaters, R.A. 2000, A&A, 356, 49
Schiminovich, D., van Gorkom, J.H., van der Hulst, J.M., Kasow, S. 1994, ApJ, 423, L101
Schwarz, U.J. 1985, A&A, 142, 273
Shane, W.W. 1980, A&A, 82, 314
Stanimirovic, S., Staveley-Smith, L., Dickey, J.M., Sault, R.J., Snowden, S.L. 1999, MNRAS, 302, 417
Swaters, R.A., Sancisi, R., van der Hulst, J.M. 1997, ApJ, 491, 140
Thilker, D.A., Braun, R., Walterbos, R.M. 1998, A&A, 332, 429
Toomre, A. 1977, in The Evolution of Galaxies and Stellar Populations, eds. B.M. Tinsley, R.B. Larson (New Haven: Yale University), 401
van der Hulst, J.M., Sancisi, R. 1988, AJ, 95, 1354
van Gorkom, J.H., Schiminovich, D. 1997, in ASP Conf. Ser. 116, The Second Stromlo Symposium: The Nature of Elliptical Galaxies, eds. M. Arnaboldi, G.S. da Costa, P. Saha (San Francisco: ASP), 310
van Gorkom, J.H., Schechter, P.L., Kristian, J. 1987, ApJ, 314, 457
Verdes-Montenegro, L., Yun, M.S., Williams, B.A., Huchtmeier, W.K., Del Olmo, A., Perea, J. 2001, A&A, 377, 812
Verheijen, M.A.W., Sancisi, R. 2001, A&A, 370, 765
Vladilo, G., Centurión, M., de Boer, K.S., King, D.L., Lipman, K., Stegert, J., Unger, S.W., Walton, N. 1994, A&A, 291, 425
Wilcots, E.M., Miller, B.W. 1998, AJ, 116, 2363
Yun, M.S., Ho, P.T.P., Lo, K.Y. 1994, Nature, 372, 530
Zaritsky, D., Rix, H.-W. 1997, ApJ, 477, 118
Zwaan, M.A. 2001, MNRAS, 325, 1142

7. THE LARGE- AND SMALL-SCALE STRUCTURE OF HVCS

ULRICH J. SCHWARZ
Kapteyn Institute, Groningen, The Netherlands, and Department of Astrophysics, Universiteit Nijmegen, The Netherlands, schwarz@astro.rug.nl

AND

BART P. WAKKER
Department of Astronomy, University of Wisconsin-Madison, USA; wakker@astro.wisc.edu

Abstract. The main question addressed in this chapter is whether or not HVCs have small-scale structure that differs from that in low-velocity gas and the IVCs. If so, this could make it possible to distinguish HVCs from galactic gas not only by their deviating velocities, but also by their structural characteristics. Further, the physical characteristics derived from the small-scale structure may provide clues to their origin. We find that the core-halo structure that was discovered in the early seventies by Giovanelli, Hulsbosch and others, is typical for all HVCs that have been observed so far. We review some methods to characterize the small-scale structure numerically. These include the autocorrelation function, low- and high-pass filtering, and the concept of fractality. In general, these methods show a tendency for HVCs to have more pronounced small-scale structure than lower-velocity gas. However, the differences are not so pronounced that it is possible to easily distinguish between high- and low-velocity gas.

1. Introduction

For as long as high-velocity clouds have been mapped, considerable structure has been found down to the scale of the resolution. This is well illustrated by the case of complex A. With a 36′ beam it appears to consist of several bright concentrations within a long filament (Wakker & van Woerden 1991). Observations with a 10′ beam (Giovanelli et al. 1973) reveal smaller cores within the concentrations. Synthesis telescopes such as the Westerbork Synthesis Radio Telescope (*WSRT*) and the Australia Tele-

H. van Woerden et al. (eds.), High-Velocity Clouds, 145–166

scope Compact Array (*ATCA*) provide maps with about $1'$ resolution, showing even more details (Schwarz & Oort 1981; Wakker & Schwarz 1991; Wakker et al. 2002). The same pattern holds for other HVCs that have been observed at arcminute resolution (cloud M I, complex H, several VHVCs, WW 187; Wakker & Schwarz 1991; Braun & Burton 1999; Wakker et al. 2002). The range of scales covers 3 decades.

Such observations were originally aimed at HVC cores, as these are the brightest, and thus are the easiest to detect. Off-core fields at random positions within HVCs have been mapped in the framework of getting distances and metallicities for HVCs from observations of interstellar lines in the directions of stars and extragalactic objects (Schwarz et al. 1995; Wakker et al. 1996, 1999; Sembach et al. 2004). These maps were made because the interpretation of the detections and non-detections of interstellar lines requires accurate H I column densities in the directions of the probes. These off-core high-resolution maps also show pronounced fine structure.

The question arises whether the observed fine structure is a special characteristic of HVCs, or a more common phenomenon of neutral hydrogen also occurring at intermediate and low velocities. To answer this, we require comparable observations at lower velocities and preferably at intermediate or high latitudes; at low latitudes many clouds at different distances contribute to a complicated emission profile for the low-velocity gas.

In early high-resolution H I studies (Schwarz & Oort 1981; Wakker & Schwarz 1991) only a limited velocity range could be observed owing to limitations of the correlator. More recent synthesis observations cover a much larger velocity range. Thus, comparisons between the HVCs and intermediate- and low-velocity gas (IVCs and LVCs) are now feasible. In this chapter we will illustrate some structural properties of HVCs, IVCs, and LVCs, using data for four selected fields observed with the same instrument and observational setup, over the full velocity range.

A great handicap for interpreting these results is the lack of knowledge of the distance, D, to the HVCs. Many of the derived physical parameters evidently depend it. For instance, density is proportional to D^{-1}, size to D, mass to D^2, and timescales to D. Some progress has been made in this area recently (see Ch. 10), but good distance estimates are still scarce, and more work is needed before the internal physics of the clouds can be understood.

In this chapter, we concentrate on the question of how to describe and characterize structure. Section 2 gives an overview of available observations, and presents a selection of four test fields for which observations of high-, intermediate- and low-velocity gas exist, made with the same telescope setup. Section 3 discusses what we can learn from data on internal motions. Section 4 describes four possible methods to characterize structure and the application of these to our test fields. Section 5 summarizes the results.

2. Observations

On the largest scales, the most detailed and sensitive HVC maps are based on the surveys of Hulsbosch & Wakker (1988) and Morras et al. (2000), which have beams of 36′ and 30′, grids of $1°\times1°$ and $0\overset{\circ}{.}5\times0\overset{\circ}{.}5$, and velocity resolutions of 16 km s^{-1}, respectively. The maps (see Ch. 2) show the presence of large HVC complexes, with different types of structure, from round (complex H, some Anti-Center clouds, and most smaller clouds) to filamentary (complexes A, C, MS, and some of the larger Anti-Center clouds).

However, these datasets do not cover the intermediate- and low-velocity gas. These are contained within the Leiden-Dwingeloo Survey (*LDS*) of Hartmann & Burton (1997), which covers the velocity range from −500 to +500 km s^{-1}, with 1.03 km s^{-1} channel spacing and a 36′ beam, at declinations more positive than −30°. The south is covered at 26 km s^{-1} and 15′ resolution by the H I Parkes All-Sky Survey (*HIPASS*; Barnes et al. 2001). A southern survey comparable to the *LDS* has been finished (Arnal et al. 2000), but as of 2004 it is not-yet generally available. These surveys allow a comparison of HVC characteristics with those of IVCs and LVCs. We will use the *LDS* data for the analyses presented here.

Studies at intermediate (10′) resolution exist for several individual HVCs or complexes, using the *Green Bank* 300-ft telescope (Giovanelli et al. 1973), the *Effelsberg* 100-m telescope (Hulsbosch 1975), and the new *Green Bank* Telescope (*GBT*; Lockman 2003). The most important general result of these observations was that they showed the presence of bright knots within the HVCs with sizes of tens of arcmin, brightness temperatures up to 30 K (compared to ∼2 K at lower resolution), velocity widths of typically ∼8 to 10 km s^{-1} (compared to ∼25 km s^{-1}), and a large scatter of velocities within the range of the 25 km s^{-1} velocity width of low-resolution maps. Thus, the cores appear to consist of cold gas, and most of the velocity width seen at lower resolution is due to random motions of these cores.

In the framework of determining column densities, Wakker et al. (2001) presented *Effelsberg* (9′ beam) observations for over a hundred directions toward extragalactic and stellar background (and foreground) targets. They derived column densities and compared these with those derived in the same directions using the *LDS* (36′ beam). Their results are discussed in Sect. 4.3.

Observations with *Arecibo* (3′ resolution) and synthesis telescopes (resolutions of 1′ or better) confirm the trend found at lower resolutions: the maps show much fine structure (Schwarz & Oort 1981; Mirabel 1981; Wakker & Schwarz 1991; Schwarz et al. 1995; Wakker et al. 1996; Tamanaha 1997; Braun & Burton 2000; Wakker et al. 2002; Sembach et al. 2004).

However, the synthesis observations are insensitive to structures larger than about 20′. Adding those requires data at intermediate resolution with

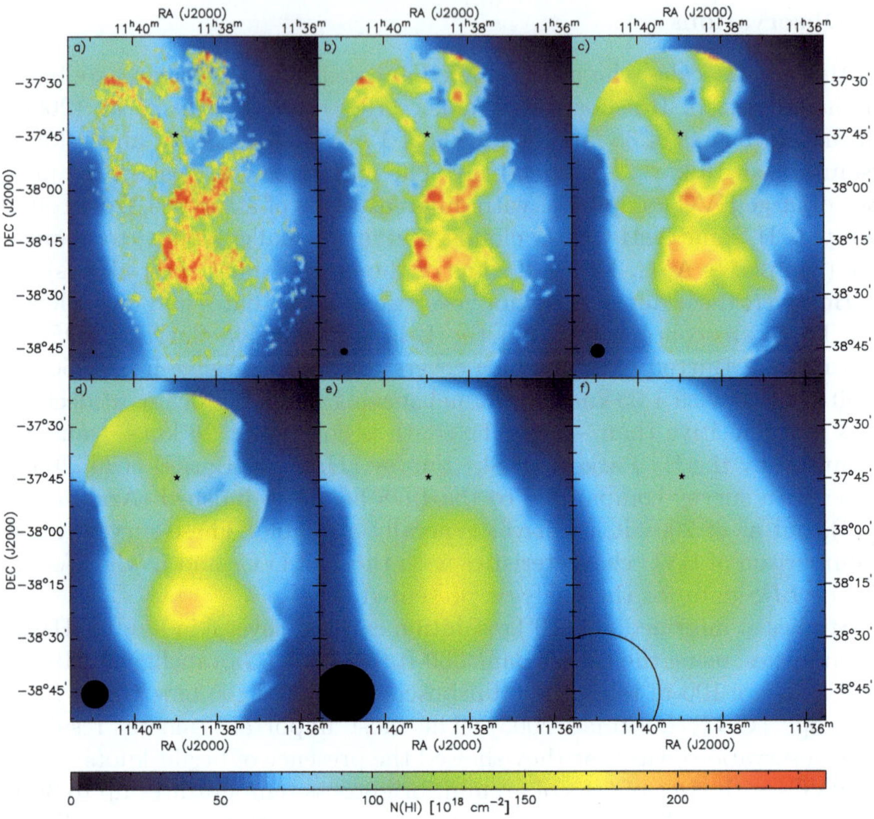

Figure 1. HVC 287+22+240 or WW 187, part of the Leading Arm of the Magellanic Stream. Shown is the total column density map at six different resolutions ($38'' \times 59''$, $2'$, $4'$, $8'$, $16'$, and $34'$). The column density scale is given at the bottom, while the size of the beam is indicated by the black dot in the bottom left corner of each panel. The symbol shows the position of NGC 3783. From Wakker et al. (2002).

similar sensitivity. Earlier studies did not include this step. Schwarz et al. (1995) and Wakker et al. (1996) described how to correct the data in a single direction toward a background target. Braun & Burton (2000) add data from the *LDS* ($36'$ beam), but this leads to some artifacts. Wakker et al. (2002) combined *ATCA* ($1'$ beam) with *Parkes* ($17'$ beam) data, to obtain the first fully-sampled map of an HVC core (shown in Fig. 1). The *ATCA* and *Parkes* data were combined in the map plane, using the method described by Schwarz & Wakker (1991), which is based on the realization that the single-dish map contains the emission filtered out by the interferometer as well as the emission seen by the interferometer.

For 12 sightlines toward extragalactic background probes, the column densities, $N(\text{H I})$, observed at high resolution (from combined single-dish and interferometer maps) can be compared to those observed with a $10'$

and 36′ beam. Wakker et al. (2001) find that for these sightlines the ratio $N(\mathrm{H\,I};1')/N(\mathrm{H\,I};10')$ lies in the range 0.75 to 1.25, a range smaller than that found for the ratio $N(\mathrm{H\,I};10')/N(\mathrm{H\,I};36')$ (0.25 to 2.5), suggesting that the cloud structure is hierarchical: the brightest knots are embedded in the brightest parts of the smoother background structure. The total flux of knots and filaments seen in the synthesis maps ranges between a few % and 30%; the remaining gas is smoothly distributed. Further study is needed.

A noteworthy feature of the structure seen at the highest resolutions is that the details show no apparent relation to the HVC as a whole, and no clear signs of shock fronts or interaction with the local gas. Further, as was the case when going from a large (36′) to an intermediate (10′) beam, the small-scale structure has random velocities within the velocity width of the profile at lower resolution. This may indicate that the small-scale features are short-lived condensations within the HVCs (see Sect. 3.1).

3. Physical characteristics

3.1. KINEMATICS

The internal motions of an HVC can give important clues about its 3-dimensional structure, its physical state, its formation history, and its future. The distribution of column density within clouds rarely or never shows systematic structures related to the HVCs as a whole. However, there are often systematic velocity patterns, which may be interpreted as evidence for rotation, expansion, contraction, or equilibrium. This is true both at large and at small scales (see e.g. Giovanelli et al. (1973) and Wakker (2001) for large-scale velocity fields; Schwarz & Oort (1981), Wakker (1991) and Braun & Burton (1999) for small-scale velocity fields).

Wakker & van Woerden (1997) discussed several timescales that can be derived from the internal motions: a) the time for cores to shift substantially with respect to each other (typically about 20 $(D/5\ \mathrm{kpc})$ Myr), b) the time a core takes to shift over its own width (typically about 0.5 $(D/5\ \mathrm{kpc})$ Myr), c) the time for a core to double its size (typically 3 $(D/5\ \mathrm{kpc})$ Myr), and d) the time for a subconcentration within a core to expand (typically 1 $(D/5\ \mathrm{kpc})$ Myr). The calculated values of these timescales are all proportional to the distance (D) of the cloud, and they thus remain uncertain to some extent. It is also necessary to assume that the clouds are not confined by some external medium or by magnetic fields; if confinement is present, expansion timescales will be much longer. Finally, projection effects make some of these timescales lower limits. The ratios of the timescales are mostly unaffected by distance estimates, however. No substantial new data have been obtained since the summary by Wakker & van Woerden (1997). The typical values listed above show that the core expansions hap-

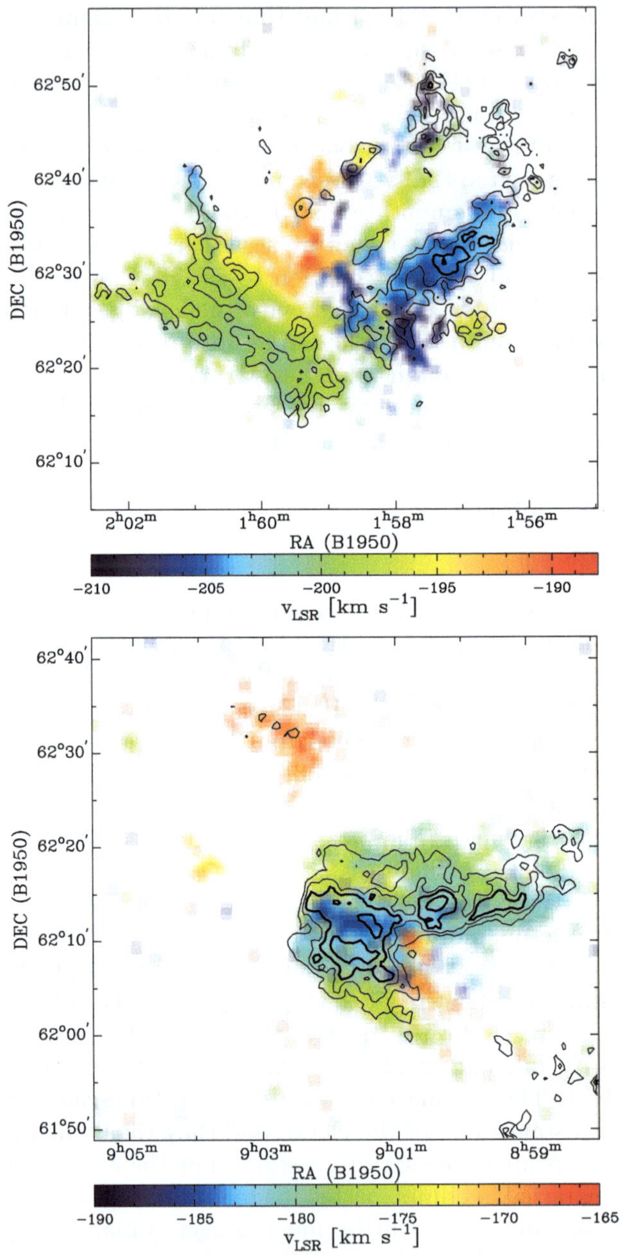

Figure 2. Velocity field of cores HI (top) and AIV (bottom) at 1′ resolution. These are representative of many HVC cores, showing multiple subcores at different velocities. Column density contours at levels of 1, 1.5, 2 and 3×10^{20} cm^{-2} have been overlaid. Adapted from Wakker (1991).

pen slower than the shifts of the cores by their own width, but much faster than the shift of the cores within a cloud. The latter timescale is comparable to that needed for the cloud to move down to the galactic plane. Thus, the fine structure will change considerably during the movement of a cloud through space, the medium-scale structure will change but remain recognizable, but the structural integrity of the complexes is mostly preserved while they travel large distances.

The cores seen at medium (10′) and high (1′) resolution could be the result of cooling gas from a fountain-type stream or cooling within a cloud. Several formation scenarios of this type are discussed by Bregman (Ch. 16).

3.2. VELOCITY GRADIENTS

Velocity gradients have been observed in a number of clouds, such as HVC 131+1−200 (core H I of complex H), MI and A IV (Wakker 1991; Wakker & Schwarz 1991; Fig. 2). Small gradients may be explained by projection effects: $\Delta v{=}\alpha\, v_r/\tan i$, where Δv is the difference in radial velocity v_r between two directions separated by a small angle α, and i is the inclination of the space velocity (i=90° in the line of sight). For small inclinations the gradient can get very large and imply a large space velocity, which may be unrealistic. For example, the core of M I (Wakker & Schwarz 1991) has a gradient of 10 km s^{-1} across 20′, which would imply a space velocity >1000 km s^{-1}, if the inclination angle were low. Brüns et al. (2001) use a similar method to argue that HVC 125+41−207 (WW 84) moves almost perpendicularly to the line of sight.

An alternative explanation for a linear velocity gradient is solid-body rotation. This is only possible if the cloud is confined by some external medium or (especially for filaments) by a magnetic field.

Parallel filaments with often different velocities are sometimes observed. Examples can be seen in H I (Wakker & Schwarz 1991) and in the Magellanic Stream (see e.g. Ch. 5). Another peculiar phenomenon was found by several authors: a correlation within a cloud between N(H I) and the infall velocity (Benjamin & Danly 1997). This could be explained by a cloud moving in an ambient medium; the cloud could then be slowed down by friction, with stronger deceleration for gas with lower column densities.

3.3. SELF-GRAVITATING CLOUDS

Blitz et al. (1999; Ch. 14) suggested that most of the HVCs are self-gravitating clouds filling the Local Group, and that they have a median distance of 1 Mpc. The large, nearby, complexes are suggested to be special examples of such clouds, which have come close to the Milky Way and are tidally disrupted. In this model, the clouds are stabilized by dark matter.

There are several problems with this model (see e.g. Chaps. 2 and 8), including one that is related to the properties of cloud structure, specifically the fact that most clouds have faint outer parts surrounding brighter cores. This complicates the derivation of total fluxes and radii. For instance, Blitz et al. (1999) used the data from Hulsbosch & Wakker (1988) to calculate cloud parameters. However, if the detection limit of the Hulsbosch & Wakker (1988) survey had been 0.20 K rather than the actual value of 0.05 K, many clouds would appear fainter and smaller (70% would not have been seen at all), and the derived neutral hydrogen fraction would have been 10% (rather than 3%). Thus, the implied neutral fraction depends strongly on the selection criteria and on the method of calculation.

Another point related to the subject of this chapter is the fact that in the Blitz et al. (1999) model the clouds are supposed to be virially stable. On the one hand this can be used to derive cloud distances, assuming a spherically symmetric cloud. On the other hand, derived masses are usually very large ($>10^8$ $M_\odot$) and then orderly rotation is more likely. Wakker & Schwarz (1991) and Braun & Burton (1999) mapped a total of 8 small ($<1°$ in diameter) HVCs with the *WSRT* at $1'$ resolution. In our opinion only some of these show weak evidence for rotation, contrary to the claim made by Braun & Burton (1999) that many show strong evidence for rotation. If such evidence were found for a cloud, a total mass could be derived from the rotation curve, but then the virial theorem would no longer be applicable.

3.4. VOLUME DENSITY CONTRAST

To estimate the H I volume density, n, of an HVC or an HVC core, one can divide the column density by the product of distance and angular size. This quotient scales as $1/D$, with D the unknown cloud distance. However, in the context of this chapter we are interested in the density contrast within a cloud, so only relative densities play a role and the distance cancels out.

We compiled the density contrast, $n/\langle n \rangle$, from several high-resolution *WSRT* datasets (Schwarz & Oort 1981 – field A I; Wakker & Schwarz 1991 – fields H I, A IV.1, A IV.2, M I). The density, n, was computed for individual cores (i.e. column density enhancements) in each *WSRT* field, assuming a distance of 1 kpc, as described in those papers. For the A I dataset, the minor and major axis of each elongated concentration had been calculated separately, and the minor axis had been used to estimate sizes, while for the other datasets an average size was listed. To make the two datasets compatible, the densities for A I were recalculated, using the average of the minor and major axis.

The average density of an HVC, $\langle n \rangle$, can be found from the column densities observed at Dwingeloo ($36'$ beam). This results in values for $\langle n \rangle$

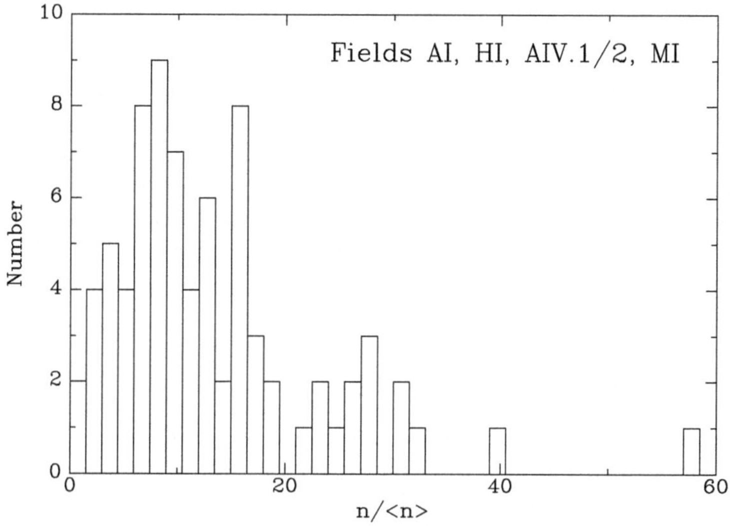

Figure 3. Density contrast, $n/<n>$, of the volume densities, n, of cores within the *WSRT* fields observed by Schwarz & Oort (1981) and Wakker & Schwarz (1991). This quantity is distance independent. Figure composed for this article.

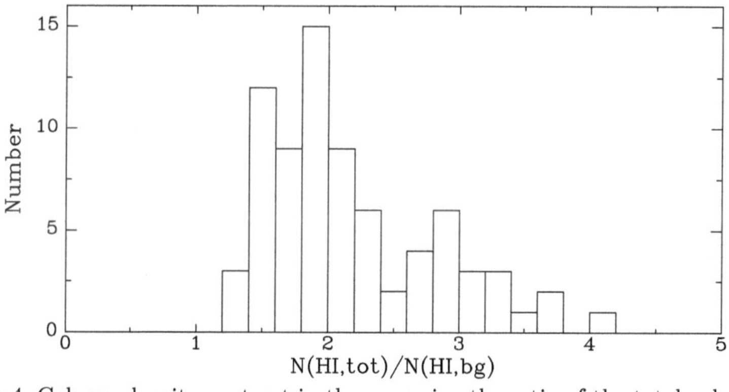

Figure 4. Column density contrast in the cores, i.e. the ratio of the total column density in each core to the background column density of the field. Figure composed for this article.

of about $1.6/D$ to $3.2/D$ cm^{-3}. In Fig. 3, we show histograms of the resulting density contrast, which most often is about 10 to 20, but can be as high as 60. Clearly, there is much small-scale structure.

3.5. COLUMN DENSITY CONTRAST

High-resolution H I data are necessary to derive accurate values of N(H I) for comparison with ionic column densities measured in absorption-line studies. In Fig. 4, we present a histogram of the ratio of total to background column density for the same five high-resolution fields as in the previous subsection. The average and rms of this ratio are 2.2±0.7. Another useful quantity is the probability to hit a concentration, which is the fraction of the field covered by the concentrations. This fraction lies between 5 and 11%. Taking into account this small probability, and taking into account that on average the increase in column density is modest, the error one makes by using intermediate- or even low-resolution H I data to derive abundances is not as severe as might be expected from the large amount of small-scale structure that is present. Nevertheless, to derive accurate abundances (within 50%), high-resolution H I data are required.

3.6. TEMPERATURES

A direct determination of the spin temperature, T_s, of H I can be made by measuring H I absorption against a background continuum source. Only a few detections of H I absorption in HVCs have been reported in the literature, giving T_s=47 K toward complex H (Wakker et al. 1991), and T_s=55 K and 147 K toward the Outer Arm (Payne et al. 1978, 1980; Akeson & Blitz 1999; for more details see Wakker 2001). The main reasons for this small number of measurements are: (1) there are few bright background sources in the direction of cores and substructures, and (2) the method is most sensitive to cold H I (<100 K), whereas the extended H I of HVCs (outside cores and substructures) is probably warm (>1000 K). Many lower temperature limits have been derived (e.g. Payne et al. 1978, 1980; Schwarz & Oort 1981; Colgan et al. 1990; Mebold et al. 1991; Akeson & Blitz 1999). However, except for three directions toward the Magellanic Stream where T>500 K was found (Mebold et al. 1991), most lower limits are only >10 K to >20 K and thus rather uninteresting.

Upper limits (T_u) for T_s can be derived from the line width W of the profile of the HVC as a whole or of cores and substructures within the HVC: $T_u = 22W^2$, where W is the velocity width (FWHM) in km s^{-1}. It is an upper limit to the spin temperature T_s, because the line width may be increased by turbulence and velocity gradients, within the beam *and* along the line-of-sight. The cores and substructures have mostly narrow lines, often implying values for T_u as low as 100 K. Mostly, though, the upper limits are in the range of several 100 K. The narrowest line was detected by Braun & Burton (2000) in the HVC 125+41−207 (WW 84): W=1.01 km s^{-1}, resulting in T_u=22 K.

In general, T_u is not a reliable estimate for the spin temperature, because of its sensitivity to many effects that can increase the observable line width W. This has implications for computing the pressure and for questions concerning the existence of pressure equilibrium within HVCs.

4. Characterization of small-scale structure

4.1. GENERAL REMARKS

It is clear that the structure of the interstellar medium is very complex, both in 3-dimensional space and in the projected observable coordinates (2 angular dimensions and velocity). The line-of-sight integration adds to the complexity, especially at low latitudes, where various features blend together because of the long path through the Galaxy. For LVCs it is important to take this into account; Lazarian & Pogosyan (2000) discuss a method to correct for the line-of-sight effects. With HVCs the pathlength effects are less severe and the signal-to-noise ratio of the high-resolution data is often poor. Therefore, we will neglect line-of-sight blending effects in the discussion below. Also, the purpose of most high-resolution maps of HVCs is to determine $N(\mathrm{H\,I})$ along the line of sight, so that we are often only interested in the projected 2-dimensional fine structure rather than in the 3-dimensional spatial distribution.

We ask: is it possible to find a quantitative measure of the structure in 2-dimensional maps or 2+1-dimensional data cubes which agrees with an intuitive concept of "fine structure"? It is interesting to note that similar questions are posed in a cosmological context, to describe the distribution of galaxies in space and the distribution of deviations from uniformity of the cosmic background radiation. In the case of the interstellar medium four main approaches have been used so far:

1. The power spectrum or the autocorrelation function.
2. Smoothing (using low-pass filters).
3. Filtering (using high-pass filters).
4. Fractality and related concepts.

We applied each of these methods to four test fields, two observed at high resolution, and two at low resolution. For the high-resolution fields we selected two *WSRT* observations not yet published elsewhere; these are fields toward the optical probes PG 1511+623 (l=99°.21, b=+47°.96, projected on complex C) and PG 0822+645 (l=151°.62, b=+34°.57, toward complex A). For PG 0822+645, a set of channel maps is shown in Fig. 5. For the low-resolution fields we chose two regions in the *LDS* (Hartmann & Burton 1997). One field is centered around complex A (l=110° to 173°, b=−13° to +50°) and the other around complex GCP (also known as the "Smith Cloud" or HVC 40−15+100; l=0° to 60°, b=−40° to −5°).

4.2. THE POWER SPECTRUM AND AUTOCORRELATION FUNCTION

The autocorrelation function (or its Fourier transform, the power spectrum) contains information on the average 2-dimensional spatial size, orientation,

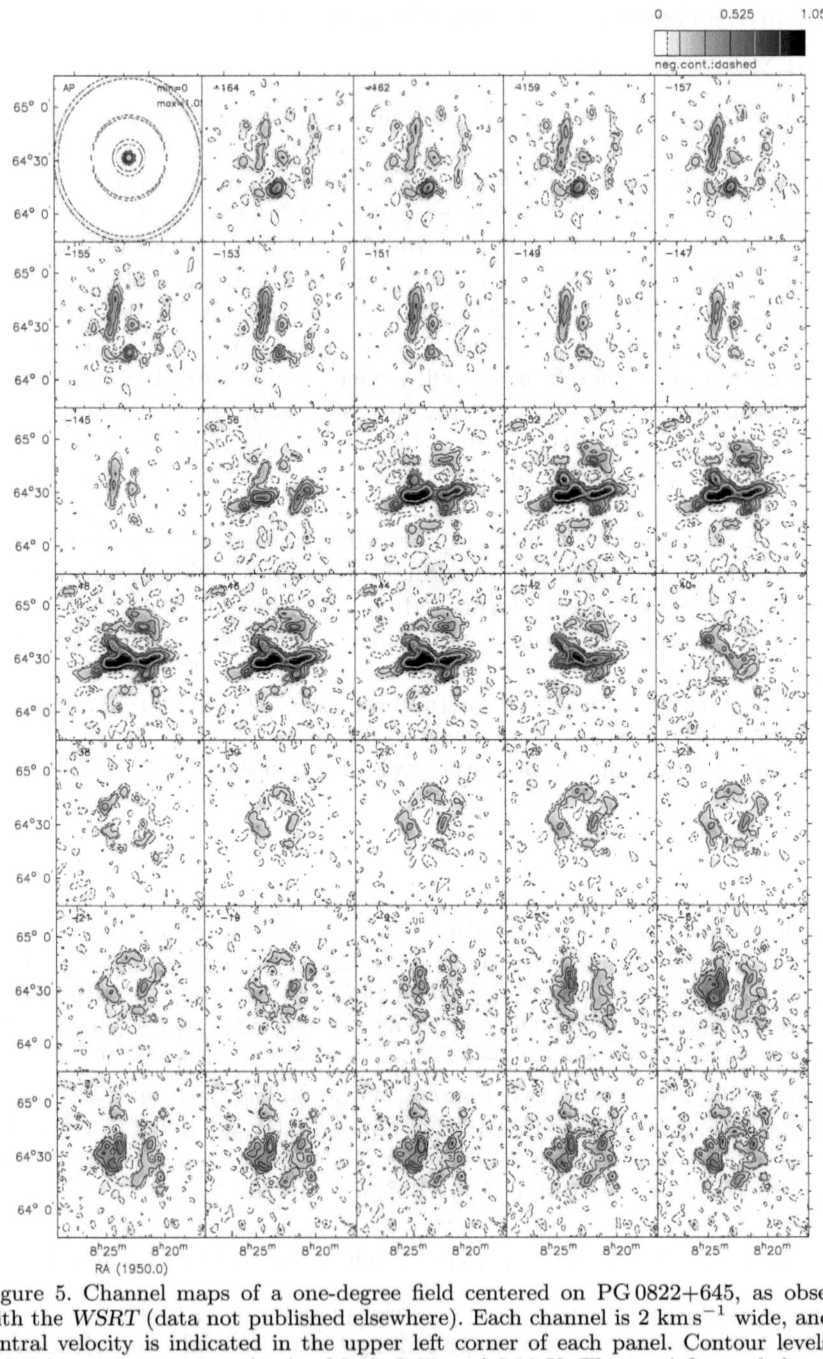

Figure 5. Channel maps of a one-degree field centered on PG 0822+645, as observed with the *WSRT* (data not published elsewhere). Each channel is 2 $\mathrm{km\,s^{-1}}$ wide, and the central velocity is indicated in the upper left corner of each panel. Contour levels are at brightness temperature levels of 0.08, 0.16, and 0.32 K. The top-left panel shows the synthesized beam.

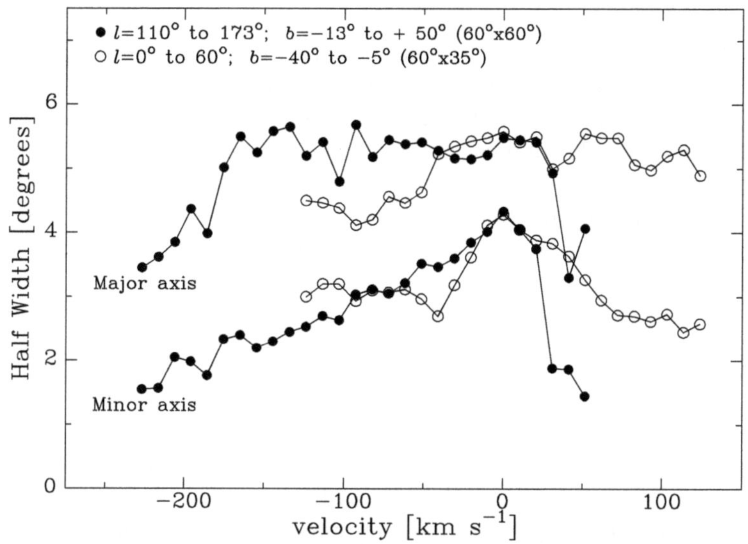

Figure 6. Fitted FWHM of minor and major axes of the autocorrelation function of the *LDS* data centered on HVC complexes A and GCP, as function of LSR velocity. Figure composed for this article.

and amplitude distribution of all features within the map. If the map consists of several isolated features, the autocorrelation function is a map with a maximum at the center and secondary maxima away from the center. We note that often just the (1-dimensional) azimuthal average of the power spectrum is used.

Many studies of the power spectrum of gas in the interstellar medium result in power laws. Almost all of these studies are based on observations with high signal-to-noise ratios. However, for HVCs the signal-to-noise ratio is in general poor. Therefore, we restrict our discussion to the central peak of the autocorrelation function, which can be parametrized in terms of moments, by a 2-dimensional Gaussian, or by other functions.

The autocorrelation of a map is most easily obtained by doing a Fourier transform and taking the square of the real part. This is usually normalized to the maximum, which occurs at the center. In practice, noise and a continuous smooth background distort the central peak. To minimize this problem, filtered data should be used, in which the background has been removed, and to which a cutoff in the signal-to-noise ratio was applied. However, clipping introduces artifacts, which need to be taken into account. In a variation of the autocorrelation technique, aimed at reducing the effects of noise and other artifacts, one can analyze the averages of parameters in a number of submaps within a large map.

Figure 6 shows the result of applying this technique to the *LDS* data of the complex A and complex GCP fields. Since the given examples are aimed at showing the principle, only channel maps separated by 10 $\mathrm{km\,s^{-1}}$ were used and a 2-dimensional Gaussian was fit to the central peak of the autocorrelation function of each selected channel. This yields the major and minor axes, which are shown as function of velocity. Two effects can be discerned. First, at the highest negative velocities (corresponding to HVC complex A), the major axis tends to be somewhat smaller than at velocities >-160 $\mathrm{km\,s^{-1}}$, implying that the typical size scale of the features in the HVC is somewhat smaller. Second, the minor axis shows a significant increase going from -220 to -10 $\mathrm{km\,s^{-1}}$. Combined with the almost constant major axis this implies that the intermediate- and low-velocity gas appears to be less filamentary than the high-velocity gas.

4.3. SMOOTHING

A simple way to characterize fine structure is to analyze the effect of angular smoothing of a peak in a map. Depending on the structure around the peak, the smoothing may have a strong effect (the maximum for an unresolved peak will decrease substantially), or a marginal effect (for a broad maximum). We define S as the ratio of a (local) maximum in a cloud after smoothing to the value in the original map. The ratio S is related to the internal structure of a feature. Low S means that there is mainly an unresolved peak, without background; S close to 1 means lots of large-scale structure with little fine-structure. For an unresolved peak, S is proportional to $1/\Omega_s$, the inverse of the solid angle of the smoothing function. Note that the effect of smoothing must be measured at the same position. Simply looking at the peaks of maps with different degrees of smoothing may give erroneous results.

Figure 7 shows an example. Not much variation with velocity is seen, indicating that to first order there is not much difference in structure within clouds between gas with high and low velocities.

A different way to approach the effects of smoothing was presented by Wakker et al. (2001), who compared column densities for HVCs, IVCs and low-velocity gas obtained from the *LDS* (36′ beam), the *Effelsberg* telescope (9′ beam), the *WSRT* (1′ beam), and Lyα absorption toward extragalactic background sources ($<0.''1$ beam). They find that there are substantial variations within a 36′ beam: for HVCs and IVCs the ratio $N(\mathrm{H\,I};9')/N(\mathrm{H\,I};36')$ (a factor 16 in beam area) lies in the range 0.67 to 1.5 in only one third of the cases, and has a full range from 0.2 to 2.5. For low-velocity gas, the range appears smaller (0.75 to 1.3), but this may be due to the fact that there are usually several different clouds with similar velocities in the line of sight, so that upward fluctuations in one cloud may be compensated by downward fluctuations in another. Within the 9′ beam

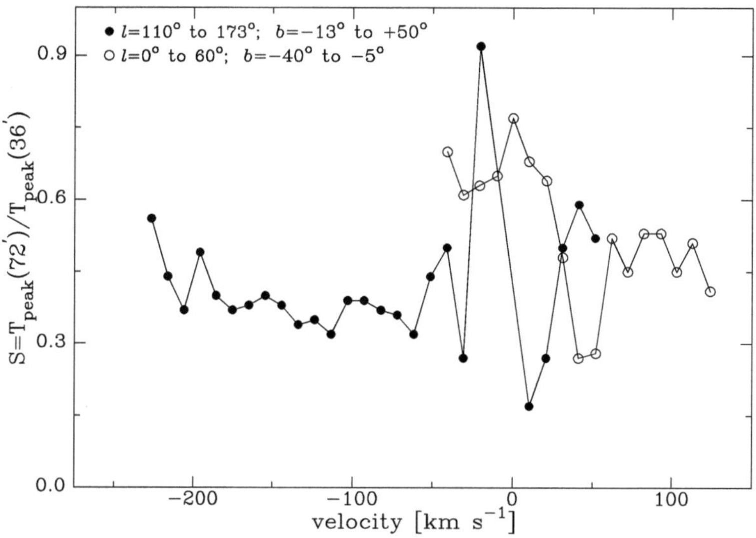

Figure 7. Ratio, S, of peak brightness before and after smoothing with a Gaussian beam twice as wide as the resolution, as function of LSR velocity, for the *LDS* data centered on HVC complex A and on complex GCP. Figure composed for this article.

fluctuations also exist. The ratio $N(\text{H I};2')/N(\text{H I};9')$ (a factor 23 in beam area) lies in the range 0.75 to 1.25. This clearly is a smaller range. Finally the ratio $N(\text{H I;Ly}\alpha)/N(\text{H I};9')$ lies in the range 0.74 to 0.98. This range is similar to the previous one, with as a notable feature that most ratios are <1. This pattern of column density ratios can be interpreted if the cloud structure is hierarchical: the relative size of fluctuations gets smaller on smaller scales, although the structure never becomes really smooth.

4.4. FILTERING

In principle, a drawback of synthesis observations is that the shortest baselines are not observed, so that large-scale features are filtered out. However, this effect can be exploited to derive a measure of fine structure, which we will call R. This is defined as the ratio of the total flux in the synthesis map to the total flux seen with a single-dish telescope with a beam that is the same as the primary beam of the synthesis telescope. The ratio R can also be calculated for maps made with single-dish telescopes, as explained below. This method allows comparisons between the amount of fine structure in different HVCs, IVCs and LVCs, although some caution is warranted. Most importantly, the non-linear processing of the synthesis map (such as applying CLEAN) may subtly influence the results. We applied this method to all four of our test fields, with results shown in Figs. 8 and 9.

To obtain the single-dish flux in the interferometer fields, the 3 closest

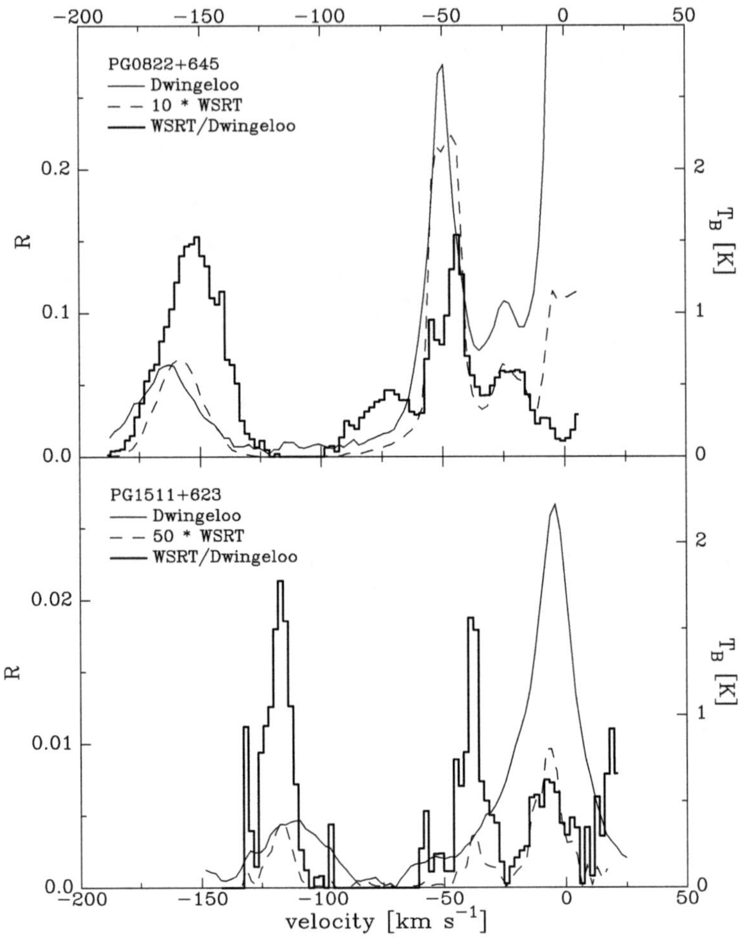

Figure 8. Ratio, R, of flux recovered in the interferometer map to total flux seen by a single-dish telescope, as function of LSR velocity. *Top:* PG 0822+645 field. *Bottom:* PG 1511+623 field. The dashed and thin lines show the observed H I profiles (scaled by a factor 10 for the *WSRT* data), with a brightness temperature scale on the right. The heavy solid line shows the ratio, with a scale on the left. Figure composed for this article.

LDS positions were averaged. The results for the high-resolution fields are displayed in Fig. 8. There appears to be a substantial dichotomy between the high-/intermediate- and the low-velocity gas, in the sense that R(low) is much smaller than R(high/intermediate). That is, the low-velocity gas is relatively smoother.

To apply a filter to the low-resolution test fields, they were split into 49 and 24 subfields of 8°×8°. Each subfield was then multiplied with a circular Gaussian "primary" beam of 8° FWHM. A filter was then applied in a manner analogous to what happens in synthesis observations. That

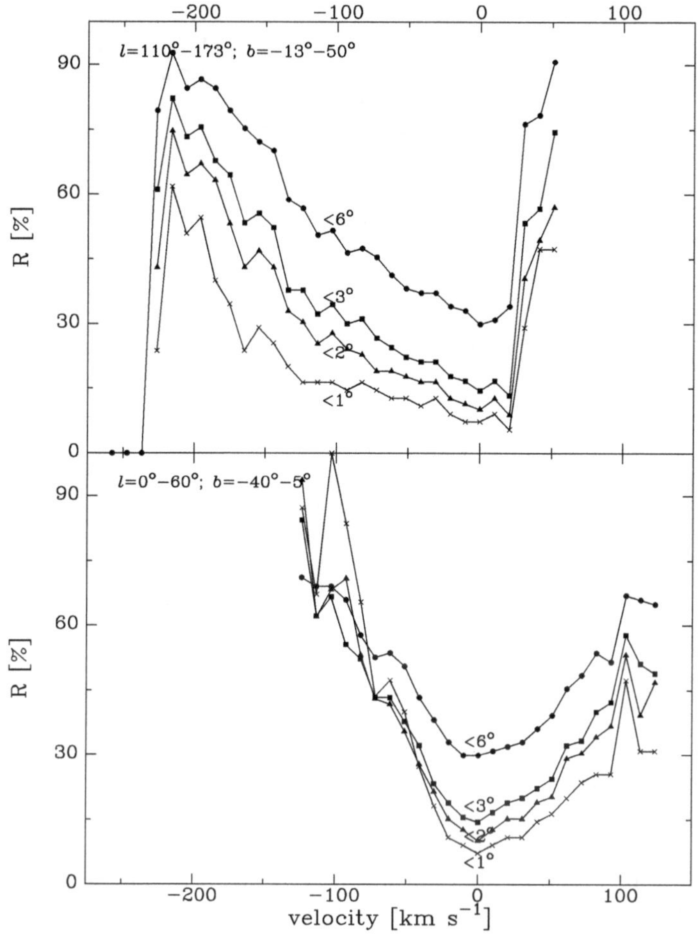

Figure 9. Ratio, R, of flux recovered after filtering to total flux, as function of LSR velocity, for the two *LDS* fields. *Top:* field centered on complex A *Bottom:* field centered on complex GCP. The four curves pertain to different filter sizes, i.e. structures at scales larger than the specified angular scale were filtered out. Figure composed for this article.

is, the Fourier transform was taken for each subfield, and the inner part of the transformed map was removed. The size of the removed area was varied to correspond to 1°, 2°, 3°, and 6° size scales. The filtered maps were then transformed again. However, instead of applying CLEAN to derive the filtered flux, we applied a simple correction factor (such that a point source gets $R = 1$). Next, R was found for each of the subfields, and averaged.

Figure 9 shows the resulting value of R as a function of velocity for the test field around complex A (l=110° to 173°, b=−13° to +50°). There is a strong correlation of R with velocity and filter size. As was the case for the high-resolution test fields, the low-velocity gas appears to be smoother.

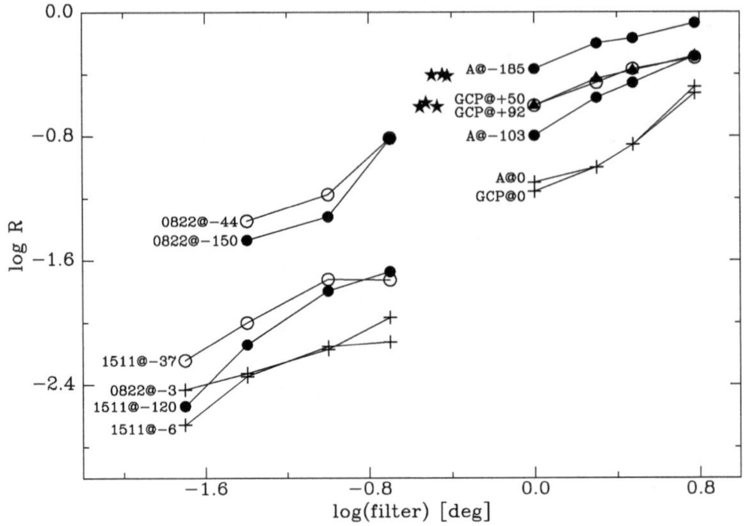

Figure 10. Ratio, R, of recovered flux after filtering to total flux, as function of filter size. Different symbols pertain to different components in the H I profile for a number of fields. The labels show the field name and, after the "@", the central velocity of the component. The five sets in the upper right were derived from the *LDS* data for the complex A and complex GCP fields, the five sets in the bottom left from the two *WSRT* fields. The six asterisks show the ratios for HVC cores found by Schwarz & Oort (1981) and Wakker & Schwarz (1991). Figure composed for this article.

For example, at low velocities only 30% of the structure is on scales <6°, and only 10% on scales <1°. On the other hand, the high-velocity gas has up to 90% of its structure on scales <6°, and 50% on scales <1°.

A different way of displaying this information is given in Fig. 10, now combining data for high- and low-resolution test fields in one plot. The horizontal axis shows the width of the filtering function, the vertical axis gives R, and different symbols correspond to three different typical velocities.

Figure 10 also contains six points for the *WSRT* data in Schwarz & Oort (1981) and Wakker & Schwarz (1991). These have R comparable to those for the larger-scale filters. This probably reflects the fact that the latter are in the direction of the cores of HVCs, which is not the case with the test fields discussed here.

The curves in Fig. 10 can also be interpreted as the probability distribution for features of a given size. For v=−100 km s^{-1} the points suggest a power-law with index −0.5. This corresponds to a power-law for the frequency distribution of −1.5, which agrees with "classical" studies of cloud sizes (e.g. Schwarz & van Woerden 1974).

The results in Fig. 10 pertain to the average R for many subfields. A question is thus whether the same behavior may be seen in each individual subfield. We inspected all subfields; although there is more scatter, the trend of R increasing to higher velocities is seen in all subfields.

This correlation of R with velocity could be genuine. However, it is possible that blending plays an important role at lower velocities (and lower latitudes), where many objects blend together on a long line of sight.

4.5. FRACTALS

The whole-sky survey of HVCs (Figs. 1, 2, 3 of Ch. 2) shows structural similarities to detailed maps of individual HVCs. This kind of similarity is the basic feature of fractals. Other constituents of the interstellar medium, such as molecular clouds, IR "cirrus", etc, have similar structural properties (e.g. Falgarone et al. 1992). Below, we discuss an example of the derivation of a fractal dimension for high-velocity clouds. The question whether or not the ISM can indeed be described by fractals is still open, and requires more detailed analysis, using methods such as those described below, but also using different approaches, such as correlation lengths and density laws.

The concept of fractals was applied to some high-resolution HVC maps by Vogelaar & Wakker (1994). They measured the length of closed contours at fixed levels, for various clouds. The idea behind this is the example of a fractal coastline. In that case the length of the coastline increases without limit when measured with rods decreasing in size. This comes about because the coastline shows more and more details, in a self-similar manner.

Do HVCs behave like this? An exact analogy to the coastline example would be to use closed contours of the cloud at different resolutions. However, the dynamic range of the data does not usually allow this; so Vogelaar & Wakker (1994) analyzed closed contours at fixed resolution, but at different contour levels within a range of velocities. They then compared the length of contours (perimeter) with the enclosed area, plotting the logarithms of the perimeters against the logarithms of the enclosed areas. In general, the perimeter P as function of the area A is given by $P = aA^b$, where a is a prefactor. For simple geometries the prefactor depends on the shape (circle: $a = \sqrt{\pi}$, square: a=4). The interpretation of the prefactor is not clear for fractals, however, although it does contain information about the kind of shape. A simply-shaped cloud would yield a scatter diagram with slope b=+0.5, i.e. the perimeter scales linearly with diameter, the area as the square of the diameter. However, if the geometry is fractal, then the slope b>0.5. The fractal dimension D equals twice the slope.

In practice there are many problems with applying this method. For instance, finite pixel size, the influence of noise, edge-effects and underlying structures, and the approximation used to measure the perimeter; for a detailed discussion see Vogelaar & Wakker (1994). Another fundamental problem is that we may have a collection of clouds with several well-defined

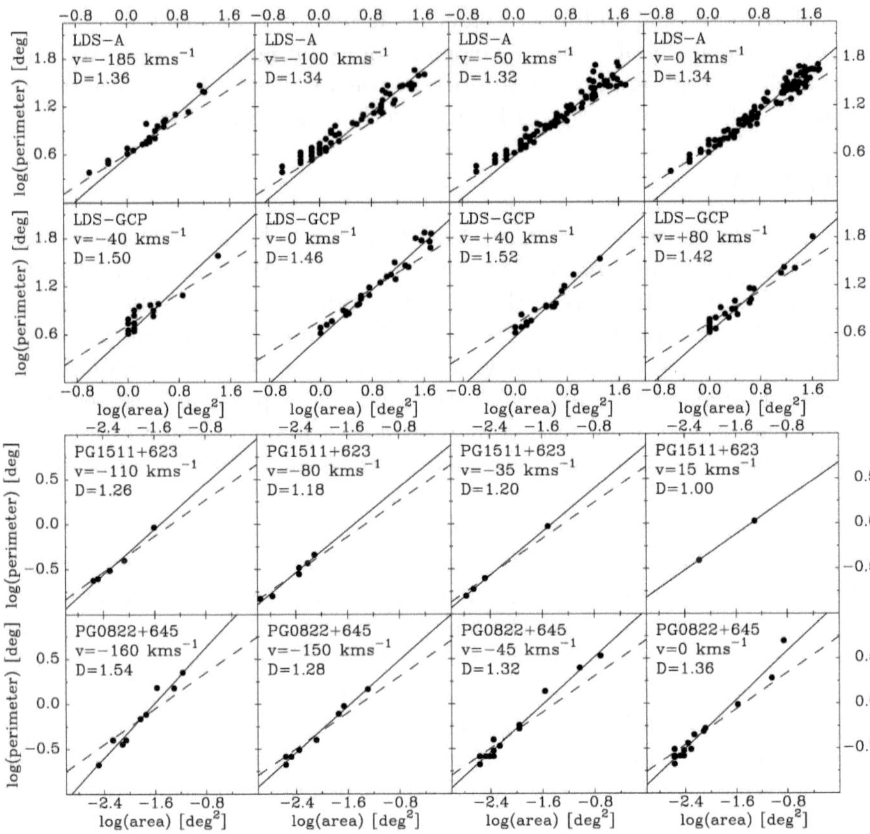

Figure 11. Fractal analysis of the four fields defined in Sect. 4.1. Plotted are the perimeter versus the area of objects ("clouds") at various velocities in the four test fields, as well as fits to the points (solid lines). For normal geometry the correlation would follow the dashed line (power law with index 0.5); for fractals the correlation follows a power-law with higher index. This has been found previously by Falgarone et al. (1992) in molecular clouds and by Vogelaar & Wakker (1994) in HVCs. Figure composed for this article.

shapes, each with a different prefactor a. The correlation diagram may then be severely distorted.

However, in general, the ingredients of the method explained above can be used to characterize the structure and the cloud's shapes: round clouds have a small ratio P/A, whereas filaments or other complicated shaped clouds have a larger ratio P/A. The slope, b, and prefactor, a, derived from the scatter diagram of P versus A can serve as indicators for the characteristics of structure as function of velocity, latitude, etc.

Vogelaar & Wakker (1994) combined all velocities within HVCs in one large correlation set, and found fractal dimensions in the range 1.2 to 1.4. We made a similar study for our 4 test fields, but for each velocity channel separately. Naturally, this results in fewer data points per scatter diagram, but as shown in Fig. 11 there is astonishingly small scatter around the best linear fit. We overplotted a line of slope 0.5, the slope for normal geometry. The observed slopes deviate significantly from this slope and agree with those found by Vogelaar & Wakker (1994).

5. Discussion and conclusions

Both the large-scale and small-scale structure of H I show rich detail. We defined various quantities which are measures of structure. We measured these quantities for four test fields and found some significant variations. However, the preliminary conclusion is that we do not find that HVCs, IVCs, and LVCs have strongly different morphologies, indicating that similar physical processes may be taking place in all clouds. This is surprising, since the various types of clouds may have very different histories and environments. If this similarity is really present, we can take advantage of it, in the sense that we can try to understand the underlying physical processes by studying HVCs, which are relatively isolated in space and velocity.

In the 1980s, the main result of the studies of HVCs at high resolution was the discovery of cores within the clouds, with narrow profiles, but with velocity ranges spanning the velocity width of the large structures. These cores were interpreted as "condensations". However, there seems to us to be a difficulty with such a simple picture. The large, smooth structure cannot simply be gas at high temperature, because condensations would result in velocities close to the systemic velocity of the cloud as a whole.

We must have gas which is highly turbulent. The spin temperature of the cells would be considerably lower than inferred from the total velocity width of the wide, smooth component, because part of the width is due to the turbulent motion. This may be checked by computing the rms velocity of the substructures within a cloud, W_s, which should be smaller than W of the HVC as a whole. Let $W_c = \sqrt{W^2 - W_s^2}$ be the intrinsic velocity width of the hypothetical cells. W_c would then provide an upper limit to the temperature, T_c, of the cells. E.g. the analysis of the 38 features found in the field A I by Schwarz & Oort (1981) gives W_s=20.4 km s^{-1}; combined with W=25 km s^{-1} this gives T_c =4600 K.

More detailed studies of medium-sized HVCs or IVCs, from low to high angular resolution, with the short-spacing problem corrected for, are necessary for a better understanding of the physical state of these clouds. Fortunately, some clouds now have good distance brackets, facilitating the interpretation. We should look into the future with patience *and* optimism regarding our knowledge of distances, which will make the study of the structure of HVCs more fruitful than what we can do today.

Acknowledgements

B.P. Wakker acknowledges financial support from NASA grant NAG5-9179.

References

Akeson, R.L., Blitz, L. 1999, ApJ, 523, 163
Arnal, E.M., Bajaja, E., Larrarte, J.J., Morras, R., Pöppel, W.G.L. 2000, A&AS, 142, 35
Barnes, D.G., Staveley-Smith, L., de Blok, W.J.G., et 36 al. 2001, MNRAS, 322, 486
Benjamin, R.A., Danly, L. 1997, ApJ, 481, 764
Blitz, L., Spergel, D., Teuben, P., Hartmann, D., Burton, W.B. 1999, ApJ, 514, 818
Braun, R., Burton, W.B. 1999, A&A, 351, 437
Braun, R., Burton, W.B. 2000, A&A, 354, 853
Brüns, C., Kerp, J., Pagels, A. 2001, A&A, 370, L26
Colgan, S.W.J., Salpeter, E.E., Terzian, Y. 1990, ApJ, 351, 503
Falgarone, E., Puget, J.-L., Pérault, M. 1992, A&A, 257, 715
Giovanelli, R., Verschuur, G.L., Cram, T.R. 1973, A&AS, 12, 209
Hartmann, D., Burton, W.B. 1997, Atlas of Galactic Neutral Hydrogen (Cambridge: Cambridge University Press)
Hulsbosch, A.N.M. 1975, A&A, 40, 1
Hulsbosch, A.N.M., Wakker, B.P. 1988, A&AS, 75, 191
Lazarian, A., Pogosyan, D. 2000, ApJ, 537, 720
Lockman, F.J. 2003, ApJ, 591, L33
Mebold, U., Greisen, E.W., Wilson, W., Haynes, R.F., Herbstmeier, U., Kalberla, P.M.W. 1991, A&A, 251, L1
Mirabel, I.F. 1981, ApJ, 247, 97
Morras, R., Bajaja, E., Arnal, E.M., Pöppel, W.G.L. 2000, A&AS, 142, 25
Payne, H.E., Dickey, J.M., Salpeter, E.E., Terzian, Y. 1978, ApJ, 221, L95
Payne, H.E., Salpeter, E.E., Terzian, Y. 1980, ApJ, 240, 499
Schwarz, U.J., Oort, J.H. 1981, A&A, 101, 305
Schwarz, U.J., van Woerden, H. 1974, in IAU Symp. 60, Galactic Radio Astronomy, eds. F.J. Kerr and S.C. Simonson III (Dordrecht: Reidel), 45
Schwarz, U.J., Wakker, B.P. 1991, in IAU Coll. 131 (ASP Conf. Ser. 19), Radio Interferometry: Theory, Techniques and Applications, eds. T.J. Cornwell, R.A. Perley (San Francisco: ASP), 188
Schwarz, U.J., Wakker, B.P., van Woerden, H. 1995, A&A, 302, 364
Sembach, K.R., Wakker, B.P., Tripp, T.M., et 13 al. 2004, ApJS, 150, 387
Tamanaha, C.M. 1997, ApJS, 109, 139
Vogelaar M.G.R., Wakker B.P. 1994, A&A, 291, 557
Wakker, B.P. 1991, A&AS, 90, 495
Wakker, B.P. 2001, ApJS, 136, 463
Wakker, B.P., Schwarz, U.J. 1991, A&A, 250, 484
Wakker, B.P., van Woerden, H. 1991, A&A, 250, 509
Wakker, B.P., van Woerden, H. 1997, ARA&A, 35, 217
Wakker, B.P., Vijfschaft, B., Schwarz, U.J. 1991, A&A, 249, 233
Wakker, B.P., Howk, J.C., Schwarz, U.J., van Woerden, H., Beers, T.C., Wilhelm, R., Kalberla, P.M.W., Danly, L. 1996, ApJ, 473, 834
Wakker, B.P., Howk, J.C., Savage, B.D., van Woerden, H., Tufte, S.L., Schwarz, U.J., Benjamin, R., Reynolds, R.J., Peletier, R.F., Kalberla, P.M.W. 1999, Nature, 402, 388
Wakker, B.P., Kalberla, P.M.W., van Woerden, H., de Boer, K.S., Putman, M.E. 2001, ApJS, 136, 537
Wakker, B.P., Oosterloo, T.E., Putman, M.E. 2002, AJ, 123, 1953

8. IONIZED GAS ASSOCIATED WITH HVCS

STEPHEN L. TUFTE
Department of Physics, Lewis & Clark College, Portland, USA
tufte@lclark.edu

Abstract. The study of emission lines from HVCs is yielding new information on their basic properties as well as on the properties of their environment. Such studies are also providing new insights into the location, origin, and role in the Galaxy of the HVCs. Very sensitive velocity-resolved measurements of Hα now exist for many HVCs, including the Magellanic Stream, complexes A and C, and some compact HVCs. These data show that ionized hydrogen is strongly associated with the neutral hydrogen, but also that the H I brightness temperature and Hα intensity are not correlated. The detection of Hα emission from many compact HVCs provides an argument against the idea that they are massive clouds distributed throughout the Local Group of galaxies – intergalactic clouds would not be sufficiently ionized to produce the observed intensities. Combined with direct distance measurements for many HVCs, the Hα data can provide measurements of the intensity of the radiation field in the environs of the Milky Way, and of the escape fraction of ionizing photons.

Emission lines also contribute to more accurate estimates of metal abundances in HVCs by providing information on the ratio of ionized to neutral hydrogen. This is exemplified by the data for complex C in the direction toward the Seyfert galaxy Mrk 290. The intensities of forbidden optical emission lines such as [S II]λ6716, [N II]λ6584, and [O III]λ5007 are sensitive to a cloud's metallicity, temperature and ambient radiation field. The ratios of the intensities of these lines to that of Hα show large variations, suggesting variations in ionization/excitation conditions. By measuring multiple emission lines on the same sightlines for which high-quality absorption-line data exist, it will be possible to put limits on the temperatures of the HVCs, as well as to characterize the spectrum of the ionizing radiation field. This review concludes by suggesting some future directions in the field of HVC emission-line studies that may prove interesting.

H. van Woerden et al. (eds.), High-Velocity Clouds, 167–182

1. Introduction

The exploration of optical emission lines from high-velocity clouds (HVCs) has emerged as an important new field of study. The vast majority of previous investigations of HVCs used radio telescopes to study the 21-cm line of hydrogen, which reveals the quantity and velocity structure of the neutral material, but gives no information on the ionized material and is mostly insensitive to the physical parameters (e.g. temperature and density) of the clouds (although rough upper and lower limits on the temperature can be gleaned from the line width and brightness temperature, respectively). $H\alpha$ measurements have shown that ionized gas is ubiquitously associated with the neutral material and are now allowing detailed studies of the spatial and kinematic relationship between the ionized and neutral material. Further, along with UV absorption-line measurements, emission-line studies are making significant progress in measuring the distance, metallicity, ionization state, and other physical parameters of the HVCs.

The two key parameters for understanding the origin and nature of HVCs are their metal abundance and their distance. Emission lines are a very important source of information for both. As I describe below, measurements of the $H\alpha$ surface brightness of HVCs give basic information on their distances. We have gathered significant evidence against the idea that the HVCs are distributed throughout the Local Group, and argue that they are more likely in the Galactic Halo. As for the metallicity, in order to measure an accurate abundance, one needs to know the total hydrogen column density and not just the neutral hydrogen column density. In cases where it has been measured, it appears that the H^+ is on the order of one third of the total hydrogen column density, but the fraction varies. Thus, to get accurate abundances, one needs to know how much ionized gas is present, and emission-line measurements play an important role in determining this.

In addition to learning about HVCs themselves, we may also use the emission lines as probes in an attempt to understand the physical conditions near their likely location in the Galactic Halo, and their connection to the disk. Assuming that the clouds are photoionized, one can hope to map out the ionizing radiation field of the Milky Way as a function of galactocentric radius and height above the plane through measurements of the $H\alpha$ line. One may also estimate the fraction of ionizing photons that escape our Galaxy altogether and are released into the intergalactic medium (IGM) in this manner. This line of inquiry may connect studies of the upper reaches of the warm ionized medium to IVCs and HVCs.

In addition to measuring the $H\alpha$ line, a small collection of measurements has been made of diagnostic forbidden lines, such as the [S II] λ6716, [N II] λ6584, and [O III] λ5007 lines. By studying their ratios to $H\alpha$, further information can be gleaned on the temperature, metallicity, and ionization conditions in and around the HVCs.

2. The observations: high-resolution Fabry-Pérot spectroscopy of emission lines

The extreme faintness of the Hα emission from HVCs made it nearly impossible to study in the past. Reynolds (1987) searched for Hα in six directions and could only place upper limits ranging from 0.2 R to 0.6 R. One Rayleigh (R) is $10^6/4\pi$ photons cm^{-2} s^{-1} sr^{-1} or 2.41×10^{-7} erg cm^{-2} s^{-1} sr^{-1} at Hα. Kutyrev & Reynolds (1989) successfully detected Hα (I(Hα)=0.08$\pm$0.02 R) from a very-high-velocity cloud ($v_{\rm LSR}$=$-$300 km s^{-1}) in Cetus, part of the Anti-Center complex. Münch & Pitz (1989) detected an intensity of 0.15 R toward the M II cloud, a part of complex M. Note that the names of HVC complexes follow Wakker & van Woerden (1991). Songaila et al. (1989) claimed a detection of Hα from complex C at the 0.03 R level; however, Tufte et al. (1998) have observed the same direction and report a significantly higher intensity. In general, these early observations involved long integration times and produced null or marginal results.

The advent of CCD detectors with both high quantum efficiency and extremely low dark noise has dramatically improved the prospects for optical emission-line studies of HVCs. Coupling these detectors to Fabry-Pérot based spectrometers, which allow high throughput at high spectral resolution for diffuse sources, has made such optical observations a burgeoning new field of study.

The author, in cooperation with Reynolds, Haffner, Madsen, and others, uses the Wisconsin H-Alpha Mapper (*WHAM*) for HVC observations. While originally designed to map out the distribution and kinematics of ionized gas in the warm ionized medium (WIM), this instrument turns out to be well suited for HVC observations. *WHAM* utilizes a 15-cm aperture, dual-etalon Fabry-Pérot spectrometer. The ring pattern passed by the Fabry-Pérot etalons is imaged onto a high quantum efficiency, low-noise CCD camera to record the spectrum (see Reynolds et al. 1998). The spectral resolution is 12 km s^{-1} over a spectral window of 200 km s^{-1}. Because typical line widths of HVCs in H I emission are 20 to 25 km s^{-1}, *WHAM* observations are unique in being able to resolve the line profiles. *WHAM* has a 1° diameter beam.

The Putman and Bland-Hawthorn team use *TAURUS*2 in conjunction with the University of Maryland Fabry-Pérot etalon (70 mm clear aperture), fed by the Anglo Australian Telescope (*AAT*) or William Herschel Telescope (*WHT*). The spectral resolution is 46 km s^{-1}, with a 45 Å (*AAT*) or 22 Å (*WHT*) spectral window. Fields of view are 10′ (*AAT*) or 5′ (*WHT*).

The team led by Weiner & Williams have used two different instruments for their work. The groundbreaking observations of the Magellanic Stream (see below) were done with the Rutgers Fabry-Pérot Interferometer coupled to the CTIO 1.5-m telescope. This instrument had a resolution of 35 km s^{-1} (FWHM) over a 22 Å window and a 7′ beam. Their second study used the Las Campanas 2.5-m telescope with the Wide Field Camera Fabry-Pérot, which had 55 km s^{-1} (FWHM) spectral resolution and a 25′ field of view.

Comparing the instruments, *WHAM* has the highest spectral resolution and the largest clear-aperture etalons, but a large field of view. The higher spectral resolution leads to more detailed velocity information and also clearer separation of components for identification with known 21-cm features. The *TAURUS* and Rutgers instruments are capable of comparable sensitivity – the smaller sizes of the etalon are partially compensated by lower spectral resolutions in terms of detection thresholds. Also, the limiting factor is usually sky subtraction issues and not Poisson statistics. The higher spatial resolution of the *TAURUS* and Rutgers instruments will allow more detailed morphological studies of individual features and also allows observations of spatially smaller features without beam dilution.

3. A brief summary of recent emission-line observations

3.1. THE MAGELLANIC STREAM GLOWS IN Hα

The first high-quality observations of optical emission lines from HVCs were of the Magellanic Stream by Weiner & Williams (1996). They detected Hα from MS II (I(Hα)=0.37 R), MS III (I(Hα)=0.21 R), and MS IV (I(Hα)=0.20 R); some non-detections were interspersed with the detections. They saw evidence of edge-brightening and proposed a model in which the clouds were ionized by ram-pressure heating due to motion through the low-density material in the Galactic Halo.

3.2. *WHAM* OBSERVATIONS OF THE M, A, AND C COMPLEXES

Tufte et al. (1998) carried out a detailed study of Hα from complex M. This study found Hα emission from various locations in the M I and M II clouds, with intensities ranging from 0.06 R to 0.20 R. This work also presented data for two directions in complex A and one in complex C, with Hα intensities ranging from 0.08 R to 0.13 R.

Figure 1 shows a map of the observation beams in complex M, overlayed on 21-cm contours from the Leiden-Dwingeloo Survey (*LDS*; Hartmann & Burton 1997), along with the Hα spectra for the direction marked 1a. This spectrum corresponds to an Hα intensity, I(Hα)=0.08±0.01 R, which we now know to be fairly typical of HVCs. Interestingly, the neighboring

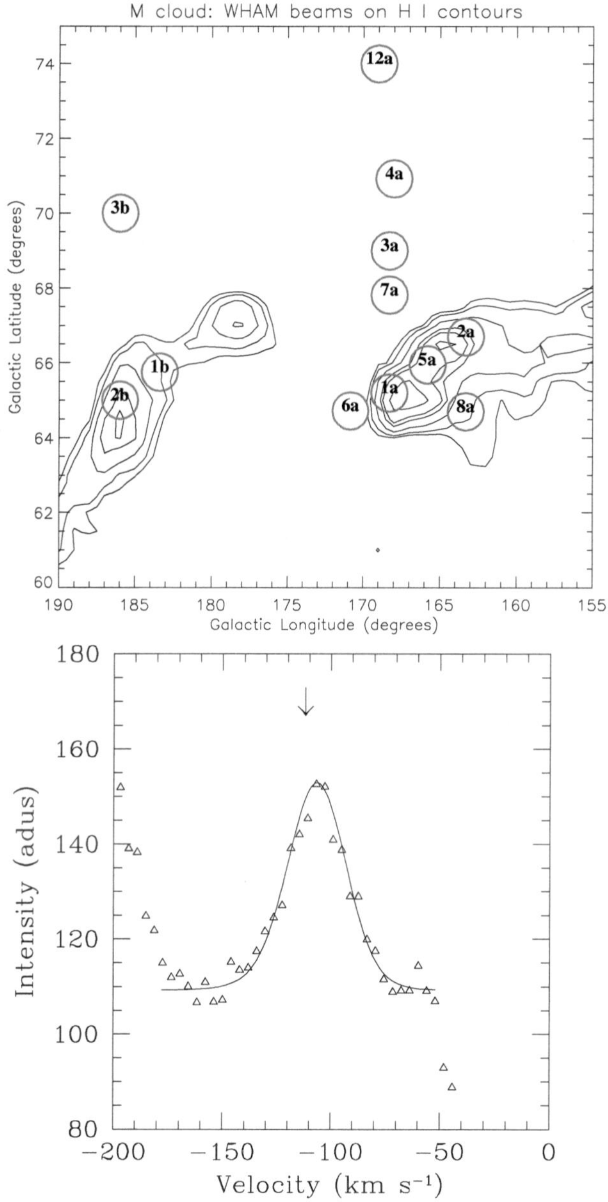

Figure 1. Hα emission from the M I cloud. *Top:* contour map of the 21-cm emission from the *LDS* (Hartmann & Burton 1997). *Bottom:* the Hα spectrum from the direction marked 1a (l=168°.3, b=+65°.2). The intensity of this emission feature is 0.08 R. The arrow shows the velocity of the 21-cm emission seen in this direction. Figure adapted from Tufte et al. (1998).

direction 6a is more than twice as bright with $I(\mathrm{H}\alpha)$=0.18±0.02 R, despite the fact that there is very little H I in this beam. This edge brightening may indicate that the ionized gas probed by the Hα envelopes the H I, but this Hα "halo", if present, is not large since beams 7a and 3a show no sign of emission. It is also interesting to note that the [S II] λ6716 line was detected for both 1a and 6a, with [S II]/Hα=0.64 and 0.11, respectively. This large difference for nearby beams indicates significant variations in ionization/excitation conditions within the M I cloud.

In this sample of Hα data, several trends were noticed that have held up more generally in subsequent surveys. The velocities of the Hα and 21-cm components match very well, leaving little doubt that the two observables are probing the same clouds. The Hα intensity, however, shows no correlation with the 21-cm column density.

The *WHAM* team has also published two additional studies of HVCs – one targeting CHVCs to test the "Local Group hypothesis", and the other in collaboration with other instruments to investigate a sightline toward complex C. These are described in more detail below.

3.3. A FLOOD OF DATA FROM DOWN UNDER

Bland-Hawthorn et al. (1998) detected the "Smith Cloud" (complex GCP) in Hα in two locations, with $I(\mathrm{H}\alpha)$=0.24 and 0.30 R. They also detected the [N II] λ6584 line with [N II]/H$\alpha\sim$0.6, somewhat elevated relative to what is found in the warm ionized medium. A measurement of the [O III] λ5007 line placed an upper limit of 0.12 R (3 σ) on the emission.

Putman (2000) and Putman et al. (2003) followed this with a major survey of emission lines from HVCs, including studies of major complexes. Complex L was found to be fairly bright. Five directions ranging in 21-cm column density from 1.6 to 3.6×10^{19} cm^{-2} showed Hα with $I(\mathrm{H}\alpha)$ ranging from 0.26 R to 0.64 R. The one non-detection was at a position beyond the H I contours and shows that this cloud has no extended ionized halo. The [N II]/Hα ratios for these clouds are very high ($\sim$2.5) but also highly uncertain owing to confusion from a bright sky line.

For complex H, Putman et al. (2003) found $I(\mathrm{H}\alpha)$=0.15 R. Since this sightline is in the galactic plane and since this cloud has a distance bound of D>5 kpc (Wakker et al. 1998), the extinction is expected to be large and in fact the extinction-corrected emission is $I(\mathrm{H}\alpha)$=3.7 R. However, the velocity of the Hα detection is -170 km s^{-1}, as compared to the 21-cm velocity of -200 km s^{-1}. This 30 km s^{-1} difference is extremely atypical and may indicate that the Hα is not from the same material, in which case the huge extinction correction may not apply and the detected 0.15 R would appear completely normal as compared to other HVCs.

The Putman team has also undertaken the important task of cross-calibrating measurements with the other observing teams by observing some of the same sightlines. Her complex M observations confirm the work of Tufte et al. (1998) by measuring nearly identical Hα intensities for two directions on the M I cloud. Actually, this is interesting since the Putman beam is so much smaller – this suggests that there is not a lot of small-scale structure in the Hα emission. Putman also observed a Magellanic Stream direction identical to one of the Weiner & Williams (1996) directions (MS IIa) and found the same result. This good agreement is not unexpected, since all three groups use nebular emission sources of known intensities for calibration, but it is reassuring nevertheless.

For the Magellanic Stream, Putman also detected several new directions in Hα, and for the direction MS IIa she found a low [N II]/Hα ratio (0.15). For the MS IIa cloud, Putman also placed an upper limit on [O III] of 0.052 R, but this was not on the same sightline as the Hα detection, so unfortunately no ratio can be derived.

Some non-detections allowed upper limits on Hα to be set. For HVCs with very high positive velocities (population EP, Wakker & van Woerden 1991), Putman found upper limits for four directions ranging from 0.039 R to 0.13 R. For material in the Leading Arm of the Magellanic Stream, she found upper limits ranging from 0.030 R to 0.078 R. For the "Sculptor clouds", she found no Hα emission, with upper limits ranging from 0.027 R to 0.074 R. This result is discrepant with a detection by Weiner et al. (2001a). The Putman survey also includes observations of a collection of compact HVCs; these are discussed below in Sect. 4.

3.4. THE WEINER, VOGEL, & WILLIAMS RESULTS

Weiner et al. (2001a) reported results of a survey of emission lines from HVCs. They find Hα detections with intensities ranging from 0.041 R to 1.68 R for 15 sightlines toward complexes ACVHV, ACHV, GCN, CR, L, and clouds in populations P and N (nomenclature from Wakker & van Woerden 1991), and also add Hα detections and non-detections toward the Magellanic Stream. As was seen in the Putman data, complex L is fairly bright; the Weiner at al. (2001a) detections have intensities ranging from 0.175 R to 1.68 R. Complexes ACHV, GCN, and ACVHV and clouds in populations N and P were all detected, but at faint levels, with Hα intensities ranging from 0.04 R to 0.15 R. For the Sculptor clouds, the Weiner group also find a non-detection for the NE lobe, but disagree with the Putman results on the SW lobe: Weiner et al. (2001a) find an intensity of 0.22 R, while the Putman group find an upper limit that is lower than this – it is unclear whether the two groups observed the same direction, because Weiner et al. (2001a) did not list the exact coordinates of the observation. It is significant that Hα was seen from all directions at levels significantly above their detection threshold of 0.01 R to 0.02 R, arguing that these clouds are unlikely to be at great distances as is asserted in the "Local Group hypothesis" (see below). They also find that the Magellanic Stream emission is more highly variable in its Hα intensity, ranging from relatively bright to not detected, which they use to argue for a different ionization mechanism for this material. This paper also presents several results for the [N II]/Hα ratio with some high values and some low.

4. Are HVCs extragalactic? The evidence from Hα

Several authors (Blitz et al. 1999; Braun & Burton 1999) have suggested that a subclass of the HVCs are dispersed throughout the Local Group of galaxies, and are the remnants of its formation. This would place them much further away (100 kpc to 1000 kpc) than other models, which place them in the Galactic Halo at 10 kpc distances. Given their angular size and column density, this would make them very massive objects.

This hypothesis has been tested by three separate studies that measure the Hα intensity for different collections of HVCs. The basic idea behind the test is that an intergalactic H I cloud will have very little ionization because the metagalactic ionizing flux is very low (Shull et al. 1999; Weymann et al. 2001; Madsen et al. 2001; Tufte et al. 2002). However, if the cloud is in the vicinity of the Milky Way, a source of ionizing photons, it will be ionized and therefore glowing in Hα (Bland-Hawthorn & Maloney 1999).

In the *WHAM* study (Tufte et al. 2002), HVCs were carefully selected as favorable candidates for being of extragalactic origin. The selection criteria included that the clouds be compact and isolated from the larger HVC complexes or galactic H I in (l,b,v) space in the *LDS*. Such clouds have been called very-high-velocity clouds (VHVCs), or compact HVCs (CHVCs).

Figure 2a shows the locations of the observations for one of these clouds, HVC WW 532, with contours showing 21-cm emission from the *LDS*. Figure 2b shows spectra from beam (A), centered on the cloud. The spectrum shows fairly bright Hα emission, with $I(\mathrm{H}\alpha)$=0.19 R. Emission was detected from 4 out of 5 of the observed CHVCs, with most of the intensities around 0.1 R, typical of the intensities measured from the large complexes thought to lie in the Galactic Halo.

Tufte et al. (2002) conclude that it is unlikely that these clouds are at 100 kpc distances from our galaxy, because the metagalactic ionizing flux level is well below that needed to produce the observed Hα surface brightness of these clouds. If they are "Local Group" objects, then either the metagalactic ionizing flux needs to be much higher than previously

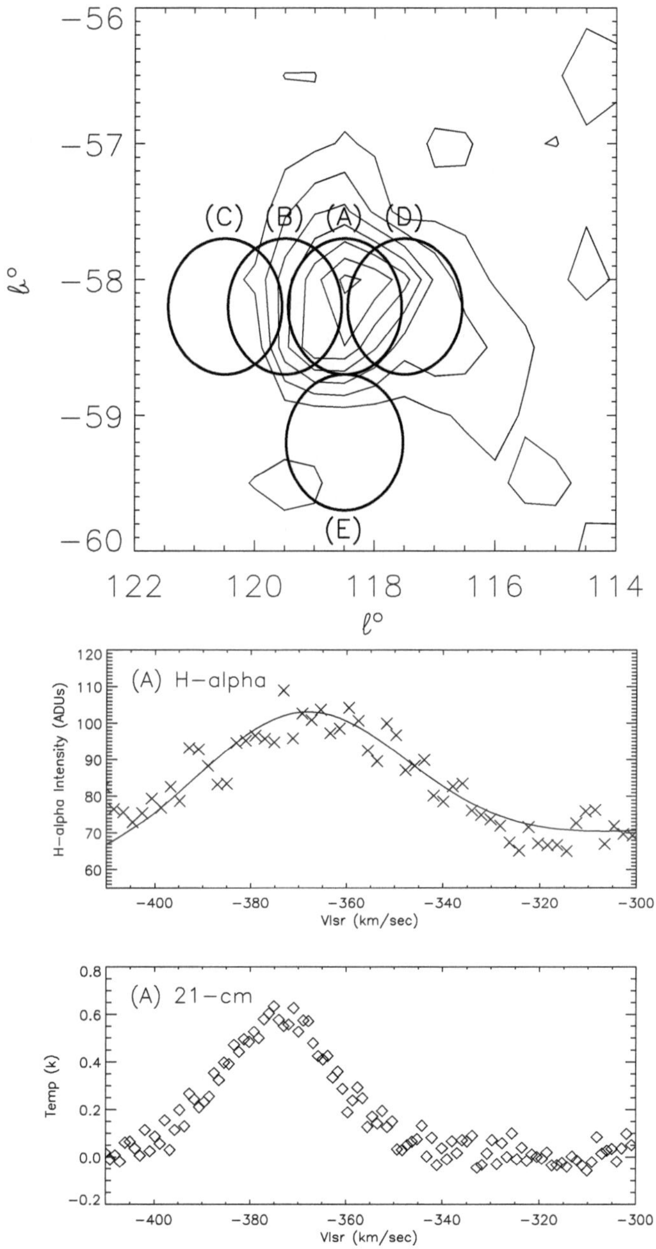

Figure 2. *Top: WHAM* observation beams for HVC WW 532 overlaid on 21-cm contours from the *LDS*. *Bottom:* Hα and 21-cm spectra for this line of sight. Figure adapted from Tufte et al. (2002).

thought (and at a level that would require an enormous number of ionizing photons, with unknown and difficult-to-explain origin), or they are ionized by some mysterious process from within. Recent observations of the outer H I disk of M 31 (Madsen et al. 2001) appear to rule out a significant external flux within the Local Group. An internal source of ionization is also difficult since no stars have ever been detected as associated with HVCs. It is more likely that these clouds are not at great distances, but are instead in the halo of the Milky Way and being ionized by it. This ionization could either be due to photoionization by UV photons leaking out of the disk, or through collisional processes as the cloud encounters ambient material in the halo at high velocity. Interestingly, there was one cloud (HVC WW 394) that had no detectable Hα emission. The upper limit at one position in this cloud, that had N(H I)= 1.6×10^{19} cm^{-2}, was I(Hα)<0.006 R, one of the lowest upper limits in existence.

An investigation by Weiner et al. (2001a) found results corroborating those of *WHAM*. All of their fifteen HVCs with well-measured spectra showed Hα, with intensities ranging from 0.04 R to 0.16 R. Further, the faintest detection was significantly brighter than their detection threshold of 0.01 R to 0.02 R. Unlike the Tufte and Putman results, these observations did not specifically target isolated compact HVCs.

In contrast to these results, Putman (2000) targeted seven compact HVCs, but did not detect Hα from any. For one direction in the galactic plane extinction prevented a meaningful upper limit. For the other six the upper limits on I(Hα) ranged from 0.037 R to 0.073 R. While it is true that two of the five *WHAM* detections are fainter than these upper limits and probably would not have been detected in the Putman study, it appears that these clouds are somehow different. The difference may be their location in the Galaxy. The Putman directions are all in the fourth quadrant, bounded by $285°<l<326°$, $-15°<b<+28°$. The Tufte et al. (2002) detections are all in the second quadrant, bounded by $100°<l<160°$, $-60°<b<-30°$, and the non-detection is in the first quadrant at l=72°, b=−22°.

Considering the observations as a whole, we conclude that many of the HVCs are ionized at levels above what would be expected if they were at extragalactic distances. The point is: extragalactic clouds emitting Hα need to be ionized somehow, and the metagalactic ionizing flux is known to be low and the gas densities in intergalactic space are too low for collisional processes to produce much Hα. However, this is not to say that the ionization of HVCs in the Galactic Halo is well understood. For example, the Magellanic Stream doesn't appear to fit neatly into the ionization models, which predict much fainter emission (Bland-Hawthorn & Maloney 1999; Bland-Hawthorn & Putman 2001; Bland-Hawthorn & Maloney 2002; Weiner et al. 2001a, 2001b).

5. Measuring metal abundances in HVCs – the ionization correction

The abundance of element X is the column density of X, N_X, divided by the *total* hydrogen column density $N_{\rm H,tot}$. However, the tradition has been to measure $N_{\rm X}/N(\rm H\,I)$ and call it the abundance, ignoring the ionized gas. In fact $N_{\rm H,tot}$=N(H I)+N(H II), and the ionized column density seems to be roughly a third of the neutral (though somewhat variable). The Hα intensity can provide important constraints on the amount of H II present and therefore improve the accuracy of the extremely important measurements of metallicity. It is interesting to note that Collins et al. (2003) found that metallicities based on [S II/H I] measured for complex C range from 0.1 to 0.6 solar but show a trend to decrease with increasing H I column density, indicating that ionization corrections may be significant in the conversion to [S/H].

As an example of how the emission lines can help unravel complicated sightlines, consider data collected toward Markarian 290, a Seyfert galaxy lying behind complex C. A team led by Bart Wakker directed an armada of instruments toward this sightline: the Goddard High Resolution Spectrograph (*GHRS*) aboard the Hubble Space Telescope (*HST*), the Wisconsin H-Alpha Mapper (*WHAM*), the *Effelsberg* 100-m Radio Telescope, the Westerbork array (*WSRT*), the *Dwingeloo* Radio Telescope (*LDS* survey data), and the William Herschel Telescope (see Wakker et al. 1999). The combined information from these telescopes allowed an unprecedented level of physical detail to be derived concerning the gas along this sightline. The results showed that for this sightline, the abundance of sulphur is A(S) = 0.089±0.024 times the solar value. Other information derived for the cloud is the hydrogen ionization fraction, $N(\rm H^+)/N_{\rm H,tot} = 0.17$, a total hydrogen density n=0.08 cm^{-3}, and a temperature T_e=7300 K (these numbers are valid for the plausible distance of 10 kpc).

Figure 3 shows the 21-cm emission from the *LDS* along with the *WHAM* measurements of Hα and [S II] λ6716 (similar to panels f, g, and h of Fig. 2 from Wakker et al. 1999). Note the close correspondence between the 21-cm data from the *LDS* survey and the *WHAM*-measured Hα spectrum. Two components with identical velocities for the Hα and the 21-cm spectra fit the data well, but the Hα components are wider than the 21-cm ones, as is typical for other HVCs (Tufte et al. 1998). The Hα intensity measured by *WHAM* is I(Hα) = 0.187±0.01 R. The [S II] λ6716 spectrum in the figure shows no signs of emission. The larger bump plotted in the [S II] spectrum corresponds to an intensity of 0.02 R, which we regard as an upper limit. The *WHAM* emission-line data are crucial to the results of the paper. Measurement of the Hα intensity provides an estimate of the amount of H$^+$, a significant correction in this case. Also, given the column

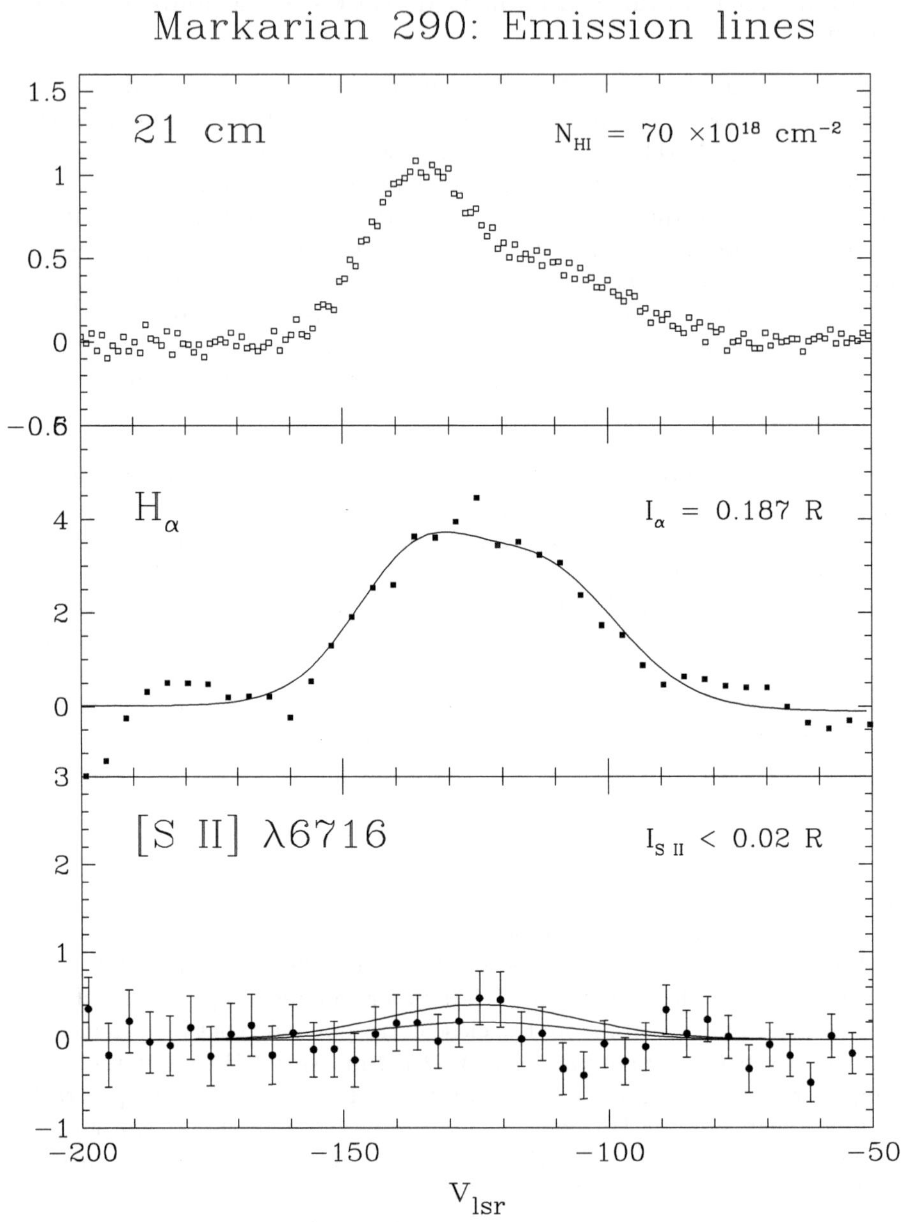

Figure 3. 21-cm emission from the *LDS* is shown along with *WHAM* measurements of Hα and [S II] for the sightline through complex C toward Mrk 290. Note the close correspondence between the ionized gas probed by the Hα and the neutral gas probed by the 21-cm emission. Figure adapted from Wakker et al. (1999).

density of S^+ from the *GHRS* data, the [S II]/Hα upper limit provides a strong constraint on the electron temperature.

The low metallicity measured for complex C toward Mrk 290 would suggest that it is falling onto the disk from without, perhaps as part of the ongoing formation of the Galaxy. If true, this would be an important result because an infall of low-metallicity gas is required by models of the chemical evolution of the Galaxy. At a plausible distance (10 kpc), the mass and velocity of complex C would correspond to a flux of low-metallicity gas that, if characteristic of past infall, would solve the long-standing "G dwarf" problem (Giovagnoli & Tosi 1995; Pagel 1997). Subsequent measurements of the metallicity of complex C on other sightlines have given conflicting values (Ch. 10). Whether this is due to real variations in the metallicity or to a failure to correct for ionization effects is yet to be determined.

6. The disk/halo connection

Evidence is accumulating that the HVCs are located high above the mid-plane of the Galaxy in the Galactic Halo. They therefore may serve as test probes of the conditions in this important, but little-understood region of the Milky Way. The Hα line gives information about the ionized hydrogen associated with the HVC, specifically: the emission measure, and the spatial and kinematic relationship to the neutral gas, already well studied through the 21-cm line. If the Hα-emitting regions are photoionized, then the Hα intensity is a direct measure of the Lyman-continuum flux incident upon the cloud. Knowledge of how this flux varies with distance above the galactic plane and with galactocentric radius can have an important bearing on more general questions regarding ionized gas in the disk and halo of the Milky Way and the leakage of ionizing radiation from the disks of spiral galaxies. If, on the other hand, the ionizations producing the Hα photons are due to collisional processes induced by the rapid motion of the HVCs (i.e. shocks), then the intensity is a measure of the ambient density through which the cloud is moving. The extent of the low-density Galactic Halo has important implication for considerations of dark matter in the Galaxy as well as for properly interpreting QSO absorption-line measurements (e.g. Sembach et al. 1995). Other potentially important ionization mechanisms that are under investigation relate to the interaction between the cloud and a hot (10^6 K) ambient medium. These models include turbulent mixing layers (Slavin et al. 1993), conductive interfaces (Breitschwerdt & Schmutzler 1994), and time-dependent non-equilibrium models of cooling, radiative (Shapiro & Moore 1976; Edgar & Chevalier 1986). These ideas seem especially relevant now that O VI has been detected from HVCs (Sembach et al. 2003). Measurements of other emission lines such as [O III]

λ5007, [S II] λ6716, and [N II] λ6584 should help to distinguish between the various ionization mechanisms.

Models of the ionization of the interstellar medium (Miller & Cox 1993; Dove & Shull 1994) can potentially connect emission lines from HVCs directly to studies of the diffuse ionized gas in the Milky Way (referred to as the warm ionized medium, or WIM). Imagine one thousand ionizing photons propagating from hot stars in the plane. Approximately 800 are captured by the local gas in the normal classical H II regions. Most of the remaining 200 that escape the immediate environment ionize the WIM layer with a scale height of 1 kpc. About 20 ionizing photons escape altogether. Exactly how many is a very important question for studies of the intergalactic medium. Of course, some hit HVCs, and in fact the brightness of the HVCs in Hα is a measure of f_{esc}, the fraction of ionizing photons originating in the plane of the Milky Way that escape the Galaxy completely. Studies of diffuse ionized gas in other galaxies have shown trends toward higher [S II]/Hα and [N II]/Hα ratios at higher distances from the midplane, with a fairly constant [S II]/[N II] ratio (Rand 1997, 1998). Haffner et al. (1999) have shown the same trends in the Milky Way and propose that they are explained by variations in the electron temperature in the WIM. A detailed model by Reynolds et al. (1999) shows that the line ratios can be explained if the temperature rises from $T_e \sim 7000$ K at $z=500$ pc to ~ 11000 K at $z=1500$ pc. Will the emission lines measured toward HVCs show the same peculiar line ratios and trends?

Several groups have made models of the ionizing radiation in the vicinity of the Milky Way (Bland-Hawthorn & Maloney 1999, 2002; Bland-Hawthorn & Putman 2001; Weiner et al. 2001a, 2001b). With such a model, one could use the Hα intensity to measure the distance to a cloud. Until a reasonable sample of clouds with independently measured Hα and distances exist, one cannot calibrate such a model without using somewhat circular arguments. Nevertheless, some interesting results are already coming out of the application of these models. For example, Weiner et al. (2001b) use the "Hα distances" to calculate $v_{\rm DEV}$, the amount by which the cloud velocity deviates from the velocity expected given its location in the differentially rotating Galaxy. They find that the HVCs with smaller $v_{\rm DEV}$ have higher Hα intensity and also higher [N II]/Hα ratios, suggesting perhaps that these clouds are an extension of galactic hydrogen into the halo, analogous to the extraplanar ionized gas seen by Rand in NGC 891 and the WIM gas above the Perseus Arm of the Milky Way studied by Haffner et al. (1999). If this trend holds as more [N II] and [S II] data are gathered, emission-line data could become a clear way to distinguish between HVCs of different origin. In a similar vein, Putman found that clouds toward the general direction of the Galactic Center tend to be brighter in Hα. This is consistent with what one would expect from the ionizing-flux models where much of the flux originates inside the solar circle. Also, interestingly, one can tune f_{esc} in these models, and the ones with the best overall match have $f_{esc}{\sim}1\%$ to 2% when averaged over solid angle (Bland-Hawthorn & Maloney 2002).

7. Conclusions and future directions

Speaking from the neutral and objective perspective of an observer, I dispassionately conclude that "We need more observations!". I expect more observations of multiple emission lines (e.g. Hα, [S II], [N II], [O III]) on the same sightlines will yield progress. This may become a key method to distinguishing halo material originating in the Galaxy from material falling in from without, since high-z halo gas tends to have high [N II]/Hα and [S II]/Hα ratios, whereas low-metallicity clouds will have low ratios due to a simple lack of emitters. Also, the pattern of emission-line ratios will be key for exploring the ionization mechanisms; and if photoionization is dominant, as seems to be the case, then we will gain information on the spectrum of ionizing photons. From a more complete observational sampling, we may also look for differences in abundance and ionization state between complexes and across complexes. Also, further observations of CHVCs in multiple emission lines will determine if this class forms a separate population with respect to the pattern of emission lines.

We must expand the sample of Hα data for the purpose of fleshing out the ionization models; especially as more clouds with distance information emerge will this be crucial. This will allow measurement of the flux and spectrum of the UV radiation escaping from the Galaxy, and physical conditions in the halo to be explored.

The author also expects much to come from combining emission-line data with the copious high-quality absorption-line data from *FUSE*, *HST*, and other space telescopes. These two relatively new windows onto HVCs can be simultaneously open, providing perspective from multiple views of the same objects.

Acknowledgements

I gratefully acknowledge support from the National Science Foundation through grants AST 02-04973 and AST 02-06349. This work was also supported by Research Corporation through a Cottrell College Science Award.

References

Bland-Hawthorn, J., Maloney, P.R. 1999, ApJ, 510, L33.
Bland-Hawthorn, J., Maloney, P.R. 2002, in ASP Conf. Ser. 254, Extragalactic Gas at Low Redshift, eds. J.S. Mulchaey, J.T. Stocke (San Francisco: ASP), 267
Bland-Hawthorn, J., Putman, M.E. 2001, in ASP Conf. Ser. 240, Gas & Galaxy evolution, eds. J.E. Hibbard, M. Rupen, J.H. van Gorkom (San Francisco: ASP), 369
Bland-Hawthorn, J., Veilleux, S., Cecil, G.N., Putman, M.E., Gibson, B.K., Maloney, P.R. 1998, MNRAS, 299, 611
Blitz, L., Spergel, D.N., Teuben, P.J., Hartmann, D., Burton, W.B. 1999, ApJ, 514, 818
Braun, R., Burton, W.B. 1999, A&A, 341, 437
Breitschwerdt, D., Schmutzler, T. 1994, Nature, 371, 774
Collins, J.A., Shull, J.M., Giroux, M.L. 2003, ApJ, 585, 336
Dove, J.B., Shull, J.M. 1994, ApJ, 428, 647
Edgar, R.J., Chevalier, R.A. 1986, ApJ, 310, L27
Giovagnoli, A., Tosi, M. 1995, MNRAS, 273, 499.
Haffner, L.M., Reynolds, R.J., Tufte, S.L. 1999, ApJ, 523, 223
Hartmann, D., Burton, W.B. 1997, Atlas of Galactic Neutral Hydrogen (Cambridge: Cambridge University Press)
Kutyrev, A.S., Reynolds, R.J. 1989, ApJ, 334, L9
Madsen, G.J., Reynolds, R.J., Haffner, L.M., Tufte, S.L., Maloney, P.R. 2001, ApJ, 560, L135
Miller III, W.W., Cox, D.P. 1993, ApJ, 417, 579
Münch, G., Pitz, E. 1989, in IAU Coll. 120 (Lecture Notes in Physics, 350), Structure and Dynamics of the Interstellar Medium (Berlin: Springer), 373
Pagel, B.E.J. 1997, Nucleosynthesis and Chemical Evolution of Galaxies (Cambridge: Cambridge University Press)
Putman, M.E. 2000, Ph.D. thesis, The Australian National University
Putman, M.E., Bland-Hawthorn, J., Veilleux, S., Gibson, B.K., Freeman, K.C., Maloney, P.R. 2003, ApJ, 597, 948
Rand, R.J. 1997, ApJ, 474, 129
Rand, R.J. 1998, ApJ, 501, 137
Reynolds, R.J. 1987, ApJ, 323, 553
Reynolds, R.J., Tufte, S.L., Haffner, L.M., Jaehnig, K., Percival, J.W. 1998, PASAu, 15, 14
Reynolds, R.J., Haffner, L.M., Tufte, S.L. 1999, ApJ, 525, L21
Sembach, K.R., Savage, B.D., Lu, L., Murphy, E.M. 1995, ApJ, 451, 616
Sembach, K.R., Wakker, B.P., Savage, B.D., Richter, P., Meade, M., Shull, J.M., Jenkins, E.B., Sonneborn, G., Moos, H.W. 2003, ApJS, 146, 165
Shapiro, P.R., Moore, R.T. 1976, ApJ, 207, 460
Shull, J.M., Roberts, D., Giroux, M.L., Penton, S.V., Fardal, M.A. 1999, AJ, 118, 1450
Slavin, J.D., Shull, J.M., Begelman, M.C. 1993, ApJ, 407, 83
Songaila, A., Bryant, W., Cowie, L.L. 1989, ApJ, 345, L71
Tufte, S.L., Reynolds, R.J., Haffner, L.M. 1998, ApJ, 504, 773
Tufte, S.L., Wilson, J.D., Madsen, G.J., Haffner, L.M., Reynolds, R.J. 2002, ApJ, 572, L153
Wakker, B.P., van Woerden, H. 1991, A&A, 250, 509
Wakker, B.P., van Woerden, H., de Boer, K.S., Kalberla, P.M.W. 1998, ApJ, 493, 762
Wakker, B.P., Howk, C., Savage, B.D., van Woerden, H., Tufte, S.L., Schwarz, U.J., Benjamin, R., Reynolds, R.J., Peletier, R.F., Kalberla, P.M.W. 1999, Nature, 402, 388
Weiner, B.J., Williams, T.B. 1996, AJ, 111, 1156
Weiner, B.J., Vogel, S.N., Williams, T.B. 2001a, in ASP Conf. Ser., 240, Gas & Galaxy Evolution, eds. J.E. Hibbard, M. Rupen, J.H. van Gorkom (San Francisco: ASP), 515
Weiner, B.J., Vogel, S.N., Williams, T.B. 2001b, in ASP Conf. Ser., 254, Extragalactic Gas at Low Redshift, eds. J.S. Mulchaey, J.T. Stocke (San Francisco: ASP), 256
Weymann, R.J., Vogel, S.N., Veilleux, S., Epps, H.W. 2001, ApJ, 561, 559

9. THE COLDEST PHASE IN HALO HIGH-VELOCITY GAS: DUST AND MOLECULES

PHILIPP RICHTER
Institut für Astrophysik und Extraterrestrische Physik
Universität Bonn, Germany
prichter@astro.uni-bonn.de

AND

KLAAS S. DE BOER
Sternwarte, Universität Bonn, Germany
deboer@astro.uni-bonn.de

Abstract. In this chapter, we discuss the dust content and the abundance and distribution of molecular gas in the Galactic Halo. During the last two decades, significant progress has been made in observing dust and molecules in halo clouds using various instruments and observing techniques. Infrared dust emission and ultraviolet absorption by molecular hydrogen, H_2, now have been detected in a number of neutral halo clouds, but only in regions where the overall metal abundance is close to that of the Milky Way or the Magellanic Clouds. The column densities of the molecular gas in the halo generally are very low, but the presence of H_2 both indicates the presence of dust and implies the existence of small-scale structure at sub-parsec/AU scales. Moreover, the H_2 measurements provide important density and temperature estimates for the absorbing structures and thus reveal important new aspects of the physical conditions in the Galactic Halo clouds.

1. Measurements of dust and molecules in the Milky Way Halo

In spite of extensive studies to explore the neutral and ionized gas content of intermediate- and high-velocity clouds in the Galactic Halo, relatively little is known about the abundance of dust and molecules in such gas. So far, emission from dust and molecules has been detected only in intermediate-velocity gas in the lower Galactic Halo close to the disk, but not in high-velocity clouds. The search for emission, however, is not a particularly sensitive method to study dust and molecules at relatively low H I

H. van Woerden et al. (eds.), High-Velocity Clouds, 183–194

column densities in gas which might be deficient in metals.

Significant progress has been made over the past years to measure molecular hydrogen and depletion of heavy elements onto dust in Galactic Halo clouds by way of absorption spectroscopy in the far ultraviolet (FUV). New instruments such as the Far Ultraviolet Spectroscopic Explorer (*FUSE*), the Orbiting and Retrievable Far and Extreme Ultraviolet Spectrometers (*ORFEUS*), and the Hubble Space Telescope (*HST*) give access to the wavelength region <1700 Å, in which the electronic transitions of many atoms and molecules allow the study of the diffuse interstellar medium at column densities as low as 10^{13} cm^{-2}. Therefore, FUV absorption spectroscopy is the most sensitive method to study the molecular content (and indirectly the dust content) in Galactic Halo clouds. FUV absorption by molecular hydrogen (by far the most abundant molecule in the Universe) has been detected in both intermediate-velocity clouds (IVCs) and high-velocity clouds (HVCs) at relatively low column densities. Although this molecular material does not contribute significantly to the mass of Galactic Halo gas, its presence provides important information about temperatures, densities and small-scale structure.

In order to understand the physical conditions in HVCs it is important to put measurements on high-velocity gas in the context of the Milky Way Halo as a whole, including disk-halo interaction processes and IVCs. This is particularly important for understanding the molecular gas phase in HVCs, since the number of measurements on dust and molecules in the Galactic Halo is still very limited. In this article, we therefore discuss the content of dust and molecules in both HVCs *and* IVCs, with the emphasis on the high-velocity gas. We regard HVCs to have radial velocities $|v_{\rm LSR}|>90$ km s^{-1} and IVCs 20 km s$^{-1}<|v_{\rm LSR}|<90$ km s^{-1}.

2. Presence of dust in halo gas

The question whether or not dust is present in HVCs is of special interest for two reasons. First, the presence of dust would require that heavy elements are also present, providing important information about the origin of the IVC and HVC gas. Second, through the well-established correlation between dust and molecules, the presence of dust gives information about the conditions in the cool interior of these clouds. We shall see, however, that the relation between dust and molecules is more useful the other way around, concluding from the presence of molecules about the presence of dust.

2.1. SEARCHES FOR HALO DUST EMISSION

Searches for thermal dust emission in Galactic Halo clouds were performed at 100μm using data from the Infra Red Astronomical Satellite (*IRAS*). A catalog of high-latitude infrared-excess clouds was presented by Désert et al. (1988), suggesting a connection between the infrared excess and intermediate-velocity H I gas for some cases. Dust emission from IVCs measured by *IRAS* was also reported in studies of individual complexes (e.g. Désert et al. 1990; Weiss et al. 1999). In contrast, several HVC complexes have been inspected for dust emission (Wakker & Boulanger 1986; Bates et al. 1988; Boulanger et al. 1996; Fong et al. 1987), but without success. One limitation of these searches lies in the combination of insufficient sensitivity of *IRAS* and the small amounts of dust expected. Another limitation concerns the larger distance of HVCs compared to that of IVCs: if dusty HVCs are relatively far from the Milky Way the absorbed energy from the Galactic Disk radiation is reduced while the (total) HVC emission is diluted as D^{-2} owing to distance.

2.2. LITTLE HYDROGEN AND THUS LITTLE DUST

Measurements of H I 21-cm emission have resulted in relatively small H I column densities for both IVCs and HVCs. The approximate maximum column density found is $N(\mathrm{H\,I}){<}3{\times}10^{20}$ cm^{-2} (see Wakker 2001). The column densities detected for molecular hydrogen (see Table 1) are $N(\mathrm{H_2}){<}10^{17}$ cm^{-2}. If the standard correlation of $N(\mathrm{H})$ with $E(B-V)$ were to hold (Bohlin et al. 1978), one would have $E(B-V){<}0.05$ mag for the Milky Way halo. For metal- and dust-deficient clouds, the expected range for $E(B-V)$ would be even lower. It therefore is extremely difficult to obtain a complete picture about the abundance and distribution of dust in such low column density gas based on direct observations.

In addition, H I 21-cm measurements that may be used to trace the dust in the halo are beam-smeared. Even if neutral and molecular material exists in relatively small, dense clumps with higher column densities, one would be unable to detect these clumps at beam sizes available for emission-line studies. The chance to detect them by absorption-line studies is strongly limited by the number of available background sources.

2.3. EXTRAPLANAR EXTINCTION IN OTHER GALAXIES

Several other galaxies seem to have dust in their halos. The best example is NGC 891 which shows very well-defined extraplanar ($|z|{>}0.4$ kpc) dust filaments in optical images (Howk & Savage 1997). These filaments are seen as far out as several kiloparsecs above the galaxy's plane, thus showing the presence of dust in regions similar to those of intermediate- and high-velocity gas in the Milky Way Halo. In a sample of 12 edge-on spiral galaxies Howk & Savage (1999) find extraplanar dust absorption in 5 systems. These structures may be typical for spiral galaxies, representing the dense phase of the multiphase medium at high z. Assuming a galactic gas-to-dust ratio, the estimated H I column densities for these features typically are several times higher than what is found for the neutral halo clouds in the Milky Way. If the extraplanar dust features in these galaxies are caused by fountain-type flows (see Sect. 2.5), the excess of extraplanar extinction in these systems may indicate an enhanced star formation activity in the underlying disk.

The topic of dusty halos in other galaxies was reviewed by Dettmar (1992, 1998), the physics of dusty halos by Ferrara (1998).

2.4. MEASUREMENTS OF DEPLETION PATTERNS

The indirect way leading to the suspicion that dust is present in the Milky Way Halo is the measurement of the metal abundances in HVCs. The methods are described in detail by Wakker & van Woerden (1997); see also Ch. 10.

The best element for determining the intrinsic metal abundance is oxygen, because it is not significantly depleted onto dust grains. Moreover, neutral oxygen and neutral hydrogen have similar ionization potentials and are strongly coupled by charge exchange reactions (Sofia & Jenkins 1998). Therefore, the O I/H I ratio is not altered by ionization effects. The FUV spectral range contains several oxygen lines with different oscillator strengths; with the new FUV instruments it is therefore possible to obtain accurate abundances of the halo clouds by combining FUV absorption-line data with H I 21-cm emission-line measurements.

Recent absorption line measurements show that IVCs typically have nearly solar oxygen abundances, while oxygen abundances in HVCs vary between ~0.1 and ~1.0 times solar (e.g. Lu et al. 1998; Bluhm et al. 2001; Wakker et al. 1999; Richter et al. 2001a; 2001b).

In contrast to oxygen, other elements such as iron and silicon tend to be incorporated into dust grains, an effect which lowers their abundance in the gas phase. Low silicon and iron abundances (in comparison to oxygen) in the local ISM are therefore interpreted as indicating the presence of dust. Depletion of iron and silicon is also seen in Galactic Halo clouds (see Savage & Sembach (1996) for a detailed review) and thus suggests the presence of dust grains beyond the disk of the Galaxy. The interpretation of the depletion pattern of Galactic IVCs and HVCs, however, is more difficult than in the local ISM, because of ionization effects and the unknown intrinsic abundances of depleted and undepleted species relative to each other. The ionization issue alone may be handled by measuring S II/H I and P II/H I ratios (e.g. Richter et al. 2001a), but finding precise gas-phase abundances of depleted elements is particularly difficult because of the unknown enrichment history (and origin) of the gas *in combination* with the ionization effects in the halo. Therefore, the relation between the presence of dust and the depletion of heavy elements remains unknown for most of the halo clouds, in particular for HVCs. An example for measuring the abundance pattern of Galactic IVC and HVC gas along one sight line is presented by Richter et al. (2001b).

2.5. ORIGIN AND NATURE OF THE DUST IN THE HALO

One major mechanism to put dust grains into the Galactic Halo is the injection by supernova explosions and/or galactic chimneys from the Disk of the Milky Way. This process would be part of the Galactic Fountain (Shapiro & Field 1976; Houck & Bregman 1990). If dust is circulating together with gas from the Galactic Disk into the halo, what is the evolution of the dust grains in a hot environment? While most of the dust grains in the ISM might be destroyed in the shock after a supernova explosion (Draine & McKee 1993), a fraction of the surviving dust grains, freed from their icy mantles, will be swept into the halo. Models for the destruction of dust grains predict that Fe grains are more easily destroyed than silicate grains (Jones et al. 1994), thus making the survival of silicate grains pumped into the halo more likely. However, Sembach & Savage (1996) concluded from absorption-line measurements that approximately 70% of the Mg, 45% of the Si, and 77% of the Fe are locked into dust grains, thus favoring the idea that the cores of dust grains are resilient and difficult to destroy completely during their ejection from the Milky Way Disk. Dust grains that have not been destroyed can be transported into the lower Milky Way Halo by way of radiation pressure in regions of intense star formation. Franco et al. (1991) have shown that this "photolevitation" mechanism may push diffuse, dusty clouds with total column densities around 10^{20} cm^{-2} above the disk.

Another (but less likely) possibility is that metal-enriched clouds fall into the Milky Way Halo from afar. If they contain dust, one would have to explain 1) where the metals came from and 2) how the dust would survive photo-destruction from the FUV background radiation.

Recently, Evans et al. (2003) detected thermal emission by dust in the core of the globular cluster NGC 7078, through observations with the Infrared Space Observatory (*ISO*). The estimated dust abundance is roughly consistent with the expected mass loss from evolved globular cluster stars. Most of this intra-cluster dust in globular clusters is likely to be stripped off when the clusters pass through the plane of the Milky Way, but some fraction of the dust may remain in the halo. As pointed out by de Boer (2004), the mass lost by red giant field stars also contributes to the gas and dust put into the halo. An estimate of the total amount of gas, based on the spatial distribution of red giant stars in the Milky Way Halo, suggests that these stars inject a significant amount of relatively metal-deficient material into the surrounding halo gas.

3. Measurements of molecules in halo gas

Since the discovery of intermediate- and high-velocity clouds one has speculated about their molecular content. The best candidate molecules are CO, observable in emission with millimeter observations from the ground, and the much more abundant H_2, observable with far-UV absorption spectroscopy from satellites.

3.1. CO AND HCO^+ OBSERVATIONS

CO emission in IVCs has been found for a handful of clouds at low z-heights <0.5 kpc (e.g. Weiss et al. 1999; Désert et al. 1990). These clouds are believed to sample galactic molecular gas in the disk-halo interface, outside the scale height of normal molecular clouds in the Milky Way Disk.

In contrast, no CO emission has been reported from HVCs so far (Hulsbosch 1978; Giovanelli 1986; Kim et al. 1989; Wakker et al. 1997), and the search for CO absorption (Akeson & Blitz 1999) also was not successful. A sensitive search for HCO^+ mm-line absorption against quasars resulted in one tentative detection (Combes & Charmandaris 2000), but this particular case needs to be confirmed.

The column density ratio of CO and H_2 is $N(CO)/N(H_2)<10^{-4}$ in Galactic Disk gas (Federman et al. 1980). The ratio is even lower for HCO^+ to H_2. Since CO emission-line measurements are not particularly sensitive, the non-detection of CO emission in HVCs gives little information about their molecular content. The unknown metallicity and the unknown dust amount in most of the HVCs underline that CO emission is not the best tracer for the general molecular content in HVCs. Moreover, because CO is easily dissociated, CO (and other molecules) need not be present in regions containing H_2.

3.2. H_2 DETECTED IN ABSORPTION

Molecular hydrogen is the most abundant molecule in space, but at low gas densities it is only measurable by FUV absorption spectroscopy, requiring satellites in space. FUV absorption-line measurements allow investigations of H_2 at column densities, $N(H_2)$, as low as 10^{14} cm^{-2}, thus being 5 orders

TABLE 1. Detections of H_2 absorption in Galactic IVCs and HVCs

Target	l (deg)	b (deg)	Cloud Name	$v_{\rm LSR}$ [km s^{-1}]	log $N(H_2)$	log f^a	Ref.^{c}
			Intermediate-velocity clouds				
Mrk 509	36.0	−29.9	complex gp	+60	14.9±0.5	−4.3	1
Mrk 876	98.3	+40.4	Draco Cloud	−30	15.6±0.2	−4.0	1
Mrk 59	111.5	+82.1	IV Arch	−44	14.7±0.2	−4.3	1
PG 1351+640	111.9	+52.0	IV Arch (IV 16)	−47	16.4±0.1	−3.3	1
HD 121800	113.0	+49.8	IV Arch	−70	14.3±0.6	−5.3	1
PG 1259+593	120.6	+58.1	IV Arch	−54	14.1±0.2	−5.1	2
PG 0804+761	138.3	+31.0	LLIV Arch	−55	14.7±0.3	−4.5	3
PG 0832+675	147.8	+35.1	LLIV Arch	−50	15.8±0.3	−3.9	1
NGC 4151	155.1	+75.1	IV Arch (IV 26)	−29	15.4±0.1	−4.5	1
NGC 3310	156.6	+54.1	IV Arch	−47	15.0±0.8	−4.5	1
HD 93521	183.1	+62.2	IV Arch	−62	14.6±0.4	−4.7	4
PG 1116+215	223.4	+68.2	IV Spur	−42	15.3±0.3	−4.3	1
HD 100340	258.9	+61.2	IV Spur	−29	16.0±0.8	−3.7	1
Sk −68 82	279.3	−32.8	IVC toward LMC	+55	present^b	...^{b}	5,6,7
Sk −60 80	279.3	−32.8	IVC toward LMC	+50	14.6±0.5	−1.2	6
3C 273	290.0	+64.4	...	+25	15.7±0.2	−3.4	1
			High-velocity clouds				
Sk −68 82	279.3	−32.8	HVC toward LMC	+120	present^b	...^{b}	5,6,7
NGC 3783	287.5	+23.0	Lead. Arm of the MS	+240	16.8±0.1	−2.9	8
Fairall 9	295.1	−57.8	Magellanic Stream	+190	$16.4^{+0.3}_{-0.5}$	−3.3	2

(a) $f=2N(H_2)/[N(\text{H I})+2N(H_2)]$. (b) H_2 is detected, but the H_2 column density is highly uncertain. (c) References: 1) Richter et al. 2003a; 2) Richter et al. 2001c; 3) Richter et al. 2001a; 4) Gringel et al. 2000; 5) Bluhm et al. 2001; 6) Richter et al. 2003b; 7) Richter et al. 1999; 8) Sembach et al. 2001.

of magnitude more sensitive than CO emission-line studies.

Several positive detections of H_2 absorption with *ORFEUS* and *FUSE* have been reported for both IVCs and HVCs (see Table 1). The first detection of H_2 in a Galactic HVC was found with *ORFEUS* by Richter et al. (1999) in the high-velocity gas in front of the Large Magellanic Cloud. With the same instrument, Gringel et al. (2000) observed H_2 absorption for the first time at intermediate velocities, related to gas in the IV Arch. So far, H_2 is seen at intermediate velocities in the IV Arch, LLIV Arch, IV Spur, complex gp, in the Draco Cloud, and in the IVC in front of the LMC. At high velocities, H_2 absorption is found in the HVC toward the LMC, in the Leading Arm of the Magellanic Stream, and in the Magellanic Stream itself. In all cases, the detected column densities are $<10^{17}$ cm^{-2}, far too weak to expect any CO emission. H_2 absorption is not seen in all sightlines through the clouds mentioned above, suggesting that the molecular gas resides in relatively confined regions. However, many of the H_2 upper limits cited in the literature are meaningless because of poor signal-to-noise ratios in the spectra. The many detections in IVCs show that H_2 is widespread in halo gas, albeit at low column density, but it is apparently not particularly abundant.

In order to explain the observations described above it is of eminent importance to consider the conditions in IVCs and HVCs for the formation and dissociation of H_2.

3.3. FORMATION AND DISSOCIATION OF MOLECULES IN THE HALO

Gas densities in high-velocity clouds are, in principle, high enough to allow the formation of molecular hydrogen (e.g. Wakker et al. 1997). H_2 forms relatively easily on dust grains: H atoms freeze out, migrate over the dust surface, meet each other, and combine to form H_2. The excess energy is partly used for heating the dust and partly for exciting the H_2 molecule ("formation pumping"; e.g. Takahashi et al. 1999). The formation rate is a function of gas density and migration rate, and in dilute gases it will be the gas density which governs the formation (see e.g. Pirronello et al. 1999). In low-metallicity HVCs it is expected that the formation rate on grains of H_2 is strongly limited by the lower dust content. This effect is probably also seen in the dust-deficient Magellanic Clouds (Richter 2000; Tumlinson et al. 2002).

Another, less efficient way to form H_2 is by associative detachment of H^- through the reaction $H^- + H \rightarrow H_2 + e^-$ (Black 1977). This process is important in regions with high fractional ionization and thus might be more relevant in the Galactic Halo than in the local ISM.

H_2 can be dissociated by UV radiation (the dissociation energy is 4.5 eV). This process involves photon excitation to levels of $\sim$12 eV and decay to high vibrational states followed by fractional (10% to 15%) dissociation. In the halo, H_2 is exposed to the local radiation field. The strength and shape of that field has been estimated by Bregman & Harrington (1986), showing that most of the UV radiation comes from hot stars in the Milky Way disk. A recent (similar) estimation by Bland-Hawthorn & Maloney (1999) aimed at finding the radiation field near the clouds of the Magellanic Stream. The UV radiation can then be absorbed by H_2 through the electronic transitions near 1100 and 1000 Å, leading to an (observable) high excitation state of H_2. Details of the effects of the so-called UV-pumping are discussed by Spitzer & Zweibel (1974). One might argue that the H_2 photo-absorption rate in halo clouds is reduced by H_2 line shielding from molecular gas in the disk. However, because of the doppler shift of the absorption lines in IVCs and HVCs, this effect probably can be neglected (Bluhm et al. 2001). A general study of the physical conditions in gas in the halo by Wolfire et al. (1995) suggests that cold and dense neutral gas clouds may exist embedded in a local warm neutral medium in an otherwise hot halo of pressure 10^3 to $10^{3.5}$ K cm^{-3}. Densities in neutral gas of order 10 to 300 cm^{-3} appear to be possible at z distances between 1 and 5 kpc (see also Wakker & Schwarz 1991), thus allowing the formation of hydrogen molecules in metal- and dust-rich IVCs and HVCs.

3.4. H_2 DIAGNOSTICS

Additional parameters of the H_2 gas, such as the molecular fractional abundance and the H_2 rotational excitation, are important diagnostic tools for the study of physical properties in the ISM. H_2 is usually assumed to be in a formation-dissociation equilibrium, involving the processes described in the previous section. With n as density, $\langle k \rangle = 0.10 - 0.15$ as the probability that the molecule is dissociated after photon absorption, β_0 as the photo-absorption rate per second, and R as the probability rate per neutral H atom to form H_2 molecules by collision with dust grains in units cm^3 s^{-1}, the formation-dissociation equilibrium (see Spitzer & Zweibel (1974) for details) is:

$$\frac{n(\mathrm{H\,I})}{n(\mathrm{H_2})} = \frac{\langle k \rangle \beta_0}{R\, n_{\mathrm{H}}}. \tag{1}$$

This equation allows to link the observed H I and H_2 column densities with physical parameters, such as the UV radiation field, the gas density, and the formation rate of H_2 on grains, where the observed $N(\mathrm{H\,I})/N(\mathrm{H_2})$ ratio has to be transformed into $n(\mathrm{H\,I})/n(\mathrm{H_2})$ assuming a certain depth structure of the cloud (see Richter et al. 2003a).

The formation rate of H_2 on grains, R_{halo}, is uncertain for IVCs and HVCs, because the unknown surface properties of the dust grains in the halo. It may be estimated by using the equation above if $n(\mathrm{H\,I})$, $n(\mathrm{H_2})$ and β_0 are known, but the number of accurate measurements is yet too small to derive a reliable formation rate of H_2 on grains for IVCs and HVCs.

In IVCs, Richter et al. (2003a, 2003b) have studied the H_2 content with *FUSE* along a large number of sight lines (see also Table 1). The *FUSE* data suggest that the chance to look through an H_2 bearing IVC cloud may be as high as 50 percent, but the molecular hydrogen fractions are low, typically $f<10^{-3}$ (Richter et al. 2003a). Using Eq. (1) together with the observed values for $N(\mathrm{H_2})$ and $N(\mathrm{H\,I})$ allows to roughly estimate typical volume densities, n_{H}, and linear diameters, L, of the regions in which the H_2 resides. The measurements imply characteristic values of $n_{\mathrm{H}}{\sim}30$ cm^{-3} and $L{\sim}0.1$ pc in IVCs. Therefore, these structures probably are related to the cold neutral medium (CNM) in these clouds. A very small ($L{\sim}40$ AU) and dense (n_H~800 cm^{-3}) IVC H_2 filament has been observed toward the LMC star Sk-68 80 (see Table 1), suggesting the detection of tiny-scale atomic structure, (Heiles 1997) in this IVC.

Gas temperatures can also be derived rather directly from detected H_2 through the distribution of the H_2 over the various excited rotational states. Once the H_2 column density exceeds a critical limit for line self-shielding (~10^{14-16} cm^{-2}), most of the H_2 in the interior of the cloud resides in the rotational ground states. The ground-state levels (J=0,1) of H_2 then are normally collisionally excited and the level population is set by the kinetic temperature (T_{01}). In contrast, the higher levels of H_2 are excited through UV radiation pumping (indicated by an equivalent excitation temperature, T_{ex}; see Spitzer & Zweibel 1974) and/or formation pumping. At column densities close to or below the self-shielding limit (as mostly found in IVCs and HVCs), the rotational states ≤ 3 can also be excited by UV pumping and thus do not necessarily reflect the kinetic temperature of the gas (e.g. Morton & Dinerstein 1976). Therefore, the interpretation of the H_2 excitation found in Galactic Halo gas is rather delicate. For the Leading Arm of the Magellanic Stream, Sembach et al. (2001) report $T_{01} = 133^{+37}_{-21}$ K at log $N(H_2)$=16.80±0.10, possibly reflecting the kinetic temperature of the gas. This temperature is slightly higher than the mean value found in the Galactic Disk (77 K; Savage et al. 1977). In all other cases, the values for T_{01} are higher, but also more uncertain because of low S/N and significantly lower column densities, and it remains unclear which process causes the H_2 rotational excitation in IVCs and HVCs. As an additional alternative, some of the excitation might be produced by shocks running through the gas, e.g. owing to cloud-cloud collisions.

4. Molecular "globules" in the outer halo?

Dense molecular clumps of AU sizes and Jovian masses have been suggested as reservoirs of cold (molecular) baryonic dark matter in the Galactic Halo (Pfenniger & Combes 1994; Kalberla et al. 1999). These molecular clumps would manifest themselves by Extreme Scattering Events (ESEs), radio flux changes of quasars occurring over several weeks (Fiedler et al. 1987; Walker & Wardle 1998). It has been speculated that collisions between such clumps might lead to halo clouds partly being molecular at much lower densities but on much larger spatial scales (Kalberla et al. 2000). Such gas thus might be connected to common metal-poor high-velocity gas in the Milky Way Halo. Observational proof of the existence of these clumps is still lacking and the implications for intermediate- and high-velocity molecular gas in the halo of the Milky Way are unclear.

5. Concluding remarks

Future absorption-line measurements in HVCs and IVCs with *FUSE* will improve our understanding of the origin and state of dust and molecules in galactic intermediate- and high-velocity gas. So far, H_2 remains still the only molecular species convincingly detected in high-velocity clouds, seen only at relatively low column densities. There is no observational evidence that molecular gas contributes significantly to the mass of IVCs and HVCs, but it cannot be excluded that more molecular material is hiding in very small, dense clumps, almost undetectable in either absorption or emission.

Acknowledgements

We like to thank Bart P. Wakker for helpful comments.

References

Akeson, R.L., Blitz, L. 1999, ApJ, 523, 163
Bates, B., Catney, M.G., Keenan, F.P. 1988, ApSpSci, 146, 195
Black, J. 1977, ApJ, 222, 125
Bland-Hawthorn, J., Maloney, P.R. 1999, ApJ, 510, L33
Bluhm, H., de Boer, K.S., Marggraf, O., Richter, P. 2001, A&A, 367, 299
Bohlin, R.C., Savage, B.D., Drake, J.F. 1978, ApJ, 224, 132
Boulanger, F., Abergel, A., Bernard, J.P., Burton, W.B., Désert, F.X, Hartmann, D., Lagache, G., Puget J.-L. 1996, A&A, 312, 256
Bregman, J.N., Harrington, J.P. 1986, ApJ, 309, 833
Combes, F., Charmandaris, V. 2000, A&A, 357, 75
de Boer, K.S. 2004, A&A, 419, 527
Désert, F.-X., Bazell, D., Boulanger, F. 1988, ApJ, 334, 815
Désert, F.-X., Bazell, D., Blitz, L. 1990, ApJ, 355, L51
Dettmar, R.-J. 1992, Fund. Cosm. Phys., 15, 14
Dettmar, R.-J. 1998, in IAU Coll. 166, (Lecture Notes in Physics, 506), The Local Bubble and Beyond, eds. D. Breitschwerdt, M.J. Freyberg, J. Trümper (Berlin: Springer), 527
Draine, B.T., McKee, C.F. 1993, ARA&A, 31, 373
Evans, A., Stickel, M., van Loon, J.Th., Eyres, S.P.S., Hopwood, M.E.L., Penny, A.J. 2003, A&A, 408, L9
Federman, S.R., Glassgold, A.E., Jenkins, E.B., Shaya, E.J. 1980, ApJ, 242, 545
Ferrara, A. 1998, in IAU Coll. 166 (Lecture Notes in Physics, 506), The Local Bubble and Beyond, eds. D. Breitschwerdt, M.J. Freyberg, J. Trümper (Berlin: Springer), 371
Fiedler, R.L., Dennison, B., Johnston, K.J., Hewish, A. 1987, Nature, 326, 675
Fong, R., Jones, L.R., Shanks, T., Stevenson, P.R.F., Strong, A.W. 1987, MNRAS, 224, 1059
Franco, J., Ferrini, F., Barsella B., Ferrara, A. 1991, ApJ, 366, 443
Giovanelli, R. 1986, Proc. NRAO workshop No. 12, Gaseous Halos of Galaxies, eds. J.N. Bregman & F.J. Lockman (Green Bank: NRAO), 99
Gringel, W., Barnstedt, J., de Boer, K.S., Grewing, M., Kappelmann, N., Richter, P. 2000, A&A, 358, L37
Heiles, C. 1997, ApJ, 481, 193
Houck, J.C., Bregman, J.N. 1990, ApJ, 352, 506
Howk, J.C., Savage, B.D. 1997, AJ, 114, 2463
Howk, J.C., Savage, B.D. 1999, AJ, 117, 2077

Hulsbosch, A.N.M. 1978, A&AS, 33, 383
Jones, A.P., Tielens, A.G.G.M., McKee, C.F., Hollenbach, D.J. 1994, ApJ, 433, 797
Kalberla, P.M.W., Shchekinov, Y.A., Dettmar, R.-J. 1999, A&A, 350, L9
Kalberla, P.M.W., Kerp, J., Haud, U. 2000, in H_2 in Space, eds. F. Combes, G. Pineau des Forêts (Cambridge: Cambridge University Press), 297
Kim, K.-T., Minh, Y.C., Hasegawa, T.I. 1989, J. Korean Astron. Soc., 22, 25
Lu, L., Savage, B.D., Sembach, K.R., Wakker, B.P., Sargent, W.L.W., Oosterloo, T.A. 1998, AJ, 115, 162
Morton, D.C., Dinerstein, H.L. 1976, ApJ, 204, 1
Pfenniger, D., Combes, F. 1994, A&A, 285, 94
Pirronello, V., Liu, C., Roser, J.E., Vidali, G. 1999, A&A, 344, 681
Richter, P. 2000, A&A, 359, 1111
Richter, P., de Boer, K.S., Widmann, H., Kappelmann, N., Gringel, W., Grewing, M., Barnstedt, J. 1999, Nature, 402, 386
Richter, P., Savage, B.D., Wakker, B.P., Sembach, K.R., Kalberla, P.M.W. 2001a, ApJ, 549, 281
Richter, P., Sembach, K.R., Wakker, B.P., Savage, B.D., Tripp, T.M., Murphy, E.M., Kalberla, P.M.W., Jenkins, E.B. 2001b, ApJ, 559, 318
Richter, P., Sembach, K.R., Wakker, B.P., Savage, B.D. 2001c, ApJ, 562, L181
Richter, P., Wakker, B.P., Savage, B.D., Sembach, K.R. 2003a, ApJ, 586, 230
Richter, P., Sembach, K.R., Howk, J.C. 2003b, A&A, 405, 1013
Savage, B.D., Sembach, K.R. 1996, ARA&A, 34, 279
Savage, B.D., Bohlin, R.C., Drake, J.F., Budich, W. 1977, ApJ, 216, 291
Sembach, K.R., Savage, B.D. 1996, ApJ, 457, 211
Sembach, K.R., Howk, J.C., Savage, B.D., Shull, J.M. 2001, AJ, 121, 992
Shapiro, P.R., Field, G.B. 1976, ApJ, 205, 762
Sofia, U.J., Jenkins, E.B. 1998, ApJ, 499, 951
Spitzer, L., Zweibel, E.G. 1974, ApJ, 191, L127
Takahashi, J., Masuda, K., Nagaoka, M. 1999, MNRAS, 306, 22
Tumlinson, J., Shull J.M., Rachford B., et 10 al. 2002, ApJ, 566, 857
Wakker, B.P. 2001, ApJS, 136, 463
Wakker, B.P., Boulanger, F. 1986, A&A, 170, 84
Wakker, B.P., Schwarz, U.J. 1991, A&A, 250, 484
Wakker, B.P., van Woerden, H. 1997, ARA&A, 35, 217
Wakker, B.P., Murphy, E.M., van Woerden, H., Dame, T. 1997, A&A, 488, 216
Wakker, B.P., Howk, J.C., Savage, B.D., van Woerden, H., Tufte, S.L., Schwarz, U.J., Benjamin, R., Reynolds, R.J., Peletier, R.F., Kalberla, P.M.W. 1999, Nature, 402, 388
Walker, M., Wardle, M. 1998, ApJ, 498, L125
Weiss, A., Heithausen, A., Herbstmeier, U., Mebold, U. 1999, A&A, 344, 955
Wolfire, M.G., McKee, C.F., Hollenbach, D., Tielens, A.G.G.M. 1995, ApJ, 453, 673

10. DISTANCES AND METALLICITIES OF HVCS

HUGO VAN WOERDEN
Kapteyn Institute, Groningen, The Netherlands
hugo@astro.rug.nl

AND

BART P. WAKKER
Department of Astronomy, University of Wisconsin-Madison, USA; wakker@astro.wisc.edu

Abstract. We review the methods for determining the distances and metallicities of high-velocity clouds. A short summary is given of several indirect methods, but we concentrate on the absorption-line method, which is based on the detection and non-detection of interstellar absorption lines in the spectra of background and foreground stars, as well as of extragalactic targets. The requirements and problems for determining both distances and metallicities are discussed, and summaries are given of current results. These include a distance bracket of 8 to 10 kpc for complex A, lower limits of ~5 kpc for complexes C and H, and an upper limit of ~4 kpc for part of complex M. We also find that several HVCs have low metallicity (~0.1 to 0.2 solar for complexes A, C, WD, and cloud WW 84), while the Magellanic Stream shows near-Magellanic values, and some other HVCs (complexes M, WB, and HVC 100−7+110) have near-solar metallicity. These results clearly show that the HVC phenomenon is diverse.

1. Introduction

Ever since the discovery of the H I high-velocity clouds (HVCs) about 40 years ago (Muller et al. 1963), it has been clear that the determination of their distances is a clue to understanding their relation to the Milky Way, their origin, and their importance for the evolution of the Galaxy. The values of many of the physical parameters that can be derived from observations are proportional to a power of the distance, D (e.g. mass $\propto D^2$, size $\propto D$, volume density $\propto D^{-1}$). Their metallicities help to pin down their origin – near-solar values indicate an origin in the Disk, values of about

H. van Woerden et al. (eds.), High-Velocity Clouds, 195–226

a quarter solar are compatible with a relation to the Magellanic Clouds, while lower values suggest an intergalactic origin.

Soon after the discovery of HVCs, Prata & Wallerstein (1967) attempted to find high-velocity Ca II toward ten early-type stars, but for over two decades such work remained unsuccessful. Given the difficulties in finding high-velocity absorption against stellar targets, many indirect methods were proposed, making different assumptions for the physical properties of the HVCs (see Sect. 2). None of these led to generally accepted distances.

Oort (1966, 1967) suggested that objects like complexes A and C may lie 1 to 2 kpc above the Disk, but that they might have come from intergalactic space. The gas would originally have primordial metallicity (i.e. Z=0), but now would be mixing with Milky Way gas, leading to $Z\sim0.7$ solar in some parts (see Oort & Hulsbosch 1978). Later it was suggested that the HVCs are part of a Galactic Fountain (Bregman 1980), in which case near-solar metallicities are expected. The fact that for a long time neither distances nor metallicities had been measured led to a proliferation of hypotheses for the origin of the HVCs, as reviewed by Wakker & van Woerden (1997).

Progress began to be made in the 1980s, with detections of Ca II absorption toward both the trailing and leading parts of the Magellanic Stream (Songaila 1981; West et al. 1985), and toward complex WB (Robertson et al. 1991). Significant distance limits for the major northern HVCs (complexes M, A, C) came only in the 1990s (Danly et al. 1993; Keenan et al. 1995; Schwarz et al. 1995; Wakker et al. 1996b; van Woerden et al. 1999a). These results also showed that HVCs contain metals, implying that the method of determining distances using stellar absorption-line spectra is in principle feasible.

As discussed below, the optical lines of Ca II can be used to determine HVC distances, but they are not suited for measuring metallicities, as calcium is strongly depleted onto dust grains, and Ca II is not the dominant ionization stage of calcium. Metallicities require ultraviolet absorption-line spectra of the dominant ionization stage for abundant, undepleted elements. After attempts with the International Ultraviolet Explorer (*IUE*), progress was made with the Hubble Space Telescope (*HST*) and, especially, the Far-Ultraviolet Spectroscopic Explorer (*FUSE*). After the first determination by Wakker et al. (1999), there now are metallicity estimates for five sightlines through complex C, and for nine other HVCs.

This chapter reviews what we know about HVC distances and metallicities as of mid-2004. First, we briefly summarize some indirect distance-determination methods (Sect. 2). In Sect. 3, we describe the absorption-line method, with its problems, followed by a discussion of current distance results (Sect. 4). In Sects. 5 and 6 we describe the methods, problems, and results of the determinations of HVC metallicities. The chapter concludes with a discussion of the implications for the origins of the HVCs (Sect. 7) and an outlook to the future (Sect. 8).

Current results confirm our earlier suspicion (Wakker & van Woerden 1997) that the HVC phenomenon is diverse, so that distances and metallicities of individual HVCs must be measured. We note that much of the discussion in Sects. 3 to 6 leans heavily on the review of absorption-line measurements by Wakker (2001), but some major new results have been added. Finally, we note that in this chapter we will only discuss HVCs, with $|v_{\mathrm{LSR}}|>90$ km s^{-1}; results for IVCs are presented in Ch. 4, while we ignore high-velocity gas that is clearly associated with stellar winds and supernova remnants.

2. Methods of distance determination

2.1. DIRECT VS INDIRECT METHODS

There are several different approaches to determining distances to HVCs. The least ambiguous results come from the "absorption-line method", described in Sect. 3. First, we summarize a few less-direct methods, restricting the discussion to developments after 1997; see Sect. 4.1 of Wakker & van Woerden (1997) for a summary of previous work. The advantage of some of these indirect methods is that they may more easily be applied to many clouds than the absorption-line method. The disadvantage is that they require making one or more key assumptions that may or may not correspond to the actual properties of the HVCs.

2.2. DISTANCES FROM Hα INTENSITIES

The idea behind this method is relatively old, but it has become practical only with the development of sensitive Hα instruments. A hydrogen cloud that is embedded in an ionizing radiation field will soak up as many photons with energy >13.6 eV as it can. Each ionization results in one recombination, yielding 0.46 Hα photons. Thus, the Hα intensity (I(Hα)) should directly measure the number of ionizing photons, which depends on the location of the cloud. Distance estimates follow from combining I(Hα) with a model for the flux of the ionizing radiation field near the Milky Way.

The Hα data are reviewed in Ch. 8. From these, we can surmise that complex M may be at lower z than complex A. Complexes GCP and L are relatively bright, probably because they lie over the Galactic Center. The few detected Anti-Center clouds may be as distant as complex A, while many of the smaller HVCs appear to be more distant than ~10 kpc.

The most detailed model of the ionizing radiation near the Galaxy was constructed by Bland-Hawthorn & Maloney (1999, 2001). However, the

Magellanic Stream is much brighter than expected from this model. The picture is further complicated by the fact that very different Hα intensities have been detected from the same cloud, implying that the internal cloud structure is important in determining I(Hα). More observations and theory are needed to make the Hα method work. Still, it remains the most promising alternative to the absorption-line method.

2.3. THERMAL AND RAM-PRESSURE EQUILIBRIUM

Estimates of the distance (D) are possible if one assumes pressure equilibrium between the HVC ($P{=}n_{\rm HVC}\,T_{\rm HVC} + P_{\rm turbulence}$) and a hot surrounding halo ($P{=}n_{\rm halo}\,T_{\rm halo}$). Observable are the column density of the HVC (N(H I)$=\ n_{\rm HVC}\,D\,\alpha$, with α its angular size) and its line width (FWHM $\propto\sqrt{T_{\rm HVC} + P_{\rm turbulence}/n}$), while the run of halo density and temperature with height above the plane can be modeled. Espresate et al. (2002) apply this to estimate the halo parameters, calibrating them using the distance to complex M (see Sect. 4). Stanimirovic et al. (2002) estimate the density of the hot halo at the distance of Magellanic Stream core MS VI. If one has confidence in the description of and parameter values for the hot halo gas, distance estimates are possible. However, an important (unexamined) complication is the contribution of magnetic pressure.

Instead of assuming that the internal outward pressure of the HVCs is balanced by the thermal pressure of a hot, confining medium, Benjamin & Danly (1997) worked out the equilibrium between the ram pressure (drag force) of an infalling cloud and the surrounding total density of halo gas (the sum of the hot, warm, and cold components). At low column density (N(H I)$<10^{19}$ cm^{-2}), the infalling gas reaches terminal velocity, which depends on column density and the halo's density and gravity structure (see also Ch. 12). Benjamin & Danly (1997) showed how this can give distances, finding that the model works for some (but not all) IVCs with known distance (including complex M).

2.4. LOCAL GROUP CLOUDS AND COMPACT HVCS

Blitz et al. (1999) assumed that to first order the virial theorem is applicable to HVCs (see Ch. 2 for the formula) and that they contain dark matter. Assuming a visible-to-dark ratio (f) of about 0.03, distances of 30 to 2000 kpc are implied (see Fig. 8 of Ch. 2). The large complexes (e.g. A, C, H) would be nearby, tidally-stretched examples for which the virial theorem is not applicable. As worked out by Zwaan & Briggs (2000) and Wakker (Ch. 2), the implied H I masses are only compatible with non-detections in other galaxy groups if $f\lesssim 0.01$, implying distances of 10 to 600 kpc. Further, when observed at high resolution, the clouds are often asymmetric, and it is doubtful that the virial theorem can be applied.

Braun & Burton (2000, 2001) mapped a small HVC (HVC 125+41−207 or WW 84) at high resolution (0$'\!.$5). Its core shows a very narrow, very bright line, tightly constraining the temperature to $T\sim85$ K and the opacity to $\tau>2$. With a model of internal conditions in an environment away from the Galaxy (e.g. Wolfire et al. 1995), they find that equilibrium requires a density of $\sim$0.7 to 3.5 cm^{-3}, so that the observed column density and angular size imply a distance between 300 and 900 kpc. Since this object has a metallicity of $\sim$0.1 solar (see Sect. 6), densities of 0.1 to 10 cm^{-3} are expected for a wide range of environmental conditions. Brüns et al. (2001) instead assume that the core is virialized and in pressure equilibrium with the warm gas surrounding it, and then derive a distance of 130 kpc. Considering the data quality and the assumptions, a distance of at least 50 kpc for this HVC seems uncontroversial, but a precise distance remains elusive. Unfortunately, this is the only known CHVC for which this conclusion can be reached, as for other objects the temperature and opacity are not well-constrained.

Braun & Burton (2000) and Burton et al. (2001) also interpret the velocity field of three other small HVCs as showing evidence for rotation. If that is correct, a dynamical mass can be estimated. Then, assuming a visible-to-dark mass ratio $f\sim0.1$, distances of about 700 kpc are implied. This interpretation remains controversial, as the velocity fields are rather irregular on small scales, and the value of f must be assumed.

2.5. KINEMATIC APPROACHES

A possible kinematic approach is to explain velocity gradients in clouds as the result of projection effects. This leads Lockman (2003) to model complex H as an object in a retrograde orbit 33 kpc from the Galactic Center. The implied H I mass (6×10^6 M$_\odot$) and size (10×5 kpc) then are comparable to those of complex C.

A completely different kinematic approach is to try to determine a distance scale for the *ensemble* of HVCs. Two recent papers (de Heij et al. 2002; Wakker 2004) both come to the conclusion that the longitude-velocity distribution of HVCs is best represented by having a population of several hundred objects distributed in a sphere of 50 to 200 kpc radius around the Galaxy (and possibly also around M 31), with an average infall velocity of about 50 km s^{-1}. See Ch. 3 for more discussion.

2.6. STARS IN HVCS?

A few attempts have been made to find stars associated with HVCs (Simon & Blitz 2002; Davies et al. 2002; Willman et al. 2002; Hopp et al. 2003).

So far, these attempts have not been successful. A convincing association might furnish a good distance determination, but it may be difficult to prove an association.

3. Distance determination from absorption lines

3.1. METHOD IN BRIEF

The absorption-line method for determining HVC distances was described in several review papers (Schwarz et al. 1995; van Woerden et al. 1999b; Wakker 2001), Applying this method requires the following steps, each with its own problems and requiring its own kind of expertise:

- Find suitable stellar candidates projected onto the HVC (using objective-prism plates or multi-color images).
- Properly classify and characterize the candidates, using high-quality photometry and intermediate-resolution spectroscopy.
- Obtain high-resolution (15 $\mathrm{km\,s^{-1}}$ or better) spectra of stars at a range of distances bracketing that of the HVC. Absorption at the velocity of the HVC in the spectrum of a background star sets an upper limit on its distance, while a significant non-detection toward a foreground star sets a lower limit. We now discuss each of these steps in some detail.

3.2. PROBE REQUIREMENTS AND SEARCHES

Suitable stars should be fairly numerous at high galactic latitude, have relatively reliable estimated distances, and their spectrum should have few stellar lines that can interfere with the interstellar absorption. In practice, this leads one to blue, evolved stars, such as Blue Horizontal Branch (BHB) stars, subdwarf B stars, and also RR Lyraes. Past programs have found thousands such stars with distances up to a few kpc (e.g. Kukarkin et al. 1970 – General Catalog of Variable Stars; Green et al. 1986 – the PG catalog; Beers et al. 1996 – the "HK" survey; Stobie et al. 1997 – the Edinburgh-Cape survey). In each of these surveys, photometric data or very-low-resolution objective-prism spectra were used to define a sample of candidates. Correlating these lists with the HVC maps yields a sample of possible HVC probes. Next, intermediate-resolution (about 1 Å) spectra and multi-band photometry are needed, which can be combined with stellar-atmosphere models to derive the stellar temperature and gravity. Assuming that the star indeed is an evolved star gives a mass estimate, and thereby the stellar radius, luminosity, and distance. Descriptions of this method are provided by e.g. Moehler et al. (1990) and Wilhelm et al. (1999). This part of the search for suitable stars is relatively well-understood. The surveys require lots of telescope time, but also serve for studies of galactic structure.

Although these surveys contain stars as faint as $V\sim15.5$ (a distance of $\sim$10 kpc for a BHB star), in practice candidate probes have $D<6$ kpc. Many HVCs may be more distant (see Sect. 4), so it is clear that samples of fainter stars will be needed to obtain a significant number of distances to HVCs. Two promising developments have occurred in the past few years.

First, the 2MASS survey obtained infra-red colors for basically all objects in the sky down to magnitude $J\sim15.5$, or a BHB distance of $\sim$10 kpc. A study of the two-color diagram (Brown et al. 2004) suggests that one can pick out the BHB stars with 50% probability. This means that there are about 65,000 BHB stars in the 2MASS survey at $|b|>15°$, of which 10,000 correlate with HVCs (not including the Magellanic Stream and the Outer Arm), 5,000 of which may lie between 5 and 10 kpc. However, confirmed classifications are still lacking for most of these stars.

Second, the Sloan Digital Sky Survey (York et al. 2000) has yielded thousands of BHB and RR Lyrae stars. The final survey will cover the 10,000 square degrees above $b=+30°$, and a small fraction ($\sim$2%) of the southern sky. This covers most of complex C, much of complex A, some of complex WA and some small positive-velocity clouds, as well as some strips crossing complexes GCN, gp, and ACHV. Most, but not all, BHB stars down to a limiting magnitude of about 18 to 20 are found. In the part of the survey that is already available (about one third) there are 200 confirmed BHBs (Sirko et al. 2004) and 400 probable RR Lyraes (York, priv. comm.) more distant than 10 kpc that lie projected on HVCs.

For HVCs not covered by the Sloan Survey, targeted probe searches are still needed. Some of these are underway.

3.3. STELLAR LINES AND VARIABILITY

Blue Horizontal Branch stars are the preferred kind of probe star, as these are in fact the evolved cores of old, usually metal-poor stars. Thus, not only does the Hϵ line at 3971 Å tend to be relatively weak (i.e. often there are no broad damping wings to confuse Ca II H at 3968 Å), and not only are there relatively few stellar metal lines (especially Fe I), but the stellar Ca II lines also are relatively weak. Still, the stellar Ca II and Fe I lines may interfere with the interstellar absorption. Therefore, when obtaining intermediate-resolution spectroscopy to confirm the stellar classification, it is also necessary to determine a stellar radial velocity to eliminate stars whose velocity is similar to that of the HVC projected onto it.

A disadvantage of the BHB stars is that the bluest (which are best for interstellar work) are also the faintest (Preston et al. 1991). Further, Lehner et al. (2001b) have pointed out that in evolved stars narrow stellar S II lines

TABLE 1. Main instruments used or usable for absorption spectra

Telescope[a]	Location[b]	Diam [m]	Instrument[c]	λ range [Å]	Cover[d] [Å]	Resolution[e] $\lambda/\Delta\lambda$	km s^{-1-1}	Years of operation
IUE	geosynchr.	0.45	*SWP*	1150–1900	750	12000	25	1978–1996
HST	600 km LEO	2.4	*GHRS*-G140M	1100–1900	26–28	20000	15	1990–1997
			GHRS-G160M	1150–2300	33–36	20000	15	1990–1997
			STIS-E140M	1150–1710	560	45800	6.5	1997–
			STIS-E230M	2000–3110	800	30000	10	1997–
ORFEUS	300 km LEO	1.0		912–1400	500	10000	30	1993/1996
FUSE	775 km LEO	0.40		912–1170	270	15000	20	1999–
WHT	La Palma	4.2	*UES*	3000–10000	700	54000	5.5	1994–2003
			ISIS	3000–6000	380	18000	17	1994–
AAT	Siding Spring	3.9	*RGO*	3100–11000	130	11500	26	1975–
VLT	Paranal	8.2	*UVES*	3000–5000	500	50000	6.0	1999–
Keck	Mauna Kea	10.0	*HIRES*	3500–11000	1600	37000	8	1994–

Notes: (a) *IUE* = International Ultraviolet Explorer, *HST* = Hubble Space Telescope, *ORFEUS* = Orbiting and Retrievable Far and Extreme Ultraviolet Spectrographs, *FUSE* = Far-Ultraviolet Spectroscopic Explorer, *WHT* = William Herschel Telescope, *AAT* = Anglo-Australian Telescope, *VLT* = Very Large Telescope. (b) Several space observatories are in Low-Earth Orbit (LEO). (c) *SWP* = Short-Wavelength Primary camera, *GHRS* = Goddard High-Resolution Spectrograph, *STIS* = Space Telescope Imaging Spectrograph, *UES* = Utrecht Echelle Spectrograph, *ISIS* = Intermediate Dispersion Spectrograph and Imaging System, *RGO* = Royal Greenwich Observatory spectrograph, *UVES* = UV-Visual Echelle Spectrograph, *HIRES* = High Resolution Echelle Spectrometer. (d) Instantaneous wavelength coverage. (e) Resolution, defined as $R=\lambda/\Delta\lambda$ or in velocity units – only resolutions relevant to HVC research are listed.

may be confused with HVC absorption.

Although RR Lyraes show more stellar lines, they may still be useful: RR Lyraes have a possible advantage, in that the stellar lines show shifts in radial velocity over time, which in principle (at the expense of more telescope time) can be used to separate stellar from interstellar lines. In practice, RR Lyrae spectra can best be used during maximum phase, when the star is both brighter and bluer. However, epochs of maximum are often only poorly known, so that new photometry is required first (c.f. van Woerden et al. 1999a).

3.4. HIGH-RESOLUTION SPECTRA OF INTERSTELLAR ABSORPTION

After identifying suitable stellar probes (see Sects. 3.2 and 3.3), high-resolution spectra of the best sample are required. In the wavelength range accessible from the ground, the strongest interstellar lines are the Ca II H and K doublet. For directions with relatively strong HVC H I emission, the

Ca II K line is found to have an optical depth $\lesssim$1 (see Wakker 2001). Often τ(Ca II) is expected to be no higher than ~0.2, so data with signal-to-noise ratio $\gtrsim$20 are needed. Furthermore, setting significant lower distance limits may require signal-to-noise ratios >100 (see Sect. 3.6).

Ultraviolet lines are much stronger, but require observatories in space. For instance, for solar abundances the Mg II λ2796.352 line is completely black for hydrogen column densities above 5×10^{17} cm^{-2}. However, in practice space observatories cannot detect stars much fainter than $V\sim15.5$ (distance ~10 kpc), and ground-based data will be needed to observe the more distant stars. Table 1 lists the properties of the telescopes that either have been used or may be used to observe stars against HVCs.

3.5. SMALL-SCALE H I STRUCTURE

The small-scale structure of HVCs is discussed in Ch. 7. It is an important factor in determining both distances and metallicities of HVCs, so we give a brief summary.

A column density (N(H I)) in the direction of a stellar or extragalactic target can easily be obtained from a 21-cm observation. However, this gives N(H I) averaged over the beam of the radio telescope, whereas the absorption-line data measure N(ion) in the pencil-beam to the target. In principle, N(H I) can be derived from Lyman-series absorption-line data toward the target, but such data are not always available, and blending with many other interstellar lines often makes a measurement impossible. The next best value is found from radio-interferometer maps (beams of $1'$ to $2'$), but these require many hours of integration time, compared to about 15 minutes for a single-dish observation with a $10'$ or $36'$ beam.

Wakker et al. (2001) show that the ratio N(H I;$36'$) to N(H I;$10'$) lies in the range 0.2 to 2.5, whereas N(H I;$10'$)/N(H I;Lyα,β) lies in the range ~0.75 to 1. In Ch. 7 it is shown that in large HVCs the column density can be up to 4 times higher in the arcminute-sized cores than in the rest of the cloud. However, the cores only cover about 10% of the observed fields. So, in a more typical situation one can assume that the H I column density measured with a $10'$ H I beam will be correct to within a factor 1.5. Note that zero-intensity *holes* in the H I distribution have never been found in synthesis maps.

Therefore, one should allow for an error in N(H I) that depends on the size of the radio beam. This error is about a factor 2.5 for N(H I;$36'$), 1.5 for N(H I;$10'$), and possibly ~5% for N(H I;$1'$–$2'$) (larger when the sightline passes close to an arcminute-sized core). Note that for small HVCs an interferometer observation is always necessary, as they clearly have structure on scales less than $10'$.

3.6. MEASURED VS. EXPECTED LINE STRENGTHS

A major complication in the process of determining distances to HVCs is that lower limits may be harder to come by than upper limits. Wakker (2001) reviews this topic. An upper limit requires detecting absorption from any ion, with the S/N ratio of the detection $(EW/\sigma(EW))>3$. An upper limit on the equivalent width of the line does not immediately yield a lower distance limit, however, as discussed below.

For any individual HVC the expected absorption strength for a given line of a given ion can be found from an abundance measurement for that ion, based on background stars or extragalactic targets. Alternatively, one can use the relations between abundance and H I column density found by Wakker & Mathis (2000) to estimate N(ion) and EW(ion). However, to obtain a lower distance limit, the ratio between expected strength and detection limit should be much larger than 3, because of small-scale structure and possible ionization (and/or abundance) variations across the cloud.

Different elements have different preferred ionization states in the diffuse ISM. Some strong ultraviolet lines, such as Mg II and C II, come almost always from the dominant stages, so the only uncertainty is the amount contained in dust particles. In the low-velocity warm ISM this is no more than about 50% for these two elements. On the other hand, the Ca II ion is not the dominant ionization stage, so that large variations in its abundance are possible. Wakker & Mathis (2000) see a range of ~1.5 dex for N(H I) between 10^{18} and 10^{20} cm^{-2}, such that sightlines with low N(H I) show only slightly lower equivalent widths than sightlines with high N(H I).

Combining these factors, Wakker (2001) argues for an ion-dependent safety factor, such that a lower distance limit only follows from a non-detection if the ratio expected/limit is larger than about 6 for the Mg II, Fe II, Si II, and C II lines, but $\gtrsim$20 for the Ca II-K line ($\gtrsim$10 if the cloud's Ca II abundance is known).

Since the typical Ca II equivalent width in HVCs like complexes A and C is found to be ~50 to 100 mÅ (Schwarz et al. 1995; Wakker et al. 1996a), detection limits of order ~5 mÅ or better are required, implying S/N ratios in the continuum >50. In fact, Wakker (2001) only finds lower distance limits based on Ca II non-detections toward a few bright OB stars at ~300 pc distance, and a probable lower limit of 6 kpc for complex C. On the other hand, since they are stronger and require a smaller safety factor, non-detections of Mg II, C II, Fe II, and Si II have yielded useful lower distance limits to complexes A, C, H, the IV Arch, IV Spur, LLIV Arch and assorted smaller HVCs and IVCs

TABLE 2. Distances, locations, and masses of HVCs

HVC/Complex name	distance [kpc]	height [kpc]	Location	log M(H I) [$M_\odot$]	Ref .	Note
complex A	8 to 10	+5 to +7	Halo	6.0±0.1	1, 2, 3	
complex M (M II/M III)	<4.0	<+3.5	Halo/Disk	<5.2	4, 5	
complex C	>6	>+4	Halo/IGM	>6.3	6	
complex G	>1.3	<−0.2		>4.6	7	
complex H	>5	∼0	Disk?	>5.8	8	a
complex ACHV	>0.3	<−0.1		>3.0	7	b
complex ACVHV, WW 507	>0.3	<−0.1		>3.0	7	c
complex GCP, WW 360	>0.3	<−0.1		>2.8	7, 9	d
complex L	>0.2	>+0.1		>1.2	7	e
complex WE	<12	>−3.2	Halo/Disk	<5.2	7, 10	
HVC 100−7+110	<1.3	>−0.2	Disk	<0	11	
Magellanic Stream	∼50	$\lesssim$−30	IGM	8.4	12	f
MS Leading Arm, WW 187	∼50	$\gtrsim$+20	IGM	7.5	12	f

References: 1: van Woerden et al. (1999a); 2: Wakker et al. (1996b); 3: Wakker et al. (2003); 4: Danly et al. (1993); 5: Ryans et al. (1997); 6: van Woerden et al. (1999b); 7: Wakker (2001); 8: Wakker et al. (1998); 9: Prata & Wallerstein (1967); 10: Sembach et al. (1991); 11: Stoppelenburg et al. (1998); 12: Putman (Ch. 5)

Notes: (a) There are estimates based on the absence or presence of interaction with Galactic Disk (see Sect. 4): D=15 kpc (Morras et al. 1998), D=27 kpc (Lockman 2003), and D=34 kpc (Blitz et al. 1999). (b) Putman et al. (2004) note that location and velocity of the Anti-Center clouds agree with the orbit of the Sagittarius Dwarf, which would imply a distance of 36 kpc (see Sect. 4). (c) Weiner et al. (2001) find Hα emission from three directions in WW 507, suggesting D=10 to 20 kpc. (d) Also called the "Smith Cloud". Bland-Hawthorn et al. (1998) detect Hα emission, indicating a distance of 26±4 kpc; they suggest a possible relation to the Sagittarius Dwarf, but this can be ruled out (see Sect. 4). (e) Upper limit of 2.5 kpc quoted by van Woerden (1993) was based on misinterpretation of stellar lines in the spectrum of HD 135485; correct interpretation due to Danly et al. (1995). (f) The distance estimates are based on the tidal models of Gardiner & Noguchi (1996) and distances of 50 and 60 kpc for the LMC and SMC, respectively.

4. Distances of high-velocity clouds

Wakker (2001) has given a comprehensive compilation of measurements of HVC absorption lines in the spectra of stellar and extragalactic probes. This paper also contains an extensive discussion of the distances and metallicities of individual HVC (and IVC) complexes, plus many maps and structural information. Much of what follows is based on that paper.

Table 2 summarizes available information on HVC distances, including upper and lower limits, plus derived quantities, such as height above or below the galactic plane and H I mass. Except for the Magellanic Stream, this information is based exclusively on results of the absorption-line method; a few other distance estimates are mentioned in the notes to the table. The order of HVCs in the table follows that in table and text of Wakker

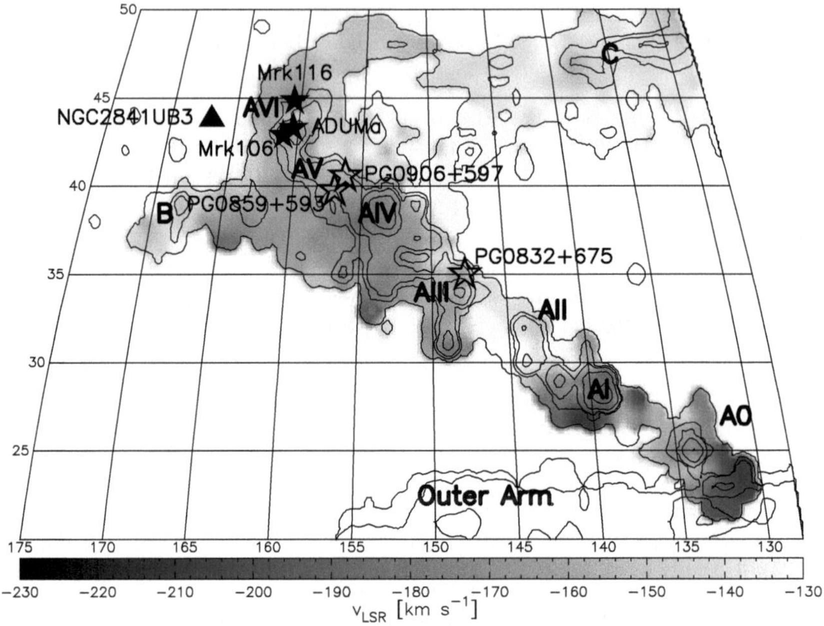

Figure 1. Map of complex A and its surroundings, based on the Hulsbosch & Wakker (1988) survey. The grey scale shows velocities, contours are at brightness temperatures of 0.05, 0.3, 1, and 2 K. Filled stars are targets showing absorption associated with complex A (the filled triangle is for a low-resolution Mg II-only detection). Open stars show the three most distant stars in whose spectrum complex A is not detected; toward AD UMa it is. Figure adapted from Wakker (2001).

(2001), again with the exception of the Magellanic Stream. Cloud names WW xxx in Col. 1 refer to the catalog of Wakker & van Woerden (1991). The H I mass in Col. 5 is obtained directly from the surveys of Hulsbosch & Wakker (1988) and Morras et al. (2000), through integration of the 21-cm line flux over the complex, assuming a velocity width of 25 km s^{-1} (FWHM). The amounts of ionized and molecular hydrogen, which in general are not known, and of helium could raise these masses by a factor of order 2; the masses given by Wakker (2001) do include these extra factors.

The complexes and clouds listed have $|v_{\rm LSR}|>\sim 100$ km s^{-1}. However, some complexes (G, H, L, GCP, WE, as well as WB in Table 5) have deviation velocities (defined in Ch. 2) below this limit, and thus may not be genuine HVCs.

4.1. DISCUSSION OF INDIVIDUAL COMPLEXES

• *Complex A:* Figure 1 shows this complex, the first HVC discovered (Muller et al. 1963), as a narrow filament with a number of strong condensations.

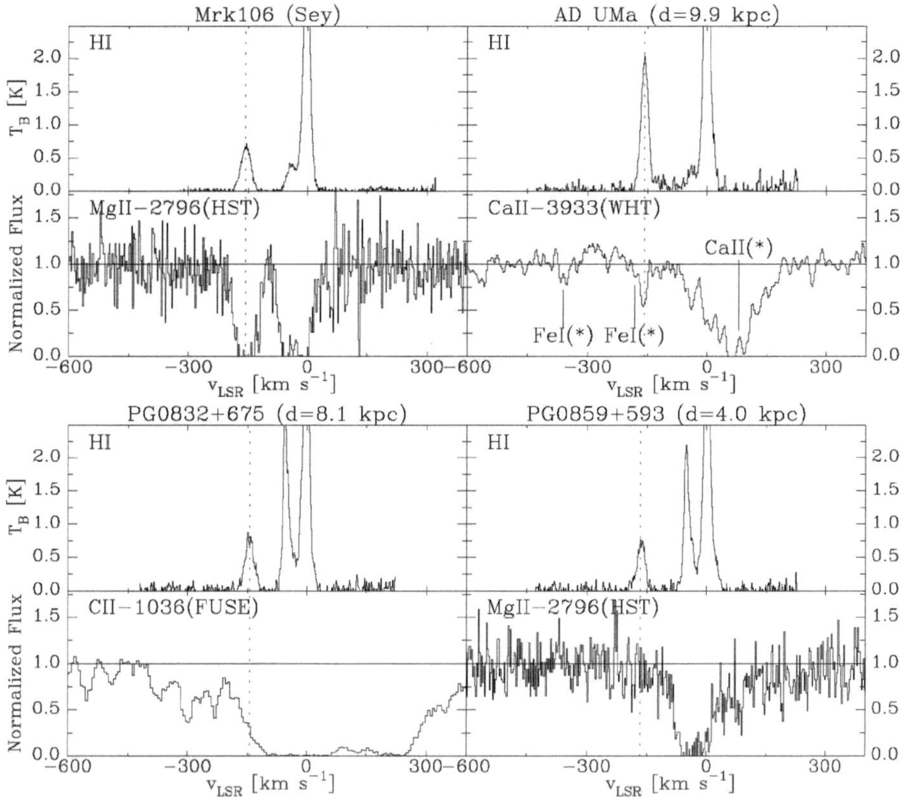

Figure 2. The spectra constraining the distance to complex A. *Top:* H I and absorption-line spectra for Mrk 106 and AD UMa, toward which complex A is detected in absorption. *Bottom:* Spectra for PG 0832+675 and PG 0859+593, toward which complex A is not detected. H I data are from *Effelsberg* (9′ beam). Dotted lines show the velocity of complex A. Absorption-line spectra have been normalized; the spectral line and instrument are given. The spectrum of AD UMa shows stellar Ca II and Fe I lines, plus interstellar Ca II absorption at velocities ~0, ~−50, and ~−160 km s^{-1}; the latter corresponds to the HVC. The several stellar Fe I and Ca II lines yield a stellar velocity $v_{\rm LSR}$=+77 km s^{-1}, corresponding to a separation of 19±3 km s^{-1} between the weak Fe I λ3930.297 line and the high-velocity Ca II. The absence of C II absorption toward PG 0832+675 sets a lower distance limit: for 1/10th solar metallicity τ(C II) is expected to be >15, given the measured value for N(H I) of 3×10^{19} cm^{-2}. Figure composed for this article.

Observations with *HST+GHRS* (Table 1) by Wakker et al. (1996b) set a lower limit to the distance: the star PG 0859+593 at 4 kpc showed no absorption by Mg II in the HVC, although the 21-cm emission along this sightline was as strong as in the direction of the Seyfert galaxy Mrk 106, which displayed strong Mg II absorption in the HVC (Fig. 2). In the spectrum of the RR Lyr star AD UMa at D=10 kpc, van Woerden et al. (1999a) found the Ca II K- and H-lines absorbed by the HVC. Potential confusion of the HVC K-line with the stellar Fe I(4) λ3930.297 line could be excluded.

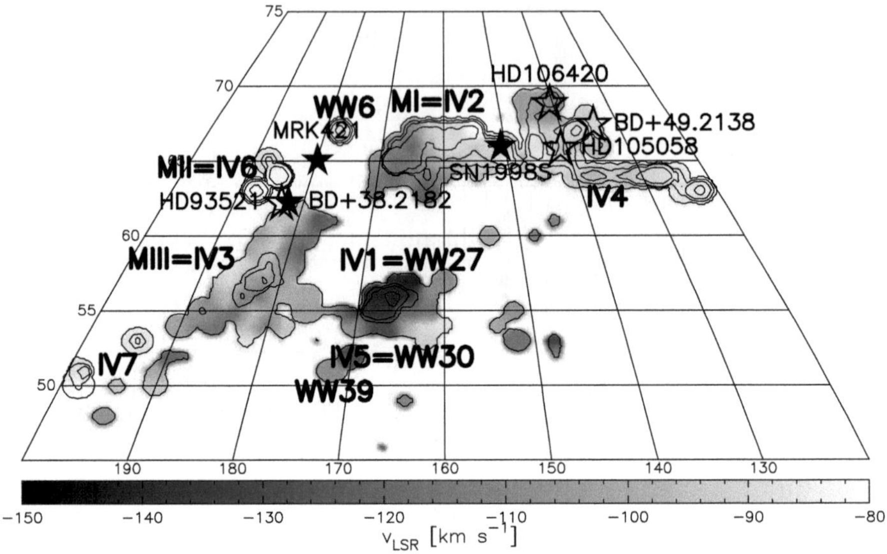

Figure 3. Map of complex M, based on the Hulsbosch & Wakker (1988) data. The grey scale shows the velocities, contours are at brightness temperatures of 0.05, 0.3, and 1 K. Filled stars show the AGN (Mrk 421), supernova (SN 1998S) and star (BD+38°2182) toward which absorption at complex M velocities has been found and published. Open stars show stellar sightlines with non-detections. Figure adapted from Wakker 2001.

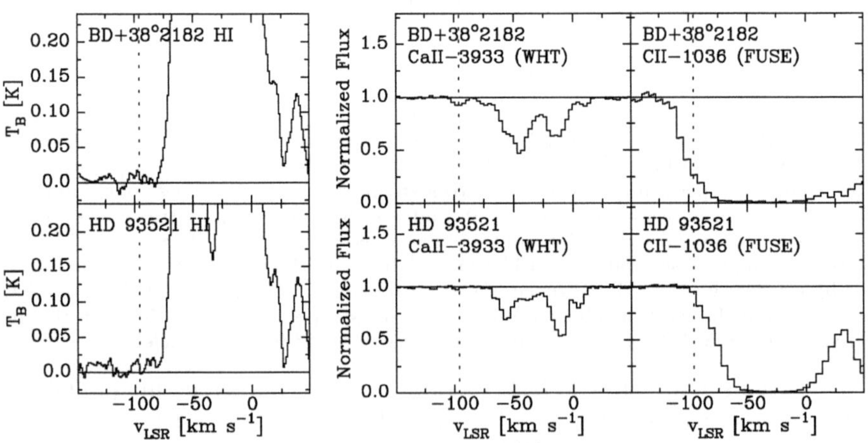

Figure 4. H I, Ca II and C II spectra for BD+38°2182 (top row) and HD 93521 (bottom row). H I and Ca II data are courtesy R.S.I. Ryans (see Ryans et al. 1997); H I data obtained with *Jodrell Bank*. The dotted lines correspond to the velocity of the Ca II absorption toward BD+38°2182. Note that the H I data for BD+38°2182 give a 3-σ upper limit for N(H I) of 8×10^{17} cm^{-2}, which would imply τ(C II)$\lesssim$5; the C II line is not completely saturated and suggests τ(C II)$\lesssim$1.5. Figure composed for this article.

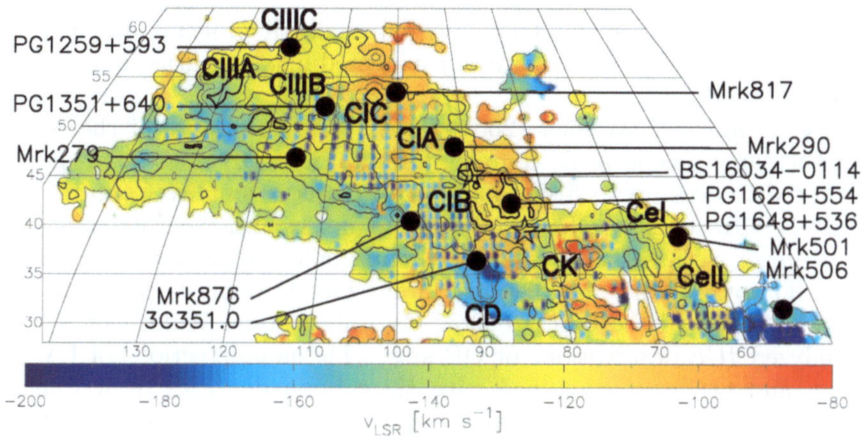

Figure 5. Map of the velocity field of complex C, based on the Hulsbosch & Wakker (1988) survey, with contours of brightness temperature at 0.05, 0.2, and 1 K overlaid. Extragalactic targets toward which absorption associated with complex C has been detected are indicated by filled circles. Open stars show the two stellar targets that set a significant distance limit (6 kpc from BS 16034-0114 and 1.2 kpc from PG 1648+536). Adapted from Wakker (2001).

The distance bracket $4<D<10$ kpc thus derived has been further tightened by Wakker et al. (2003) with *FUSE*: absorption in the intrinsically strong C II λ1036.337 line is not detected for the star PG 0832+675 (D=8 kpc).

• *Complex M:* This loose collection of clouds (Fig. 3), with velocities $v_{\rm LSR}$ around -100 km s^{-1}, may not all be at the same distance. The complex may, at least in part, be related to the Intermediate-Velocity Arch, discussed in Ch. 4, which lies at distances between 0.8 and 1.8 kpc (Wakker 2001).

The star BD+38°2182, which lies in a gap between clouds M II and M III, sets a limit $D<4$ kpc to the distance of these clouds by showing UV absorption lines (Danly et al. 1993) and Ca II K absorption (Keenan et al. 1995; Ryans et al. 1997; see Fig4) at the velocity of the clouds. Although these absorptions are absent in the spectrum of HD 93521, only 0°.5 away, the lack of measurable H I in both directions (Fig. 4) makes it impossible to estimate expected line strengths; hence HD 93521 does *not* set a lower distance limit to complex M (Ryans et al. 1997).

• *Complex C:* The structure of this largest HVC, covering 2000 deg^2 (Fig. 5), is highly complex; at many positions it shows 2, sometimes 3, profile components. The lack of absorption in many probe spectra sets a hard limit $D>1.2$ kpc to its distance (Wakker 2001). Van Woerden et al. (1999b) found no absorption in several stars at distances 3 to 8 kpc; the significant (10-σ) non-detection against BS 16034-0114 sets a lower limit of 6 kpc to the distance of complex C, implying an H I mass exceeding 2×10^6 M$_\odot$. In view of its large size, distance determinations for different parts of complex C are badly needed. The many extragalactic probes available (Fig. 5) allow making reliable estimates for expected line strengths toward stars in many directions.

• *Complex H:* Wakker (2001, Fig. 7) gives a picture of the structure of this large, low-latitude complex. Many bright OB stars lie projected on it, at distances up to 5 kpc and more. Observations by Centuríon et al. (1994) with the Isaac Newton Telescope and by van Woerden and Peletier (in prep.) with *WHT+UES* at La Palma failed to show Ca II absorption, but do not yield a lower distance limit, since the only background quasar available proved too faint to show Ca II absorption. Wakker et al. (1998) analyzed *IUE* observations of OB stars. The non-detections of UV Mg II, C II, and O I lines, combined with H I column densities from *Effelsberg*, imply *either* upper limits on abundances ranging from 0.03 to 0.0003 times solar, *or* a location of complex H behind the stars, at D>5 kpc.

The fact that complex H straddles the galactic plane raises the question of a possible interaction with the Galactic Disk. Blitz et al. (1999, and Ch. 14) found no signs of such interaction in the Leiden-Dwingeloo Survey, and hence put the HVC at galactocentric distance $R \gtrsim 40$ kpc. However, Morras et al. (1998) reported disturbances in the Disk, suggesting shock phenomena, and estimated R=22 kpc. Recent observations by Lockman (2003) with the new Green Bank Telescope (beam 9′) clearly show an ablation pattern around the brightest core of complex H. Lockman interprets the velocity gradients in terms of the collision with the Galactic Disk of a satellite in a retrograde, inclined circular orbit at R=33±9 kpc, i.e. D=27 kpc. This distance depends, of course, on model assumptions. Verification of such a great distance with the absorption-line method will be hard to obtain.

• *Complexes ACHV and ACVHV:* For the Anti-Center clouds (see Wakker 2001, Fig. 9), as for some other complexes, only uninteresting lower limits on distance and mass are available. However, some other pieces of information deserve mention.

Putman et al. (2004) point out that the AC clouds lie along the orbit followed by the Sagittarius (Sgr) dwarf galaxy; also the velocities (−100 to −200 km s^{-1} relative to LSR) appear to be consistent. The implied distance of the clouds would be about 36 kpc. The clouds may represent gas swept out of the Sgr Dwarf by its collision with the outer Milky Way disk, some 200 to 300 Myr ago.

In the cloud WW 507 (l,b,v=165−44−285), which is part of the ACVHV complex, Weiner et al. (2001) detected Hα emission at 3 positions, indicating a distance of order 10 to 20 kpc. Sensitive absorption-line studies of the AC clouds would be of great importance.

• *Complex GCP:* Around the time of the first HVC discovery, Smith

(1963) found a "peculiar object" at (l,b,v=40−15+100). Wakker (2001, Fig. 18) gives a map of complex GCP, of which the Smith cloud is the brightest core. Bland-Hawthorn et al. (1998) derived a distance 26±4 kpc from Hα emission, and suggested a possible relation to the Sgr Dwarf. However, the orbit shown by Putman et al. (2004) is essentially parallel, at 30° distance, to the GCP complex (which lies along the celestial equator). Hence, a relationship seems to be ruled out.

• *Complex WE:* The upper limit D<13 kpc for this complex (Wakker 2001, Fig. 11) is based on several UV absorption lines observed by Sembach et al. (1991) in the distant O-star HD 156359, which lies near WW 364 = HVC 323−16+115.

• *Magellanic Stream (MS) and Leading Arm (LA):* These extended systems of HVCs are discussed in detail in Ch. 5. No distances from absorption lines are available. However, both MS and LA are well represented by models for the tidal interactions between the Galaxy and both Magellanic Clouds (Gardiner & Noguchi 1996; Gardiner 1999). The 50 kpc distance listed here is a rough value, consistent with the distances of LMC and SMC; parts of the tidal features may deviate by some 50% from this value.

• *HVC 100−7+110:* This tiny cloud was first found by absorption lines in *IUE* spectra of the O9-star 4 Lac at 1.2 kpc distance (Bates et al. 1990); next it was mapped in 21-cm emission at Jodrell Bank (Bates et al. 1991). The H I mass (<1 $M_\odot$) was derived from a Westerbork map by Stoppelenburg et al. (1998). The cloud clearly lies in the Galactic Disk, less than 200 pc from the plane, and moves toward the star (which has v_{LSR}=−26 km s^{-1}) with at least 100 km s^{-1} relative to its immediate surroundings; but its small size (less than 1 pc) may save it for some time.

5. Metallicity determination – requirements and problems

Determining metallicities of HVCs requires a different approach than determining their distances. Instead of taking spectra of the optical Ca II lines or very strong ultraviolet lines (Mg II, C II, Si II) toward faint blue stars, one needs spectra of background targets for ions which have lines with optical depths between about 0.1 and 2 (which of course depends on N(H I)). The best targets are AGNs, such as quasars and Seyfert galaxies, which have flat intrinsic spectra that rise toward shorter wavelengths. Starburst galaxies (containing many bright O stars) may also be useful, but for these disentangling stellar and interstellar features can be more complicated. A few items need to be taken into account when measuring metallicities.

• *Small-scale structure.* To convert an ionic column density into an abundance requires a hydrogen column density. As discussed in Sect. 3.5, a good determination requires a 21-cm observation with a beam of 2′ or better. This is not always practical, so 10′ beam data are often used. However, one should keep this in mind when interpreting abundance results.

• *Molecular hydrogen.* In principle, a derivation of the hydrogen column density also requires taking H_2 into account. However, its column density is generally found to be low in the HVCs. Richter et al. (2001b, 2003) find a fraction $N(H_2)/N(H\,I)<10^{-3}$ toward IVCs and the Magellanic Stream, and no H_2 in complex C.

• *Ionization.* At typical temperatures in the warm ISM (~7000 K), most elements will be either neutral or once-ionized (see Sutherland & Dopita 1993). However, photoionization caused by UV radiation produced by hot stars may change the balance. As a result, many elements can exist in multiple ionization stages, and the balance differs in different parts of a cloud. For some elements lines of more than one ionization stage are accessible, which then provide information on the physical conditions. Photoionization also affects hydrogen, and to derive abundances one needs to know the sum of $N(H\,I)$ and $N(H^+)$. The latter is difficult to measure, but one can get a handle on it from Hα emission, even though its intensity, $I(H\alpha)$, is not directly proportional to $N(H^+)$. Tufte (Ch. 8) discusses one way to deal with this problem.

• *Depletion.* In the diffuse ISM almost all elements are to some extent incorporated in dust grains, reducing their abundance in the gas phase. This effect is referred to as "depletion". Savage & Sembach (1996) review the depletions of many elements, and conclude that the amount of an element locked up in dust depends on the environment: in dense, cold clouds more atoms are in dust grains than in the warmer, more diffuse ISM. The depletion is least in halo gas (such as IVCs). Wakker & Mathis (2000) give a slightly different possible interpretation, showing that the apparent abundance of Ca II, Fe II, Mg II, and Mn II anti-correlates with $N(H\,I)$ (which is highest for cold clouds and lowest for halo gas). The more refractory elements (Ni, Fe, Cr) are more depleted than Mg and Si, while S and P are almost undepleted. C and O make up a substantial fraction of the dust particles, but because they are so abundant, their depletion is low.

• *Reference abundances.* To compare the abundances measured in the ISM, it is useful to divide them by a standard reference. Solar abundances are most commonly used, since these are relatively well-measured for all elements. However, recent work on the subject shows that even these need to be used with care. The tables of Anders & Grevesse (1989) (updated for several elements by Grevesse & Noels (1993) and Grevesse et al. (1996)) have been widely used. However, Holweger (2001) and Allende-Prieto et al. (2001, 2002) find that a more careful analysis of the solar spectrum gives solar C, N, O, Mg, Si, and Fe abundances that differ by up to 0.22 dex from values published earlier. Fox et al. (2004) show that for complex C homogenizing the different determinations of O/H and S/H removes some apparent discrepancies, see Sect. 6.1. For the discussion of other HVCs in Sect. 6.2 published values have not been corrected, however.

• *Intrinsic variations.* If there are no data to the contrary, the usual assumption is that the abundance of an ion is constant across a cloud. For heavily depleted and/or non-dominant ions (e.g. N I, Ca II) this may not be a good assumption, but for lightly-depleted ions that are the dominant ion (e.g. O I, S II, Si II) it probably is. However, even in the latter case there may be abundance differences in a cloud, as is shown by the range of about a factor 2 found for S II/H I in complex C (see Table 4); some of this variation may be related to unknown corrections for hydrogen ionization and small-scale structure, but some may be real.

• *Abundance vs metallicity.* The existence of element depletion and ionization effects (see below) leads to a subtle difference in the common usage of the terms "abundance" and "metallicity". The observed ratio N(ion)/N(H I) is commonly called the ion's abundance, or (better) its gas-phase abundance. Sometimes a correction for H^+ is attempted. However, the metallicity of a cloud is to be understood as the *total* abundance of a heavy element relative to *total* hydrogen, divided by its solar abundance: $A(\mathrm{X})/A_\odot(\mathrm{X}) = \{N(\mathrm{X})/N(\mathrm{H})\}\ /\ \{N(\mathrm{X})/N(\mathrm{H})\}_\odot$. The total abundances imply corrections for ionization effects and depletion, or: $N(\mathrm{X}) = \delta^{-1}$ $\Sigma_{\mathrm{i.s.}} N$(element in i.s.), where "i.s." stands for "ionization stages", and δ is the fraction of the element in the gas-phase. As δ is a-priori unknown, only elements for which δ=1 is expected are suitable for measuring cloud metallicities. This includes oxygen, sulphur, and phosphorus (see Table 3), although phosphorus may show some depletion if $N(\mathrm{H\,I})>5\times10^{19}$ cm^{-2} (Lehner et al. 2004). For S and P ionization effects still are important. Collins et al. (2003) find that in complex C N(S II)/N(H I) decreases with increasing N(H I). The least ambiguous results are obtained when using the O I ion, as its column density is coupled to that of H I through a charge-exchange reaction (Osterbrock 1989), because of the tiny difference in ionization potential between O I and H I (13.618 vs 13.598 eV). Thus, in almost all environments, N(O I)/N(H I) is equivalent to an ionization-corrected metallicity, with just a small uncertainty associated with possible oxygen depletion.

There are a final few considerations to take into account. First, the ISM is opaque between about 200 Å and 912 Å because of Lyman-continuum absorption. Second, in the diffuse ISM most ions occur only in the ground state – the density is too low to collisionally excite the ions. Third, most ground-state lines are in the ultraviolet part of the spectrum. Fourth, turbulent velocities in the ISM are at most tens of km s^{-1}, so that typically a resolution of $\Delta\lambda/\lambda$>30,000 is required. With these considerations, determining

TABLE 3. Interstellar lines useful for metallicities and physical conditions

Ion	Lines [Å]	N(H I;τ=0.1)[a] [10^{18} cm^{-2}]	Measures
C II	1036.337, 1334.523	1.7, 1.2/depl(C)	extended H I
C II*	1037.018, 1335.708		heating-cooling balance
C III	977.020	0.03/frac(C III)[b]	hot or photoionized envelope
N I	1200, 1134, 963, 953[c]	(0.6 to 10)/frac(N I)	metallicity/radiation field[d]
N II	1083.994	0.8/frac(N II)	N I/N II: ionization correction
O I	1302.169, 1039.230, 948.686, 936.630, 929.517, 950.885, 937.841, 930.256, ...	0.22 to 30	metallicity[e]
Si II	1260.422, 1193.290, 1190.416, 1526.707, 1304.370, 1020.699	0.32 to 26/depl(Si)	(metallicity), depletion[f]
Si III	1206.500	0.22/frac(Si III)	hot or photoionized envelope
P II	1152.818, 963.801	20 to 105	(metallicity[g])
S II	1259.519, 1253.811, 1250.584	19 to 55/frac(SII)	metallicity[h]
Ar I	1048.220, 1066.660	(36 to 144)/frac(Ar I)	photoionization balance[i]
Fe II	2382.765, 2600.173, 1144.938, 1608.451, 2374.461, 1063.176, 1096.877, 1121.975, ...	(0.24 to 19)/depl(Fe)	depletion onto dust
Ni II	1317.217, ...	24/depl(Ni)	depletion onto dust
Zn II	2062.664, 2026.136	180, 360	metallicity[j]

Notes: (a) H I column density at which the optical depth (τ) of the line is 0.1, assuming an FWHM of 20 km s^{-1} (b-value 12 km s^{-1}), a metallicity of 0.1 times solar, and using oscillator strengths from Morton (2000).
(b) frac(ion) and depl(element) are, respectively, the fraction of the element in the appropriate ionization stage, and the fraction remaining in the gas phase after depletion onto dust.
(c) N I has several triplets near the listed wavelengths. However, the lines of HVCs and low-velocity gas often blend, and many other ISM lines are present near 963 and 953 Å.
(d) Most of the nitrogen is in the gas phase, not in dust grains. Sofia & Jenkins (1998) show that photoionization tends to favor N^{+1} over N^0.
(e) O I is the best ion for a metallicity determination, as it is coupled to H I via a charge-exchange reaction (Osterbrock 1989).
(f) Silicon tends to be lightly depleted onto dust grains, making metallicities uncertain by a factor up to 2.
(g) Phosphorus tends to be lightly or not at all depleted at low N(H I); however, the lines are relatively weak and have not been seen in HVCs.
(h) The S II triplet gives the second most-reliable metallicities, as sulphur is barely if at all present in the dust. However, an ionization correction seems necessary at low N(H I).
(i) Like N I, Ar I is not depleted onto dust, but strongly affected by photoionization (Sofia & Jenkins 1998).
(j) Zinc is an iron-group element that is not depleted onto dust, and in principle can provide the ratio between iron-group elements (mostly formed in type Ia supernovae) and α-elements such as O, Mg, Si, and S (mostly formed in type II supernovae).

abundances requires space observatories with high-resolution spectrographs sensitive to wavelengths between 900 and 3000 Å. Two observatories (*HST* and *FUSE*) have been useful in this respect (see Table 1). Table 3 summarizes accessible absorption lines that contain useful information about HVC metallicities and physical conditions.

6. Ion abundances and metallicities of HVCs

The far-UV spectrographs onboard *HST* and *FUSE* have allowed rapid progress in the study of metal ions in HVCs. We first discuss (Sect. 6.1) the many results obtained on the largest HVC, complex C; in Sect. 6.2 we summarize work on other HVCs.

6.1. COMPLEX C

Combining *HST*/*GHRS* observations of the S II absorption lines at 1250 to 1259 Å with 21-cm H I measurements taken with the *Westerbork*, *Effelsberg* and *Dwingeloo* telescopes (resolutions 1′, 9′, and 36′), and observations of Hα and [S II] λ6716 with the Wisconsin H-Alpha Mapper (*WHAM*; resolution 1°), Wakker et al. (1999) derived a value of $A(\mathrm{S})/A_\odot(\mathrm{S})$=0.09±0.03 for the sulphur abundance in complex C on the sightline to the Seyfert galaxy Mrk 290. The ratio $N(\mathrm{H}^+)/N(\mathrm{H}^0)$ was derived from a model of the H II region adjoining the HVC; the ratio $N(\mathrm{S}^{+2})/N(\mathrm{S}^+)$ was estimated. With its mass exceeding 10^6 M$_\odot$ (Table 2) and downward velocity, complex C was found to provide a major influx of low-metallicity gas, with significant effects on the Galaxy's future chemical evolution.

Since then, about a dozen AGN sight lines through complex C have been studied. Figure 6 offers a small selection of the many absorption lines found in these rich UV spectra. Table 4 summarizes the metallicities $A/A_\odot$ derived from these spectra. The values given in the original publications were adjusted by Fox et al. (2004) to a common set of standard (solar) abundances. Since the errors quoted for $A/A_\odot$ are often asymmetric, metallicity ranges spanned by these errors are given instead. Note that some overlap exists in the material used in the different papers; hence the different $A/A_\odot$ values quoted for the same sightline are not always independent!

Except for the ratio S/H that was derived by Wakker et al. (1999) toward Mrk 290, no ionization corrections were applied to the measured ratios $N(\mathrm{S\,II})/N(\mathrm{H\,I})$ and $N(\mathrm{O\,I})/N(\mathrm{H\,I})$. For the ratio O I/H I no corrections are necessary (Sect. 5), but for low H I column densities the corrections to S II/H I may be significant. Collins et al. (2003) have calculated such corrections, "for illustrative purposes only"; Tripp et al. (2003) have calculated and applied them (to other ion ratios than those mentioned here). However, on every sightline the derived ratios for oxygen and sulphur appear consis-

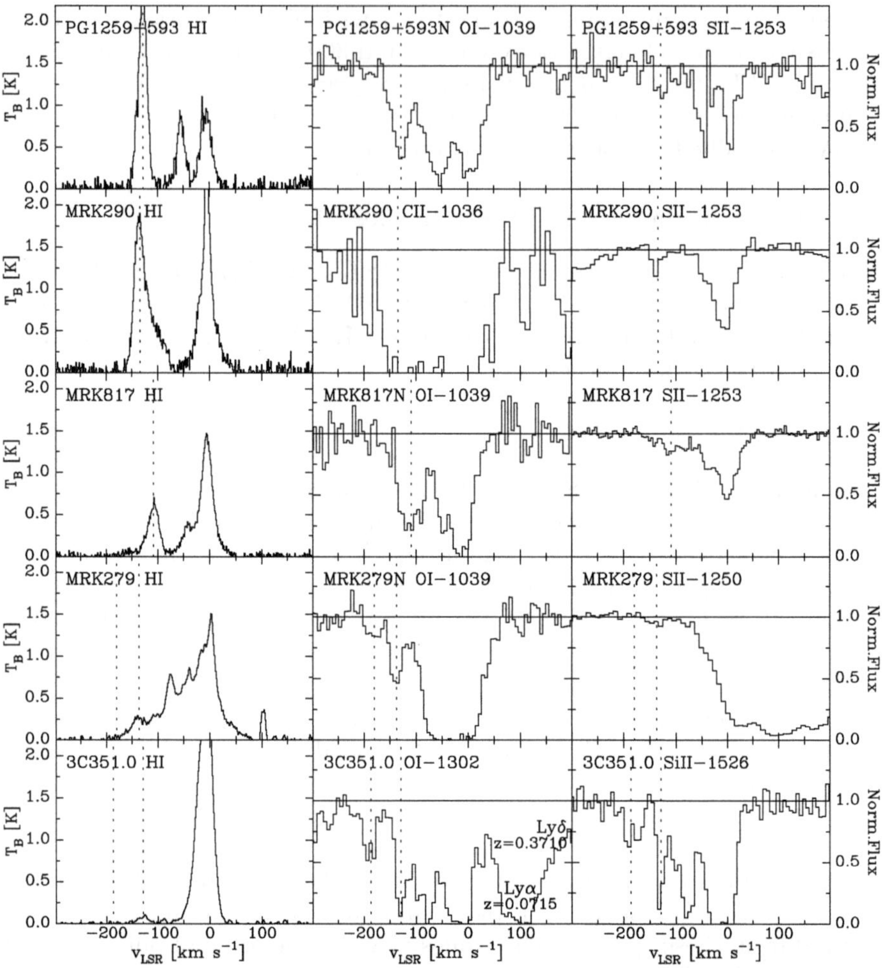

Figure 6. Representative H I and ionic spectra for five sightlines through complex C, sorted by N(H I) in complex C. For each sightline the *Effelsberg* H I spectrum is shown, as well as two good ionic lines (O I and S II when available). In all panels dotted vertical lines indicate the velocity of complex C; a second dotted line sometimes indicates a second component, the "high-velocity ridge" (Tripp et al. 2003). Some target names are followed by "N" when *FUSE* orbital-night-only data were used, to avoid contamination by geocoronal O I emission. Figure composed for this article.

tent. While the sightlines with high N(H I) show lower metallicities (0.09 toward Mrk 290, 0.15 toward PG 1259+593) than the values (around 0.4) on sightlines with more modest column densities (Mrk 279 and Mrk 817), the case of 3C 351.0 breaks this possible trend.

Taken together, the metallicities found for complex C span a range from 0.1 to 0.3, possibly even 0.5 solar. However, the lowest values, 0.1 to 0.2,

TABLE 4. Metallicity determinations for complex C

Probe	l [°]	b [°]	N(H I)^a [cm^{-2}]	H I^a source	Ion ratio^b	$A/A_\odot{}^c$	Ref.	Notes
Mrk 279	115.04	46.86	30.7×10^{18}	Eff	SII/HI	0.30–0.68	1	(d)
					SII/HI	0.34–0.78	2	(d)
					OI/HI	0.15–0.62	2	(d,e)
Mrk 290	91.49	47.95	91.7×10^{18}	WSRT+Eff	S/H	0.07–0.12	3	(d,f)
					SII/HI	0.07–0.09	1	(d)
					SII/HI	0.07–0.14	2	(d)
Mrk 817	100.30	53.48	30.5×10^{18}	Eff	SII/HI	0.30–0.39	1	(d)
					SII/HI	0.38–0.55	2	(d)
					OI/HI	0.24–0.63	2	(d,e)
PG 1259+593	120.56	58.05	90.0×10^{18}	WSRT+Eff	SII/HI	0.10–0.19	4	(d)
					SII/HI	0.13–0.25	2	(d)
					OI/HI	0.07–0.33	4	(d,e)
					OI/HI	0.08–0.21	2	(d,e)
					OI/HI	0.11–0.21	5	(d)
3C 351.0	90.08	36.38	4.2×10^{18}	Eff	OI/HI	0.12–0.33	6	(d,e)

References: 1: Gibson et al. (2001); 2: Collins et al. (2003); 3: Wakker et al. (1999); 4: Richter et al. (2001a); 5: Sembach et al. (2004); 6: Tripp et al. (2003).
Notes: (a) H I column density and source of H I data – Eff means spectrum was measured with *Effelsberg* (9′ beam, see Wakker et al. 2001), WSRT+Eff means *Effelsberg* data and *WSRT* data were combined (see Wakker et al. 1999 and Sembach et al. 2004). (b) Measured ion ratio from which metallicity is derived. (c) $A/A_\odot$ values derived from S II/H I have *not* been corrected for ionization effects; those from O I/H I are insensitive to ionization. (d) Metallicity, $A/A_\odot$, adjusted by Fox et al. (2004) to use standard solar reference abundances. (e) Fe II, Si II also measured. (f) S II measured in absorption and (undetected) in emission, H I and Hα both measured, ionization corrections applied.

appear to be the most reliable; and while Gibson et al. (2001) and Collins et al. (2003) claim that "Complex C cannot be characterized by a single metallicity", Tripp et al. (2003) consider this still possible.

The UV lines provide several other important abundance ratios. Collins et al. (2003) note that the abundance ratios of S, Si, and O in complex C are essentially solar; since S and O are always essentially undepleted, this shows that in this HVC silicon suffers little depletion onto dust. On the other hand, the ratio N/O is strongly subsolar. This suggests that the gas producing complex C may have been removed from its parent galaxy after production of the α-elements O, Si, and S in supernovae of type II, but before enrichment of N by evolved low-mass stars took place.

From the spectrum of PG 1259+593, Sembach et al. (2004) have derived the deuterium abundance in complex C. In view of the extreme weakness of the D I lines compared to the H I Lyman series this is a tricky job, but it appears to have been successfully done. Sembach et al. (2004) find an abundance $N(\mathrm{D})/N(\mathrm{H})=(2.2\pm0.7)\times10^{-5}$ for complex C, consistent with the primordial value of $(2.6\pm0.2)\times10^{-5}$, but probably higher than the Galactic value, which lies between 0.8 and 1.5×10^{-5}. This suggests an extragalactic, rather than a galactic, origin for complex C. Note that this is the *only* determination of D/H in the local universe outside the Milky Way Disk!

Sembach et al. (2003) have found O VI absorption on 9 sightlines through complex C. In view of the close agreement between H I, Si II, and O VI velocities on 5 sightlines, Fox et al. (2004) argue that complex C is surrounded by an envelope of highly-ionized material. The ratios of high-ion line strengths appear to be inconsistent with isothermal collisional equilibrium; the highly-ionized high-velocity gas may rather originate at the conductive or turbulent interfaces between the neutral or lightly-ionized parts of complex C and a surrounding hot medium (Galactic Halo or intergalactic Local Group gas). Tripp et al. (2003) find that the strength of O VI rises toward the lower (l and b) end of complex C, suggesting that this "low-metallicity, high-velocity cloud plunges into the Milky Way".

6.2. OTHER HVCS

Table 5 summarizes the metallicity values $A/A_\odot$ derived for HVCs. It also mentions the Ca II abundances, discussed in Sect. 6.3 below. The appropriate H I column densities and references are given separately for the metallicities and for the Ca II abundances.

For complex C the metallicity could be derived on 5 sightlines (Table 4); for most other HVCs only one determination is available.

• *Complex A.* This HVC has the best-determined distance (Table 2), but its metallicity is only poorly known. The value given, 0.1 with an uncertainty of a factor of order 3, is based on a 3-σ detection (Kunth et al. 1994) of an O I line whose structure and width are uncertain (Wakker 2001). The saturated Mg II absorption measured in Mrk 106 (Fig. 2; Wakker et al. 1996b) just gives a lower limit $A(\mathrm{Mg})>0.035\ A_\odot(\mathrm{Mg})$. Toward a number of other AGNs observed with *FUSE* the O I λ1039 absorption associated with complex A is contaminated by an H_2 line. However, if the metallicity of complex A were larger than about 0.3 solar, there would be a saturated feature, which is not observed to be the case.

• *Complex M.* In cloud M I, at -120 km s^{-1}, Tufte et al. (1998) have measured the red [S II] emission line, yielding $I([\mathrm{S\,II}])/I(\mathrm{H}\alpha)=0.64\pm0.14$. Making proper assumptions about geometry, temperature, and the ratio S^{+2}/S^{+} this gives an abundance $A(\mathrm{S})$ of order 0.8 $A_\odot(\mathrm{S})$, with a wide error range (Wakker 2001).

• *Magellanic Stream (tail).* Against Fairall 9 (Fig. 7a) Gibson et al. (2000) find $A(\mathrm{S\,II})=(0.15$ to $0.48)\ A_\odot(\mathrm{S})$ without ionization correction. Two other

TABLE 5. Metallicities and Ca II abundances

complex or cloud name	$A/A_\odot$	ion	N(H I) [10^{18}cm^{-2}]	Ref.	A(Ca II) [10^{-9}]	N(H I) [10^{18}cm^{-2}]	Ref.
complex A	0.03–0.4	O I	20	1, 2	19±5	38, 80	3, 4
complex M (M I)	0.4–1.8	[S II]	<0.5?	5, 2			
complex C[a]	0.10–0.5?	O I,S II	4–92	6	17±5	33–92	7, 2
Magellanic Stream	0.3±0.2	S II	94	8	27±2	89	9, 2
Leading Arm, WW187	0.25±0.08	S II	80	10, 11	6.7±0.8	83	12
WW 84	0.10–0.32	O I	11	13	<89	8±5	14, 2
complex WB, WW92[b]	$\gtrsim$0.3[c]	O I	11	13	280±120		15, 2
complex WD, WW226	0.05–0.35	O I	51	13			
HVC 224−83−197	<0.46	O I	2.2	16			
HVC 100−7+110	0.1–4	O I	5	17, 18			
HVC against LMC,+120	0.1–0.5	O I	6–12	19	20–40	6–7	20[d]

References: 1: Kunth et al. (1994), 2: Wakker (2001), 3: Schwarz et al. (1995), 4: van Woerden et al. (1999a), 5: Tufte et al. (1998), 6: see Table 4, 7: Wakker et al. (1996a), 8: Gibson et al. (2000), 9: Songaila & York (1980) and Songaila (1981), 10: Lu et al. (1998), 11: Sembach et al. (2001), 12: West et al. (1985), 13: Wakker et al. (in prep.), 14: Bowen et al. (1991a, b), 15: Robertson et al. (1991), 16: Sembach et al. (2002), 17: Stoppelenburg et al. (1998), 18: Bates et al. (1990, 1991), 19: Bluhm et al. (2001), 20: Molaro et al. (1993). Notes: (a) Based on several sightlines, see Table 4. (b) Metallicity and Ca II abundance based on different sightlines. (c) The absorption has a relatively small and uncertain b-value, making the ionic column densities uncertain: $A/A_\odot$=0.28 if b=10 km s^{-1}, 0.58 if b=6 km s^{-1}. (d) For this HVC absorption by Ca II and other ions has been published in many papers, but for only two stars are reliable values for both N(H I) and N(Ca II) known (see Wakker 2001).

sightlines provide lower limits of 0.03 and 0.05 times $A_\odot$(Mg) for A(Mg II). From low-resolution (220 km s^{-1}) spectra on several sightlines, Savage et al. (2000) derive 0.10 solar for A(Mg II)/$A_\odot$(Mg), indicating mild depletion of Mg. Wakker (2001) derives an Fe/S ratio of 0.18±0.06 solar, showing depletion of iron onto dust.

• *Leading Arm.* Against NGC 3783, Lu et al. (1998) and Sembach et al. (2001) find an S II/HI ratio of 0.25±0.08 solar in the cloud WW 187 (Fig. 7b). The Si II/S II ratio of 0.61±0.23 solar shows that Si is only slightly depleted. The ratio Fe II/S II=0.13±0.04 (Fe/S)$_\odot$ indicates strong depletion of iron.

The metallicity values of both Magellanic Stream and Leading Arm agree with that of the Small Magellanic Cloud, which is about 0.3 solar. In the Magellanic Bridge a metallicity of only 0.08 solar has been found (Lehner et al. 2001a), consistent with the metallicities of stars in the Bridge.

• *WW 84 = HVC 125+41−207.* For this HVC, which lies projected on Mrk 205 at l=125°, b=+42°, and is part of Population EN, Wakker et al. (in prep.) find O I/H I between 0.1 and 0.4 solar. Si II (0.05±0.02) and Fe II (0.03 to 0.12 solar) appear depleted, and the abundance of N I is also low:

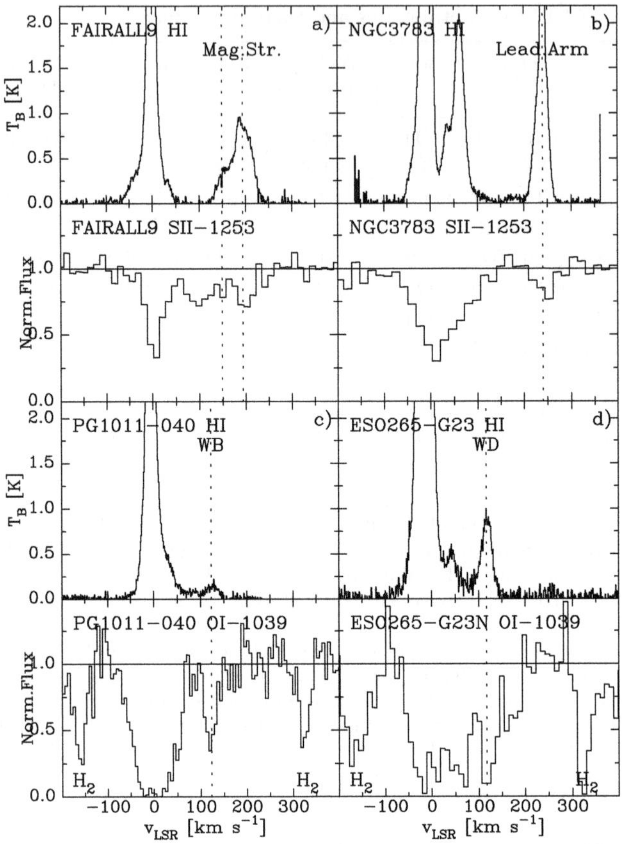

Figure 7. H I and metal-line spectra for four objects that yield useful metallicity measurements for HVCs other than complex C. *Top two rows:* (a, b) *Parkes* and *Green Bank* H I 21-cm data, and *HST* S II absorption for the two sightlines through the trailing and leading arms of the Magellanic Stream. *Bottom two rows:* (c, d) *Green Bank* and *Villa Elisa* H I 21-cm data and (previously unpublished) *FUSE* O I absorption for two sightlines through two positive-velocity HVC complexes near l=270°, b=+30°. Figure composed for this article.

0.03 to 0.12 solar.

- *WW 92 = HVC 244+41+113.* This HVC lies in complex WB. Fig. 7c shows the O I absorption in PG1011−040 from a *FUSE* spectrum obtained by Wakker, Gibson et al. (in prep.). Abundances of O I, Si II and Fe II all are ill-determined, with lower limits only of 0.1 solar, and probable values around 0.7 solar. N I also has a lower limit of 0.1, but 0.2 solar as probable value.
- *WW 226 = HVC 284+15+117.* This HVC lies in complex WD, not far from WW 187 (which lies in the Leading Arm, but at a quite different velocity). The spectrum of ESO 265−G23 (Fig. 7d) yields an O/H ratio

between 0.05 and 0.35 solar, Si II/H I between 0.08 and 0.35, Fe II/H I between 0.08 and 0.19 solar, and again a low N I abundance: 0.02 to 0.11 solar (Wakker et al., in prep.).

• *HVC 224−83−197.* This compact object has been observed by Sembach et al. (2002) in absorption against Ton S210. The upper limit for its metallicity, O/H<0.46 solar, would allow a relationship with the Magellanic Stream, which also lies nearby.

• *HVC 100−7+110.* This tiny HVC, at less than 1.3 kpc distance and within 200 pc from the plane (Sect. 4), has wide limits on its O/H abundance: between 0.1 and 4 times solar. Its Mg II, Fe II and Al II abundances are consistent with this roughly solar value.

• *The HVC at $v_{\rm LSR}$=+120 km s^{-1} in front of the LMC.* The nature and location of this cloud are controversial (Wakker & van Woerden 1997). It may lie close to the LMC and belong to it (but would then be moving *away* from it), or it may lie in the Milky Way Halo. Absorption lines have been measured in *IUE* spectra of many LMC stars, and especially in SN 1987A. However, accurate abundances are lacking, as are high-resolution H I observations in the directions of the background targets, so no reliable metallicity values can be given (see Wakker 2001). Note that for the HVC at $v_{\rm LSR}\sim$+165 km s^{-1} seen against the LMC even fewer reliable H I column densities are known.

6.3. ABUNDANCES OF Ca II

Table 5 also mentions the abundances of Ca^{+} ions relative to hydrogen in several HVC complexes. In complex A, the two values measured against Mrk 106 (Schwarz et al. 1995) and AD UMa (van Woerden et al. 1999a; Wakker 2001) are consistent. Also, in complex C the values for the sightlines to Mrk 290 and PG 1351+640 (Wakker et al. 1996a) differ little. These two large HVCs have almost the same Ca II abundance.

The value reported for the Magellanic Stream (Songaila 1981) is not much larger than those for complexes A and C. On the other hand, the value 6.7±0.8 for the Leading Arm (West et al. 1985) is a factor 4 lower than that of the Stream. (Note that the value reported earlier by Wakker (2001) is a factor 10 too large.)

The high value measured for complex WB by Robertson et al. (1991) raises the range of A(Ca II) to a factor 40. As discussed in Sects. 3 and 4, usage of the Ca II lines for distance determination does require measuring a fiducial value against an extragalactic probe behind the same HVC!

The values found for A(Ca II) show that its lines are useful for distance determinations, but may require long exposures (c.f. Sect. 3.6).

7. Origins

What do the available distances (Table 2) and metallicities (Table 5) of HVCs tell us about their origins? Bregman (Ch. 16) discusses several possible origins in detail.

Locations in the intergalactic medium (IGM) or high in the Halo ($|z|>5$ to 10 kpc, say) are compatible with origins through tidal disruption or capture of Milky Way satellites, and with origins in the galaxy-formation process. The only HVCs with accurate locations: complex A, the Magellanic Stream, and the Leading Arm, fall in this category. HVCs in the lower Halo or in the Disk (cloud M II–III; complex WE; complex H) share the same possible extragalactic origins, but may also originate in a Galactic Fountain. The same wide range of options is open to several HVCs for which only a low lower limit to distance is available. Only the dwarf-HVC 100−7+110 (Sect. 4) almost certainly must have a local origin, in the Milky Way Disk. For any HVC with $D>100$ kpc, the galaxy-formation process (Ch. 14) would be the most likely origin. But except in the latter case, a distance alone can hardly betray an HVC's roots. And with velocity information limited to one dimension, orbit calculations can only be done statistically.

Metallicities give a somewhat stronger handle. HVCs originating in the Disk, via a Galactic Fountain, should have near-solar metallicities. Tidal disruption of the Magellanic Clouds should entail metallicities of order 0.3 (for origins in the SMC) to 0.5 (LMC), or somewhat lower values if the origin dates back several Gyr. Disruption of smaller Milky-Way satellites may imply metallicities of 0.01 to 0.2. An intergalactic (or galaxy-formation) origin allows even more leeway; Tripp et al. (2002) find average C/H and Si/H abundances around 0.05 to 0.2 solar for Lyman-α clouds in the Virgo Supercluster; Nestor et al. (2003) derive average metallicities of 0.05 to 0.2 solar for zinc (which is known to show little depletion) in damped Lyman-α gas at redshifts $0.9<z<2.0$ from Sloan Digital-Sky Survey spectra; and Prochaska et al. (2003) find metallicities between 0.002 and 0.5 from 100 damped Lyman-α systems at redshifts $0.5<z<5$. In addition to the "standard" metallicity values based on O, S, Si, or C, the N/H underabundances may carry information on HVC origins (cf. Sect. 6.1).

Clues to the origins of HVCs may also be obtained through possible relationships to the stellar streams discovered in the Galactic Halo. And another stellar source of information which has remained largely overlooked is the gas (and dust) lost by evolving red giant stars (de Boer 2004): to what extent can this source produce high-velocity gas falling toward the Disk?

Combining the data on distance and metallicity, we find that complex C must have an extragalactic (dwarf or intergalactic) origin, and complex A probably is a similar case. Complex M appears to be part of a Galactic

Fountain; and so may complex WB if its metallicity is indeed high. The origin of Magellanic Stream and Leading Arm clearly lies in tidal disruption of the Magellanic Clouds. As suggested by Putman et al. (2004), the AC clouds may be tidal debris of the Sgr Dwarf, but neither distance nor metallicity information is available to support this. HVC 100−7+110 is a minor local event. For the other HVCs and complexes mentioned in Tables 2 and 5, available information is too meager to derive an origin.

From this summary of only scarce data, it is clear that the HVC phenomena are diverse in terms of distance, metallicity, and origins.

8. Needs and prospects

Although important progress has been made in recent years, information on distances, metallicities, and origins of HVCs is still scarce. We badly need an upper limit on the distance of complex C; or rather: distance brackets for different parts of this large object, which may be strongly interacting with the Milky Way Halo. For complex A also, brackets at both the high-b and the low-b end will help to understand the geometry (and possibly the dynamics). A good distance bracket for cloud M I could be readily obtained, since suitable stellar and extragalactic probes are at hand. Distances for the ACHV and ACVHV complexes would be of great interest, in view also of their possible relation to the Sgr Dwarf. A distance bracket for complex H would be important, but hard to obtain in this region of high extinction. The distances of GCP and GCN, of complexes WA-WD, and of the many small HVCs are fully open questions. For the small HVCs, suitable probes will be rare. At distances of 10 to 50 kpc, probes of magnitudes 16 to 21 must be found, and observed. The *SDSS* data bank (Sect. 3.2) offers hope.

Recent metallicity studies have strongly emphasized complex C and the Magellanic tidal filaments, with considerable success. Current needs lie elsewhere: especially complex A and the AC clouds, but also GCN and GCP, the WA-WD complexes, and especially the many small HVCs. Since for metallicities extragalactic probes suffice, some of the small HVCs should be possible; but they may require very long exposures on fainter AGNs.

Large optical telescopes, such as *Keck* and the *VLT* (and others), have so far hardly been used for HVC distances. The faint magnitudes just mentioned will require time on these great instruments. The strong current interest in the HVC problem may help time-allocation committees to decide in favor.

Metallicities can only be measured in space. How long will *FUSE* and *HST* last, and will there be suitable successors? The importance of HVCs to problems of galaxy formation and evolution calls for one.

Progress in this field will profit from collaboration of astronomers working on various aspects of the problem. The website created by Wakker (http://www.astro.wisc.edu/~wakker/dist/dist.html), contains a wealth of information on HVCs and recommended observing programs, and may foster such collaboration.

Acknowledgements

Van Woerden thanks Reynier Peletier and Ulrich Schwarz for their years of collaboration on the distance-determination program.

References

Allende-Prieto, C., Lambert, D.L., Asplund, A. 2001, ApJ, 556, L63

Allende-Prieto, C., Lambert, D.L., Asplund, A. 2002, ApJ, 573, L137

Anders, E., Grevesse, N. 1989, Geochim, Cosmochim Acta, 53, 197

Bates, B., Catney, M.G., Keenan, F.P. 1990, MNRAS, 242, 267

Bates, B., Catney, M.G., Gilheany, S., Keenan, F.P., Davies, R.D., Hummel, E. 1991, MNRAS, 249, 282

Beers, T.C., Wilhelm, R., Doinidis, S.P., Mattson, C.J. 1996, ApJS, 103, 433

Benjamin, R.A., Danly, L. 1997, ApJ, 481, 764

Bland-Hawthorn, J., Maloney, P.R. 1999, ApJ, 510, L33; erratum in ApJ, 550, 231

Bland-Hawthorn, J., Veilleux, S., Cecil, G.N., Putman, M.E., Gibson, B.K., Maloney, P.R. 1998, MNRAS, 299, 611

Blitz, L., Spergel, D.N., Teuben, P.J., Hartmann, D., Burton, W.B. 1999, ApJ, 514, 818

Bluhm, H., de Boer, K.S., Marggraf, O., Richter, P. 2001, A&A, 367, 299

Bowen, D.V., Pettini, M., Penston, M.V., Blades, J.C. 1991a, MNRAS, 248, 153

Bowen, D.V., Pettini, M., Penston, M.V., Blades, J.C. 1991b, MNRAS, 249, 145

Braun, R., Burton, W.B. 2000, A&A, 354, 853

Braun, R., Burton, W.B. 2001, A&A, 375, 219

Bregman, J.N. 1980, ApJ, 236, 577

Brown, W.R., Geller, M.J., Kenyon, S.J., Beers, T.C., Kurtz, M.J.,Roll, J.B. 2004, AJ, 127, 1555

Brüns, C., Kerp, J., Pagels, A. 2001, A&A, 370, L26

Burton, W.B., Braun, R., Chengalur, J.N. 2001, A&A, 369, 616; erratum in A&A, 375, 227

Centuríon, M., Vladilo, G., de Boer, K.S., Herbstmeier, U., Schwarz, U.J. 1994, A&A, 292, 261

Collins, J.A., Shull, J.M., Giroux, M.L. 2003, ApJ, 585, 336

Danly, L., Albert, C.E., Kuntz, K.D. 1993, ApJ, 416, L29

Danly, L., Lee, Y.P., Albert, C.E. 1995, BAAS, 27, 860

Davies, J., Sabatini, S., Davies, L., Linder, S., Roberts, S., Smith, R., Evans, Rh. 2002, MNRAS, 336, 155

de Boer, K.S. 2004, A&A, 419, 527

de Heij, V., Braun, R., Burton, W.B. 2002, A&A, 392, 417

Espresate, J., Cantó, J., Franco, J. 2002, ApJ, 575, 194

Fox, A.J., Savage, B.D., Wakker, B.P., Richter, P., Sembach, K.R., Tripp, T.M. 2004, ApJ, 602, 738

Gardiner, L.T. 1999, in ASP Conf. Ser. 166, Stromlo Workshop on High-Velocity Clouds, eds. B.K. Gibson, M.E. Putman (San Francisco: ASP), 292

Gardiner, L.T., Noguchi, M. 1996, MNRAS, 278, 191

Gibson, B.K., Giroux, M.L., Penton, S.V., Putman, M.E., Stocke, J.T., Shull, J.M. 2000, AJ, 120, 1830

Gibson, B.K., Giroux, M.L., Penton, S.V., Stocke, J.T., Shull, J.M., Tumlinson, J. 2001, AJ, 122, 3280
Green, R.F., Schmidt, M., Liebert, J. 1986, ApJS, 61, 305
Grevesse, N., Noels, A. 1993, in Origin and Evolution of the Elements, eds. N. Prantzos, E. Vangioni-Flam, M, Cassé (Cambridge: CUP), 15
Grevesse, N., Noels, A., Sauval, A.J. 1996, in ASP Conf. Ser. 99, Cosmic Abundances, eds. S.S. Holt, G. Sonneborn (San Francisco: ASP), 117
Holweger, H. 2001, in AIP Conf. Proc. 598, Solar and Galactic Composition, ed. R.F. Wimmer-Schweingruber (New York: AIP), 23
Hopp, U., Schulte-Ladbeck, R.E., Kerp, J. 2003, MNRAS, 339, 33
Hulsbosch, A.N.M., Wakker, B.P. 1988, A&AS, 75, 191
Keenan, F.P., Shaw, C.R., Bates, B., Dufton, P.L., Kemp, S.N. 1995, MNRAS, 272, 599
Kukarkin, B.V., et al. 1970, General Catalogue of Variable Stars, Third Edition, Volumes I and II (Moscow: Sternberg Institute)
Kunth, D., Lequeux, J., Sargent, W.L.W., Viallefond, F. 1994, A&A, 282, 709
Lehner, N., Sembach, K.R., Dufton, P.L., Rolleston, W.R.J., Keenan, F.P. 2001a, ApJ, 551, 781
Lehner, N., Trundle, C., Keenan, F.P., Sembach, K.R., Lambert, D.L. 2001b, A&A, 370, 996
Lehner, N., Wakker, B.P., Savage, B.D. 2004, ApJ, submitted
Lockman, F.J. 2003, ApJ, 591, L33
Lu, L., Savage, B.D., Sembach, K.R., Wakker, B.P., Sargent, W.L.W., Oosterloo, T.A. 1998, AJ, 115, 162
Moehler, S., Heber, U., de Boer, K.S. 1990, A&A, 239, 265
Molaro, P., Vladilo, G., Monai, S., d'Odorico, S., Ferlet, R., Vidal-Madjar, A., Dennefeld, M. 1993, A&A, 274, 505
Morras, R., Bajaja, E., Arnal, E.M. 1998, A&A, 334, 659
Morras, R., Bajaja, E., Arnal, E.M., Pöppel, W.G.L. 2000, A&AS, 142, 25
Morton, D.C 2000, ApJS, 130, 403
Muller, C.A., Oort, J.H., Raimond, E. 1963, C.R. Acad. Sci. Paris, 257, 1661
Nestor, D.B., Rao, S.M., Turnshek, D.A., Vanden Berk, D. 2003, ApJ, 595, L5
Oort, J.H. 1966, BAN, 18, 421
Oort, J.H. 1967, in IAU Symp. 31, Radio Astronomy and the Galactic System, ed. H. van Woerden (London: Academic Press), 279
Oort, J.H., Hulsbosch, A.N.M. 1978, in Astronomical Papers Dedicated to Bengt Strömgren, eds. A. Reiz, T. Anderson (Copenhagen: Copenhagen University), 409
Osterbrock, D.P. 1989, Astrophysics of Gaseous Nebulae and Active Galactic Nuclei (Mill Valley CA: University Science Book)
Prata, S., Wallerstein, G. 1967, PASP, 79, 202
Preston, G.W., Shectman, S.A., Beers, T.C. 1991, ApJ, 375, 121
Prochaska, J.X., Gawiser, E., Wolfe, A.M., Castro, S., Djorgovski, S.G. 2003, ApJ, 595, L9
Putman, M.E., Thom, C., Gibson, B.K., Staveley-Smith, L. 2004, ApJ, 603, L77
Richter, P., Sembach, K.R., Wakker, B.P., Savage, B.D., Tripp, T.M., Murphy, E.M., Kalberla, P.M.W., Jenkins, E.B. 2001a, ApJ, 559, 318
Richter, P., Sembach, K.R., Wakker, B.P., Savage, B.D. 2001b, ApJ, 562, L181
Richter, P., Wakker, B.P., Savage, B.D., Sembach, K.R. 2003, ApJ, 586, 230
Robertson, J.G., Schwarz, U.J., van Woerden, H., Murray, J.D., Morton, D.C., Hulsbosch, A.N.M. 1991, MNRAS, 248, 508
Ryans, R.S.I., Keenan, F.P., Sembach, K.R., Davies, R.D. 1997, MNRAS, 289, 83
Savage, B.D., Sembach, K.R. 1996, ARA&A, 34, 279
Savage, B.D., Wakker, B.P., Jannuzi, B.T., et 10 al. 2000, ApJS, 129, 563
Schwarz, U.J., Wakker, B.P., van Woerden, H. 1995, A&A, 302, 364
Sembach, K.R., Savage, B.D., Massa, D. 1991, ApJ, 372, 81
Sembach, K.R., Howk, J.C., Savage, B.D., Shull, J.M. 2001, AJ, 121, 992
Sembach, K.R., Gibson, B.K., Fenner, Y., Putman, M.E. 2002, ApJ, 572, 178

Sembach, K.R., Wakker, B.P., Savage, B.D., Richter, P., Meade, M., Shull, J.M., Jenkins, E.B., Sonneborn, G., Moos, H.W. 2003, ApJS, 146, 165
Sembach, K.R., Wakker, B.P., Tripp, T.M., et 13 al. 2004, ApJS, 150, 487
Simon, J.D., Blitz, L. 2002, ApJ, 574, 726
Sirko, E., Goodman, J., Knapp, G.R., Brinkmann, J., Ivezic, Z., Knerr, E.J., Schlegel, D., Schneider, D.P., York, D.G. 2004, AJ, 127, 899
Smith, G.P. 1963, BAN, 17, 203
Sofia, U.J., Jenkins, E.B. 1998, ApJ, 499, 951
Songaila, A. 1981, ApJ, 243, L19
Songaila, A., York, D.G. 1980, ApJ, 242, 976
Stanimirovic, S., Dickey, J.M., Krco, M., Brooks, A.M. 2002, ApJ, 576, 773
Stobie, R.S., Kilkenny, D., O'Donoghue, D., et 24 al. 1997, MNRAS, 287, 848
Stoppelenburg, P.S., Schwarz, U.J., van Woerden, H. 1998, A&A, 338, 200
Sutherland, R.S., Dopita, M.A. 1993, ApJS, 88, 253
Tripp, T.M., Jenkins, E.B., Williger, G., et 11 al. 2002, ApJ, 575, 697
Tripp, T.M., Wakker, B.P., Jenkins, E.B., Bowers, C.W., Danks, A.C., Green, R.F., Heap, S.R., Joseph, C.L., Kaiser, M.E., Linsky, J.L., Woodgate, B.E. 2003, AJ, 125, 3122
Tufte, S.L., Reynolds, R.J., Haffner, L.M. 1998, ApJ, 504, 773
van Woerden, H. 1993, in ASP Conf. Ser. 45, Luminous High-Latitude Stars, ed. D.D. Sasselov (San Francisco: ASP), 11
van Woerden, H., Schwarz, U.J., Peletier, R.F., Wakker, B.P., Kalberla, P.M.W. 1999a, Nature, 400, 138
van Woerden, H., Peletier, R.D., Schwarz, U.J., Wakker, B.P., Kalberla, P.M.W. 1999b, in ASP Conf. Ser. 166, Stromlo Workshop on High-Velocity Clouds, eds. B.K. Gibson, M.E. Putman (San Francisco: ASP), 1
Wakker, B.P. 2001, ApJS, 136, 463
Wakker, B.P. 2004, ApJ, submitted
Wakker, B.P., van Woerden, H. 1991, A&A, 250, 509
Wakker, B.P., van Woerden, H. 1997, ARA&A, 35, 217
Wakker, B.P., Mathis, J.S. 2000, ApJ, 544, L107
Wakker, B.P., van Woerden, H., Schwarz, U.J., Peletier, R.F., Douglas, N.G. 1996a, A&A, 306, L25
Wakker, B.P., Howk, J.C., Schwarz, U.J., van Woerden, H., Beers, T.C., Wilhelm, R., Kalberla, P.M.W., Danly, L. 1996b, ApJ, 473, 834
Wakker, B.P., van Woerden, H., de Boer, K.S., Kalberla, P.M.W. 1998, ApJ, 493, 762
Wakker, B.P., Howk, J.C., Savage, B.D., van Woerden, H., Tufte, S.L., Schwarz, U.J., Benjamin, R., Reynolds, R.J., Peletier, R.F., Kalberla, P.M.W. 1999, Nature, 402, 388
Wakker, B.P., Kalberla, P.M.W., van Woerden, H., de Boer, K.S., Putman, M.E. 2001, ApJS, 136, 537
Wakker, B.P., Savage, B.D., Sembach, K.R., et 18 al. 2003, ApJS, 146, 1
Weiner, B.J., Vogel, S.N., Williams, T.B. 2001, in ASP Conf. Ser., 240, Gas & Galaxy Evolution, eds. J.E. Hibbard, M. Rupen, J.H. van Gorkom (San Francisco: ASP), 515
West, K.A., Pettini, M., Penston, M., Blades, J.C., Morton, D.C. 1985, MNRAS, 215, 481
Wilhelm, R., Beers, T.C., Sommer-Larsen, J., Pier, J.R., Layden, A.C., Flynn, C., Rossi, S., Christensen, P.R. 1999, AJ, 117, 293
Willman, B., Dalcanton, J., Ivezic, Z., Schneider, D.P., York, D.G. 2002, AJ, 124, 2600
Wolfire, M.G., McKee, C., Hollenbach, D.J., Tielens, A.G.G.M. 1995, ApJ, 453, 673
York, D.G., et 143 al. 2000, AJ, 120, 1579
Zwaan, M.A., Briggs, R.H. 2000, ApJ, 530, L61

11. THE HOT HALO

KLAAS S. DE BOER
Sternwarte, Universität Bonn, Germany
deboer@astro.uni-bonn.de

Abstract. The detection in the 1950s of halo high-velocity clouds in Ca II absorption lines led to the prediction of the presence of a "Galactic Corona". Starting in 1973, the hot phase of the local ISM was probed with the *Copernicus* satellite, while X-ray sensors detected a diffuse X-ray background of, at that time, unknown origin. Since 1978, the hot gas in the vast stretches of halo has become accessible through absorption spectra obtained with the *IUE*, *HST* and *FUSE* satellites. This chapter reviews the essentials of both absorption and emission by hot halo gas as well as theories for origin and properties of the hot phase. The large body of results on the hot Galactic Halo obtained with *FUSE* using extragalactic sight lines is summarized at the end.

1. Toward the concept of a hot halo

Ca II absorption lines were discovered by Münch (1952, 1957) at "high" velocities in spectra of stars at high galactic latitude. This absorption implied that clouds of gas exist outside the Milky Way Disk, and that there had to be a medium confining these clouds. Because that medium did not show up in absorption lines, it presumably was well-ionized and thus hot, and so received the name Galactic Corona (Spitzer 1956), borrowed from the solar corona. Today we use not so much the word corona but rather the word halo. In fact, the word corona rather emphasizes the temperature aspect, while "halo" points more at the spatial nature of the gas, as with halo stars. Further details can be found in the chapter on history (Ch. 1).

How does one detect hot gas in the Universe? When background light sources of sufficient brightness are available, one can detect gas through absorption lines. Hot gas is supposedly also well-ionized, so one looks for background light sources with sufficient flux in the spectral range where species dominant in gas of a given high temperature have absorption lines. Such absorption lines are those of ions in the isoelectronic sequence C IV,

H. van Woerden et al. (eds.), High-Velocity Clouds, 227–250

N V, and O VI. These elements have a fairly large abundance in the cosmos (see Table 1 for solar abundances). Alternatively, one attempts to detect continuum emission in the X-ray spectral domain.

Instruments became available to record spectra in the UV at a resolution fit to study interstellar absorption lines. Especially the lines given in Table 1 have been looked for in the spectra of stars in general, and of objects outside the Milky Way Disk in particular. With the *Copernicus* satellite the first extensive measurements could be made of Si IV, C IV, N V, and O VI in interstellar gas in the solar vicinity. Ubiquitous O VI (Jenkins & Meloy 1974; Jenkins 1978a, b) was discovered.

Using rocket-borne detectors diffuse X-ray emission from high galactic latitudes was discovered (see Williamson et al. 1974). The combined detections of O VI and X-ray radiation led to a complete overhaul of the models for the general interstellar medium (Shapiro & Field 1976). A hot and rarefied component was required (in addition to the molecular, cool, and H II phases), with a filling factor of anywhere between 0.2 and 0.8. The hot gas probably comes from supernova explosions, and overpressurized hot gas then may flow into the halo. Theories of the hot interstellar gas have been reviewed by Spitzer (1990).

A large body of data at wavelengths between 1150 and 3000 Å is available from measurements with the International Ultraviolet Explorer (*IUE*) and the Hubble Space Telescope (*HST*). Space Shuttle-borne instruments like the Hopkins Ultraviolet Telescope (*HUT*) and the Orbiting Retrievable Far- and Extreme Ultraviolet Spectrometer (*ORFEUS*) on the shuttle satellite *ASTRO-SPAS* briefly permitted far-UV measurements. The Far Ultraviolet Spectroscopic Explorer (*FUSE*) satellite was launched in 1999 and is now contributing extensive measurements between 900 and 1200 Å.

The *FUSE* satellite has contributed tremendously to our observational knowledge of the hot phase of the ISM. Since this text was conceived before the bulk of the *FUSE* results became known, some aspects of those results are mentioned in the relevant sections, while for the O VI halo and HVC absorption two separate sections are included.

Diffuse radiation at X-ray wavelengths was explored further with rocket and balloon experiments, the *EINSTEIN* satellite and the Röntgen Satellite (*ROSAT*). In particular the *ROSAT* All Sky Survey (*RASS*) provided a wealth of data.

2. Absorption by highly-ionized species in the halo

The first detection of high ions along a sight line through a substantial stretch of Milky Way Halo was made in C IV toward stars in the Large Magellanic Cloud (LMC) (Savage & de Boer 1979). In addition to absorp-

TABLE 1. Data on species with ionization potential (I.P.) above 50 eV having observable resonance absorption lines

Ion	[M/H][a]	Range in I.P. [eV]	Absorption lines [Å]	Strength[a] log $f\lambda$	Temp.[b] [K]	max. frac.[b]	log P^c
C^{+3}	−3.44	47.9−64.5	1548.195	2.470	1.0×10^5	0.35	9.57
			1550.770	2.169			
N^{+4}	−3.95	77.5−97.9	1238.821	2.289	1.8×10^5	0.25	8.48
			1242.804	1.988			
O^{+5}	−3.07	113.9−138.1	1031.926	2.137	2.9×10^5	0.25	9.01
			1037.617	1.836			

Notes: (a) Logarithmic solar abundances [M/H] and f-values from the compilation by Morton (1991); in case of depletion onto dust different values apply (see de Boer et al. 1987b). (b) Temperature of gas in thermodynamic equilibrium at which the ion has the maximum of the ionic relative abundance, "max. frac." (from Arnaud & Rothenflug 1985). (c) Relative logarithmic probability $\log P = \log$ ((abundance+12)×max.frac.×$f\lambda\times n_e$) to find absorption from equal volumes at T, n_e, assuming constant pressure $nT{=}10^4$ cm^{-3} K.

tion at velocities pertaining to gas in the solar vicinity, absorption by C IV was found at +60 and +130 km s^{-1}, associated with high-velocity clouds on the line of sight.

In many other directions C IV has also been found in absorption. These directions include those using extragalactic background light sources, such as quasars, nuclei of galaxies, or extragalactic supernovae. These lines of sight extend far beyond the Milky Way and its local environment. Only the lines of sight to galactic stars and to the Magellanic Clouds will give significant limits to the distance of the hot gas.

The predominance of C IV detections can be explained in a rather mundane way: the C^{+3} ion is intrinsically easier to observe than N^{+4} and O^{+5} (see Table 1). The detection of N V is more difficult due to two factors. The abundance of nitrogen is a factor of 3 below that of carbon and the f-value of the absorption line is a factor of 1.5 smaller than that of C IV. Thus, normally N V has an optical depth a factor of 5 smaller than C IV. In addition, the spectral range near 1240 Å contains complicating spectral features, among them interstellar lines of Mg II which blend with N V.

Detection of O VI is in principle less easy, because the O^{+5} gas should be more rarefied than C^{+3} gas (see Table 1) on a given sight line (assuming thermodynamic equilibrium and constant pressure in the ISM). In addition, the absorption lines lie in a technically more difficult part of the spectrum.

In principle, lines from S VI near 940 Å should also be detectable but with a relative probability log$P$=7.76, almost two orders of magnitude below that of C IV. No report on that line in absorption through halo gas is known, probably because in most stars the spectral continuum is weak and

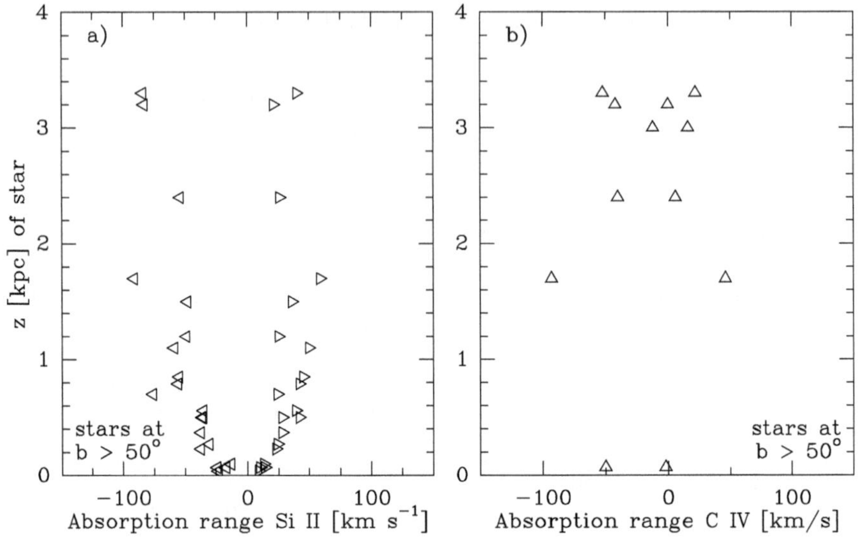

Figure 1. To investigate the motion of the gas in the halo, absorption data toward stars with $|b|>50°$, eliminating a good portion of the effects of galactic rotation, have been collected from the literature (for details see Sect. 3). The panel at left shows absorption data for Si II. The absorption width increases to reach a maximum at $|z|\sim0.8$ kpc, and the average velocity is ~-10 km s^{-1}, suggesting a slow descent of neutral gas toward the Milky Way Disk. The panel at right shows the same information for C IV. C IV absorption is reported only toward stars with $|z|>1$ kpc. Moreover, the velocity of absorption shows a large scatter, in width as well as in average velocity. Figure from de Boer (1998a).

dominated by densely-packed hydrogen lines.

Finally, Si IV is also easily observable. However, its ionization potential is such (33.5 eV) that it can be produced through ionization by stellar photons. Detections of Si IV may be partly due to the H II region gas in the stellar vicinity, and the absorption by halo gas is difficult to separate.

Due to the low far-UV intensity of stellar radiation, the ions of Table 1 normally cannot be produced through stellar photoionization. Rather, they exist in the ISM under conditions related to thermodynamic equilibrium, or derivates from that. Therefore, most attention is given to C IV, N V, and O VI, because they most likely represent the truly hot gas phase.

3. Large-scale distribution of high ions in the halo

The absorption seen in the spectra of high-latitude stars shows a varying width in velocity. This width may be related to the distance of the stars, or to their galactic latitude. Clearly, since galactic rotation causes a velocity spread, its effects must be taken into account. It is therefore mandatory to restrict the study of true halo gas to high-latitude lines of sight with, roughly, $|b|>50°$.

As a reference for the study of the high ions one can use Si II, an ion representing neutral gas. Absorption by Si II shows a width which does not increase for stars with $|z|>0.8$ kpc (see Fig. 1). This follows from the data in the sample collected by Danly (1989) and Danly et al. (1992). Newer *HST* data (Savage et al. 1997) for the same stars confirm these values.

3.1. C IV

Inspection of the data available for C IV shows that absorption is present only for stars with $|z|>1.5$ kpc. The width in velocity is on the order of 100 $\mathrm{km\,s^{-1}}$. However, the width values scatter and, in contrast to what is seen for Si II, the average velocity of absorption per star shows a large scatter (see Fig. 1). A review of the C IV data is available from Savage et al. (1997).

3.2. N V

The absorption by N V is more difficult to detect because of the smaller optical depth (as discussed above). At lower latitudes as well as in directions where the shear due to galactic rotation is less, sufficient optical depth may be reached. Thus, Savage & Massa (1985, 1987) found N V in absorption in directions around the bulge of the Milky Way.

Further excellent data are available to a number of high-latitude stars and to extragalactic objects using the *HST* (see Savage et al. 1997).

3.3. O VI

Between 1973 and 1979 O VI was detected only with *Copernicus*, and thus very few data toward halo stars were available (Jenkins 1978a, b). The five lines of sight reaching into the halo were rediscussed by Spitzer (1996). Since then O VI data from *ORFEUS* have become available (in 1996).

The *ORFEUS* Tübingen-Heidelberg echelle spectrograph (Kraemer et al. 1990) has a resolution near ~15000 and O VI from halo gas has been measured toward several stars in the flight of 1996 (Widmann et al. 1998). They detected O VI toward 14 stars, of which 9 are in the halo at $z<1$ kpc and 5 are in the Magellanic Clouds. A few of the *ORFEUS* stars are in common with those measured by *Copernicus* and the results are mutually consistent. A few *ORFEUS* low-dispersion spectra of ~0.3 Å resolution were analyzed by Hurwitz & Bowyer (1996).

Since 1999 *FUSE* has enabled measurements of O VI on numerous galactic and extragalactic sight lines (see Sects. 8 and 9).

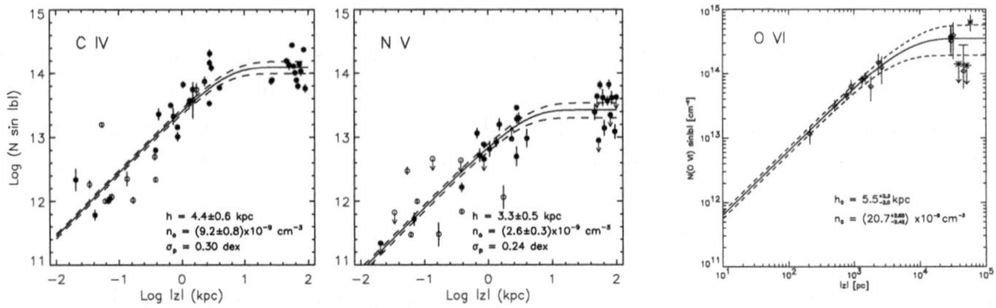

Figure 2. Data for absorption by C IV, N V, and O VI toward halo objects, collected from the literature, are plotted versus z-height. The solid lines show the fit to hydrostatic models resulting in a scale height, h, and a mid-plane density, n_0, as indicated. For details see Sect. 3.4. This figure shows the panels for C IV and N V from Fig. 4 of Savage et al. (1997), and for O VI from Fig. 3 of Widmann et al. (1998). For details about the symbols see the original papers. Comments on such an interpretation are given in Sect. 3.4.2.

3.4. HYDROSTATIC HALO AND SCALE HEIGHT

3.4.1. *The data*

The observations of high ions toward halo stars indicate that the latitude-corrected column density through halo gas has a maximum (the values read from Fig. 2 are given in Table 2). However, there is substantial scatter among the lines of sight.

The distribution of gas perpendicular to the Milky Way Disk is, in a stationary situation, the result of the balance between the buoyancy of the gas and the containing gravitational force of the Milky Way (mostly the disk). In most analyses the density distribution in z-direction is assumed to follow hydrostatic equilibrium (Spitzer 1956; Jenkins 1978b), given by

$$n_z = n_0 \, e^{-z/h},$$

in which n_0 [cm^{-3}] is the mid-plane gas density (or ion density) and h [kpc] the scale height. The column densities observed along halo sight lines are transformed to the equivalent column density in z-direction, $N \sin|b|$, and then plotted against z-distance of the background light source (both parameters logarithmic). Then an exponential is fitted (for examples see Fig. 2). In this manner one can derive that, e.g., the scale height for Milky Way neutral hydrogen is just h(H I)=0.30±0.03 kpc (Savage et al. 1997).

The values for the scale height of the high ions found in the literature are based on two methods for the fit: either one fits the scale height h and the mid-plane density n_0 simultaneously, or one assumes the midplane density from some line of reasoning and then derives h. Since it is not clear that the high ions are distributed according to the hydrostatic formula, one should regard values for the scale height as just numbers convenient for comparing various studies of halo gas.

TABLE 2. Average parameters for the high ions found on halo lines of sight (see Fig. 2 and Sect. 3.4).

Species	disk density [cm^{-3}]	scale height [kpc]	number of sight lines	source	$\log(N \sin\|b\|)_{max}$ [cm^{-2}]
C IV	9.2×10^{-9}	4.4 ± 0.6	40	S(*IUE,HST*)	~14.5
N V	2.6×10^{-9}	3.3 ± 0.5	37	S(*IUE,HST*)	~13.9
O VI	2.1×10^{-8}	5.5 ± 2.2	16	W(*ORFEUS*)	~14.6
O VI	$1.7\times10^{-8\,a}$	2.3 ± 0.4	100	S(*FUSE*)	~14.8

Notes: (a) adopted value. S(*IUE,HST*) = Savage et al. (1997); W(*ORFEUS*) = Widmann et al. (1998); S(*FUSE*) = Savage et al. (2003).

The available accurate C IV and N V data for halo lines of sight, including extragalactic sight lines, have been discussed by Savage et al. (1997). They present the currently most-complete plots of the normalized column density of these ions versus z-distance of the star (see Fig. 2). They find a C IV scale height of h(C IV)=4.4±0.6 kpc. For N V the number of upper limits for the absorption is fairly large, and for the fit shown in Fig. 2 these upper limits were regarded as detections. A statistical analysis treating the upper limits properly leads to h(N V)=3.3±0.5 kpc (Savage et al. 1997).

For O VI two data sets are available: one from *ORFEUS* (Widmann et al. 1998) and one from *FUSE* (Savage et al. 2000). Widmann et al. (1998) have mostly halo stars as background light sources, plus a few stars in the Magellanic Clouds. They fit the hydrostatic relation in order to find both h and n_0 simultaneously. The *ORFEUS* data indicate an O VI scale height h(O VI)=5.5±2.2 kpc. Savage et al. (2003) have only extragalactic background light sources. They adopt a midplane density for O VI from general arguments but relying mostly on the value from Jenkins (1978b). They then find h(O VI)=2.3±0.4 kpc, a value substantially smaller than that of Widmann et al. (1998), in spite of using essentially the same n_0 as Widmann et al. (1998) had found. The difference of a factor of 2 is caused by the average of the O VI column densities found for objects with z>10 kpc, i.e. the Magellanic Cloud objects of Widmann et al. (1998) versus the extragalactic objects of Savage et al. (2003).

3.4.2. *Comments on the scale-height concept*

The maximum column density of O VI found in both the *ORFEUS* and the *FUSE* sample, is larger than the maximum value for N V and C IV. For N V this is expected, but for O VI column densities much smaller than those for C IV are expected (Table 1).

The fact that the observed mean maximum column density of O VI is similar to or even larger than that of C IV (see Table 2), in contrast to the expectations of Table 1, indicates that *at least one* of the assumptions for the usual interpretation of such data is not valid.

First, the gas need not be in thermodynamic equilibrium. In fact, on such sight lines it is never clear *a-priori* that the gas seen in different species absorbs on the same portion of the line of sight. One always detects the portion where the particular ion has the most favorable conditions to be present. This implies that it would never be correct to compare O VI and C IV directly and that the column density ratios are not expected to be in line with thermodynamic equilibrium anyway.

Secondly, the ion-gas need not be in hydrostatic equilibrium, or, the z-distribution of the ion-gas is not given by the hydrostatic equation. Another way of describing this situation is that ionization conditions vary widely between sight lines, so that there are large pockets of O VI gas on one line of sight and only small amounts on other lines of sight. This, in a way, is equal to the regional version of deviation from thermodynamic equilibrium.

Thirdly, the ionization balance in the halo favors the high-ion gas, whereas in the disk there is still a substantial amount of neutral gas. In this case it is clear that n_0(ion) has no meaning for the hydrostatic fits normally used. In fact, one would expect column densities smaller than the hydrostatic case over the first, say, 1 kpc and larger column densities from the rest of the line of sight. Further aspects are discussed in relation with models for the nature of the halo gas (Sect. 7).

Additionally, in the hydrostatic case the gas will be less dense toward larger z (Spitzer 1956). This reduces the possibility for detection of most constituents of the gas at higher z. Above some height the prevalent radiation density will keep the gas ionized (Bregman & Harrington 1986). Then the only good representative of the gas distribution would be the distribution of free electrons, representing all ionic matter. The scale height of free electrons is 1.5 to 2.5 kpc (Reynolds 1989). Since the scale heights of in particular C IV and O VI are rather larger, this indicates that C IV and O VI are underrepresented in the lower part of the halo. In short, the hydrostatic formula is not applicable for these ions.

Given the large probability that the hot halo is not homogeneous nor in thermodynamic equilibrium, it is very unlikely that the study of a small number of sight lines will bring meaningful values for the scale height or averages for the conditions of the halo gas.

3.5. ANALYZING THE ABSORPTION VELOCITIES

3.5.1. *Profile analyses and* $|z|$*-extent of ionized gas*

Analyzing the rather smooth C IV absorption profiles toward LMC stars ($l{\sim}270°$, $b{\sim}{-32}°$), Savage & de Boer (1981) found that, when assuming co-rotation of the halo gas, the C IV gas extends to ~10 kpc outside the disk of the Milky Way, well in line with the postulations by Spitzer (1956). However, later observations of halo gas using two globular cluster stars as background light source (de Boer & Savage 1983, 1984) led to the recognition that the halo gas rotates at a slower pace than the disk, which fits to models of a Galactic-Fountain-type flow (Bregman 1980).

Further analyses of this kind mostly based on *HST* data are collectively interpreted in a careful and excellent way by Savage et al. (1997). They explicitly show the example for the galactic C IV absorption on the Mrk 509 sight line (see also Sect. 6.3). When assuming co-rotation, the velocities indicate that the bulk of the C IV absorption is due to gas with a scale height of ~4.5 kpc, in line with the hydrostatic value.

A drawback of extragalactic sight lines is the absence of a distance limit to the absorbing gas. Without such a limit, deviations from co-rotation will mostly stay undetected. In fact, halo rotation velocities smaller than the ones of disk gas lead to predicted velocities very similar to those resulting from downward halo flows. Examples of these effects can, for various galactic directions, be inferred from Fig. 2 of Kaelble et al. (1985).

3.5.2. *Flow of high ions*

In the concept of a fountain-type flow, very hot gas leaves the disk, cools and recombines in some way, and descends toward the disk. The end product may be the neutral HVCs, which indeed on average flow toward the disk (Bregman 1980; Kaelble et al. 1985; Wakker 1991). It is thus of relevance to investigate the velocities of the high ions.

The C IV velocities found from stellar spectra show a large scatter (see Fig. 1). Including newer results from *HST* spectra of stars, but in particular of extragalactic objects, Savage et al. (1997) find that the C IV gas has predominantly a diskward velocity, as does the N V gas. Most sight lines analyzed are in the northern sky, where also the high-velocity complexes of neutral gas move toward the disk. However, since there is no distance limit to the C IV gas on most of the analyzed halo lines of sight, no hard numbers can be derived for the flow velocities, unless one compares these with a specific spatial model (z-distribution) for the halo gas (c.f. the remarks at the end of Sect. 3.5.1).

Radial velocities and absorption widths of N V are hard to get because its lines lie in parts of the spectrum with a lot of other spectral structure. The O VI lines show no systematic trend of in- and outflow (see Sect. 8). If the Milky Way has a fountain-like flow, one would expect to see outflow of very hot gas. It is not clear, however, if such an outflow would contain measurable amounts of C IV and O VI, nor if these species would manifest themselves in absorption or in emission. Both C IV and O VI have been detected in emission from halo directions (see Sect. 4). The spectral resolution of those instruments used was too poor to determine a velocity.

3.5.3. *High ions and HVCs*

In several cases high ions (C IV and O VI) have been detected at the same velocity as HVCs. These cases include the lines of sight to the Magellanic Clouds (Savage & de Boer 1979, 1981; Widmann et al. 1998; Sembach et al. 2000). In all these cases one may regard the high-ion gas to represent an ionized envelope around the HVC or perhaps the ionization front at the head of the HVC. The interpretation is related to the details of the various theoretical considerations reviewed in Sect. 7.

3.6. DISCUSSION

Do the various ions represent the same gas? Since N V is expected to exist in gas of higher temperature and thus of lower density than C IV, smaller column densities of N V are expected. However, O VI has rather large column densities, in spite of possibly being from gas at yet higher temperatures.

When trying to extract the meaning from scale-height type plots, one tends to overemphasize the data points toward extragalactic light sources, because these determine the leveling-off of the column densities. The high column densities to stars in excess of the fit subjectively get less weight. When a large number of lines of sight is available for the scale-height determination, this kind of interpretation may be too quick a generalization and is still prone to statistical effects (compare the O VI scale heights from *ORFEUS* and *FUSE* in Table 2). But remember also the comments on non-equilibrium conditions in Sect. 3.4.2.

Given the available data, one must conclude that the scale heights for C IV, N V and O VI are, within the respective uncertainties, similar. However, the variations in (normalized) column densities between the sight lines indicate that the ions are distributed in a patchy way. Thus, studies of the bulk nature of the hot halo gas clearly should include all possible particularities of the individual lines of sight (see some special cases in Sect. 6).

4. Emission by high ions

Emission by halo gas was reported by Martin & Bowyer (1990) using the Berkeley EUV/FUV Shuttle Telescope. A number of positions at high galactic latitude (most of them at $|b|>30°$) show well-defined C IV emission. Possibly also O III at 1661 Å was detected. Dixon et al. (1996) give (3 σ) evidence for O VI emission by halo gas in *HUT* spectra with 7 Å resolution. This emission shows as a shoulder to the geocoronal Lyβ emission profile. One significant detection is in a direction just about 10° off the line of sight to 3C 273, a line of sight for which Davidsen (1993) had reported strong O VI absorption. It coincides with bright diffuse emission in the X-ray M-band.

With *FUSE* two detections of O VI emission are reported. Dixon et al. (2001) have seen O VI toward the Coma and Virgo clusters (North Galactic Pole region). The emission is seen near zero $v_{\rm LSR}$ and thus from gas in the Milky Way. Combining the detected intensity with a column density from absorption toward a target in the same direction they arrive at $n_e{\sim}0.05$ cm^{-3}, assuming thermodynamic equilibrium with $\log T_e{\sim}5.5$. Shelton et al. (2001) detected O VI emission toward l=315° and b=−41°. They favor from their models (in which the Local Bubble is accounted for) halo gas with $n_e{\sim}0.01$ cm^{-3} at $T_e \sim 10^5$ to 10^6 K.

Under equilibrium conditions O VI emission is strongest from gas with T=3×10^5 K (which is the temperature at which the fraction of O^{+5} is at its maximum), with an energy emission coefficient of $10^{-22.5}$ erg cm^3 s^{-1} (Landini & Monsignori Fossi 1990). For C IV these values are T=10^5 K and $10^{-22.2}$ erg cm^3 s^{-1}. If one considers pressure equilibrium, the higher temperature for the existence of O VI at the same time implies a lower gas density, so that the C IV emission line may be expected to be one order of magnitude stronger than the O VI line. On the other hand, the diffuse background light near 1035 Å is most likely much weaker than near 1550 Å (see e.g. de Boer et al. (1973) for data on the average interstellar radiation field), so that the diminished confusion near the O VI lines may compensate for its smaller flux. However that all may be, since the optimum temperatures differ considerably, the C IV and O VI emissions probably come from different pockets of gas anyway, which thus hampers the derivation of meaningful values for the conditions in the gas.

5. Diffuse X-ray emission

There must be more than one source of the diffuse X-ray radiation. Since the O VI gas was detected locally, the diffuse soft X-ray emission was originally interpreted as entirely due to hot gas in the "Local Hot Bubble". However, when inspecting the maps of the rocket-observed diffuse X-ray emission sky, larger intensities are seen away from the Milky Way Disk (see the review by McCammon & Sanders 1990). The high-energy flux, however, comes rather more from regions of the disk, and it was speculated that a hot halo exists with a temperature of perhaps 4×10^6 K (Rocchia et al. 1984).

A first indication of interaction of HVCs with the ambient halo medium was found by Hirth et al. (1985), who noted a local intensity enhancement coincident with the Draco HVC. Later, using *ROSAT* All Sky Survey data, it was discovered that the maps of diffuse soft X-ray emission show neutral clouds at high galactic latitude as shadows (Snowden et al. 1991), requiring diffuse background halo emission. This proposition fueled a heated debate about the size of the halo contribution, since this aspect has consequences for the models to explain the high-latitude soft X-ray emission (e.g. Hirth et al. 1991; Snowden et al. 1998; Pietz et al. 1998). The debate about this topic has not subsided yet.

The temperature of the emitting gas normally is inferred from a fit of a modeled spectrum to the soft X-ray spectral energy distribution. Here lies the biggest problem in the interpretation of the measurements. There are many free parameters in such models: the number of components and the n, T combination for each of these. Moreover, absorption by neutral and ionized gas with unknown spatial distribution has to be accounted for, too. In fact, a detailed radiation-transfer analysis is required.

The intensity of diffuse X-ray emission is, if thermal, governed by the emission measure and the temperature. The emission detected comes predominantly from the portions of the line of sight where $\int n_e^2/\sqrt{T_e}\ dl$ is maximal, thus from the high-density parts. The interpretation of the diffuse X-ray emission must also cope with the fact that the filling factor of the particular gas is unknown. Note that the cooling function of such gas has a local minimum near $4{\times}10^6$ K (see e.g. Böhringer & Hensler 1989) so that this temperature is the more likely one for the diffuse hot medium. This is an order of magnitude above the thermodynamic-equilibrium temperature for O VI.

ROSAT All Sky Survey data confirmed and substantiated the presence of a widespread diffuse emission component with 1 to $2{\times}10^6$ K (Kerp 1994; Snowden et al. 1998). Of the locally-detected diffuse soft X-ray flux it was thought that roughly half may come from a hot halo. Further shadows by foreground neutral gas clouds were discovered (Snowden et al. 1991; Lilienthal et al. 1992; Snowden et al. 2000).

Later deep X-ray surveys showed that in the soft (1 to 10 keV) band 90% of the flux is resolved (Hasinger et al. 1998). This has been confirmed recently in measurements with *CHANDRA* (Rosati et al. 2002) and *XMM-NEWTON* (Hasinger et al. 2001).

The diffuse flux was recently measured again at high spectral resolution, showing emission lines. The temperature derived is about 2 to $3{\times}10^6$ K (McCammon et al. 2002). Evidently, HVCs exist in a diffuse hot medium (which does produce soft X-rays) of still poorly-known temperature and density.

TABLE 3. Properties of Galactic Halo gas in front of the MCs

Milky Way gas in front of:	LMC (12 l.o.s.)	SMC (11 l.o.s.)
Angular scales probed	$0\overset{\circ}{.}5$ to $5\overset{\circ}{.}0$	$0\overset{\circ}{.}05$ to $3\overset{\circ}{.}3$
Range in log N(O VI)	14.22 to 14.67	13.77 to 14.39
Mean log N(O VI)	$14.52^{+0.10}_{-0.14}$	$14.13^{+0.14}_{-0.20}$

6. Special lines of sight

6.1. HIGH IONS TOWARD THE MAGELLANIC CLOUDS

Numerous spectra have been obtained of stars in the Magellanic Clouds. A deep analysis of the first set of *IUE* spectra showed (Savage & de Boer 1981) that C IV is present on this line of sight only outside the disk of the Milky Way, since C IV was not seen in absorption toward stars with distance <1 kpc in the same direction. For additional data on these sight lines see, e.g. Savage et al. (1989) and, based on *HST* spectra (e.g. Bomans et al. 1996; Wakker et al. 1998). FUV data, available from Hurwitz et al. (1995), Widmann et al. (1998) as well as Friedman et al. (2000) and Howk et al. (2002), show clear O VI absorption by the Galactic Halo (as well as in each of the MCs). The LMC sight line has the great advantage that the LMC velocity itself is well separated from that of the Galaxy and the halo. Toward the SMC this is much less so.

It has been suggested that the HVCs detected *in front* of the LMC are part of the Magellanic System (Blades et al. 1988), that they are stripped in some way from the LMC. The argument is that the abundance of the elements seen in absorption in the spectra of SN 1987A (de Boer et al. 1987a, Blades et al. 1988) are very similar to the ones of the LMC itself (approximately a factor 2 less than in the Milky Way). The claim about this association is not justified; the observed HVCs extend in space beyond the boundaries of the LMC (de Boer et al. 1990). Moreover, the LMC space motion is close to perpendicular to the line of sight (see Kroupa & Bastian 1997; de Boer 1998b), so that a dynamic relation of these clouds at $v_{\rm rad}{\sim}{+}130$ km s^{-1} with the LMC at $v_{\rm rad}{\sim}{+}250$ km s^{-1} is extremely unlikely. And if gas were dynamically torn off the Magellanic System it would be SMC gas, having abundances much lower than LMC gas. Toward the SMC the halo absorption takes place over the much more limited velocity range of 0 to +150 km s^{-1} (the SMC velocity is ${\sim}{+}150$ km s^{-1}). C IV has been detected in the spectrum of HD 5980, but the separation between disk and halo absorption is tricky. Galactic O VI is detected in *ORFEUS* spectra of HD 5980 (Widmann et al. 1998) and in numerous further SMC stars by Howk et al. (2002) with *FUSE*.

The *FUSE* data allow on the MC lines of sight (l.o.s.) the probing of "small-scale" variations (Table 3). Howk et al. (2002) find that the O VI absorption extends on the LMC sight lines from about -20 km s^{-1} to $+180$ km s^{-1}, thereby often blending with the O VI absorption by gas in the LMC itself at higher velocities. The largest optical depth (a shallow maximum) is found near $+70$ km s^{-1} (heliocentric velocities), the same as that of the IVC there, with a secondary maximum near $+130$ km s^{-1}, the velocity of the HVC, the first HVC detected in H_2 (Richter et al. 1999).

6.2. NORTHERN-SKY CLOUDS AND LOOPS

The northern galactic sky is full of clouds and loops of gas. These structures have to be known and incorporated in an analysis of absorption by high ions. When background light sources lie *behind* such clouds and loops, the absorptions may be related to these clouds themselves and then the absorption shown might not be due to diffuse halo gas. Since the northern sky is well studied, the most prominent features are briefly reviewed.

The Intermediate-Velocity Arch, running from $(l,b)=(110°,+30°)$ via $(l,b)=(150°,+65°)$ to $(l,b)=(200°,+40°)$, has been studied in H I in great detail by Danly & Kuntz (1993) and Kuntz & Danly (1996). In its direction lie several stars, among them HD 121800 and HD 93521, to which C IV and O VI absorption is detected. C IV is also seen in emission. The distance to this arc is estimated to be z=0.6 to 1.2 kpc (Wakker 2001) and the metal content is essentially solar (Richter et al. 2001), facts which have to be included in an interpretation of the high ion absorptions seen.

The Radio Loop I extends from $l=20°$, $b=+30°$ via $l=0°$, $b=+70°$ to $l=310°$, $b=+50°$. Breitschwerdt et al. (1996) suggest that this Loop is in fact a ring of denser gas at the intersection of two bubbles. They estimate the distance to the neutral gas of the loop to be 70 pc. Both C IV and O VI are seen in emission as well as in absorption for directions inside Loop I.

In the north pole region of the Milky Way (see Fig. 3) high ions are detected both in emission and in absorption in many directions. There was no particular association with either Radio Loop I or with the Intermediate-Velocity Arch in the pre-2000 data. However, the number of sight lines was too small to prove or disprove a real association. The *FUSE* data from Savage et al. (2003) do not indicate clear correlations either.

The high-velocity cloud complex C extends over a large portion of the northern sky. The distance to this complex is still relatively poorly determined, since the number of spectra for relevant lines of sight is small. Wakker & van Woerden (1997) give in their review a summarizing limit of z>1.7 kpc. The metal content is $\sim$0.1 times solar (Wakker et al. 1999; Richter et al. 2001). With *FUSE* O VI has been detected in absorption

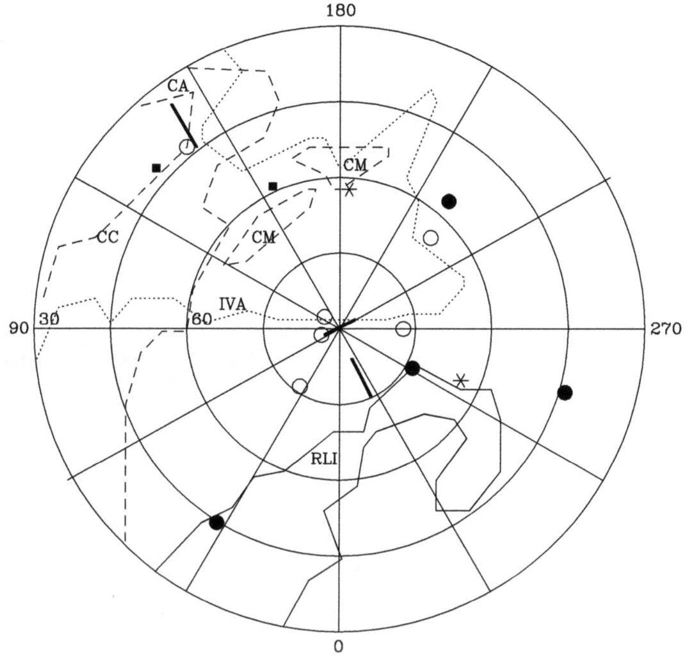

Figure 3. In the direction of the Northern Polar Cap of the Milky Way numerous observations pertaining to hot gas are available. The positions marked show: 1) O VI: •=detected in emission, ∘=not detected in emission, ∗=O VI in absorption to halo objects with $z>1$ kpc; 2) C IV: detected in emission at two positions (filled square) and along scans (solid lines). For references see Sect. 4. The contours sketch: Radio Loop I, RLI, with a distance of $d\sim70$ pc (Breitschwerdt et al. 1996); the Intermediate-Velocity Arch (dotted contour), IVA, (Kuntz & Danly 1996) at $z=0.6$ to 1.2 kpc; the large-z high-velocity clouds (dashed contours) of complex A, CA, complex M, CM, and complex C, CC. Data from before 1998.

at complex C velocities using extragalactic objects as background light sources.

6.3. HIGH-ION ABSORPTION WITHOUT NEUTRALS

6.3.1. *C IV without neutral gas*

In most of the absorption spectra C IV is present at the same velocity as Si II, albeit normally with a smoother profile (see e.g. toward SN 1987A, Savage et al. 1989). There are now three cases known in which C IV is present *without* detection of Si II or of H I.

• Mrk 509. Sembach et al. (1995) found high-velocity C IV at -270 km s^{-1} in the spectrum of Mrk 509 (l=36°, b=−30°), no Si II at that velocity, and H I at that velocity only at a position 2° away. The velocity is exceptional; from galactic rotation one would expect velocities ranging from 0 to $+80$ km s^{-1}

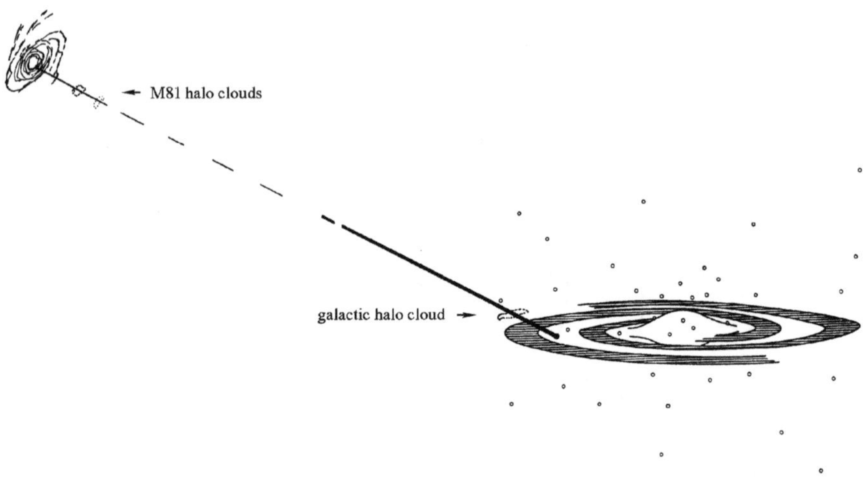

Figure 4. The line of sight from the Milky Way to SN 1993J in M 81 contains numerous clouds. Absorption near 0 km s^{-1} is due to gas in the solar vicinity, absorption near -50 km s^{-1} to a local halo cloud, absorption near -120 km s^{-1} to gas in M 81. The absorption complex at velocities between +120 and +240 km s^{-1} is due to clouds with very small spatial extent but with high column densities of Ca II and Mg II, as well as of C IV. These must lie somewhere along the 3 Mpc line of sight, but most likely near M 81. For the exceptional absorption profiles see Vladilo et al. (1993) and de Boer et al. (1993).

for gas with d<16 kpc, or at most down to -100 km s^{-1} at larger distances. The low-velocity C IV absorption is seen between 0 and +80 km s^{-1} indeed. Both clouds have considerable amounts of O VI absorption, too (Sembach et al. 2003). The -270 km s^{-1} gas may be an infalling ionized HVC. However, without a limit on the distance, the cloud need not be in the *Galactic* Halo.

• HD 119608. Toward stars seen in the sky inside galactic Radio Loop I and IV Sembach et al. (1997) detected C IV and N V at -45 km s^{-1} (l=320°, b=+43°) without Mg II at 1239 Å and H I at 21 cm. This gas is part of the Galaxy since it is seen against a star. Its velocity is similar to that of other absorptions found in this general direction. However, a final check on the absence of neutral gas should be made by obtaining a spectrum of the much more sensitive line of Mg II at 2795 Å.

• SN 1993J. On the 3.25 Mpc long sight line toward SN 1993J in M 81 (l=142°, b=41°) absorption was found in C IV at the expected velocities (see Fig. 4). In addition, strong absorption was found at +130 km s^{-1} in Ca II, in Mg II, and in C IV (Vladilo et al. 1993, 1994; de Boer et al. 1993; Bowen et al. 1994). Most extraordinarily, strong C IV absorption was also found at velocities up to +200 km s^{-1} (de Boer et al. 1993) with indications for absorption at up to +240 km s^{-1}. These velocities are not expected with any reasonable model for the dynamics of gas in the Milky Way. In addition, there is no trace of H I in *Effelsberg* 21-cm spectra related to the Ca II and Mg II absorption velocities. This implies that the gas can occupy only a very small fraction of the radiotelescope beam, hence that it probably is far away from us. The complex most likely represents HVCs falling toward M 81, or moving somewhere through intergalactic space. The SN 1993J line of sight is an example of a hot halo of another galaxy (see also Marggraf & de Boer 2000).

6.3.2. *O VI without neutral gas*
Up to half of the lines of sight studied with *FUSE* show indications for O VI absorption at $|v_{\rm LSR}|$>150 km s^{-1} *without* H I detected at 21 cm (see Savage et al. (2003) and the profiles presented by Wakker et al. (2003)). However, since the structure of the spectra in the region of the O VI lines is complicated and the S/N ratio not always large, while the detection limit in 21-cm emission is not always at a significant level, most of these lines of sight need further scrutiny. In many cases broad absorption wings to the 0 km s^{-1} gas are present reaching into the high-velocity range.

Clear cases of O VI *without* H I are those of NGC 3310, PG 0947+396, PG 1001+291, PG 1116+215, and ESO 572-G34 in the Northern Galactic Cap and PHL 1811, PKS 2155−304 and HE 0226−4110 (as well as Mrk 509), in the southern (see Wakker et al. 2003). The northern ones (excluding NGC 3310) have positive absorption velocities and are in the general direction of the northern third galactic quadrant. Sembach et al. (2003) suggest these may be O VI clouds within the Local Group.

7. Models for the presence of high ions

The absorption and emission in the high-ion lines cannot be explained in a simple manner: The absorption (or emission) seen in these species need not be produced by the same gas, since the optimum temperatures for the presence of these ions differ, in thermodynamic equilibrium, considerably.

7.1. PROCESSES TO MAKE HOT HALO GAS

Various models have been developed and tested in attempts to reproduce the observed column densities. These can be grouped into several larger categories as given below. Most models share, of course, fundamental physical processes.
∘ EQUILIBRIUM WITH RADIATION FIELD: Bregman & Harrington (1986) investigated the nature of the average radiation field in the halo. They are able to reproduce the observed column density ratios of C IV and Si IV for acceptable temperature and density values with photoionization. The models can not explain the N V and O VI column densities. Thus these ions

must reside in gas under different conditions.

∘ RADIATIVE COOLING: Radiative cooling was explored by Edgar & Chevalier (1986) using the best atomic data available at that time. They calculated cooling from $T=10^6$ K at constant pressure (isobaric), constant density (isochoric), or a combination of both. The resulting integrated column densities of O VI and N V for a halo line of sight are in fair agreement with observations, while those of C IV (and Si IV) fall short. Shapiro & Benjamin (1993) performed similar calculations, including the full radiative transfer, implying the inclusion of the self-emitted X-ray continuum and of self-ionization. The column densities in the isochoric case are consistent with the observations. This would imply that halo gas is rather homogeneous on kiloparsec scales.

∘ CONDUCTIVE HEATING: Numerous calculations are available for cool clouds embedded in a hot medium (see review by Spitzer 1996). The clouds may be spherical or treated as a slab and may or may not contain a magnetic field. The column density ratio $N(\text{C IV})/N(\text{O VI})$ is calculated and it appears not to be very sensitive to varying conditions, being in the range from 0.1 to 0.4. Most of these aspects can be found in Slavin & Cox (1993).

∘ TURBULENT MIXING: When neutral warm clouds flow through hot gas, a turbulent-mixing layer will occur (Begelman & Fabian 1990). Detailed modeling is presented by Slavin et al. (1993). For one mixing layer they expect column densities log $N(\text{C IV})$ from 10.4 to 12.8, log $N(\text{N V})$ from 9.4 to 11.5, and log $N(\text{O VI})$ from 10.2 to 11.9, where the first values are for a small velocity difference between the clouds and the 10^5 K gas, the other values are for large velocity differences and $10^{5.5}$ K gas. Emission-line strengths are also predicted. The absorption C IV to O VI ratio does not fit the observations.

∘ ADIABATIC COOLING: When very hot gas expands rapidly, it cools without recombination. The result would be a well ionized gas with "frozen-in" ionization structure at moderate temperatures. This possibility has been explored by Breitschwerdt & Schmutzler (1994) to explain the soft X-ray emission from gas at larger z-distances. With adiabatic cooling any mixture of ions may exist on any line of sight, and the ratio of the C IV to O VI column density depends on the details of the original gas conditions and the relative speeds of cooling and recombination.

7.2. HVCS IN HOT HALO GAS

Wolfire et al. (1995) calculated that two-phase stable (high-velocity) clouds can exist embedded in the hot ($T{\sim}1$ to $2{\times}10^6$ K) halo. According to their models, the neutral clouds can exist only within a narrow range of pressures at a given z-height.

Stable primordial gas clouds can exist only at $z<2$ kpc (Wolfire et al. 1995). However, all HVCs contain metals (Savage & de Boer 1981; Wakker & van Woerden 1997; Wakker et al. 1999; Richter et al. 1999, 2001), indicating they are not made of primordial material.

For many HVCs the z-distance is observed to be larger than 2 kpc (Wakker & van Woerden 1997; Ch. 10). They may be stable and the distance is in line with the Wolfire et al. (1995) calculations for clouds that contain heavy elements.

8. FUSE and the hot halo

In 1999 the *FUSE* satellite was launched. *FUSE* spectra are obtained at $\sim$20 km s^{-1} resolution (Moos et al. 2000), or more coarse, depending on the available signal. In a first paper with results on the Milky Way Halo, spectra of 11 extragalactic sources were analyzed by Savage et al. (2000). Most sight lines show the presence of O VI covering a large range of column densities. Further data are available from Sembach et al. (2000) and Friedman et al. (2000). In this first analysis Savage et al. (2000) arrived at a scale height for O VI of 2.7 kpc, much smaller than that found by Widmann et al. (1998) from *ORFEUS*. As mentioned in Sect. 3.4.2, since the halo is patchy, a small number of sight lines cannot give a generally valid result, also not in terms of scale height (if at all).

A full analysis of the data *for 100 extragalactic sources* taken in the first several years of *FUSE* reveals a complicated picture for the structure of the halo. The relevant spectra are presented by Wakker et al. (2003), the interpretation by Savage et al. (2003) and Sembach et al. (2003). The many sight lines observed toward the LMC and SMC allow a study of small-scale structure of halo gas (Howk et al. 2002; see also Sect. 6.1). The overall results of these studies may be summarized as follows.

The most important finding from the *FUSE* observations is that the O VI absorption is *widespread and highly irregular.* Note, however, that the *FUSE* targets were unevenly distributed in galactic longitude and obviously concentrated toward higher latitudes.

COLUMN DENSITIES: The highest column densities found are log N(O VI) $\sim$14.7, with the lowest detected ones near 13.3 (from high S/N data). On average, column densities toward the North Galactic Cap are larger than those toward the South Galactic Pole (by $\sim$0.25 dex). The average column density toward the LMC is similar to that toward southern extragalactic sources.

SCALE HEIGHT: Since all sources analyzed are extragalactic (except two), the scale height plot by Savage et al. (2003) does contribute little further knowledge beyond the earlier Savage et al. (2000) results (see Sect. 3.4).

DISTRIBUTION OVER THE SKY: The distribution of O VI is *not correlated* with the distribution of the other tracer of hot gas, the diffuse X-rays. Furthermore, no correlation is found with other gas indicators such as H I and diffuse Hα emission.

VELOCITY DISPERSION: The O VI absorption shows a velocity dispersion (based on b-values derived) ranging from 30 to 100 km s^{-1}. This is much wider than what would be expected solely from thermal broadening. Clearly, at least one of four effects must play a role for the observed lines of sight: multiple velocity structure, non-thermal conditions, effects of galactic rotation, and motions perpendicular to the disk.

RADIAL VELOCITIES: The radial velocities of O VI on sight lines toward higher latitudes ($|b|$>45°) range from about −40 to +80 km s^{-1}. The sample average is close to 0 km s^{-1}, with a standard deviation of 22 km s^{-1}.

VELOCITIES AND GALACTIC ROTATION: In the Northern Galactic Cap there is a crude correlation of observed radial velocity with direction: negative $v_{\rm LSR}$ in the region $30<l<150°$, more strongly positive velocities toward $210<l<330°$, both as expected from galactic rotation. However, in both directions velocities of opposite sign occur as well. Clearly gas is moving *in and out* from our vantage point in the Milky Way. Toward the Southern Galactic Cap a larger fraction of positive velocities was found, notably in the longitude range $240<l<60°$. There is *no overall correlation with galactic rotation.*

SPECIAL REGIONS: The loops seen in the radio continuum are regarded as regions related to previous energetic events. There is no clear correlation of O VI with the radio loops. The LMC and SMC lines of sight allow an investigation of fine structure (see Sect. 6.1).

Summarizing the findings on diffuse halo gas from these extragalactic sight lines, Savage et al. (2003) state that *"a combination of .... radiative cooling of hot fountain gas, the cooling of supernova bubbles, and the turbulent mixing of warm and hot halo gases is required to explain the presence of* O VI *in the halo"*.

9. Association of O VI with HVCs

The association of HVCs with O VI was investigated by Sembach et al. (2003). They identify 85 individual O VI features at high velocity, defined as $|v_{\rm LSR}|$>100 km s^{-1}, in the extragalactic sight-line sample. With few exceptions, the O VI absorption is confined to $|v_{\rm LSR}|$<400 km s^{-1}, indicating that most features are associated with the Milky Way.

The sample shows several clear-cut associations of detected O VI absorption with known high-velocity 21-cm emission (apart from those already mentioned toward the Magellanic Clouds in Sect. 6.1).

Complex C: 9 sight lines with O VI absorption have been found. The velocity ranges from -200 to -100 km s^{-1}, and closely aligns with the H I. The metal content in complex C is only 1/10 solar (Richter et al. 2001), so the detected O VI indicates plentiful gas of high-ionization state.

Complex A: One of the three available sight lines shows O VI absorption. PG 0832+675 shows no high-velocity O VI and hence provides a lower limit for the distance to complex A of 8 kpc (see Wakker et al. 2003).

Magellanic Stream: Detection of O VI is foremost toward Fairall 9, but it is found on 5 more lines of sight.

Some of the associations of O VI with HVCs on individual sight lines are not terribly clear for the reasons mentioned in Sect. 6.3.2. The relevant profiles can be found in Wakker et al. (2003).

10. Concluding remarks

The observational information reviewed above shows that hot gas exists in the halo. Its presence is demonstrated by the detection of absorption and emission by ions such as C IV, N V, and O VI, as well as by the nature of the X-ray emission and its distribution over the sky.

It is not clear if the ions represent pockets of gas in thermal equilibrium, of partly-recombined gas originally at very high temperatures, or of gas in mixing layers. Some H I HVCs are associated with O VI, others are not, and O VI is also present without detected H I.

Most likely, the halo gas is not in thermodynamic equilibrium. The observed C IV, N V, and O VI thus demonstrate the need for the existence of a very hot gas phase at $T > 10^6$ K, from which these ions are the products through a variety of steps from high to lower energetic states. With *CHANDRA* possibilities to detect yet higher ions in the X-ray regime have become available (see Nicastro et al. 2001; Fang et al. 2003).

HVCs exist in the halo of the Milky Way, at distances sufficiently large that they are not associated with disk gas. Also the high ions and the X-ray emitting gas both exist away from the disk. Evidently, the HVCs are located in a *hot halo*, but it is not clear whether extensive interaction between HVCs and hot halo occurs.

The scale-height concept fails to describe the distribution of the O VI halo gas properly.

The motion of the hot gas appears to be modest and its v_{LSR} is hardly correlated with galactic rotation (in contrast to the halo H I 21-cm emission). In some cases the absorption profile of C IV seems to be similar to the absorption profile of neutral species. These cases indicate a relation of the high-ion gas with H I high-velocity clouds. The O VI absorptions are so broad that a detailed correlation with H I velocity is not always clear.

Acknowledgements

I like to thank Blair Savage and Ken Sembach for providing me with the subset of the C IV and N V scale height plots from the figure in Savage et al. (1997) and Hansjörg Widmann for redrafting the O VI scale height plot. Blair Savage provided early versions of the newest *FUSE* papers. I thank Jürgen Kerp for advice and Philipp Richter for a critical reading of the manuscript.

References

Arnaud, M., Rothenflug, R. 1985, A&AS, 60, 425
Begelman, M.C., Fabian, A.C. 1990, MNRAS, 244, 26P
Blades, J.C., Wheatley, J.M., Panagia, N., Grewing, M., Pettini, M., Wamsteker, A. 1988, ApJ, 334, 308
Böhringer, H., Hensler, G. 1989, A&A, 215, 147
Bomans, D.J., de Boer, K.S., Koornneef, J., Grebel, E.K. 1996, A&A, 313, 101
Bowen, D., Roth, K.C., Blades, J.C., Meyer, D.M. 1994, ApJ, 420, L71
Bregman, J.N. 1980, ApJ, 236, 577
Bregman, J.N., Harrington, J.P. 1986, ApJ, 309, 833
Breitschwerdt, D., Schmutzler, T. 1994, Nature, 371, 774
Breitschwerdt, D., Egger, R., Freyberg, M.J., Frisch, P.C., Vallerga, J.V. 1996, Space Science Rev., 78, 183
Danly, L. 1989, ApJ, 342, 785
Danly, L., Lockman, F.J., Meade, M.R., Savage, B.D. 1992, ApJS, 81, 125
Danly, L., Kuntz, K.D. 1993, in Star Formation, Galaxies and the Interstellar Medium, eds. J. Franco, F. Ferrini, G. Tenorio-Tagle (Cambridge: Cambridge University Press), 86
Davidsen, A.F. 1993, Science, 259, 327
de Boer, K.S. 1998a, in IAU Coll. 166, The Local Bubble and Beyond (Lecture Notes in Physics, 506), eds. D. Breitschwerdt, M.J. Freyberg, J. Trümper (Berlin: Springer), 433
de Boer, K.S. 1998b, in The Magellanic Clouds and other Dwarf Galaxies, eds. T. Richtler, J.M. Braun (Aachen: Shaker Verlag), 125
de Boer, K.S., Savage, B.D. 1983, ApJ, 265, 210
de Boer, K.S., Savage, B.D. 1984, A&A, 136, L7
de Boer, K.S., Koppenaal, K., Pottasch, S.R. 1973, A&A, 28, 145
de Boer, K.S., Grewing, M., Richtler, T., Wamsteker, W., Gry C., Panagia, N. 1987a, A&A, 177, L37
de Boer, K.S., Jura, M.A., Shull, J.M. 1987b, in Exploring the Universe at Ultraviolet Wavelengths, eds. Y. Kondo et al. (Dordrecht, Reidel), 485
de Boer, K.S., Morras, R., Bajaja, E. 1990, A&A, 233, 523
de Boer, K.S., Rodriguez Pascual, P., Wamsteker, W., Sonneborn, G., Fransson, C., Bomans, D.J., Kirshner, R.P. 1993, A&A, 280, L15
Dixon, W.V.D., Davidsen, A.F., Ferguson, H.C. 1996, ApJ, 465, 288
Dixon, W.V.D., Sallmen, S., Hurwitz, M., Lieu, R. 2001, ApJ, 552, L69
Edgar, R.J., Chevalier, R.A. 1986, ApJ, 310, L27
Fang, T., Sembach, K.R., Canizares, C.R. 2003, ApJ, 586, 49
Friedman, S.D., Howk, J.C., Andersson, B.-G., Sembach, K.R., Ake, T., Roth, K., Sahnow, D.J., Savage, B.D., York, D.G., Sonneborn, G., Vidal-Madjar, A., Wilkinson, E. 2000, ApJ, 538, L39
Hasinger, G., Burg, R., Giacconi, R., Schmidt, M., Trumper, J., Zamorani, G. 1998,

A&A, 329, 482
Hasinger, G., Altieri, B., Arnaud, M., et 25 al. 2001, A&A, 365, L45
Hirth, W., Mebold, U., Müller, P. 1985, A&A, 153, 249
Hirth, W., Mebold, U., Dahlem, M., Müller, P. 1991, ApSpSci, 186, 211
Howk, J.C., Savage, B.D., Sembach, K.R., Hoopes, C.G. 2002, ApJ, 572, 264
Hurwitz, M., Bowyer, S. 1996, ApJ, 465, 296
Hurwitz, M., Bowyer, S., Kudritzki, R.-P., Lennon, D.J. 1995, ApJ, 450, 149
Jenkins, E.B. 1978a, ApJ, 219, 845
Jenkins, E.B. 1978b, ApJ, 220, 107
Jenkins, E.B., Meloy, D.A. 1974, ApJ, 193, L121
Kaelble, A., de Boer, K.S., Grewing, M. 1985, A&A, 143, 408
Kerp, J. 1994, A&A, 289, 597
Kraemer, G., Barnstedt, J., Eberhardt, N., et al. 1990, in Observatories in Earth Orbit and beyond, ed. Y. Kondo (Dordrecht: Kluwer), 177
Kroupa, P., Bastian, U. 1997, New Astronomy, 2, 77
Kuntz, K.D., Danly, L. 1996, ApJ, 457, 703
Landini, M., Monsignori Fossi, B.C. 1990, A&AS, 82, 229
Lilienthal, D., Hirth, W., Mebold, U., de Boer, K.S. 1992, A&A, 255, 323
Marggraf, O., de Boer, K.S. 2000, A&A, 363, 733
Martin, C., Bowyer, S. 1990, ApJ, 350, 242
McCammon, D., Sanders, W.T. 1990, ARA&A, 28, 657
McCammon, D., Almy, R., Apodaca, E., et 17 al. 2002, ApJ, 576, 188
Moos, H.W., Cash, W.C., Cowie, L.L., et 54 al. 2000, ApJ, 538, L1
Morton, D.C. 1991, ApJS, 77, 119
Münch, G. 1952, PASP, 64, 312
Münch, G. 1957, ApJ, 125, 42
Nicastro, F., Zezas, A., Drake, J., Elvis, M., Fiore, F., Fruscione, A., Marengo, M., Mathur, S., Bianchi, S. 2001, ApJ, 573, 157
Pietz, J., Kerp, J., Kalberla, P.M.W., Burton, W.B., Hartmann, D., Mebold, U. 1998, A&A, 332, 55
Reynolds, R.J. 1989, ApJ, 339, L39
Richter, P., de Boer, K.S., Widmann, H., Kappelmann, N., Gringel, W., Grewing, M., Barnstedt, J. 1999, Nature, 402, 386
Richter, P., Sembach, K.R., Wakker, B.P., Savage, B.D., Tripp, T.M., Murphy, E.M., Kalberla, P.W.M., Jenkins, E.B. 2001, ApJ, 559, 318
Rocchia, R., Arnaud, M., Blondel, C., Cheron, C., Christy, J.C., Rothenflug, R., Schnopper, H.W., Delvaille, J.P. 1984, A&A, 130, 53
Rosati, P., Tozzi, P., Giacconi, R., et 16 al. 2002, ApJ, 566, 667
Savage, B.D., de Boer, K.S. 1979, ApJ, 230, L77
Savage, B.D., de Boer, K.S. 1981, ApJ, 243, 460
Savage, B.D., Massa, D. 1985, ApJ, 295, L9
Savage, B.D., Massa, D. 1987, ApJ, 314, 380
Savage, B.D., Jenkins, E.B., Joseph, C.L., de Boer, K.S. 1989, ApJ, 345, 393
Savage, B.D., Sembach, K.R., Lu, L. 1997, AJ, 113, 2158
Savage, B.D., Sembach, K.R., Jenkins, E.B., et 10 al. 2000, ApJ, 538, L27
Savage, B.D., Sembach, K.R., Wakker, B.P., Richter, P., Meade, M., Jenkins, E.B., Shull, J.M., Moos, H.W., Sonneborn, G. 2003, ApJS, 146, 125
Sembach, K.R., Savage, B.D., Lu, L., Murphy, E.M. 1995, ApJ, 451, 616
Sembach, K.R., Savage, B.D., Tripp, T.M. 1997, ApJ, 480, 216
Sembach, K.R., Savage, B.D., Shull, J.M., et 13 al. 2000, ApJ, 538, L31
Sembach, K.R., Wakker, B.P., Savage, B.D., Richter, P., Meade, M., Shull, J.M., Jenkins, E.B., Sonneborn, G., Moos, H.W. 2003, ApJS, 146, 165
Shapiro, P.R., Field, G.B. 1976, ApJ, 205, 762
Shapiro, P.R., Benjamin, R.A. 1993, in Star Formation, Galaxies and the Interstellar Medium, eds. J. Franco, F. Ferrini, G. Tenorio-Tagle (Cambridge: Cambridge Uni-

versity Press), 275
Shelton, R.L., Kruk, J.W., Murphy, E.M., et 22 al. 2001, ApJ, 560, 730
Slavin, J.D., Cox, D.P. 1993, ApJ, 417, 187
Slavin, J.D., Shull, J.M., Begelman, M.C. 1993, ApJ, 407, 83
Snowden, S.L., Mebold, U., Hirth, W., Herbstmeier, U., Schmitt, J.H. 1991, Science, 252, 1529
Snowden, S.L., Egger, R., Finkbeiner, D.P., Freyberg, M.J., Plucinsky, P.P. 1998, ApJ, 493, 715
Snowden, S.L., Freyberg, M.J., Kuntz, K.D., Sanders, W.T. 2000, ApJS, 128, 171
Spitzer, L. 1956, ApJ, 124, 20
Spitzer, L. 1990, ARA&A, 28, 71
Spitzer, L. 1996, ApJ, 458, L29
Vladilo, G., Centurión, M., de Boer, K.S., King, D.L., Lipman, K., Stegert, J., Unger, S.W., Walton, N. 1993, A&A, 280, L11
Vladilo, G., Centurión, M., de Boer, K.S., King, D.L., Lipman, K., Stegert, J., Unger, S.W., Walton, N. 1994, A&A, 291, 425
Wakker, B.P. 1991, A&A, 250, 499
Wakker, B.P. 2001, ApJS, 136, 463
Wakker, B.P., van Woerden, H. 1997, ARA&A, 35, 217
Wakker, B.P., Howk, J.C., Chu, Y.-H., Bomans, D.J., Points, S.D. 1998, ApJ, 499, L87
Wakker, B.P., Howk, J.C., Savage, B.D., van Woerden, H., Tufte, S.L., Schwarz, U.J., Benjamin, R., Reynolds, R.J., Peletier, R.F., Kalberla, P.M.W. 1999, Nature, 402, 388
Wakker, B.P., Savage, B.D., Sembach, K.R., et 18 al. 2003, ApJS, 146, 1
Widmann, H., de Boer, K.S., Richter, P., Krämer, G., Appenzeller, I., Barnstedt, J., Gölz, M., Grewing, M., Gringel, W., Mandel, H., Werner, K. 1998, A&A, 338, L1
Williamson, F.O., Sanders, W.T., Kraushaar, W.L., McCammon, D., Borken, R., Bunner, A.N. 1974, ApJ, 193, L133
Wolfire, M.G., McKee, C.F., Hollenbach, D., Tielens, A.G.G.M. 1995, ApJ, 453, 673

12. HVCS INTERACTING WITH THEIR ENVIRONMENT

CHRISTIAN BRÜNS, ULRICH MEBOLD
Radioastronomisches Institut, Universität Bonn, Germany
cbruens@astro.uni-bonn.de, umebold@astro.uni-bonn.de

Abstract. We present observational evidence and review theoretical considerations for an interaction of high-velocity clouds (HVCs) with their ambient medium. HVCs showing either a head-tail structure or a horseshoe shape are likely to be interacting. Such HVCs are observed in complexes in the gaseous Galactic Halo (e.g. complex A), in HVC complexes in the outer Galactic Halo (e.g. the Magellanic Stream), and in a sample of compact HVCs that might be located in the intergalactic medium of the Local Group. The overall morphology of the head-tail HVCs is the same for clouds located in different environments: cold, dense condensations are found near the sharp (leading) edge of the HVC, embedded in a warm, neutral medium forming a head-tail structure. The observed morphology – decreasing column density, increasing velocity dispersion and decreasing radial velocity in the direction of the tail – is consistent with gas having been stripped off the main body of the HVC. The detections of Hα emission, O VI absorption, and possibly X-ray emission associated with HVCs provide further evidence for a ram-pressure interaction. Moreover, recent numerical simulations modeling the ram-pressure interaction of HVCs with their ambient medium predict the existence of head-tail structures.

1. Introduction

In this chapter, we focus on the interaction of HVCs with their ambient medium. We will review both observational evidence and numerical simulations for interacting HVCs.

Shortly after the detection of HVCs by Muller et al. (1963), it was evident that a large fraction of HVCs show negative radial velocities (Oort 1966). As a consequence, it was expected that they will reach the disk after some time and merge with Milky Way gas. The density of the gas near the galactic plane is quite high and ram pressure will slow down the HVCs significantly. Gravitational acceleration and ram-pressure deceleration will

H. van Woerden et al. (eds.), High-Velocity Clouds, 251–271

probably balance each other, leading to a constant velocity, also known as terminal velocity (Benjamin & Danly 1997).

High-velocity clouds are known to be located in various environments. First, there are HVCs in the gaseous Galactic Halo, located a few kpc above the galactic plane (e.g. complex A, van Woerden et al. 1999a); second, there are HVCs located in the outer Galactic Halo at distances on the order of 50 kpc (e.g. the Magellanic Stream, see Ch. 5); and third, there are HVCs thought to be located in the local intergalactic medium (CHVCs, see Ch. 15).

Section 2 gives an estimate of the influence of ram pressure on the dynamical evolution of HVCs. Section 3 shows H I observations that indicate an interaction of HVCs with their ambient medium. Section 4 summarizes observations in the high-energy bands (UV, X-rays, γ-rays) that associate excess emission with HVCs. Section 5 summarizes recent numerical simulations of interacting HVCs. Section 6 gives a summary and conclusions.

2. The influence of ram pressure

A significant fraction of HVCs show negative velocities (see Ch. 3). They fall down toward the Galactic Disk and are anticipated to merge with the Milky Way gas. HVCs are embedded in a diffuse, low-density medium for most of their lifetime, up to the time that they impact the disk, where the density of the ambient medium is quite high. The density of the ambient halo medium decreases from $n_{\mathrm{h}}{\sim}10^{-3}$ cm^{-3} in the lower Galactic Halo to $n_{\mathrm{h}}{\sim}10^{-6}$ cm^{-3} in the intergalactic medium (Kalberla & Kerp 1998). The ambient medium helps to stabilize HVCs, as the external pressure acts in the same direction as the gravitational forces in the HVC. For HVC complexes for which an approximate distance is known, observed cloud masses are orders of magnitude too low to provide gravitational stabilization. By definition HVCs move with a high velocity relative to the Milky Way gas. In this section, HVCs will be treated as solid bodies, where momentum is transferred to the cloud as a whole, in order to gain a first impression of the influence of ram pressure. In Sect. 5, more realistic models are presented. The decelerating force due to ram-pressure interaction can be calculated by a simple 1-D version of the equation of motion (Benjamin & Danly 1997):

$$m_{\mathrm{c}}\frac{dv_{\mathrm{c}}}{dt} = m_{\mathrm{c}}g(z) - \frac{1}{2}C_{\mathrm{D}}\rho_{\mathrm{h}}(z)A_{\mathrm{c}}\left[v_{\mathrm{c}} - v_{\mathrm{h}}(z)\right]^2 . \quad (1)$$

A_{c}, m_{c}, v_{c}, ρ_{h}, v_{h}, and g represent the surface area of the HVC, the HVC mass, the HVC velocity, the halo density, the velocity of the halo gas, and the gravitational acceleration, respectively. The drag coefficient, C_{D}, indicates the efficiency of momentum transfer to the cloud, the factor $\frac{1}{2}$ is conventionally included. Usually, $C_{\mathrm{D}} = 1$ is assumed. The equa-

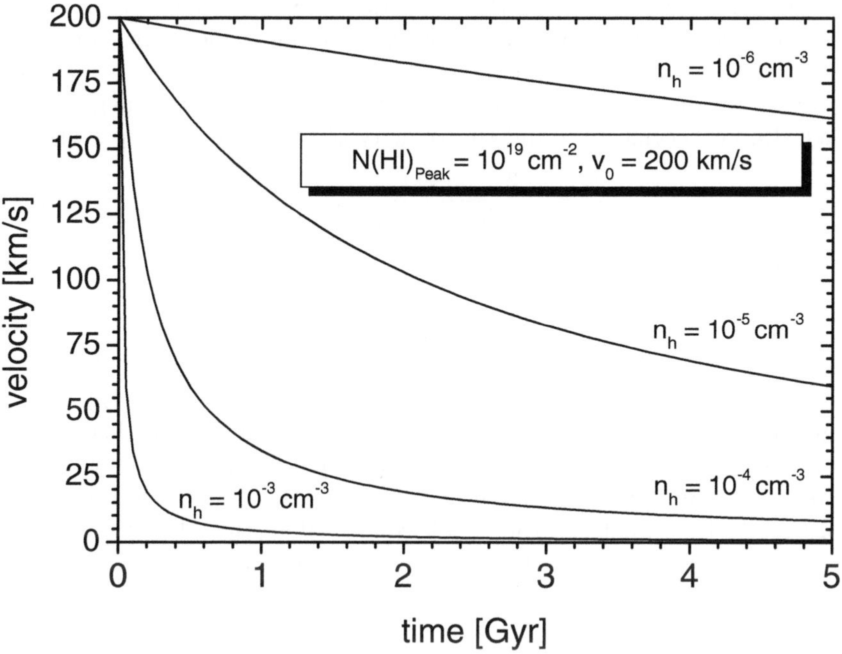

Figure 1. The diagram shows the evolution of the velocity of an HVC decelerated solely by ram pressure as a function of time for different values of the density, $n_{\rm h}$, of the ambient medium. Since $v(t)$ depends only on the ratio of $n_{\rm h}$ and $N(\mathrm{H\,I})_{\rm peak}$ (see Eq. 4), the curve for $N(\mathrm{H\,I})_{\rm peak} = 10^{19}$ cm^{-2} and $n_{\rm h} = 10^{-5}$ cm^{-3} is identical to the curve for $N(\mathrm{H\,I})_{\rm peak} = 10^{20}$ cm^{-2} and $n_{\rm h} = 10^{-4}$ cm^{-3}. Figure from Brüns (2003).

tion can be further simplified by approximating the HVC as a homogeneous sphere with constant density. It follows that $m_{\rm c} = m_{\rm u}\overline{n_{\rm c}}\ \frac{4}{3}\pi r^3$, $\overline{n_{\rm c}} = N(\mathrm{H\,I})_{\rm peak}\ (2r)^{-1}$ and $A_{\rm c} = \pi r^2$, where $m_{\rm u}$ represents the mean mass per particle and $N(\mathrm{H\,I})_{\rm peak}$ the peak column density. Furthermore, $\rho_{\rm h} = m_{\rm u} n_{\rm h}$ is assumed. Benjamin & Danly (1997) investigated the dynamical evolution of clouds falling to the Galactic Disk along the vertical (z) axis. They assumed that HVCs are moving with terminal velocity, i.e. the case where locally $\mathrm{d}v/\mathrm{dt} = 0$. It follows that

$$v_t = \sqrt{\frac{2\ g(z)\ N(H\,\mathrm{I})_{\rm peak}}{C_{\rm D}\ f_{\rm c}\ n_{\rm h}(z)}}. \quad (2)$$

Here $f_{\rm c}$ is the ratio of the H I column density to the total column density (H I+H II). Benjamin & Danly (1997) assumed that the HVC is moving only in the direction of the z-axis. They calculated the space velocity from the radial velocity and the galactic latitude. They derived a distance to the HVCs assuming values for $C_{\rm D}$ and $f_{\rm c}$, and taking $g(z)$ and $n_{\rm h}(z)$ from a Milky Way model.

Although it is unlikely that HVCs fall down exactly along the z-axis, and it is unclear whether they move with terminal velocity or not, the model is useful as a first reconnaissance.

The deceleration due to ram pressure alone can be estimated, neglecting the gravitational acceleration and the velocity of the halo gas. Eq. 1 yields

$$\frac{dv_c}{dt} = -\frac{3}{4}\frac{n_h(z)}{N(H\,\mathrm{I})_{\mathrm{peak}}}\,[v_c]^2\,. \tag{3}$$

The solution of Eq. 3 is

$$v_c(t) = \frac{v_0}{1 + v_0\, t\, \frac{3}{4}\frac{n_h(z)}{N(H\,\mathrm{I})_{\mathrm{peak}}}}, \tag{4}$$

where v_0 represents the initial velocity at t = 0. Figure 1 shows Eq. 4 for different values of the density of the ambient medium, for $N(\mathrm{H\,I})_{\mathrm{peak}} = 10^{19}$ cm^{-2} and v_0= 200 km s^{-1}. The solution depends only on the ratio of the halo density and the peak column density of the cloud. Figure 1 can therefore easily be scaled to other peak column densities. The time scale for the deceleration can be expressed by $t_{1/2}$, the time needed to reduce the initial velocity by a factor of 2:

$$t_{1/2} = 2.1 \times 10^7 \mathrm{yr} \left[\frac{N(H\,\mathrm{I})_{\mathrm{peak}}}{10^{19}\ \mathrm{cm}^{-2}}\right] \left[\frac{n_h}{10^{-3}\ \mathrm{cm}^{-3}}\right]^{-1} \left[\frac{v_0}{200\ \mathrm{km\ s}^{-1}}\right]^{-1}. \tag{5}$$

An HVC with a peak column density of 10^{19} cm^{-2}, embedded in an ambient medium with a density of 10^{-3} cm^{-3}, decelerates from 200 km s^{-1} to 100 km s^{-1} in 21 Myr. For comparison, the free-fall timescale of clouds in complex A is of the same order of magnitude. The density of the ambient medium surrounding the Magellanic Stream is expected to be two orders of magnitude lower. The corresponding timescale of $t_{1/2}$ = 2.1 Gyr is similar to the age of the Magellanic Stream predicted by numerical simulations (Gardiner 1999 and references therein). As a consequence, ram pressure is expected to play a significant role in the dynamical evolution of HVCs.

3. Evidence for interaction with the ambient halo medium

In the following subsections, we present examples for HVCs in HVC complexes located at different heights above the Galactic Disk that are probably interacting with their ambient medium.

3.1. HVCS IN THE GALACTIC HALO

HVC complex M has a distance to the galactic plane of z<3.5 kpc (Danly et al. 1993). Giovanelli et al. (1973) observed the cloud M I with the 300-ft *Green Bank* telescope and found a “wing” of diffuse gas that connects the HVC with gas at intermediate velocities. The region with the highest column density has the most negative velocity. While there are cores with narrow line width in the high column density region (the head), there are no such cores in the wing (or tail).

HVC complex A is the only HVC complex with a distance bracket, placing it within the gaseous Galactic Halo at $4.5<z<7$ kpc (van Woerden et al. 1999a; Wakker et al. 2003). Giovanelli & Haynes (1977) observed HVC 139+28−191 (also known as A I) with the 300-ft *Green Bank* telescope and found cold cores embedded in a warmer gas phase. The warmer phase showed strong velocity and column density gradients. Giovanelli & Haynes (1977) were the first to call this kind of structure a tail.

Hulsbosch(1978) mapped the H I in HVC 132+23−211 (also known as A0), using the 100-m *Effelsberg* telescope. This HVC is located in the same line of sight as the Outer Arm complex. Like other clouds in complex A, it shows strong velocity gradients and cold components near the edge of the fastest part. He argued that the cold cores are shock instabilities formed during the interaction with the ambient medium. Meyerdierks (1991, 1992) performed new observations with the 100-m *Effelsberg* telescope and analyzed the structure of this cloud in more detail. He argued that HVC 132+23−211 is currently interacting with low-velocity gas. A problem with the suggested interaction of complex A with low-velocity gas is the apparently large distance of that HVC. However, the distance determination is valid for the high-latitude end of the complex, whereas Meyerdierks (1991, 1992) studied a low-latitude core. A determination of the distance to this core is required to settle this question.

Pietz et al. (1996) used the Leiden-Dwingeloo Survey of galactic neutral atomic hydrogen (*LDS*, Hartmann & Burton 1997) to analyze the H I distribution of complex C, looking for signs of interaction. They found a number of clouds where a faint bridge of gas connects complex C with IVCs (so-called velocity bridges). Follow-up observations with the 100-m *Effelsberg* telescope enhanced the recognizability of these velocity bridges.

Figure 2 shows two examples for tails connecting HVCs with gas at lower velocities near complex M (HVC 166+56−140) and toward complex H (HVC 129+4−210). While the H I emission clearly connects the HVCs with gas at lower velocities, it remains unclear whether these clouds are physically connected or not. IVCs have usually much smaller distances than HVCs, making a direct connection of these clouds unlikely. These might just represent cases where the low-velocity end of a tail accidentally superimposes with IVC gas.

Observational indications for interactions of HVCs with the Galactic Disk have been claimed for several HVC complexes. The review by Mebold et al. (1990) focused on the interaction of HVCs with the Galactic Disk.

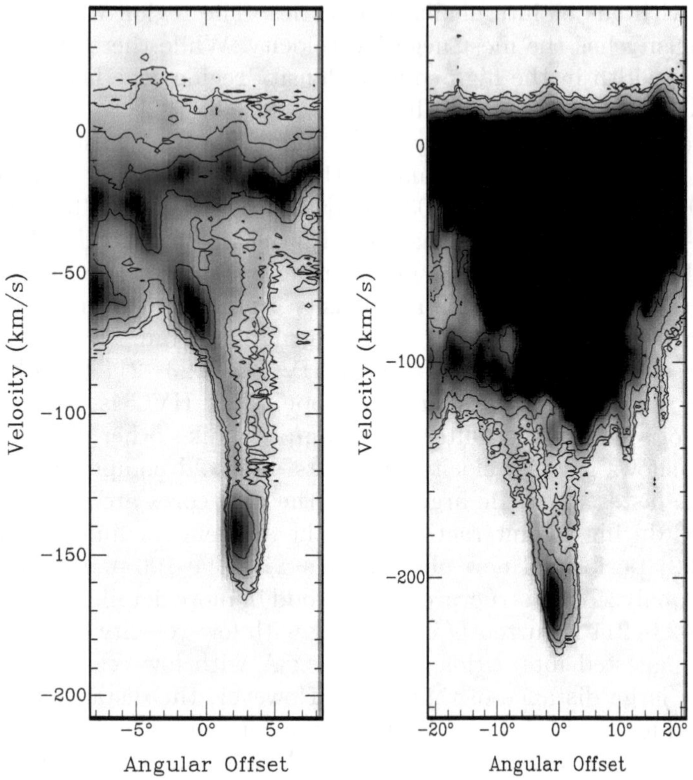

Figure 2. Position-velocity slices from the *LDS*. *Left*: position-velocity slice through HVC 166+56−140 (near complex M). *Right*: slice through HVC 129+4−210 (in complex H). The slices are oriented along the direction of the tail. The contour lines represent for both figures 0.14, 0.21, 0.42, 0.84, 1.68, and 3.36 K. There is a velocity bridge connecting the HVCs with gas at lower velocities. Nevertheless, this does not necessarily imply a physical connection. Figure from Brüns (1998).

One prominent example is the interaction of an HVC with the so-called Draco Nebula (Mebold et al. 1985), where observed structures in the Draco Nebula seemed to show some relationship to those in complex C; however, the two clouds also seem to be at different distances: 0.3 to 2.5 kpc for the Draco Nebula (Goerigk & Mebold 1986) and >5 kpc for complex C (van Woerden et al. 1999b). A more recent example was presented by Tamanaha (1997), who found evidence for an interaction of the Anti-Center complex with Milky Way gas. Another example was presented by Morras et al. (1998), who investigated the interaction of complex H with the Milky Way disk. The H I data of complex H show evidence for an interaction in form of steep velocity and column density gradients. In the same line of sight, there is a hole in the H I distribution of the disk gas. The kinematic distance

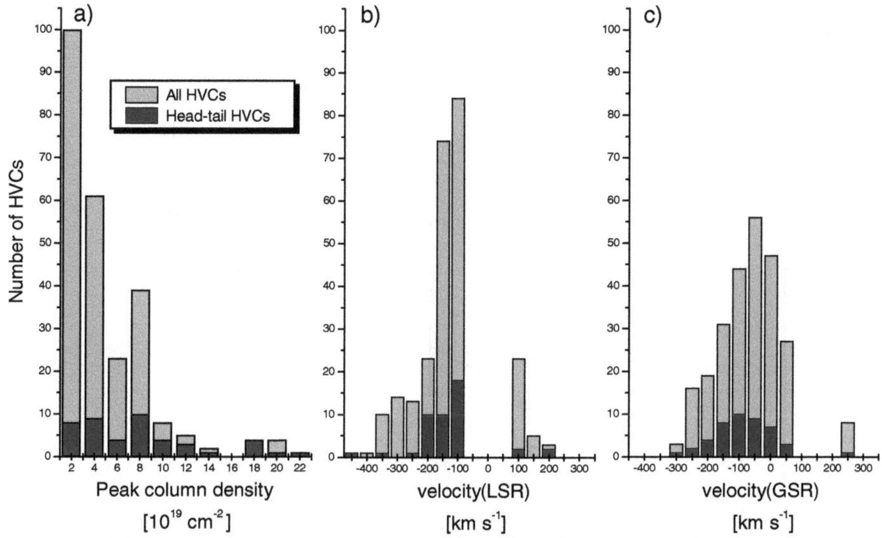

Figure 3. These histograms show the number of HVCs in the *LDS* versus (a) peak column density, (b) v_{LSR} and (c) v_{GSR}. The entire HVC sample is plotted in light gray, the head-tail HVCs are plotted in dark gray. ($\Sigma_{\mathrm{HVC}} = 252$, $\Sigma_{\mathrm{head-tail}} = 45$). Figure from Brüns et al. (2000a).

of the disk gas is 15 kpc, corresponding to a galactocentric distance of 22 kpc, i.e. complex H is likely interacting with the outermost Galactic Disk. Lockman (2003) observed a part of complex H using the *Green Bank* Telescope (*GBT*), confirming the evidence for an interaction with the Milky Way. He models complex H as a satellite galaxy at a distance of 33 kpc from the Galactic Center. There is also evidence for an interaction of the Leading Arm of the Magellanic System with the gas in the Galactic Disk (see Sect. 3.3).

3.2. HEAD-TAIL HVCS IN THE LEIDEN-DWINGELOO SURVEY

Brüns et al. (2000a) analyzed the HVCs in the whole sky covered by the *LDS*. As a first step, they compiled a "bright HVC catalog" containing all HVCs north of declination $-30°$ with peak column densities larger than 10^{19} cm^{-2} and diameters larger than $1°$. This catalog contains 252 clouds and forms a complete, flux-limited sample (Brüns 1998).

In a second step, each cloud was analyzed by eye with respect to velocity gradients and asymmetries in the column density distribution (more details can be found in Brüns et al. 2000a). Altogether, about 20% of the clouds showed both a significant velocity gradient and an asymmetric H I distribution, a so called head-tail structure.

Head-tail HVCs were found in *all* HVC complexes except the very faint

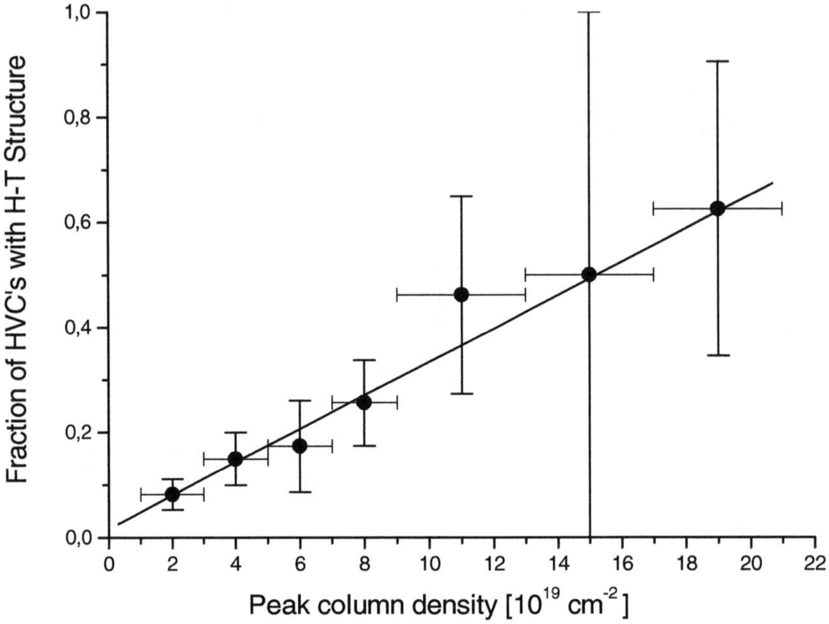

Figure 4. Correlation between the fraction of head-tail HVCs in the *LDS* among all HVCs of the sample in a certain column density interval and the peak column density. The solid line marks the result of a weighted linear regression. Figure from Brüns et al. (2000a).

complex L. The analysis showed a clear trend in that a larger fraction of clouds with higher column density show a head-tail structure than clouds with a lower column density (Figs. 3 and 4). However, there might be a selection effect for the faintest clouds in the sample, where a tail might not be detected because of a small angular extent or a low column density. There is also a trend that a larger fraction of clouds with high radial velocities $|v_{\rm GSR}|$ are head-tail HVCs. The mean radial velocity $v_{\rm GSR}$ of the head-tail HVCs is more negative than the mean radial velocity of the entire HVC sample (Fig. 3c). A gaussian fit to the histograms yielded mean velocities of $\overline{v_{\rm head-tail\ HVC}} = -86.1\pm3.8$ km s^{-1} and $\overline{v_{\rm all\ HVC}} = -62.3\pm7.5$ km s^{-1}, the difference being significant. It is so far unclear whether this trend is influenced by selection effects or not. The southern sky contains a number of clouds with positive radial velocities. A full-sky survey might change the results significantly.

The identification of head-tail HVCs in the *LDS* is difficult, because of the relatively low angular resolution of 36′ and grid spacings of 30′. Especially within large HVC complexes like complex C a clear identification is difficult. While clear cases of head-tail HVCs exist, some head-tail structures may be explained by random superpositions of unresolved clouds with similar velocities, while "real" head-tail structures might be hidden.

3.3. MAGELLANIC HVCS

The gaseous tidal arms of the Magellanic System (Ch. 5) consist of H I gas with radial velocities in the HVC regime. The *Parkes* narrow-band multibeam survey of the entire Magellanic System (Brüns et al. 2000b) allows for the first time a detailed analysis of Magellanic HVCs with respect to velocity gradients and asymmetries in the column density distribution. The survey comprises full sampling at an angular resolution of $14'$ and high velocity resolution ($\Delta v = 0.8$ km s^{-1}).

The Magellanic Stream forms a continuous stream of gas that starts at the Magellanic Clouds and trails over 100°, passing the South Galactic Pole. Figure 5 shows part of the Magellanic Stream, centered on MS IV. In a crowded region like the Magellanic Stream, a detailed analysis of single clouds is extremely difficult. Some clouds have a horse-shoe shape (e.g. near α=23^{h}44^m, δ=−12°) or a head-tail structure (e.g. near α=23^{h}54^m, δ=−12°). They may be indicators for interactions of gas clumps within the Stream, rather than for an interaction with the surrounding medium. Nevertheless, the overall structure shows a sharp column density drop-off toward the eastern side, while the western side drops off more smoothly.

A plausible explanation for these asymmetries is an interaction with the ambient medium. Another hint for an interaction with the ambient medium is the detection of Hα associated with the Magellanic Stream (Weiner & Williams 1996; Bland-Hawthorn & Maloney 1999, 2001). The Hα intensity is much brighter than expected from photoionization due to Lyα photons that escape from the Galactic Disk. The high Hα brightness could be explained if the Magellanic Stream is about 5 times nearer than expected from numerical simulations. Alternatively, the excess Hα emission may be explained by ram-pressure interactions with the ambient medium. Konz et al. (2001) explained the Hα emission from the Magellanic Stream considering the interaction of a neutral gas cloud with an ambient magnetized plasma. In cases where the relative velocity is high (v_{rel}∼300 km s^{-1}), the magnetic field is on the order of B∼10 μG, and the ambient density is on the order of n∼10^{-5} cm^{-3}, efficient ionization of the outer layers of the Magellanic Stream is possible. While photoionization leads to a homogeneous illumination of the cloud, interaction processes produce much brighter emission at the leading edge of the clouds. More Hα observations covering the entire Magellanic Stream, rather than single lines of sight, are needed to discriminate between the alternatives.

The Leading Arm (Putman et al. 1998) is a clumpy structure containing a number of unresolved, dense condensations. Figure 6 shows one example of a head-tail HVC in the Leading Arm, presenting both a column density map and a position-velocity slice though the H I data cube. Associated

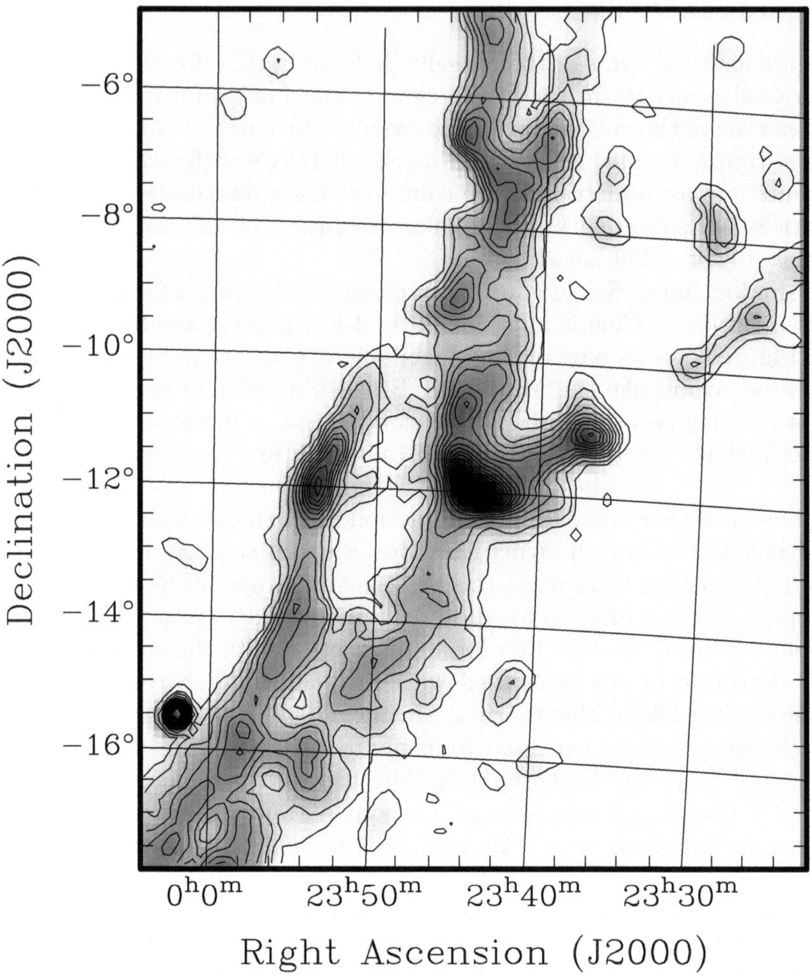

Figure 5. H I column density distribution of a part of the Magellanic Stream observed with the multi-beam facility of the *Parkes* telescope. The contour levels represent 0.1, 1, 2, 3, ...$\times 10^{19}$ cm^{-2}. This part of the Magellanic Stream was formerly known as MS IV (Mathewson et al. 1974). Figure from Brüns et al. (2000b).

with the column density gradient of the HVC is a velocity gradient. Lower column densities appear at lower velocities, suggesting that the gas in the tail may be stripped off. Similar structures were observed for a number of clouds in this region.

The Leading Arm abruptly changes in orientation as well as in radial velocity where it crosses the galactic plane and continues on the other side of the galactic plane shifted in the direction of lower galactic longitude (see left part of Fig. 5). A number of clouds near the galactic plane show a

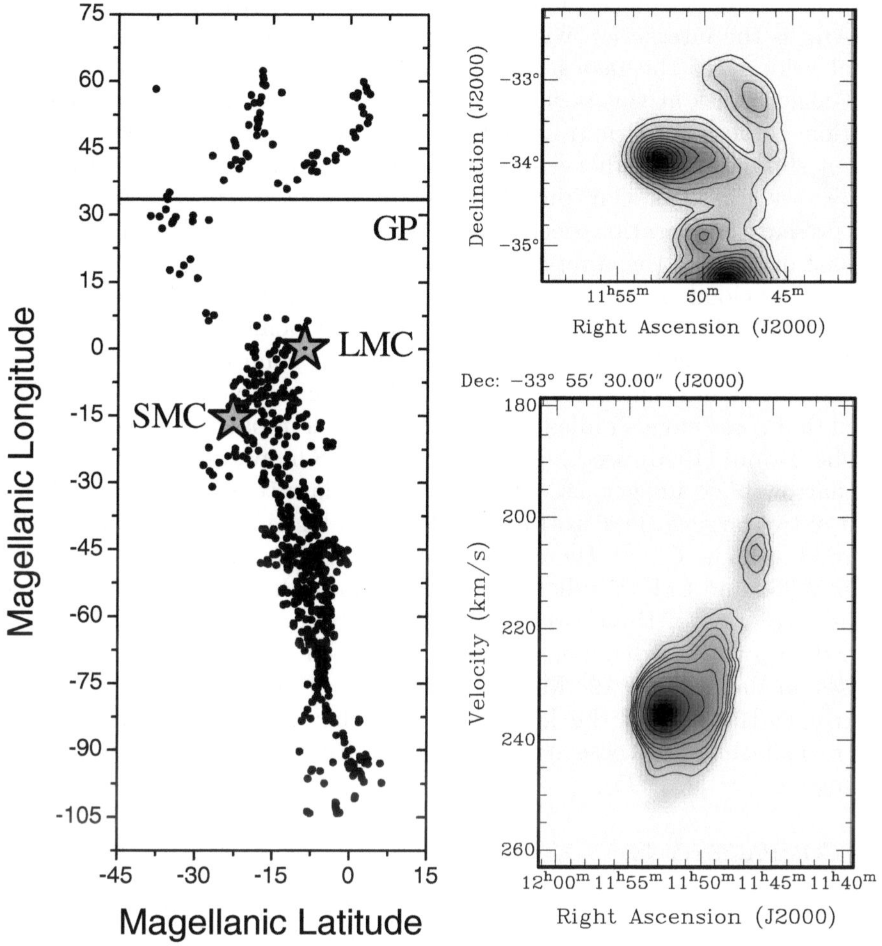

Figure 6. *Left:* The distribution of Magellanic HVCs in Magellanic coordinates (the line of zero latitude is defined as the line from galactic longitude l=90° across the South Galactic Pole to l=270° (zero longitude is defined by the position of the LMC). Each marker represents one entry in the catalog of clouds in the tidal arms of the Magellanic System. The Magellanic Clouds are marked by asterisks. The line at Magellanic longitude 33°.4 represents the location of the galactic plane. Note that the Leading Arm bends over where it crosses the line of sight to the galactic plane (GP).
Right: a head-tail HVC of the Leading Arm: the top image shows the column density distribution (contours represent 0.5 and 1×10^{19} cm^{-2}, increasing in steps of 1×10^{19} cm^{-2}), the lower image shows a position-velocity slice though the data cube at a declination of −33°.92 (the contour lines represent 0.2, 0.3, 0.4, 0.5, 0.7, 1, 2, 3 K; 0.2 K corresponds to $\sim 5\ \sigma$). This cloud is clearly asymmetric both in the column density distribution and in velocity. Figure from Brüns et al. (2000a).

head-tail structure. A possible explanation for the turnover of the Leading Arm is the interaction with gas layers in the outer Galactic Disk. The radial velocity of the fastest (and therefore most distant) component of the Galactic Disk at this position is $v_{\rm LSR}$=+122 km s^{-1}. Extrapolating the rotation curve from Brand & Blitz (1993) yields a distance of ~30 kpc. Recent simulations predict a similar distance for the Leading Arm in this region (see Fig. 2 of Gardiner 1999). This reinforces the possibility of a ram-pressure acceleration/deceleration in the relatively high density of the ambient medium in the outer disk. Further work is necessary to make plausible that the clouds in the Leading Arm receive sufficient momentum during their passage through the Galactic Disk without being destroyed.

The Magellanic Bridge is a complex structure connecting the LMC and the SMC. The internal velocity structure is quite complex, and it is anticipated that some clouds pulled out of the SMC will impact on the LMC after a while. Braun (1996) used high-resolution *IRAS* maps, optical broad-band and narrow-band images, as well as H I 21-cm data to analyze the environment of two large-scale distortions (LMC 4 and DEM 268) in the disk of the Large Magellanic Cloud. He argued that the cone-like structures connected to DEM 268 and LMC 4 indicate HVC impacts as the most probable source of the formation of these two distortions. In both cases, the energy input needed to form the distortions might have been provided by HVCs with masses on the order of 10^5 $M_\odot$, assuming an initial velocity of 100 km s^{-1} relative to the disk of the LMC. On the other hand, there are alternative explanations for these structures, e.g. self-propagating star formation (Domgörgen et al. 1995).

3.4. COMPACT HVCS

Compact high-velocity clouds (CHVCs) are thought to be located in the intergalactic medium of the Local Group (see Ch. 15). The fact that they are compact *and* isolated makes them the perfect candidates for a detailed analysis of the H I distribution. The analysis of CHVCs with respect to head-tail structures is very interesting, as these observations may help to prove the existence of an ambient medium around CHVCs. If CHVCs are indeed located in intergalactic space, the observations may help to constrain the physical parameters of the intergalactic medium of the Local Group.

So far, the best studied CHVC is HVC 125+41−207 (also known as WW 84; Braun & Burton 2000, Brüns et al. 2001). Figure 7 shows the H I column density distribution of HVC 125+41−207, as observed with the 100-m *Effelsberg* telescope. This cloud has a distinct head-tail structure. More than 10% of the H I gas is located in the tail. A significant part of the tail has column densities below 10^{19} cm^{-2}, i.e. deep integrations are

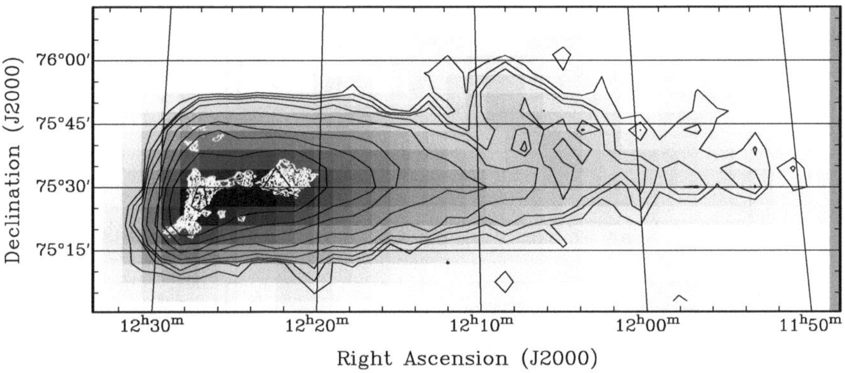

Figure 7. H I column density distribution of HVC 125+41−207. The black contours represent 1, 2, 3, 5, 10, 20, 30, 50, 100×10^{18} cm^{-2}, as observed with the 100-m *Effelsberg* telescope. The H I distribution shows a distinct head-tail structure. The white contour levels represent the high column density part of the Westerbork data from Braun & Burton (2000), starting from 10^{20} cm^{-2} in steps of 0.5×10^{20} cm^{-2}. Figure from Brüns et al. (2001).

needed to detect tails of this kind.

In general, the spectra show two components, one with a low and one with a high velocity dispersion (i.e. a cold and a warm gas phase). White contours in Fig. 7 represent the cold gas phase as observed with the Westerbork interferometer (Braun & Burton 2000). Cold cores were only observed in the head of HVC 125+41−207. Figure 8 shows the result of a gaussian decomposition of the *Effelsberg* H I spectra from HVC 125+41−207 for two representative slices through the HVC (see Brüns et al. (2001) for a detailed discussion). Figure 8a shows that the position of the maximum column density of the warm component is shifted relative to the cold component. The radial velocities $|v_{\rm GSR}|$ of the warm component are always less negative than the velocities of the colder gas (Fig. 8e), indicating that the two gas phases are separating from each other. There is a small velocity gradient for the cold gas phase. This gradient can be used to disentangle projection effects and estimate the projection angle between the line of sight and the 3-D HVC velocity vector (Brüns et al. 2001). The result is that HVC 125+41−207 moves almost perpendicularly to the line of sight. The radial velocity of the warm component does not show a regular gradient. The run of the curve can be explained by an oscillating tail, caused by turbulent flows along its trajectory (Santillán et al. 1999).

The velocity dispersion of the warm H I gas is increasing from FWHM $\sim$10 km s^{-1} to FWHM $\sim$20 km s^{-1} across the extent of the cloud and remains approximately constant in the tail. This corresponds to a Doppler temperature increasing from $T_{\rm D}$=2000 K to $T_{\rm D}$=10000 K in the tail. There is also a trend of higher velocity dispersions toward the outskirts of the

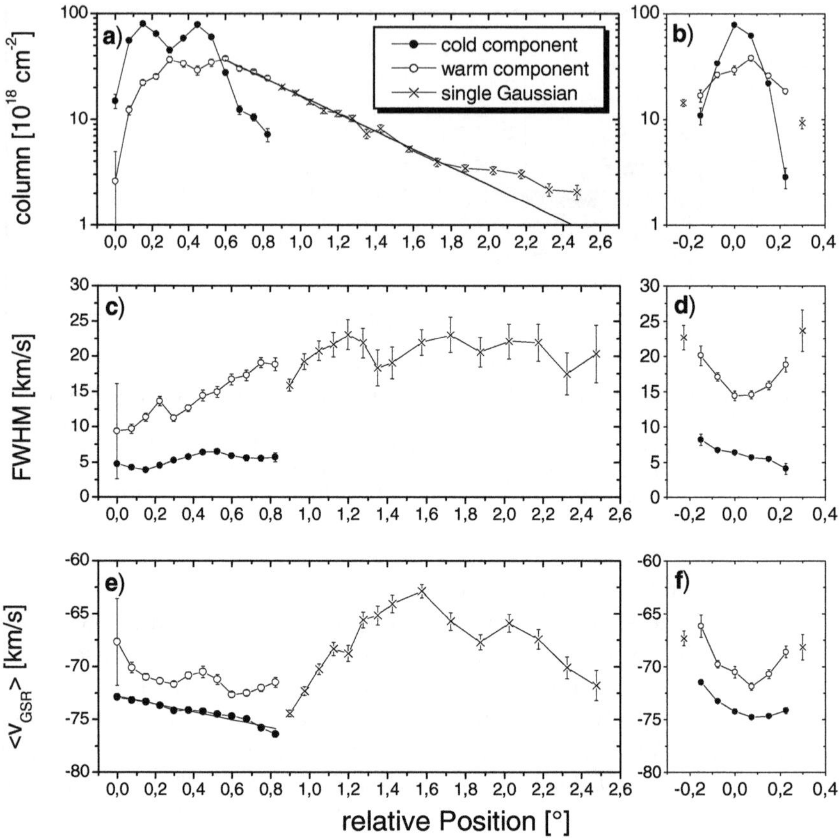

Figure 8. Showing the results of the gaussian decomposition of the H I spectra of HVC 125+41−207 along two perpendicular slices. The cold gas phase is represented by filled circles, the warm gas phase by open circles. Results from a single gaussian fit are represented by crosses. The diagrams a, c and e show the results for a slice in the direction of the tail (at $\delta_{2000} = +75\overset{\circ}{.}5$), the diagrams b, d and f represent a slice perpendicular to the first one (at $\alpha_{2000} = 12^{\mathrm{h}}23^{\mathrm{m}}$). The line in diagram a is an exponential fit to the data. The line in diagram e is a fit to the velocity field. Figure from Brüns et al. (2001).

cloud. These results give strong evidence that the warm gas is in the process of being stripped off the main body of the HVC.

Brüns (2003) performed deep H I observations of 8 CHVCs using the 100-m *Effelsberg* telescope. Most of them show up with an asymmetric column density distribution. Figure 9 shows two examples of CHVCs with a head-tail structure, HVC 070+50−145 and HVC 024−02−285. The tails of these CHVCs have only a small velocity gradient. Figure 10 shows two examples of CHVCs with a horse-shoe shape, also indicative for ram-pressure interaction. The four examples of Figs. 9 and 10 illustrate typical shapes for compact, isolated HVCs.

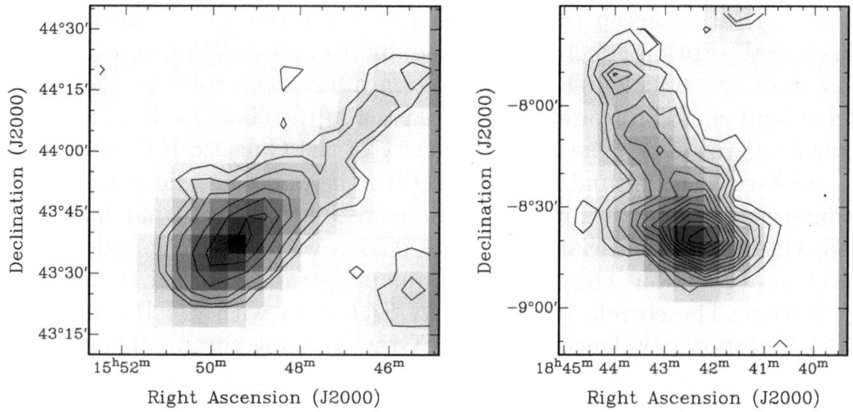

Figure 9. Two examples of head-tail CHVCs. *Left:* H I column density distribution of HVC 070+50−145. The black contours represent 2, 5, 10, 20, 30, 40×10^{18} cm^{-2} as observed with the 100-m *Effelsberg* telescope. The H I distribution shows a faint head-tail structure with average column densities in the tail on the order of 5×10^{18} cm^{-2}. *Right:* H I column density distribution of HVC 024−02−285, observed with the 100-m *Effelsberg* telescope. The black contours represent 2 and 5×10^{18} cm^{-2}, further increasing in steps of 5×10^{18} cm^{-2}. The tail has an average column density on the order of 1.5×10^{19} cm^{-2}. Figure from Brüns (2003).

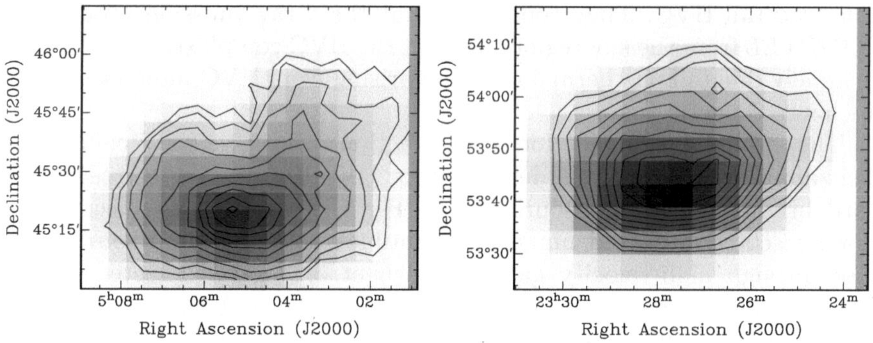

Figure 10. Two examples for horse-shoe shaped CHVCs. *Left:* H I column density distribution of HVC 162+03−180. *Right:* H I column density distribution of HVC 110−7−465. Both were observed with the 100-m *Effelsberg* telescope. The black contours represent 2 and 5×10^{18} cm^{-2}, further increasing in steps of 5×10^{18} cm^{-2}. Figure from Brüns (2003).

4. Evidence from UV, X-ray and γ-ray observations

Kerp et al. (1999) searched for enhanced soft X-ray emission associated with HVC complexes using the *ROSAT* all-sky survey. They studied the transfer of 0.25-keV photons through the interstellar medium, in order to distinguish variations in the observed soft X-ray intensity caused by photoelectric absorption effects from those due to excess X-ray emission. The X-ray data are modeled as a combination of emission from the Local Hot Bubble and emission from a distant plasma in the Galactic Halo and from extragalactic sources. The X-ray intensity of the Galactic Halo and extragalactic X-ray background is modulated by the photoelectric absorption of the intervening galactic interstellar matter. They showed that large- and small-scale intensity variations of the 0.25-keV data are caused by photoelectric absorption which is predominantly traced by the total $N(\text{H I})$ distribution. The correlation of the *ROSAT* data with the H I data from the *LDS* supports evidence for a hot, X-ray emitting corona. They showed that their method leads to a good representation of the soft X-ray observations. In four large areas on the sky, they searched for regions where the modeled and observed X-ray emission differed significantly and found that there is significant excess X-ray emission near HVC complexes C, D, and GCN. It was suggested that the excess X-ray emission is correlated with HVCs. However, the situation is complicated, as some lines of sight toward HVCs also pass through significant amounts of intermediate-velocity gas.

Using *COMPTEL* observations, Blom et al. (1997) found evidence for enhanced γ-ray emission between 0.75 and 3 MeV, related to the complexes M and A. The peak of the γ-ray emission is located at the position of a head-tail HVC. They concluded that the γ-ray emission observed by *COMPTEL* from the sky region between the HVC complexes M and A may be largely due to non-thermal emission arising from HVC interactions with the lower Galactic Halo.

The strongest evidence for an interaction of high-velocity clouds with their ambient medium was derived from UV absorption spectroscopy toward bright background sources. The Far Ultraviolet Spectroscopic Explorer (*FUSE*) observed numerous sightlines to background sources that pass through high-velocity gas. Sembach et al. (2003) detected 85 O VI HVCs in absorption on 100 sightlines. The correlation between the derived O VI column densities and those derived from other highly-ionized species (C IV and N V) suggests collisional ionization as the most probable source for the O VI. The detection of O VI absorption associated with the Magellanic Stream indicates the existence of an extended Galactic Corona even at distances beyond 50 kpc. This is supported by the strong Hα detections associated with the Magellanic Stream discussed in Sect. 3.3.

In summary, observations give clear evidence for an interaction of HVCs with their ambient medium. Observations using the new generation X-ray telescopes are needed to further explore the possibility of a correlation between HVCs and X-ray emission.

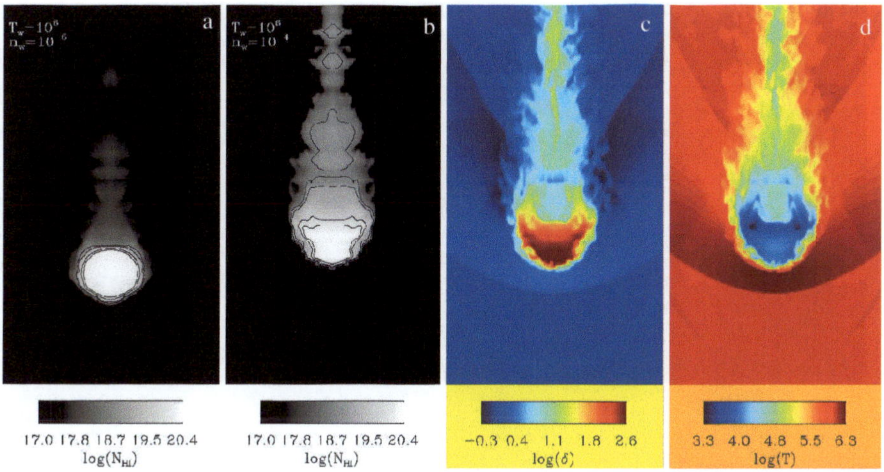

Figure 11. Figure 2 from Quilis & Moore (2001). The column density for the dark-matter dominated HVC after 300 Myrs of evolution in a wind of velocity 200 $\mathrm{km\,s^{-1}}$. (a) density 10^{-5} $\mathrm{cm^{-3}}$ and (b) density 10^{-4} $\mathrm{cm^{-3}}$. The contours correspond to column densities 5, 10, 50, 100 $\times 10^{18}$ $\mathrm{cm^{-2}}$. Panels (c) and (d) show the volume density ($\delta = \rho/\rho_w$) and temperature within a central slice through the HVC in panel (b). ρ and ρ_w denote the volume densities of the cloud and of the ambient medium, respectively.

5. Numerical simulations

A few years after the detection of HVCs by Muller et al. (1963), the first theoretical calculations regarding HVC interactions with their environment were performed. Savedoff et al. (1966) investigated the effects that may occur when extragalactic gas falls down to the Milky Way. They concluded that the observed HVCs may be generated as a secondary effect during the interaction of extragalactic material with the Milky Way gas. In the following years, most papers focused on the collision of clouds with the Galactic Disk. E.g. Tenorio-Tagle (1981) studied the impact of low-, intermediate-, and high-velocity clouds on the Galactic Disk, and concluded that the energy transfer to the disk gas is high enough to create holes and trigger star formation. Lesch & Birk (1998) modeled the impact of a magnetized, partially-ionized cloud onto the Galactic Disk. They claim that the X-ray emission associated with HVCs (Kerp et al. 1994) could be explained by magnetic reconnection.

Recent numerical simulations to investigate the dynamical evolution of HVCs embedded in an ambient medium were performed by Santillán et al. (1999), Vieser (2001), Quilis & Moore (2001), and Konz et al. (2002).

The work of Santillán et al. (1999) concentrates on the impact of HVCs on the Galactic Disk. They performed 2-D magneto-hydrodynamic (MHD) numerical simulations of the interaction of HVCs with both magnetic and

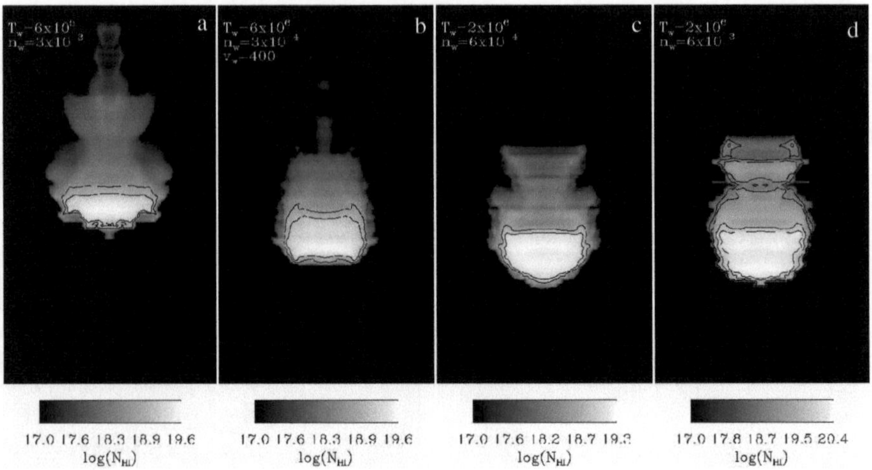

Figure 12. Figure 3 from Quilis & Moore (2001). Column density maps ($N(\mathrm{H\,I})$) for several of the pure-gas HVCs after 3 Myr of evolution (compare to 300 Myr for Fig. 11 – they evolve much faster than dark-matter-dominated HVCs). The temperature and density of the external medium are indicated in each panel, and in each case the wind velocity is 200 $\mathrm{km\,s^{-1}}$, except for panel (b) where it is 400 $\mathrm{km\,s^{-1}}$. The initial peak column density in HVCs (models a, b, c) is 2×10^{19} $\mathrm{cm^{-2}}$, whereas for model (d) it was increased to 2×10^{20} $\mathrm{cm^{-2}}$. The contours correspond to column densities 5, 10, 50, 100 $\times10^{18}$ $\mathrm{cm^{-2}}$.

non-magnetic thick galactic gaseous disks. They found that the magnetic field lines are distorted and compressed during the collision, increasing the field pressure and tension. This prevents the cloud material from penetrating into the disk and may even transform a high-velocity inflow into an outflow, moving away from the disk. The perturbation creates a complex, turbulent pattern of MHD waves, which are able to traverse the disk of the Galaxy and induce oscillations on both sides of the plane. They concluded that the magnetic field efficiently transmits the perturbation over a large volume, but also acts as a shield that inhibits the mass exchange between the halo and the disk. For the non-magnetized case they found that the evolution of the shocked layer generates a tail that oscillates, creating vorticity and turbulent flows along its trajectory.

Vieser (2001) focused on 2-D hydrodynamical simulations of clouds embedded in a hot plasma. He analyzed clouds at rest and clouds moving with a high velocity relative to the ambient medium, and found that the inclusion of heat conduction in the numerical model is essential to the stability of the cloud. The simulation demonstrated that a head-tail structure is a natural consequence if a cloud moves with a high velocity relative to the ambient medium. Moreover, the model showed that there is some mass exchange between the ambient medium (i.e. the gas in the disk) and the head of the HVC, leading to an increasing metallicity of the HVC.

Quilis & Moore (2001) were the first to perform 3-D hydrodynamical simulations for HVCs, both with and without dark-matter content. The results for HVCs with dark matter are displayed in Fig. 11; the results for HVCs without dark matter are displayed in Fig. 12. Note that the HVCs without dark-matter content evolve on much shorter timescales than HVCs that are dominated by dark matter. Quilis & Moore (2001) conclude that all HVCs show low column density tails ($N(\mathrm{H\,I})<10^{18}$ cm^{-2}). Tails with higher column densities ($N(\mathrm{H\,I})>10^{19}$ cm^{-2}) are expected if the density of the surrounding medium is greater than 10^{-4} cm^{-3}.

The observational fact that clouds in the Magellanic Stream and CHVCs show high column density tails indicates that the formation of significant tails must be possible even in low-density environments. The problem might be solved by the inclusion of magnetic fields in the simulation. Konz et al. (2002) performed 2-D magneto-hydrodynamical simulations to analyze the stability of fast-moving clouds in a hot ambient medium. They found that the magnetic field is very important for the stability of the HVCs. Moreover, the simulations showed the existence of long-lived tails surrounded by a magnetic barrier.

3-D MHD simulations for various physical parameters are needed to analyze the dynamical evolution of clouds in the intergalactic medium in detail. These simulations could be used to derive the basic physical parameters of the intergalactic medium of the Local Group.

6. Summary and conclusions

In this chapter, we presented observational evidence and review theoretical considerations for an interaction of HVCs with their ambient medium. HVCs showing either a head-tail structure or a horse-shoe shape are likely interacting with their ambient medium.

Head-tail HVCs are observed in complexes in the gaseous Galactic Halo (e.g. complex A), in HVC complexes located in the outer Galactic Halo (e.g. the Magellanic Stream), and in a sample of compact HVCs that might be located in the intergalactic medium of the Local Group. The overall morphology of the head-tail HVCs is the same for clouds located in different environments.

A simple illustration of the structure of HVCs, reconstructed from observations and theoretical considerations, yields an onion-like structure: dense condensations are embedded in a warm neutral medium. Both gas phases are traced by the 21-cm line emission of H I. The neutral cloud is surrounded by a layer of ionized gas that is traced by Hα emission. Outside this layer, there is the Galactic Halo or the intergalactic medium, i.e. a hot ionized gas. It is expected that these gas phases are in pressure equilibrium. The observations demonstrate that the diffuse H I component forms a head-tail structure with decreasing column density, increasing velocity dispersion, and decreasing radial velocity in the direction of the tail, consistent with gas having been stripped off the main body of the HVC. Cold, dense condensations are located near the sharp (leading) edge of the HVC, demonstrating that the more diffuse components of the HVC are much more affected by the interaction with the ambient medium than the dense condensations. Detailed observations of the outer ionized layer are not available at the moment, but it is expected that this layer is even more deformed during the interaction.

All recent simulations regarding the interaction of HVCs with their ambient medium predict the existence of head-tail structures. In the near future, realistic simulations can be made for HVCs located in various environments. Deep H I observations of a large sample of HVCs are in progress to derive statistically relevant properties like the mass fraction in the tail, the orientation of the tails and so forth. The observations, in combination with new 3-D MHD simulations, will for the first time enable the derivation of physical parameters for the gas in the outer Galactic Halo and the intergalactic medium of the Local Group.

Acknowledgements

Part of this work was supported by the German *Deutsche Forschungsgemeinschaft, DFG* project number ME 745/19. We thank Vicent Quilis for providing Figs. 11 and 12.

References

Benjamin, R.A., Danly, L. 1997, ApJ, 481, 764
Bland-Hawthorn, J., Maloney, P.R. 1999, ApJ, 510, L33
Bland-Hawthorn, J., Maloney, P.R. 2001, ApJ, 550, 231
Blom, J.J., Bloemen, H., Bykov, A.M., et 8 al. 1997, A&A, 321, 288
Brand, J., Blitz, L. 1993, A&A, 275, 67
Braun, M. 1996, AN, 317, 369
Braun, R., Burton, W.B. 2000, A&A, 354, 853
Brüns, C. 1998, Diploma thesis, Universität Bonn
Brüns, C. 2003, Ph.D. thesis, Universität Bonn
Brüns, C., Kerp, J., Kalberla, P.M.W., Mebold, U. 2000a, A&A, 357, 120
Brüns, C., Kerp, J., Staveley-Smith, L. 2000b, in ASP Conf. Ser. 218, Mapping the Hidden Universe, eds. R.C. Kraan Korteweg,P.A. Henning, H. Andernach (San Francisco: ASP), 349
Brüns, C., Kerp, J., Pagels, A. 2001, A&A, 370, L26
Danly, L., Albert, C.E., Kuntz, K.D. 1993, ApJ, 416, L29
Domgörgen, H., Bomans, D.J., de Boer, K.S. 1995, A&A, 296, 523
Gardiner, L.T. 1999, in ASP Conf. Ser. 166, Stromlo Workshop on High-Velocity Clouds, eds. B.K. Gibson, M.E. Putman (San Francisco: ASP), 292
Giovanelli, R., Haynes, M.P. 1977, A&A, 54, 909

Giovanelli, R., Verschuur, G.L., Cram, T.R. 1973, A&AS, 12, 209
Goerigk, W., Mebold, U. 1986, A&A, 162, 279
Hartmann, D., Burton, W.B. 1997, Atlas of Galactic Neutral Hydrogen (Cambridge: Cambridge University Press)
Hulsbosch, A.N.M. 1978, A&AS, 33, 383
Kalberla, P.M.W., Kerp, J. 1998, A&A, 339, 745
Kerp, J., Lesch, H., Mack, K.H. 1994, A&A, 286, L13
Kerp, J., Burton, W.B., Egger, R., Freyberg, M.J., Hartmann, D., Kalberla, P.M.W., Mebold, U., Pietz, J. 1999, A&A, 342, 213
Konz, C., Lesch, H., Birk, G.T., Wiechen, H. 2001, ApJ, 548, 249
Konz, C., Brüns, C., Birk, G.T. 2002, A&A, 391, 713
Lesch, H., Birk, G.T. 1998, in IAU Coll. 166 (Lecture Notes in Physics, 506), The Local Bubble and Beyond, eds. D. Breitschwerdt, M.J. Freyberg, J. Trümper (Berlin: Springer), 451
Lockman, F.J. 2003, ApJ, 591, L33
Mathewson, D.S., Cleary, M.N., Murray, J.D. 1974, ApJ, 190, 291
Mebold, U., Cernicharo, J., Velden, L., Reif, K., Crezelius, C., Goerigk, W. 1985, A&A, 151, 427
Mebold, U., de Boer, K.S., Wennmacher, A. 1990, XIth European Regional Astronomical Meeting of the IAU, eds. F. Sanchez, M. Vazquez (Cambridge: Cambridge University Press), 413
Meyerdierks, H. 1991, A&A, 251, 269
Meyerdierks, H. 1992, A&A, 253, 515
Morras, R., Bajaja, E., Arnal, E.M. 1998, A&A, 334, 359
Muller, C.A., Oort, J.H., Raimond, E. 1963, C.R. Acad. Sci. Paris, 257, 1661
Oort, J.H. 1966, BAN, 18, 421
Pietz, J., Kerp, J., Kalberla, P.M.W., Mebold, U., Burton, W.B., Hartmann, D. 1996, A&A, 308, L37
Putman, M.E., Gibson, B.K., Staveley-Smith, L., et 23 al. 1998, Nature, 394, 752
Quilis, V., Moore, B. 2001, ApJ, 555, L95
Santillán, A., Franco, J., Martos, M., Kim, J. 1999, ApJ, 515, 657
Savedoff, M.P., Hovenier, J.W., van Leer, B. 1966, BAN, 19, 107
Sembach, K.R., Wakker, B.P., Savage, B.D., Richter, P., Meade, M., Shull, J.M., Jenkins, E.B., Sonneborn, G., Moos, H.W. 2003, ApJS, 146, 165
Tamanaha, C.M. 1997, ApJS, 109, 139
Tenorio-Tagle, G. 1981, A&A, 94, 338
Tenorio-Tagle, G., Bodenheimer, P., Rozyczka, M., Franco, J. 1986, A&A, 170, 107
van Woerden, H., Schwarz, U.J., Peletier, R.F., Wakker, B.P., Kalberla, P.M.W. 1999a, Nature, 400, 138
van Woerden, H., Peletier, R.D., Schwarz, U.J., Wakker, B.P., Kalberla, P.M.W. 1999b, in ASP Conf. Ser. 166, Stromlo Workshop on High-Velocity Clouds, eds. B.K. Gibson, M.E. Putman (San Francisco: ASP), 1
Vieser, W. 2001, Ph.D. thesis, Universität Kiel
Wakker, B.P., Savage, B.D., Sembach, K.R., et 18 al. 2003, ApJS, 146, 1
Weiner, B.J., Williams, T.B. 1996, AJ, 111, 1156

13. WARPS, POLAR RINGS AND HIGH-VELOCITY CLOUDS

LINDA S. SPARKE
Department of Astronomy, University of Wisconsin-Madison, USA; sparke@astro.wisc.edu

Abstract. Both in warped disk galaxies and in polar ring systems, matter at the largest radii appears to have settled into near-circular orbits, that are tilted away from the equatorial plane of the inner galaxy. In both warps and polar rings, the outer material is largely gaseous, and seems to be less advanced in its star forming life than the inner parts of the system. The disks of most spiral galaxies are warped. Recent work shows no easy path to exciting warps to the amplitude that we now observe, so perhaps the outer warped region is a later addition. In polar ring systems, the outer ring is always more gas-rich than the inner galaxy, and appears to have formed a larger fraction of its stars more recently. The rings are probably a relic of infall or accretion, which perhaps took place many gigayears ago, when gas was more plentiful in the cosmos. An astronomer in a forming polar ring system would have observed an amazingly profuse system of HVCs, bringing in as much baryonic matter as the original galaxy contained.

1. Introduction

How might HVCs be related to the warps that are often observed in the outer disks of spiral galaxies, or to the polar rings of gas, dust and stars seen orbiting some early-type galaxies? Both the outer warped regions of a disk, and the encircling polar rings, involve a large fraction of the galaxy's gas, but relatively few of its stars. So it makes sense to ask whether they could be built up by the accretion of cool neutral gas, like that of HVCs.

The answer depends on what we believe to be the nature of these clouds. Bregman, in this volume (Ch. 16), reviews two leading hypotheses. HVCs may represent a "Galactic Fountain": disk gas heated by supernovae expands into the halo; the clouds condense out of this gas as it cools, and fall back onto the disk. The other possibility is that gas is entering the Milky Way from outside, either as gas streams torn from dwarf satellite systems, or incoming pregalactic material.

H. van Woerden et al. (eds.), High-Velocity Clouds, 273–296

In both galactic warps and polar rings, the outermost gas of the system is "anomalous"; it does not share the rotation axis of the inner disk, and has far more angular momentum. Gas accreted from outside a galaxy can scarcely avoid having a great deal of angular momentum about an axis unrelated to that of the inner system; but the fountain model has no obvious mechanism for pushing gas onto tilted orbits at large radii. So here we discuss only the question: if HVCs represent gas now entering the Milky Way from outside, could similar accretion have given rise to galactic warps and polar rings?

We begin with warps in galactic disks, and then consider polar rings and other structures within galaxies that show abnormal kinematics. Binney (1992) has given a comprehensive review of galactic warps, while Rubin (1994) describes the "multi-spin" galaxies, including polar rings. This chapter concentrates on more recent developments, and aspects relevant to a connection with HVCs. We assume $H_0 = 75$ km s^{-1} Mpc^{-1} throughout.

2. Warps in the outer disks of galaxies

2.1. OBSERVED WARPS IN THE GASEOUS AND STELLAR DISK

Figure 1 shows the "superthin" disk galaxy UGC 7170, which has a classical warp: the outer parts of the gas layer curve away from the plane of the inner disk, into an "S" or "integral-sign" shape. Warping is very common. We see it in the outer disk of all three spirals in the Local Group – our Milky Way, the Andromeda galaxy M 31, and M 33 (see e.g. Burton et al. 1992). Some of the warps are spectacularly huge: in NGC 4013 (Bottema 1995, 1996; Fig. 2) the gas curves in a very symmetric manner so that its outer parts lie in a plane inclined more than 20° from the inner gaseous and stellar disk. Of 26 edge-on disk galaxies that were mapped in neutral hydrogen for the *WHISP* survey, García-Ruiz et al. (2002a) found that in all 20 systems where the H I disk extended beyond the stellar disk, the gaseous layer was warped. Eleven of these showed clear integral-sign warps, in rough agreement with the earlier study of Bosma (1991). The true fraction of warped disks must be higher than 50%, since whether or not we can detect a warp depends on its orientation.

Usually, we think of a warped disk as composed of gas clouds on near-circular orbits about the galaxy center. These orbits are concentric but not coplanar; the direction of the orbital angular momentum changes smoothly with galactocentric radius. This "tilted-ring" picture ignores the slow precession that such orbits would undergo in the galaxy's aspherical gravitational field. It generally provides an excellent fit to H I observations of galaxy disks, and has become the standard description (e.g. Begeman 1987). Generally, the inner part of the H I disk is fairly flat, and appears coplanar

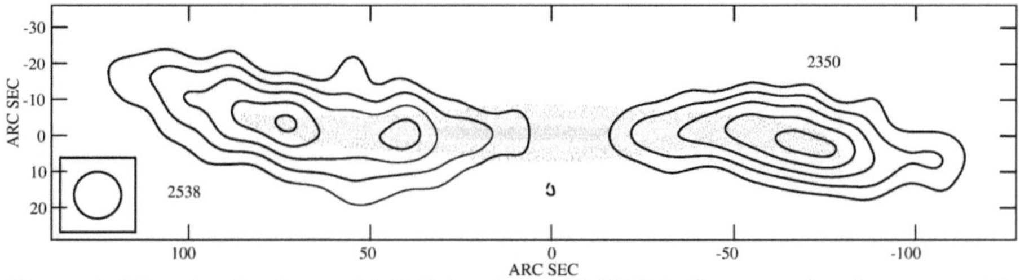

Figure 1. Warp in the "superthin" disk galaxy UGC 7170. Contours for the extreme H I channels, labeled by the velocity in km s^{-1}, are superposed on a B-band image. Circle at lower left shows the beam size. Figure from Cox et al. (1996.)

with the stellar disk. So we can describe the shape of the warped outer disk by giving the tilt angle at each radius with respect to the inner disk, and the azimuth of the *line of nodes* at which the gas orbit cuts the plane of the inner disk.

When a warped disk is seen close to edge-on, it will appear most strongly curved when the line of nodes points close to our direction, as it does in Figs. 1 and 2. If instead the line of nodes lies close to the plane of the sky, we see a thickened gas layer, but little bending. Only close examination of the velocity field can distinguish between such a warped disk and a general thickening of the gas layer. In M 31, the line of nodes lies within 20° of the sky plane; multiply-peaked velocity profiles reveal that our line of sight passes more than once through the warped gas layer (Brinks & Burton 1984). Swaters et al. (1997) went to considerable trouble to show that the disk of NGC 891 is thickened rather than warped.

Although the H I layer warps to greater heights, de Grijs (1997) showed that warps in the *stellar* disks of galaxies seen edge-on are as common as those in the gas. Schwarzkopf & Dettmar (2001) agree: in the latter study, 45% of non-interacting edge-on galaxies show warps in the R-band images. Reshetnikov & Combes (1998) argued that almost all of the optical warps must be true warps, rather than the effect of spiral arms, while Reshetnikov et al. (2002) find that warps are also common in disks at redshifts $z{\sim}1$.

If warps are a gravitational phenomenon, we expect stars and gas to follow the same warped plane. Magnetic stresses, or the ram pressure of gas through which the galaxy travels, should instead act on the gas disk while leaving the stellar disk undisturbed. Since warps are characteristic of the outer disk, it is more difficult to trace them in stars than in the H I gas, which usually extends to larger radii. But wherever stars are seen at radii where the gaseous disk is bent (see Fig. 1), they follow the same warped plane. Stars have been found in the outer Milky Way, following the curve of the warped gas disk (e.g. Carney & Seitzer 1993; López-Corredoira et al. 2002).

Even where stars and gas follow the same warp, the stellar disk will ap-

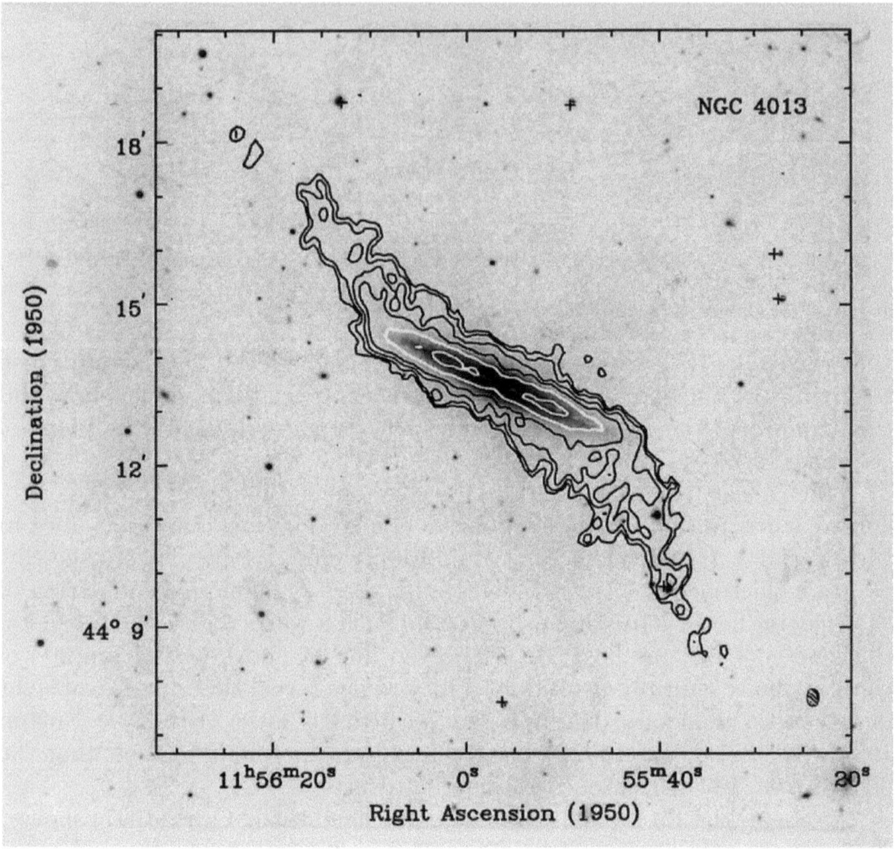

Figure 2. The prodigious warp in the galaxy NGC 4013. Contours show the total H I emission, superposed on an optical image of this edge-on disk galaxy. After Fig. 4 of Bottema (1995).

pear less strongly curved in projection, because its density falls faster with radius than the H I. If the disk becomes bluer in its outer parts, the warp should also be more pronounced in blue light than in the red, as was found by Florido et al. (1991). Making allowance for this effect, Sasaki (1987) concluded that the stars of the disk of NGC 5907 follow the same warped plane as the gas; Casertano et al. (1987) came to the same conclusion for NGC 4565. In UGC 7170, the gas layer does not extend far beyond the stars, and the optical warp follows that of the gas (Cox et al. 1996).

2.2. HOW LONG DO GALACTIC WARPS LIVE?

Could the Milky Way's warp have been built up from gas that arrived in the disk as HVCs? The midplane of the disk begins to curve just beyond the solar circle, and the warp involves about half the disk gas, or 2 to 3×10^9 $M_\odot$ of H I. Wakker et al. (1999) estimate that HVCs now bring in no more than 1 $M_\odot$ of new material per year, and so could replace the gas of the outer disk over 2 to 3 Gyr. The Galactic Warp could be made from HVC gas only if it can persist for at least that long.

The line of nodes of the warp of the Milky Way's H I disk remains constant to within about 20° over a range from the solar circle at radius R_0 to $\sim 2.5 R_0$ (see e.g. Fig. 7.22 of Burton 1988). This is at first sight surprising. If we consider a gas cloud following a tilted circular orbit of radius R in a galaxy with gravitational potential $\Phi(R, z)$, it oscillates up and down through the galaxy's equatorial plane with frequency $\nu = \sqrt{\partial^2\Phi/\partial z^2}$, where z is the height above the midplane. At the same time, the cloud circles the galaxy with frequency $\Omega(R)$, where $R\Omega^2(R) = \partial\Phi/\partial R$. Since the galaxy is flattened, we have $\nu > \Omega$; the vertical oscillation is faster, and the line of nodes regresses with frequency $\nu - \Omega$. Both ν and Ω tend to zero at large radii, so the line of nodes of the inner warp regresses faster than that in the outer parts, twisting the warp into a leading spiral. Binney & Tremaine (1987; Sect. 6.6) show that the Milky Way's warp should have become severely spiral inside a gigayear. A warp that began with a straight line of nodes would develop a pitch angle of less than a radian within 100 Myr: we call this the *winding time*. Either the warp in the outer disk has a very recent origin, or something stops it from winding up.

This "winding problem" is common to other galaxies. In a study of 12 warped galaxies that were well mapped in the 21-cm line, Briggs (1990) found a number of regularities. The gas disks tend to be fairly flat within the radius R_{25}, where the B-band surface brightness falls to 25 mag arcsec^{-2}. Generally, but not invariably, warping develops between this point and the Holmberg radius $R_{\rm Ho}$, where the surface brightness is 26.5 mag arcsec^{-2} in the B-band. The warp has a straight line of nodes within $R_{\rm Ho}$; outside that radius, it curves into a loosely wound leading spiral. The *WHISP* survey (Kamphuis et al. 1996; van der Hulst et al. 2001) will provide a much more extensive database for testing these regularities.

At least some warps are likely to be transient. Reshetnikov & Combes (1998) noted that galaxies with an optically prominent "S"-warp tend to lie in regions that are denser than average, suggesting that recent accretion or interaction may be important. Verdes-Montenegro et al. (2002) propose that the galaxy NGC 3642, where the warped outer disk is actively forming stars, has recently eaten a dwarf companion. Schwarzkopf & Dettmar (2001) found that the stellar disks in 93% of their 61 edge-on interacting disk galaxies were warped. Roughly one third of the interacting systems showed "U"-shaped warps, rather than an "integral-sign"; corrugations with a short radial wavelength were also common. However, these authors concluded that recent tidal interactions and mergers could not entirely explain the observed abundance of large-scale warps. By contrast, García-Ruiz et al. (2002a) found that galaxies in poorer environments were *more* likely to show H I warps, perhaps because their gas disks had not been truncated by interactions.

2.3. DYNAMICAL MODELS FOR WARPS

Despite four decades of work, we lack a complete understanding of the physics of galactic warps. Lynden-Bell (1965) proposed that warps might represent a discrete normal mode of bending for a self-gravitating disk. Sparke & Casertano (1988), adapting Hunter & Toomre's (1969) treatment to include the background potential of an oblate "dark halo", found that the disk can indeed have discrete bending modes. These correspond to a warped disk with a straight line of nodes, precessing slowly about the halo's symmetry axis. Varying the disk's exponential scale length in relation to the core radius, within which the halo density is roughly constant, changes the shape of the warp: see also Kuijken (1991). To obtain bending like that in the disk of NGC 4013 requires that the halo has a large core. Hofner & Sparke (1994) showed that, when a disk of surface density $\Sigma(R)$ is given some arbitrary initial tilt, bending waves propagate outward with the group velocity $c_g \sim \pi G \Sigma(R)/\Omega(R)$. The central disk settles rapidly toward the mode shape, with its straight line of nodes, while in the outer, unsettled, parts the line of nodes winds into a leading spiral, just as Briggs (1990) had found. Ideta et al. (2000) showed that warps survive exceptionally well if the halo is prolate rather than oblate, though the line of nodes should then twist in a trailing sense.

This explanation assumes that the dark halo will steadfastly retain its shape as the disk precesses within it. But gravitational N-body experiments by Dubinski & Kuijken (1995), in which both the disk and an oblate halo were represented by "live" simulation particles, showed that their halo reacted to the precessing disk. Starting the disk from a warped mode in the potential corresponding to their initial model halo, they found that the halo would align with a heavy disk within only a few orbital periods, and would remove the off-axis angular momentum via dynamical friction from a light disk. Nelson & Tremaine (1995) concluded from an analytic calculation that dynamical friction could either excite a warp or damp it. In a model like that of Dubinski & Kuijken (1995), where halo particles stream rapidly in the direction of the disk's rotation, the warp should damp; but Nelson & Tremaine's Fig. 4 implies that warping persists for most of a Hubble time. A prolate halo like that of Ideta et al. (2000), or a counter-streaming halo, could amplify the warp.

Binney et al. (1998) performed a slightly different N-body experiment. Their oblate halo consists of "live" particles, while the disk is a set of concentric massive rings that are free to tilt. They found that a disk started from the appropriate warped mode did not settle, but wound rapidly into a spiral; the warp gains energy from the halo, rather than losing it. They speculate that galactic warps might represent a joint oscillation mode of the whole galaxy, in which the disk configuration might resemble the bending mode derived for the "frozen" halo, while the state of the halo could be very different. Decades of work on spiral structure have shown that stability calculations are always delicate; May & James (1984) warned that slightly anisotropic interparticle forces in an N-body code can play havoc with tilted-disk problems. So it is unsurprising that the role of the halo is not yet fully understood.

Appeals to viscosity to prolong the life of a warp have had no better luck. The recent efforts of New et al. (1998) reach the same conclusion as Steiman-Cameron & Durisen (1988) and Colley & Sparke (1996): if viscous forces are strong enough to prevent the warped disk from winding into a tight spiral, they also force the disk gas inward, on the same timescale as the warp would otherwise wind up. Since the winding time is no more than 100 Myr at the Sun's position, while the outer disk contains 2 to 3×10^9 $M_\odot$ of gas, we would then require 20 to 30 solar masses of gas per year to flow inward through the solar circle. This is significantly more than what is now arriving as HVCs. There is no indication that gas in the Milky Way's disk, or that of any other spiral, is draining so rapidly into the center; we conclude that warps are probably not stabilized by viscous forces.

Perhaps the dark halo itself is not axisymmetric. A triaxial halo, tumbling about its short axis, can induce warping in a gas disk circling that axis, if the figure of the halo rotates retrograde to the disk (Binney 1978, 1981; Sparke 1984; Tremaine & Yu 2000). This is unlikely to be the case in most galaxies, since the particles making up the halo would probably have to counter stream with respect to the disk material (Vietri 1986). If the halo tumbles about an axis in the plane of the disk, we have the situation envisaged by Ostriker & Binney (1989), who suppose the symmetry axis of the halo to shift as additional dark matter joins it. The disk warps, with its line of nodes along the halo's rotation axis. Jiang & Binney (1999) perform gravitational N-body experiments, adding particles to the halo to change its shape and angular momentum over the course of a run. As the halo gains 30% in mass, their disk of tilted rings indeed develops a 7° warp. Unfortunately, if the halo's smooth reorientation is ever halted, we are left with the original problem of how to preserve the warp. Debattista & Sellwood (1999) show that if the halo's angular momentum, rather than its figure rotation, is misaligned with the disk, dynamical friction aligns the disk's spin with that of the halo. This happens faster for the inner disk, so again a warp develops. Their model halo is nearly spherical, so that the "winding problem" does not arise.

Masset & Tagger (1997) suggest instead that warps are generated by spiral waves in the disk. As two-armed (m=2) spiral disturbances deposit their energy at the outer Lindblad resonance, part of it is re-radiated via nonlinear coupling, as m=1 bending waves that correspond to an "integral-sign" warp. These waves propagate into the outer disk. This model predicts that large-scale warps should be accompanied by the short-wavelength corrugations that are often observed (e.g. Florido et al. 1991; Schwarzkopf & Dettmar 2001). In this model, we would expect that galaxies with integral-sign warps should also show a strong two-armed spiral. But observationally, there is no clear relation between these phenomena; for example, the flocculent spiral galaxy NGC 2841 is quite strongly warped (Begeman 1987).

Weinberg (1998), using a method based on linear perturbation theory, investigated warping caused by the repeated close passage of a satellite galaxy. He found that the galaxy halo should develop a strong wake, which exerts enough force on the disk to cause a substantial warp. The N-body experiments of Tsuchiya (2002) agree, although the form of the induced warp does not match that of the Milky Way's gas disk, and the line of nodes may be in the wrong position. However, a similar calculation by García-Ruiz et al. (2002b) fails to reproduce this strong bending. Bailyn (2003) points out that the off-axis angular momentum in the warped region of the Milky Way's disk is aligned with that of the Sagittarius Dwarf galaxy, and suggests that this smaller system might have induced the warp.

2.4. WARPS FROM HIGH-VELOCITY CLOUDS?

Although theorists are in some confusion, observations give a simpler answer to the question of warp longevity. Maps in the near infrared with the COBE satellite (Freudenreich 1998, Porcel et al. 1997) suggest that the Milky Way's old stellar disk has the same warp as the gas. Warps have been found in nearly isolated galaxies, such as NGC 5907 (Sancisi 1976) and UGC 7170 (Cox et al. 1996). Deep H I and optical imaging of NGC 5907 reveals only a tiny companion, with less than 1% of the main galaxy's mass (Shang et al. 1998); for comparison, the Large Magellanic Cloud has about 10% of the Milky Way's mass and luminosity. The warped systems UGC 7170 and UGC 7321 are "superthin" galaxies. As Uson & Matthews (2003) argue, the vertically thin disks of these objects, and the fact that they have turned relatively little of their gas into stars, imply that they have never been strongly perturbed, since interactions would have heated the disk, and triggered star formation. Even if some warps are tidally induced by galaxy interactions, the bending itself must last for many orbital times after the perturber has left the vicinity.

Thus we can conclude: if HVCs represent gas falling into the Galaxy from outside, that gas will have a great deal of angular momentum. It will also be added preferentially to the outer disk. Like meteoric cratering in the solar system, the infall of HVC gas would have been more rapid at earlier times. If we have faith that Nature can solve the winding problem, even if galaxy theorists cannot, and that warps are persistent features of galaxies, then the bending in the outer disk of the Milky Way and of other galaxies could well reflect the capture of high-velocity gas clouds from orbits tilted with respect to the plane of the inner disk.

3. Galaxies with polar rings

Polar ring systems are "multi-spin galaxies" (Rubin 1994): different components have highly discrepant axes of rotation. In this, they represent a more severe departure from "normality" than the warped disks. Warps are common, while polar rings are rare; they are found in <1% of S0 galaxies (Whitmore et al. 1990), and are even rarer in spirals. In most polar rings, essentially all the galaxy's gas, not merely the outer portion, follows orbits that lie nearly orthogonal, rather than just slightly tilted, to the plane of the inner stellar disk. The inner galaxy is an early-type system, generally an S0 or an elliptical, that contains little gas, and consists largely of old stars. The outer gas-rich material in polar orbit can resemble the disk of an extreme late-type galaxy, upended over the host system; it is probably a late acquisition.

3.1. NGC 4650 A: A PROTOTYPICAL POLAR RING GALAXY

Figure 3 shows a view from the Hubble Space Telescope of one of the closest and best studied polar rings, NGC 4650 A, from Gallagher et al. (2002). The redshift of this system is cz=2910 km s^{-1}, or $v_{\rm LG}$ = 2625 km s^{-1} relative to the centroid of the Local Group, placing it at a distance of 35 Mpc (Sackett et al. 1994), where $1'$ corresponds to 10 kpc, and $1''$ to 170 pc. The ring is much larger than the central galaxy, and clearly bluer. The ring is not exactly polar, but appears tilted about 15° from the minor axis of the inner stellar body. Diffuse young stellar associations are prominent in its outer parts. They outline two "spiral arms" to the top right and lower left of the ring, earlier identified by Arnaboldi et al. (1997). Hα emission is strong, indicating recent vigorous star formation.

The small central galaxy, classified as S0, consists of densely-packed red stars. Strong absorption at Hγ (Whitmore et al. 1987) and red optical colors suggest that star formation was abruptly truncated 3 to 5 Gyr ago

Figure 3. The polar ring galaxy NGC 4650 A = PRC A-5, observed in the F450W band by *WFPC2* on the Hubble Space Telescope. The long dimension of the image is 2.′1; the upward direction is slightly west of north.

(Gallagher et al. 2002); the near-infrared colors of Iodice et al. (2002) imply significant starbirth within the past 3 Gyr. Absorption-line spectroscopy (Whitmore et al. 1987; Sackett et al. 1994) shows that its stars stream about the minor axis with $V/\sigma \sim 1$, as expected in the disk of a galaxy. Kinematic measurements of both stars and ionized gas in the ring (Swaters & Rubin 2003) show that this is indeed a "multi-spin" system: the ring material is in near-circular rotation about its own minor axis, which is perpendicular to that of the inner galaxy. The ionized gas forms a slightly warped disk, with a central hole (Laustsen & West 1980; Nicholson 1989). In Fig. 3, we see a dark dust lane on the right, where the ring passes in front of the galaxy to the West – the ring gas is not primordial, but contains heavy elements. Figure 4 shows the dark lanes extending continuously inward to the central galaxy; Iodice's infrared photometry reveals some bright polar structures in the same region. Some of the ring material apparently has managed to reach the inner body.

The S0 galaxy has a bright compact nucleus; the *WFPC2* images, with $0\farcs1$ pixels, do not resolve it. Its measured luminosity of $L_I = 8 \times 10^7$ $L_\odot$ within a 20 pc radius is a lower limit to the true value, since the nucleus is probably reddened by dust. This is unlikely to be even a low-powered active nucleus. The optical emission-line spectrum shows low excitation (Laustsen & West 1980), and Arnaboldi et al. (1997) detected 20-cm continuum radio emission from star forming regions in the ring, but no nuclear source. More likely, the central object is a nuclear star cluster. It would be 35 times brighter than that in M 33 ($L_V = 2 \times 10^6$ $L_\odot$ and core radius $r_c \sim 0\farcs1$ or 0.4 pc; Matthews et al. 1999), and slightly brighter than the most luminous young super star-clusters (Whitmore et al. 1999). Around this unresolved source is a small nuclear disk or central bulge, with a core radius of $0\farcs36$ or 60 pc.

Recent mapping shows 7×10^9 $M_\odot$ of H I gas in the ring (Arnaboldi et al. 1997), along with roughly a tenth as much molecular gas (Watson et al. 1994): this galaxy is about as gas-rich as the Milky Way. As is typical for polar ring systems (e.g. van Gorkom et al. 1987), the central galaxy is almost empty of gas, while the ring contains at least as many baryons as the central S0. For the polar disk alone, $M/L_B \sim 5$, a high value typical of late-type or low-surface-brightness disks (Matthews & Gallagher 1997). Like extreme late-type galaxies, this disk has been inefficient at turning its gas into stars.

The H I maps barely resolve a central hole in the disk, probably less than $10''$ in size, and coincident with the S0 galaxy's nucleus. Most of the gas lies within $1\farcm5$ of the center, although H I filaments stretch to $4'$ north and south of the galaxy, while the Hα emission is found only within $1'$ of the central galaxy. The averaged surface density of H I within radius $1\farcm5$ exceeds 8 $M_\odot$ pc^{-2}. We can compare this with the density $\Sigma_{\rm crit} = \kappa v_s / \pi G$

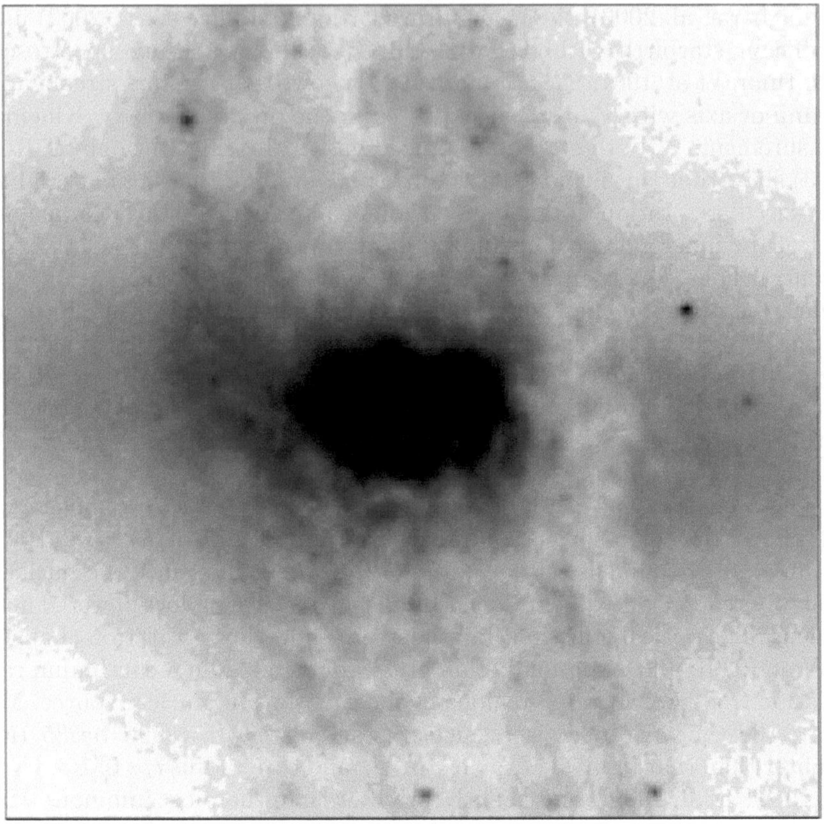

Figure 4. The inner regions of NGC 4650 A, from the same observation as Figure 3. Dust lanes cross in front of the galaxy on both sides, and extend inward to the stellar body.

at which the gas disk should become unstable to axisymmetric disturbances (Binney & Tremaine, Sect. 6.2.3). Taking the gas sound speed v_s=6 km s^{-1} and calculating the epicyclic frequency κ at $1'$, or 10 kpc from the center, yields $\Sigma_{\rm crit} = 5\ M_\odot\ {\rm pc}^{-2}$, so the average gas density within $1\farcm5$ is $\sim$1.6 $\Sigma_{\rm crit}$. By contrast, Kennicutt (1989) finds in a sample of giant spiral galaxies that star formation (as measured by the Hα intensity) remains high until the gas density has fallen below about $0.63\Sigma_{\rm crit}$. Despite the large surface density of gas, the outer regions of this polar ring are unusually reluctant to make stars.

The polar H I disk shares the orbital motion of the ionized material. The rotation speed remains roughly constant at about 120 km s^{-1}, with no indication of a decrease even in the outermost filaments. This indicates an extended distribution of non-luminous "dark matter". Sackett and collaborators (Sackett & Sparke 1990; Sackett et al. 1994) infer from a comparison

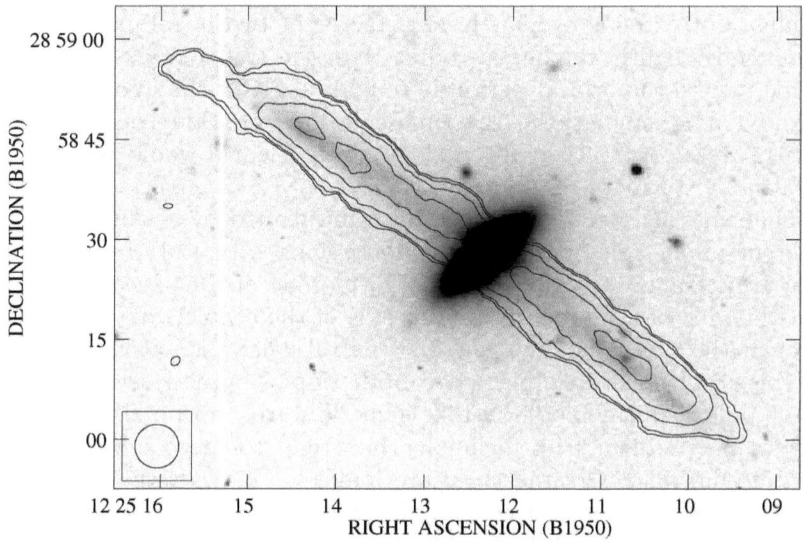

Figure 5. V-band image of UGC 7576 = PRC A-4; the inner S0 galaxy is encircled by a thin polar ring of dust, gas and stars. Contours show 21-cm emission of H I mapped at the Very Large Array (*VLA*); the circle at left shows the beam. Figure from Cox (1996).

of observed rotation speeds in the disk and ring that the dark matter does not form a spherical halo, but instead is flattened toward the equatorial plane of the inner galaxy. Combes & Arnaboldi (1996) show that the observed motions can also be explained if the dark matter lies in the polar ring itself. Curiously, despite its unusual structure, NGC 4650 A lies right on the K-band Tully-Fisher relation (Gallagher et al. 2002).

3.2. AGES AND TIMESCALES

Both the morphology and the stellar colors can provide a clue to the age of a polar ring. Figure 5 shows a thin, almost planar, stellar ring around the S0 galaxy UGC 7576, seen close to edge-on. Again, dust lanes are visible as the ring crosses in front of the central galaxy. This ring has a slight "C-shaped" or "banana" warp (Whitmore 1991). Hα emission is weak (Cox 1996), and there are no clumps of blue stars. Mould et al. (1982) conclude from broadband colors that the stars of the ring are neither very young nor extremely old; most have ages between 0.1 Gyr and 3 Gyr. While accreted gas can dissipate energy to settle into a ring, the stars cannot; so they must have been born since the ring gas found its present configuration. Hence the ring is probably at least 3 Gyr old.

The H I map in Fig. 5 does not resolve the gas layer perpendicular to its length. A position-velocity cut along the ring shows that it is not a thin annulus, but extends over at least a factor of two in radius. The atomic gas extends slightly further in radius than the stellar ring, to about $1'$ or 25 kpc, where the orbital period is roughly 0.7 Gyr. In smoothed-particle hydrodynamic simulations, the timescale for smoothing out an irregular ring is several orbits (e.g. Rix & Katz 1991), which would put the ring's age at 5 Gyr or more.

The rings in Figs. 3 and 5 are both misaligned by about 15° with the apparent minor axis of the central body. This is typical of objects in the Polar Ring Galaxy Catalog (PRC; Whitmore et al. 1990, see below). Most lie within 20° of the apparent minor axis of the central galaxy (Whitmore 1991). If the ring is circular and the central galaxy is axisymmetric, then the rings in these two objects are both tipped by at least 10° from the galaxy pole (Whitmore 1984). Like galactic warps, a tipped polar ring has a "winding problem": its inclined orbits tend to precess in the galaxy's gravitational field. Orbital times are longer at larger radii, so precession will be fastest at the inner edge of the ring, and a twist should develop.

The stellar ring in Fig. 5 appears so nearly planar that it cannot be warped by more than about a tenth of a radian. To estimate a winding timescale, we can assume that the potential in the outer parts of the galaxy deviates by about 10% from sphericity, and that the rotation speed is constant. For near-polar orbits, the precession rate is proportional to the misalignment angle (e.g. Sparke 1986). At an inclination of 10° from the pole, the outer edge of the ring will have turned by about 0.1 radians after one rotation, while an orbit at half that radius has moved twice as far. We have just argued that the ring of UGC 7576 is probably several orbits old. So the galaxy potential must be very closely spherical, or the ring is stabilized in some way: for example, by its self-gravity (Sparke 1986, 1990). If polar rings *have* been stabilized, some of them might be very old indeed.

Whitmore et al. (1990) used photographic survey plates to compile a Polar Ring Galaxy Catalog (PRC), listing over 100 polar ring galaxies and related objects. Category A comprised objects confirmed as polar ring systems, by measuring both stellar velocities in the inner galaxy and gas motions in the ring. Good candidates were placed in category B, possible polar rings in category C, and related objects in category D. About a dozen systems have now been confirmed as polar rings; Table 1 gives a list. The luminosity in Col. 6 includes both the central galaxy and the polar ring. The host galaxy for B-19 and C-45 is an elliptical; in the others, it is an early-type disk galaxy, S0 or Sa. All of the disk systems except UGC 4261 (an asymmetric object that is classified only as S?) are small galaxies, with $L_B < 8 \times 10^9$ $L_\odot$. The H I masses, by contrast, are comparable to that of a large spiral galaxy; the median value for their ratio is $M(\text{H I})/L_B \sim 0.7$. Bettoni et al. (2001), using a larger sample from the PRC, find a normal

TABLE 1. Kinematically confirmed polar rings

	PRC		cz km s^{-1}	D Mpc	L_B 10^9 L$_\odot$	M(H I) 10^9 M$_\odot$	inner galaxy	polar ring
r	A-1	A 0136−0801	5590	74	4	3	SWR	S91, C96
n	A-2	ESO 415−G26	4560	60	8	5	WMS	RSS, Sch01
tw	A-3	NGC 2685	870	13	4.6	1.6	Si98	M92, S02
r	A-4	UGC 7576	7028	95	6	5	PRC	S91, C96
i	A-5	NGC 4650 A	2910	35	4	7	S94	A97
i	A-6	UGC 9796	5570	75	3.4	5	PRC	S91, C96
n	B-3	IC 1689	4570	63	7	< 0.6	Si98	vGKS
r	B-16	NGC 5122	2725	35	3.9	2	S04	C96
m	B-17	UGC 9562	1371	20	1.3	0.6	RC94[b]	S91, C01
n	B-19	AM 2020−504	5058	65	11	9	A93	S91
tw	C-13	NGC 660	823	13	7	5[a]	vD95[b]	vD95
m	C-24	UGC 4261	6420	86	17	8	RC94[b]	RSS
m	C-27	UGC 4385	1950	25	1.7	1.2	RC94[b]	RSS
tw	C-45	NGC 5128	540	3.3	20	0.8	W86	vG90, N92, Sch95

References: A93: Arnaboldi et al. 1993; A97: Arnaboldi et al. 1997; C01: Cox et al. 2001; C96: Cox 1996; M92: Mahon 1992; N92: Nicholson et al. 1992; RC94: Reshetnikov & Combes 1994; RSS: Richter et al. 1994; S90: Sackett & Sparke 1990; S91: Sackett 1991; S94: Sackett et al. 1994; Sch95: Schiminovich et al. 1995; Sch01: Schiminovich 2001; S02: Schinnerer & Scoville 2002; SWR: Schweitzer et al. 1983; Si98: Sil'chenko 1998; S04: Sparke et al. 2004; vD95: van Driel et al. 1995; vGKS: van Gorkom et al. 1987; vG90: van Gorkom et al. 1990; WMS: Whitmore et al. 1987; W86: Wilkinson et al. 1986; PRC: Whitmore et al. 1990.
Notes: (a) A quarter of this gas is in the central galaxy, the rest in the polar ring: vD95. (b) Rotation measured from emission of gas in the central body, not stellar absorption lines. The leftmost column gives the ring morphology, as described in the text.

ratio of atomic to molecular gas, with both about 10 times overabundant compared with normal galaxies of the same type as the host.

A number of authors (e.g. van Gorkom et al. 1987; Whitmore 1991; Reshetnikov & Sotnikova 1997) have pointed out that polar rings can be ordered by the regularity of their appearance, in both starlight and the H I gas. In Table 1, the smoothest and most orderly rings are those of A 0136-0801, UGC 7576 and NGC 5122, labeled "r". The rings around UGC 9796 and NGC 4650 A are slightly less regular and form an intermediate class "i", while the gas around UGC 4261, UGC 4385 and UGC 9562 can only be described as a mess (m). The rings around ESO 415−G26, IC 1689 and AM 2020−504 are narrow (n), while those in NGC 2685, NGC 660 and NGC 5128 extend over a substantial radial range and appear highly twisted (tw), but their velocity fields are strikingly bisymmetric. Van Gorkom et al. (1987) propose that the regular rings are the oldest, and the messy rings the

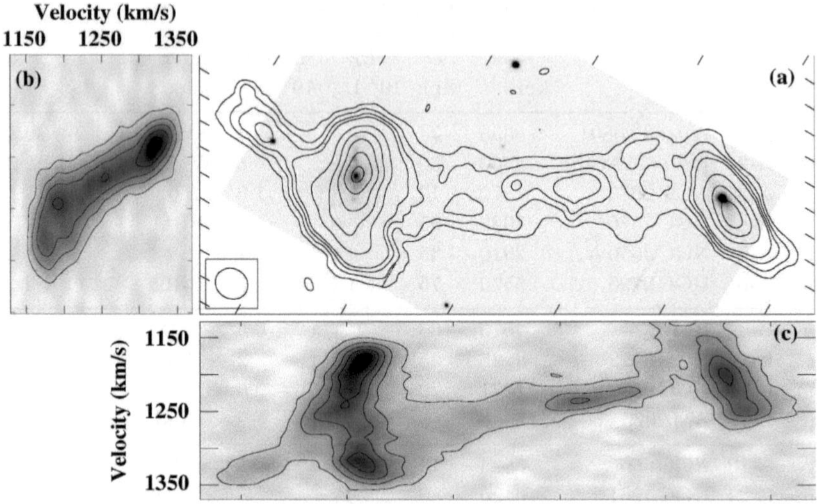

Figure 6. (a) B-band image of the polar ring system UGC 9562 = PRC B-17 = II Zw 71 (left) and nearby blue dwarf galaxy II Zw 70 (right), with contours showing H I emission from the polar ring, the dwarf, and a gas streamer between them; (b) position-velocity cut along the apparent major axis of the polar ring, in rotation about its minor axis; (c) position-velocity cut along the quiescent gas of the streamer. Figure 3 of Cox et al. (2001).

youngest. Reshetnikov & Combes (1994) suggest that the ring of UGC 4261 is still being formed in a galaxy merger. But since products of the most recent starbirth will dominate the ring's output of visible light, it will be hard to distinguish between a recently formed ring and one where, for example, a passing galaxy has triggered a starburst. The ring in UGC 9562 (Fig. 6) may have suffered such a perturbation.

An indication that many polar rings are fairly old and stable is that "confirmed" (category A) and "good-candidate" (B) objects from the PRC typically have weak far-infrared emission for their H I gas mass and their blue luminosity (Richter et al. 1994; Schinnerer & Scoville 2002). This contrasts with interacting galaxies, which are typically bright in the far-infrared. A survey of 40 polar ring galaxies in the radio continuum (Cox & Sparke 2004) shows that they are only slightly brighter at 20 cm than normal elliptical and S0 galaxies. Where emission is resolved, it comes from the polar structure, rather than the central galaxy. This absence of strong far-infrared and radio emission suggests that the polar ring gas is in stable orbits; if it were flowing rapidly into the central galaxy, we would expect to see evidence of star formation there. Brocca et al. (1997) find that polar ring systems are no more likely than normal galaxies to have companions close enough to the two systems to have interacted within the past gigayear.

3.3. HOW ARE POLAR RINGS FORMED?

While polar rings are thought to represent captured material, it is not clear what process would lead to their formation. Accretion of gas from a passing spiral seems unlikely, since the rings themselves contain as much H I gas as a large spiral, and often substantial molecular material as well. Polar rings fall close to the Tully-Fisher relation for normal disk galaxies, but are slightly *underluminous* for the measured rotation speeds (Iodice et al. 2003) – this is not what we would expect in systems that have gained a large amount of extra gas.

Bekki (1997, 1998) suggests that polar rings might be produced in a way very similar to (planar) ring galaxies such as the Cartwheel (AM 0035−33); a small galaxy collides head-on with a larger disk system, sending out a shock wave into the disk, which triggers star formation. If the smaller system loses enough energy, it will be captured by the larger one to form a dense central body. This model scores some successes. The larger galaxy supplies both the dark matter and the majority of the baryonic material, so it is natural that these systems follow the Tully-Fisher relation. If the dark matter is concentrated to the plane of the polar ring (which was originally the equatorial plane of the larger system), that would increase the ring rotation speeds, explaining the deviations seen by Iodice et al. (2003). If in addition the dark halo is triaxial and tumbles about its shortest axis, it could induce spiral structure in the ring, similar to what we see in NGC 4650 A (Bekki & Freeman 2002).

However, there is no reason why the inner body should come to rest with its rotation axis almost perpendicular to that of the outer ring gas. Numerical simulations by Bournaud & Combes (2003) show that only a minority of collisions will result in such an orientation. Once the collision-induced burst of star formation is over, the system will have *less* H I gas per unit luminosity than before it. Since Table 1 gives much higher values for the ratio $M(H\,\textsc{i})/L_B$ than for a normal spiral at the present day, this model requires the initial disk to be even more gas-rich.

Another possibility is that the rings represent the remains of a gas-rich dwarf irregular galaxy, which has been captured and torn apart by differential rotation. However, the H I content of present-day dwarf irregulars is generally 10^8 to 10^9 $M_\odot$ (e.g. Matthews & Gallagher 1996), well below the masses in Table 1. Further, the ratios $M(\mathrm{H\,I})/L_B$ computed from Table 1 include the central galaxy, which contributes most of the light but almost none of the gas. Thus the accreted object must have been unusually gas-rich, but with an optical luminosity that is small even compared to the relatively dim central galaxies of most polar rings. If the rings are old, they could have been made earlier in the Universe's history, when neutral gas was several times more plentiful than it is now. Consistent with this picture, Reshetnikov (1997) found two candidate polar rings among galaxies of the Hubble Deep Field, implying that polar rings were ten times more common at $z \sim 1$ than they now are.

Could polar rings, like some of the HVCs, represent the delayed infall of primordial or intragroup material? One might question whether such an ordered structure as a polar ring can form out of intragroup gas. But the giant ring in Leo (Schneider et al. 1989), a structure 200 kpc in diameter, containing at least 10^9 $M_\odot$ of H I that appears to be in coherent rotation, may be a larger-scale example of such a phenomenon. The Magellanic Clouds follow a near-polar orbit around our Milky Way, along with several dwarf galaxies and the HVCs of the Magellanic Stream (see Haud 1988; Majewski 1994). The gas of the Stream has probably been torn from the Small Magellanic Cloud by a close encounter with the Large Cloud (Ch. 5). If the material that makes up these small galaxies had instead remained gaseous, spreading around their orbital plane, the Milky Way would have had a true polar ring. Perhaps polar rings are best viewed as instances of failed satellite formation, just as the solar system's asteroid belt contains material that was never incorporated into one of the planets.

Like NGC 4650 A, many polar ring galaxies have bright compact nuclei, suggesting that gas inflow and dissipation were important at earlier times, to build these dense inner structures. Sil'chenko (1998) found that the nuclei of NGC 2685 (A-3) and IC 1689 (B-3) are chemically distinct, with stronger Mg absorption than the surrounding stellar population, implying that the centers have undergone multiple bursts of star formation. Reshetnikov et al. (2001) found LINER and even Seyfert spectra in half of their sample of polar ring systems.

As with merging and interacting galaxies (e.g. Hibbard & van Gorkom 1996; Hibbard & Yun 1999), we might expect to find outlying gas around polar ring galaxies, which has escaped incorporation into the ring itself, and may point to a possible source. Schiminovich et al. (1995) found H I gas associated with the outer stellar shells of NGC 5128 (PRC C-45). Single-dish H I surveys often find considerably more gas around a polar ring than lies within the system in an interferometric map (Richter et al. 1994; van Driel et al. 2000), pointing to a gas-rich environment. Figure 6 of Cox et al. (2001) shows the distribution of H I gas in and around UGC 9562. A streamer of gas connects the polar ring to a small nearby star-forming dwarf galaxy; streamer and dwarf each contain about a third as much gas as is in the ring. Insets show that the gas velocity varies smoothly with position along the streamer, suggesting a physical connection between the galaxies. But the fact that a "tag end" also appears to the east of the ring indicates that perhaps the streamer passes behind or in front of the ring,

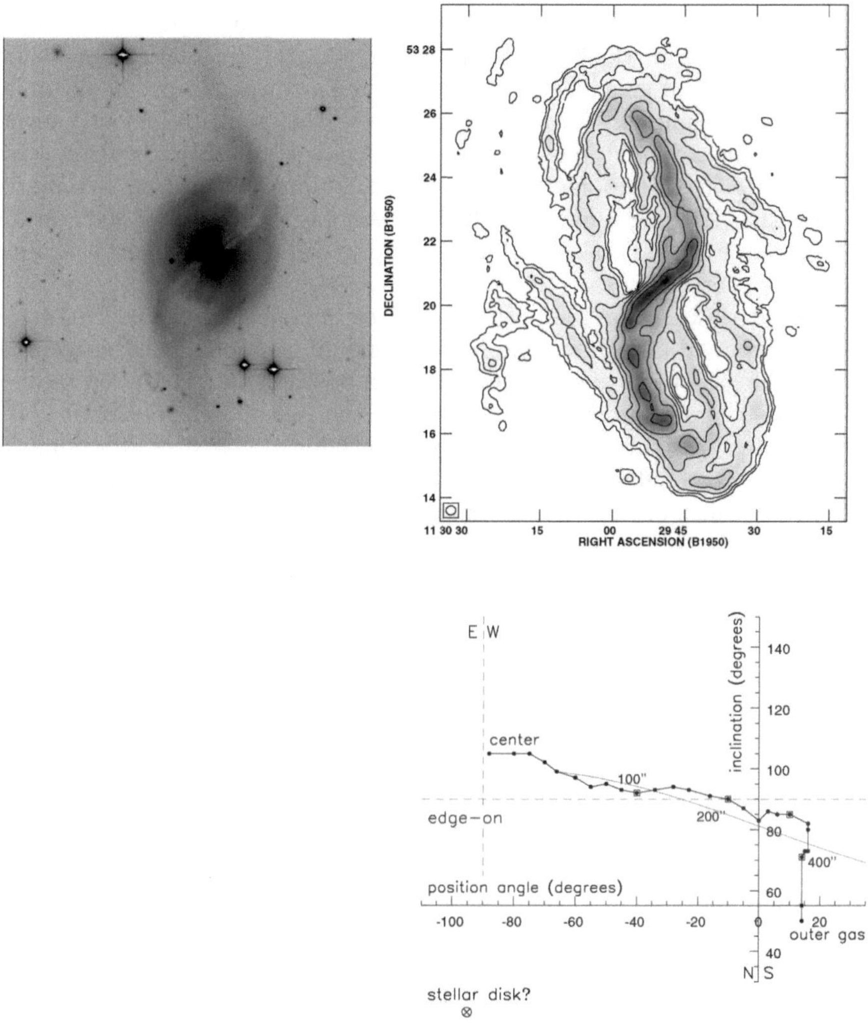

Figure 7. NGC 3718 = PRC D-18. *Top left: R*-band image, taken at the Michigan-Dartmouth-MIT Observatory on Kitt Peak, by K.R. Sembach and J.S. Gallagher. The vertical extent of this image is $8' \times 40''$. *Top right:* total H I map, from van Moorsel et al., (in preparation); note that the twisted disk extends further in the gas than in stars. *Lower right:* Position angle and inclination for the tilted-ring fit to the H I layer; the curve shows the locus of material inclined at 80° to the galaxy's disk.

rather than flowing into it. The ring itself is blue, and star forming; it may have been disturbed by the streamer gas and "rejuvenated".

The galaxy NGC 3718 (PRC D-18) has an even more complex gas structure around it. The top left panel of Fig. 7 shows a peculiar symmetric dust lane, and concentrations of dust and stars in the "spiral arms" of a galaxy seen close to face-on. The neutral hydrogen map to the right shows a concentration of gas in the "spiral". However, the H I kinematics show that this is not a spiral galaxy, but a system with a warped and twisted polar ring. We see a complex bisymmetric structure, with multiple velocity peaks where our line of sight passes more than once through the twisted disk (Schwarz 1985; van Moorsel et al., in preparation). The strong symmetry suggests that the ring is old enough that even gas at a radius of 500″, in the outer parts of the H I disk, has had time to make at least one orbit around the galaxy center.

The gas disk can be modeled as a series of concentric tilted rings; the lower right panel shows how their inclination and position angle change with radius. The gas orbits twist roughly 120° in position angle between radii of 30″ and 500″, warping through edge-on to the line of sight at about 200″; they are always highly inclined to the underlying stellar disk. The smooth curve shows the angles for a twisted disk with a constant inclination of 80° to the stellar disk of the host, assuming that this is seen with its equatorial plane tipped by 20° to 30° to face-on, and the major axis is close to north-south (so that its pole lies close to PA 90°).

We can interpret this twisting as differential precession in the galaxy's gravitational potential, and construct a dynamical model like that of Sparke (1996) for NGC 5128. A crude mass model (Sparke 2002; based on Peletier & Willner's (1993) H-band photometry), puts roughly 60% of the mass in an exponential disk: $M_d=7\times10^{10}$ $M_\odot$ with scale length $h_R=27''$. The central Plummer bulge has core radius $r_P=10''$ and $M_b=2\times10^{10}$ $M_\odot$, while the pseudo-isothermal halo has a core $r_H=150''$, and the circular speed rises to $V_H=300$ km s^{-1} at large radii. In this gravitational potential, we can compute how fast the gas layer twists up through differential precession. Even when the halo is taken to be spherical, the model gas disk twists by 120° at $r\sim50''$ during a single orbital period of the outer gas at 500″. A substantially flattened dark halo would cause even stronger differential precession. Unlike NGC 4650 A (Sackett et al. 1994; Combes & Arnaboldi 1996) and A 0136−0801 (Sackett & Pogge 1995), but like NGC 4753 where we see a complex warped dust structure (Steiman-Cameron et al. 1992), this galaxy seems to have a near-spherical distribution of "dark matter".

Measuring chemical abundances in the rings, by optical emission-line spectroscopy of H II regions, might establish whether the ring gas has low metallicity, similar to that in the HVCs of the Milky Way. Eskridge & Pogge (1997) obtained spectra of several H II regions in the ring of NGC 2685; from these, they derive oxygen abundances 0.8 to 1.1 times the solar value! Such gas is far too metal-rich to have come from a dwarf galaxy, and the small central galaxy is unlikely to have been able to strip metal-rich gas from the inner parts of a large spiral. Eskridge & Pogge (1997) suggest instead that the ring is old, and has been enriched in metals since its formation; the ejecta of supernovae must have been "smothered" and prevented from escaping by the ring gas. Since some of their observed H II regions were in the outer parts of the ring, this retention of heavy elements is a little surprising.

The spectra of H II regions in the ring of NGC 4650 A (Sparke et al., in preparation) suggest a lower abundance. Compared to the spectrum in Fig. 2a of Eskridge & Pogge (1997), the [O III] line at 5007 Å is stronger relative to Hβ, and the [N II] lines are weaker relative to Hα, indicating that the gas of NGC 4650 A is poorer in metals. The abundance is probably similar to that of gas in the Large Magellanic Cloud, about a third of the solar level. But this is still significantly higher than for the infalling HVC complex C, where Wakker et al. (1999) measure 0.1 times solar metallicity.

It is far from clear how polar ring systems are related to the HVCs of our Milky Way. Observers living in a galaxy with a polar ring would see the ring as a giant band of high-velocity gas, stretching completely around the sky, and with a distinctive kinematic signature. A twisted polar disk like that of NGC 3718, or gas streamers like those of UGC 9562, would be seen as a more complex system of HVCs. Some of our own HVCs, notably those of the Magellanic Stream, do lie on near-polar orbits at radii of tens of kiloparsecs; but others follow very different orbits, and may be much more local to the Galactic Disk. While some aspects of the Milky Way's HVC system resemble a polar ring, the correspondence is severely incomplete.

References

Arnaboldi, M., Capaccioli, M., Cappellaro, E., Held, E.V., Sparke, L.S. 1993, A&A, 267, 21
Arnaboldi, M., Oosterloo, T., Combes, F., Freeman, K.C., Koribalski, B. 1997, AJ, 113, 585
Bailyn, J. 2003, ApJ, 583, L79
Begeman, K. 1987, Ph.D. thesis, Rijks Universiteit Groningen
Bekki, K. 1997, ApJ, 490, L37
Bekki, K. 1998, ApJ, 499, 635
Bekki, K., Freeman, K.C. 2002, ApJ, 574, L21
Bettoni, D., Galletta, G., García-Burillo, S., Rodríguez-Franco, A. 2001, A&A, 374, 421
Binney, J.J. 1978, MNRAS, 183, 779
Binney, J.J. 1981, MNRAS, 196, 455
Binney, J.J. 1992, ARA&A, 30, 51
Binney, J.J., Tremaine, S. 1987, Galactic Dynamics (Princeton: Princeton University Press)
Binney, J.J., Jiang, I.-G., Dutta, S.N. 1998, MNRAS, 297, 1237
Bosma, A. 1991, in Warped Disks and Inclined Rings Around Galaxies, eds. S. Casertano, P. Sackett, F. Briggs, 181
Bottema, R. 1995, A&A, 295, 605
Bottema, R. 1996, A&A, 306, 345
Bournaud, F., Combes, F. 2003, A&A, 401, 817
Briggs, F.H. 1990, ApJ, 352, 15
Brinks, E., Burton, W.B. 1984, A&A, 141, 195

Brocca, C., Bettoni, D., Galletta, G. 1997, A&A, 326, 907
Burton, W.B. 1988, in Galactic and Extragalactic Radio Astronomy, 2nd edition, eds. G.L. Verschuur, K.I. Kellerman (Berlin: Springer), 295
Burton, W.B., Elmegreen, B.G., Genzel, R. 1992, in The Galactic Interstellar Medium, eds. D. Pfenniger, P. Bartholdi (Berlin: Springer)
Carney, B., Seitzer, P. 1993, AJ, 105, 2127
Casertano, S., Sancisi, R., van Albada, T.S. 1987, in IAU Symp. 117, Dark Matter in the Universe, eds. J. Kormendy, G.R. Knapp (Dordrecht: Reidel), 82
Colley, W.N., Sparke, L.S. 1996, ApJ, 471, 748
Combes, F., Arnaboldi, M. 1996, A&A, 305, 763
Cox, A.L. 1996, Ph.D. thesis, University of Wisconsin-Madison, USA
Cox, A.L., Sparke, L.S. 2004, AJ, submitted
Cox, A.L., Sparke, L.S., van Moorsel, G., Shaw, M. 1996, AJ, 111, 1505
Cox, A.L., Sparke, L.S., Watson, A.M., van Moorsel, G. 2001, AJ, 121, 692 (C01)
Debattista, V., Sellwood, J.A. 1999, ApJ, 513, L107
de Grijs, R. 1997, Ph.D. thesis, Rijks Universiteit Groningen, Chapter 9
Dubinski, J., Kuijken, K. 1995, ApJ, 442, 492
Eskridge, P.B., Pogge, R.W. 1997, ApJ, 486, 259
Florido, E., Prieto, M, Battener, E., Mediavilla, E., Sanchez-Saavedra, M.L. 1991, A&A, 242, 301
Freudenreich, H.T. 1998, ApJ, 492, 495
Gallagher, J.S., Sparke, L.S., Matthews, L.D., Frattare, L.M., English, J., Kinney, A.L., Iodice, E., Arnaboldi, M. 2002, ApJ, 568, 199
García-Ruiz, I., Sancisi, R., Kuijken, K. 2002a, A&A, 394, 769
García-Ruiz, I., Kuijken, K., Dubinski, J. 2002b, MNRAS, 337, 459
Haud, U. 1998, A&A, 198, 125
Hibbard, J.E., van Gorkom, J.H. 1996, AJ, 111, 655
Hibbard, J.E., Yun, M.S. 1999, AJ, 118, 162
Hofner, P., Sparke, L.S. 1994, ApJ, 428, 466
Hunter, C., Toomre, A. 1969, ApJ, 155, 747
Ideta, M., Hozumi, S., Tsuchiya, T., Takizawa, M. 2000, MNRAS, 311, 733
Iodice, E., Arnaboldi, M., De Lucia, G., Gallagher, J.S., Sparke, L.S., Freeman, K.C. 2002, AJ, 123, 195
Iodice, E., Arnaboldi, M., Bournaud, F., Combes, F., Sparke, L.S., van Driel, W., Capaccioli, M. 2003, ApJ, 585, 730
Jiang, I.-G., Binney, J.J. 1999, MNRAS, 303, L7
Kamphuis, J.J., Sijbring, D., van Albada, T.S. 1996, A&AS, 116, 15
Kennicutt, R.C. 1989, ApJ, 344, 685
Kuijken, K. 1991, ApJ, 376, 467
Laustsen, S., West, R.M. 1980, JApA, 1, 177
López-Corredoira, M., Cabrera-Lavers, A., Garzón, F., Hammersley, P.L. 2002, A&A, 394, 883
Lynden-Bell, D. 1965, MNRAS, 129, 299
Mahon, E. 1992, Ph.D. thesis, University of Florida, Gainesville, USA
Majewski, S.R. 1994, ApJ, 431, L17
Masset, F., Tagger, M. 1997, A&A, 322, 442
Matthews, L.D., Gallagher, J.S. 1996, AJ, 111, 1098
Matthews, L.D., Gallager, J.S., van Driel, W. 1999, AJ, 118, 208
May, A., James, R.A. 1984, MNRAS, 206, 691
Mould, J., Balick, B., Bothun, G., Aaronson, M. 1982, ApJ, 260, L37
Nelson, R.W., Tremaine, S. 1995, MNRAS, 275, 897
New, K.C.B., Tohline, J.E., Frank, J., Väth, H.M. 1998, ApJ, 503, 632
Nicholson, R.A. 1989, Ph.D. thesis, Sussex University, UK
Nicholson, R.A., Bland-Hawthorn, J., Taylor, K. 1992, ApJ, 387, 503
Ostriker, E.C., Binney, J.J. 1989, MNRAS, 237, 785

Peletier, R.F., Willner, S.P. 1993, ApJ, 418, 626
Porcel, C., Battener, E., Jimenez-Vincente, J. 1997, A&A, 322, 103
Reshetnikov, V.P. 1997, A&A, 321, 749
Reshetnikov, V.P., Combes, F. 1994, A&A, 291, 57
Reshetnikov, V.P., Sotnikova, N. 1997, A&A, 325, 933
Reshetnikov, V.P., Combes, F. 1998, A&A, 382, 513
Reshetnikov, V.P., Faúndez-Abans, M., de Oliveira-Abans, M. 2001, MNRAS, 322, 689
Reshetnikov, V.P., Battener, E., Combes, F., Jiménez-Vincente, J. 2002, A&A, 382, 513
Richter, O.-G., Sackett, P.D., Sparke, L.S. 1994, AJ, 107, 99
Rix, H.-W., Katz, N. 1991, in Warped Disks and Inclined Rings Around Galaxies, eds. S. Casertano, P. Sackett, F. Briggs, 112
Rubin, V.C. 1994, AJ, 108, 456
Sackett, P.D. 1991, in Warped Disks and Inclined Rings Around Galaxies, eds. S. Casertano, P. Sackett, F. Briggs, 73
Sackett, P.D., Sparke, L.S. 1990, ApJ, 361, 408
Sackett, P.D., Pogge, R.W. 1995, in AIP Conf. Ser., 336, Dark Matter, eds. S.S. Holt, C.L. Bennett (New York: AIP), 141
Sackett, P.D., Rix, H.-W., Jarvis, B.J., Freeman, K.C. 1994, ApJ, 436, 629
Sancisi, R. 1976, A&A, 53, 159
Sasaki, T. 1987, PASJ, 39, 849
Schiminovich, D. 2001, in ASP Conf. Ser., 240, Gas and Galaxy Evolution, eds. J.E. Hibbard, M.P. Rupen, J.H. van Gorkom (San Francisco: ASP), 147
Schiminovich, D., van Gorkom, J.H., van der Hulst, J.M., Malin, D.F. 1995, ApJ, 444, L77
Schinnerer, E., Scoville, N. 2002, ApJ, 577, L103
Schneider, S., Skrutskie, M.F., Hacking, P.B., et 8 al. 1989, AJ, 97, 666
Schwarz, U.J. 1985, A&A, 142, 273
Schwarzkopf, U., Dettmar, R.-J. 2001, A&A, 373, 402
Schweitzer, F., Whitmore, B.C., Rubin, V.C. 1983, AJ, 88, 909
Shang, Z., Brinks, E., Zheng, Z. et 22 al. 1998, ApJ, 504, L23
Sil'chenko, O.K. 1998, A&A, 330, 412
Sparke, L.S. 1984, MNRAS, 211, 911
Sparke, L.S. 1986, MNRAS, 219, 657
Sparke, L.S. 1990, in Dynamics and Interactions of Galaxies, ed. R. Wielen (Berlin: Springer)
Sparke, L.S. 1996, ApJ, 473, 810
Sparke, L.S. 2002, in The Shapes of Galaxies and Their Dark Matter Halos, ed. P. Natarajan (Singapore: World Scientific), 178
Sparke, L.S., Casertano, S. 1988, MNRAS, 234, 873
Sparke, L.S., Shoemaker, W., Cox, A.L. 2004, in preparation
Steiman-Cameron, T.Y., Durisen, R.H. 1988, ApJ, 325, 26
Steiman-Cameron, T.Y., Kormendy, J., Durisen, R.H. 1992, AJ, 104, 1339
Swaters, R.A., Rubin, V.C. 2003, ApJ, 587, L23
Swaters, R.A., Sancisi, R., van der Hulst, J.M. 1997, ApJ, 491, 140
Tremaine, S., Yu, Q. 2000, MNRAS, 319, 1
Tsuchiya, T. 2002, NewA, 7, 293
Uson, J.M., Matthews, L.D. 2003, AJ, 125, 2455
van der Hulst, J.M., van Albada, T.S., Sancisi, R. 2001, in ASP Conf. Ser., 240, Gas and Galaxy Evolution, eds. J.E. Hibbard, M.P. Rupen, J.H. van Gorkom (San Francisco: ASP), 451
van Driel, W., Combes, F., Casoli, F., et 16 al. 1995, AJ, 109, 942
van Driel, W., Arnaboldi, M., Combes, F., Sparke, L.S. 2000, A&AS, 141, 385
van Gorkom, J.H., Schechter, P.L., Kristian, J. 1987, ApJ, 314, 457
van Gorkom, J.H., van der Hulst, J.M., Haschick, A.D., Tubbs, A.D. 1990, AJ, 99, 1781
Verdes-Montenegro, L., Bosma, A., Athanassoula E. 2002, A&A, 389, 825
Vietri, M. 1986, ApJ, 306, 48

Wakker, B.P., Howk, J.C., Savage, B.D., van Woerden, H., Tufte, S.L., Schwarz, U.J., Benjamin, R., Reynolds, R.J., Peletier, R.F., Kalberla, P.M.W. 1999, Nature, 402, 388
Watson, D.M., Guptill, M.T., Buchholz, L.M. 1994, ApJ, 420, L21
Weinberg, M.D. 1998, MNRAS, 299, 499
Whitmore, B.C. 1984, AJ, 89, 618
Whitmore, B.C. 1991, in Warped Disks and Inclined Rings Around Galaxies, eds. S. Casertano, P. Sackett, F. Briggs, 60
Whitmore, B.C., McElroy, D.B., Schweitzer, F. 1987, ApJ, 314, 439
Whitmore, B.C., Lucas, R.A., McElroy, D.B., Steiman-Cameron, T.Y., Sackett, P.D., Olling, R.P. 1990, AJ, 314, 439
Whitmore, B.C., Zhang, Q., Leitherer, C., Fall, S.M., Schweizer, F., Miller, B.W. 1999, AJ, 118, 1551
Wilkinson, A., Sharples, R.M., Fosbury, R.A.E., Wallace, P.T. 1986, MNRAS, 218, 297

14. HIGH-VELOCITY CLOUDS: THE MISSING LINK?[1]

LEO BLITZ
Department of Astronomy, University of California, Berkeley, USA; blitz@astro.berkeley.edu
DAVID N. SPERGEL
Department of Astrophysical Sciences, Princeton University, USA; dns@astro.princeton.edu
PETER J. TEUBEN
Department of Astronomy, University of Maryland, USA teuben@astro.umd.edu
AND
DAP HARTMANN
Center for Astrophysics, Cambridge, USA; Delft University of Technology, the Netherlands; dap@tbm.tudelft.nl

Abstract. Hierarchical structure formation models predict the existence of large numbers of low velocity dispersion dark halos. Galaxy surveys find far fewer galaxies than predicted by analytical estimates and numerical simulations. In this paper, we suggest that these dark halos are not missing, but have been merely misplaced in the galactic astronomy section of the journals: they are the high-velocity clouds (HVCs). We review the predictions of our model for the Local Group origin of the HVCs and its implications for the formation and the evolution of our Galaxy. We describe recent observations that confirm many earlier predictions and discuss future tests of the model.

1. The missing galaxy problem

A generic prediction of hierarchical structure formation models is the existence of large numbers of low-mass halos. The Press-Schechter formalism (Press & Schechter 1974) predicts that the galaxy mass function,

$$n(M) \propto M^{\alpha} \exp(-M/M_*) \tag{1}$$

[1]This chapter has remained essentially unchanged since its submission (January 1999).

H. van Woerden et al. (eds.), High-Velocity Clouds, 297–312

has a steep faint-end slope, $\alpha \simeq -2$. Numerical simulations (Efstathiou et al. 1988; Gelb & Bertschinger 1994) are consistent with the Press-Schechter approach: they also predict copious low-mass halos.

Most galaxy surveys, however, do not seem to find large numbers of low-luminosity, low-velocity-dispersion galaxies. Loveday et al. (1999) summarize a number of recent field surveys that find a faint-end slope in the range $-1.2 \lesssim \alpha \lesssim -0.7$. Groups have similar "flat" galaxy luminosity functions with slopes typically ~ -1 (Muriel et al. 1998). While surveys that reach lower surface-brightness limits find more dwarf galaxies (Bothun et al. 1997, Dalcanton et al. 1997), even the inclusion of these systems does not appear to increase the faint-end slope enough to reconcile theory and observation.

In our own Local Group, where the galaxy inventory is thought to be essentially complete, the discrepancy is even more severe. Simulations at the appropriate scale suggest that the Local Group should contain roughly 1000 objects with velocity dispersions larger than 10 $\mathrm{km\,s^{-1}}$ (Klypin et al. 1999). Observers however have only been able to find ~30 galaxies in the Local Group (Mateo 1998).

Where are the missing dark halos? There is either something wrong with hierarchical structure formation, the numerical simulations, or there are a host of unidentified bound systems in the Local Group.

What are the likely properties of these low-velocity-dispersion halos? Star formation is likely to be inefficient in these low-luminosity systems, because the cosmological ultraviolet background can prevent or at least delay the formation of atomic and molecular hydrogen (Babul & Rees 1992; Kepner et al. 1997; Barkana & Loeb 1999). If these missing galaxies have not formed stars, they likely persist as small bound objects containing mostly ionized hydrogen and possibly a handful of stars. These dark halos may be the high-velocity clouds (HVCs).

The HVCs are clouds of atomic hydrogen, detected primarily by means of their 21-cm emission, that cannot be in circular rotation about the Galactic Center. Because they are largely found at high galactic latitude, and because the H I layer of the Milky Way is so thin, the characteristic distance to the HVCs can, in principle, be anywhere between several hundred parsecs and 1 Mpc. In this paper, we will argue that the evidence points to a Local Group origin for the HVCs. Oort's (1964) original idea that the clouds represent infall onto the Milky Way was abandoned long ago, because of the discovery of HVCs with positive galactocentric velocities. Nevertheless, we will argue that his insight that the HVCs represent the unaccreted remnants of galaxy formation is largely correct. We will also offer some speculations on the implications of the Local Group origin. A fuller, more detailed account of the arguments presented in this paper may

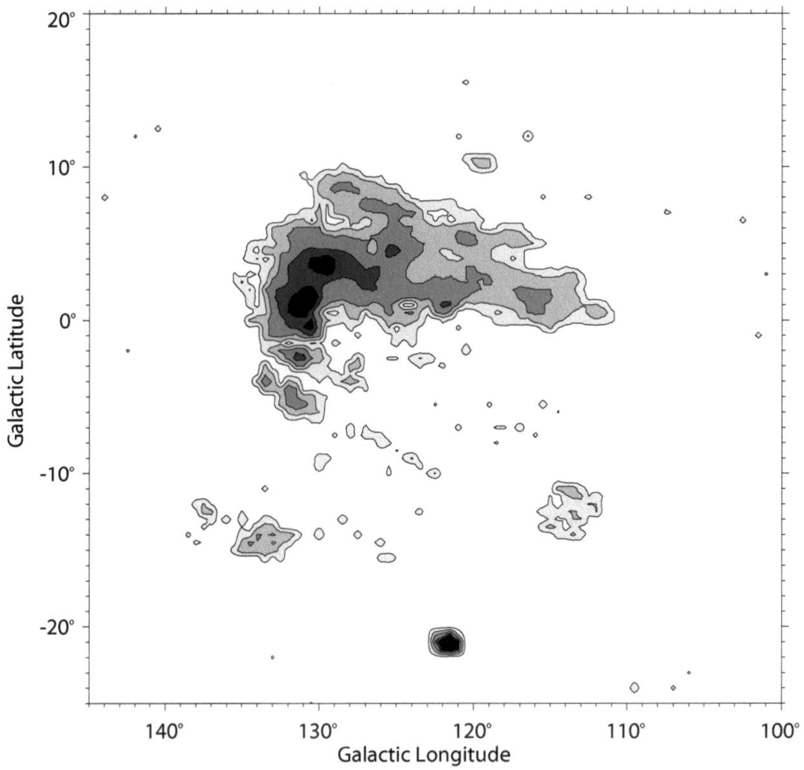

Figure 1. H I emission from complex H, as seen in the Leiden-Dwingeloo Survey, integrated over the velocity range $-240 \leq v_{\rm LSR} \leq -170$ km s^{-1}, effectively excluding the conventional Galactic gaseous disk. The bright object at l=122°, b=−21° represents the portion of M 31 emitting within the chosen velocity range. Figure from Blitz et al. (1999a).

be found in Blitz et al. (1999a). Some of the arguments in this article have already appeared as Blitz et al. (1999b).

2. Evidence for a Local Group origin

2.1. KINEMATIC DATA

The flux from HVCs is less than 10^{-4} of the normal galactic emission. In the outermost parts of the Milky Way, a longitude-velocity plot of the normal galactic emission shows that the contours are very nearly sinusoidal. The radial velocity, v_r of the gas in circular orbit around the center obeys:

$$v_r = (\Theta \frac{R_0}{R} - \Theta_0) \sin l \ \cos b \tag{2}$$

where Θ is the circular velocity of a gas parcel at a distance R from the Galactic Center, and where R_0 is the distance of the Sun from the Center.

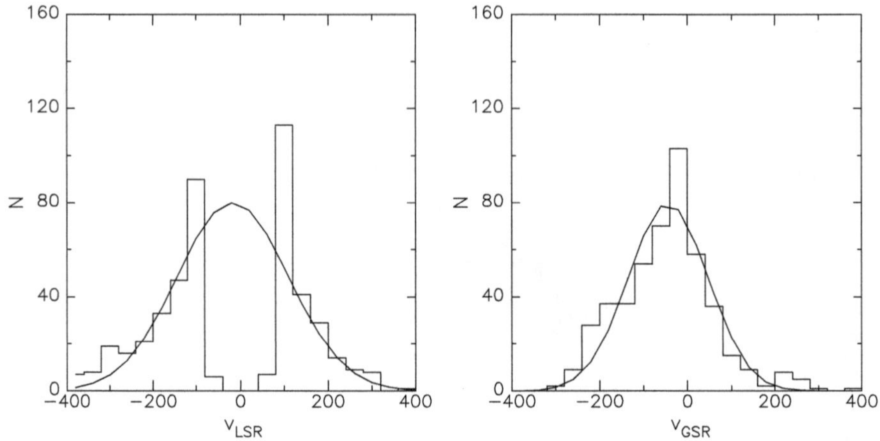

Figure 2. *Left:* Histogram of the distribution of HVC velocities relative to the LSR. HVCs which might have a $v_{\rm LSR}$ near zero would not be separable from conventional-velocity galactic emission. *Right:* Distribution of HVC velocities relative to the GSR. A gaussian profile was fit to the wings of both histograms. The GSR distribution of the HVCs is more narrowly confined than that relative to the LSR, suggesting that the GSR system is the more appropriate inertial reference frame. The data in both panels are from the Wakker & van Woerden (1991) catalog of HVCs (from Blitz et al. (1999a)).

Thus, for gas close to the plane ($\cos b \simeq 1$), sinusoidal profiles for a flat rotation curve are indicative of gas at constant radius. Conversely, the galactocentric distance of gas along a sinusoid can be inferred from its maximum velocity ($\sin l \cos b = 1$). Longitude-velocity plots from the Leiden-Dwingeloo survey (Hartmann & Burton 1997) exhibit H I emission with maximum velocities of about ± 170 km s^{-1} (Blitz et al. 1999a), implying that for a flat rotation curve, the H I disk of the Milky Way extends to a distance of $\simeq 37$ kpc from the Galactic Center.

Although most of the HVCs are far from the galactic plane, one large cloud known as complex H (after Aad Hulsbosch who has spent years observing and categorizing the HVCs) lies directly in the plane and is shown in Fig. 1. The velocity centroid of this cloud is -194 km s^{-1}. If it were within the H I disk, the velocity difference between this cloud and the gas in normal galactic rotation would be between 30 and 200 km s^{-1}, giving rise to a very large region of highly shocked gas with an energy of $\simeq 10^{54}$ erg, nearly independent of distance over a large portion of the disk. This much energy would give rise to a region of disturbed gas in the H I disk at least 25 degrees in extent, and also to strong Hα and perhaps X-ray emission, but none of these is evident. Complex H must therefore lie beyond the H I disk at a distance of at least 40 kpc from the center (see Lockman (2003) for a recent confirmation of this distance). If it is at a distance of 50 kpc, the cloud has a diameter of 20 kpc, and an H I mass of 9×10^7 M$_\odot$. It would be a huge H I cloud, comparable in mass to a dwarf spheroidal galaxy. If this cloud is typical of the others, then the typical distance of the HVCs will be $\simeq$25 times larger (since the median angular size of the HVCs is $\simeq$25 times smaller), and the reservoir of H I locked up in the HVCs is enough to make a Milky Way-sized galaxy.

The velocity dispersion and the velocity centroid of the cloud ensemble can also be used to determine what the most appropriate inertial frame of reference is. A non-inertial frame always gives rise to a larger velocity dispersion than an inertial frame, because the former adds a position-dependent velocity in quadrature to each observed radial velocity. A non-zero velocity centroid suggests that the ensemble is moving relative to the observer. Figure 2 shows the distribution of the velocities of the HVCs in the LSR and GSR frames of reference; velocities in the latter frame are much smaller, suggesting that the Galactic Center is a better inertial frame than the Local Standard of Rest. The mean velocity of -46 ± 7 km s^{-1} implies that the Galactic Center is moving with respect to the barycenter of the ensemble. If we concentrate on the clouds with negative velocities seen in the lower panel of Fig. 3, the centroid of these clouds is close to the Local Group barycenter (l=147°, b=−25°, v=−82 km s^{-1}). Relative to the LSR, the mean velocity of this group of clouds is −173 km s^{-1}; relative to the GSR, it is −88 km s^{-1}. However, relative to a frame of reference centered on the barycenter of the Local Group, the LGSR, the mean velocity is only -28 ± 10 km s^{-1}. These numbers suggest that the barycenter of the Local Group is the proper inertial frame for the HVCs and that the Milky Way is approaching the barycenter at a velocity of about 60 to 90 km s^{-1}.

Burton et al. (Ch. 15) and Braun & Burton (1999) have suggested that there is a separate class of HVCs, which they call "Compact HVCs" or CHVCs, that are different from those discussed and analyzed above, but somehow share the kinematic properties of the conventional HVCs. In fact, the Burton & Braun (1999) clouds are simply a subset of the HVCs compiled by Wakker & van Woerden (1991), with a few additional clouds that fall between the latter's sampled points. Even though the clouds listed in Burton & Braun (1999) are chosen from the smallish end of the Wakker & van Woerden sample, the former have a median surface area $\lesssim$1.0 deg^2 compared to 1.5 deg^2 for the Wakker & van Woerden sample (Blitz et al. 1999a), and are not more compact than the typical HVCs. Even this difference may result largely because Braun & Burton (1999) use the beam-deconvolved area and Blitz et al. (1999a) do not. In any event, the small difference hardly warrants a new designation. Apparently, Burton & Braun (1999) have succeeded in showing that a representative subset of HVCs has properties of and behaves similarly to the HVC population as a whole.

2.2. DYNAMICAL SIMULATION

The Local Group is dynamically simple, thus it should be possible to simulate the dynamical history of the HVCs and reproduce both the spatial and kinematic distribution of the clouds. Since 95% of the mass of the Local Group is in the Milky Way and M 31, we modeled the Local Group in Blitz et al. (1999a) as a modified, restricted 3-body system, with the HVCs as test particles in a potential defined by the Galaxy and M 31. The simulation begins with the HVCs on a regular grid and no initial velocity dispersion. The Milky Way and M 31 are separated initially by 100 kpc and expand with the Hubble flow; M 31 is taken to have twice the mass of the Milky Way. Enough mass is put into the two galaxies so that they are turned around from the Hubble flow until they reach their present separation and velocity of approach. After the start of the simulation, if a particle comes within 100 co-moving kpc of either galaxy, we assume that the cloud is accreted. This is a somewhat larger interaction radius than the geometric cross-section of the galaxies, to allow for gravitational focusing, tidal disruption and dynamical friction.

The basic results of the model are insensitive to the exact value of the relative masses of the two galaxies, or the accretion radius. The model is exceedingly simple and contains no hydrodynamics (the test particles are non-interacting), though it is not self-consistent in that all of the mass is placed in the two galaxies at t=0 and accretion does not increase the masses of the galaxies. These shortcomings are compensated by the simplicity of the model. It contains no free parameters and no fine-tuning is done to improve the comparison with the observations.

The results are shown in Fig. 3 and 4. Figure 3 is a comparison of the simulated and observed spatial distributions of the HVCs, with the Magellanic Stream and the A, C and M complexes removed. The comparison shows a rather good agreement considering the simplicity of the model. The model reproduces the two concentrations of clouds, the separation into positive and negative LSR velocities, and the tilt in the positive- and negative-velocity cloud groups relative to the galactic plane. No extraneous groups of clouds are produced. No other model considered to date reproduces all of these features of the HVC spatial distribution. The separation into two groups occurs naturally in the model because the clouds are distributed along a wide filament along the line connecting the Milky Way and M 31. The negative LSR velocity clouds are seen along the filament toward the Local Group barycenter, and the positive LSR velocity clouds are seen primarily in the anti-barycenter direction. Both groups are falling toward the LGSR.

The Magellanic Stream was removed from the comparison because it is

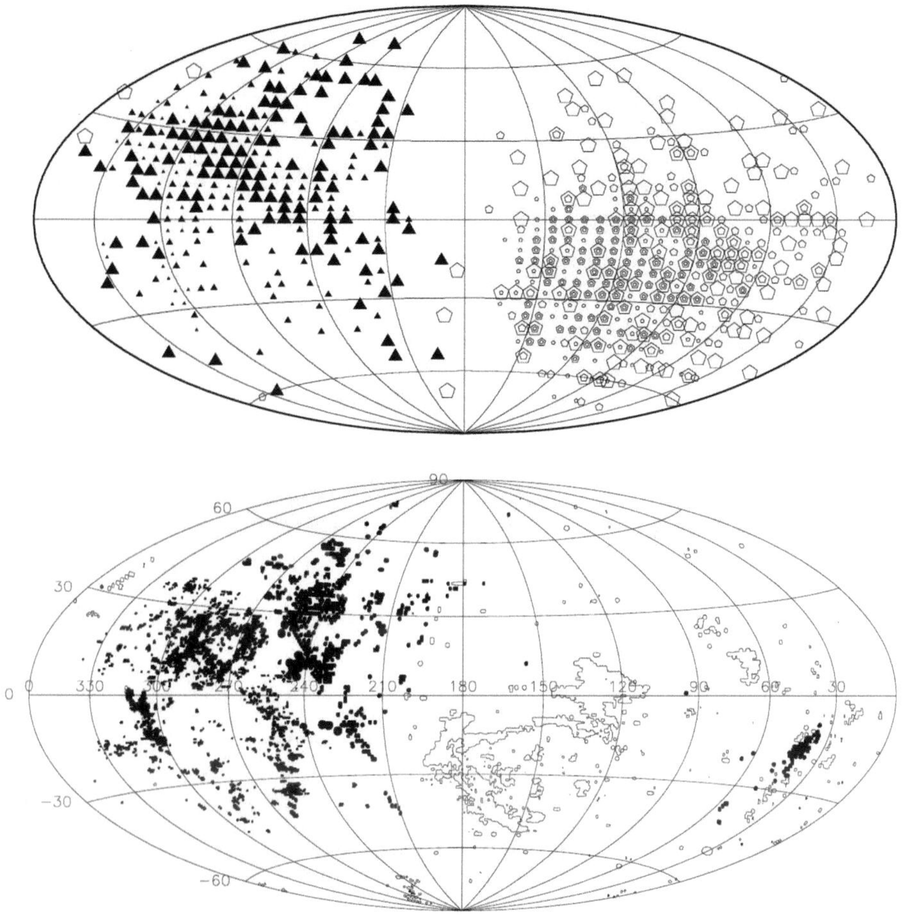

Figure 3. Comparison of simulated and observed distributions of the HVC ensemble on the sky. *Top*: Distribution of all simulated clouds having H I column densities greater than 3×10^{18} cm^{-2} and $|v_{\rm LSR}|$ greater than 100 km s^{-1}. The size of the symbols is proportional to column density and ranges from 3×10^{18} to greater than 3×10^{19} cm^{-2}. Strictly speaking, these simulated column densities are total ones, i.e. including the dark-matter content. The triangles represent clouds with positive LSR velocities; the pentagons, clouds with negative LSR velocities. This figure represents the distribution of HVCs if the clouds have not been destroyed by passage through a hot intergalactic medium and if collisions between HVCs are rare. *Bottom*: Distribution of observed HVCs but excluding the Magellanic Stream and the northern-hemisphere complexes A, C, and M, which are evidently relatively nearby and thus not representative of the angular size of individual clouds in the Local Group ensemble. Positive LSR velocities are denoted by filled contours, negative LSR velocities by open contours. The lower panel was kindly provided by Bart Wakker. Figure from Blitz et al. (1999a).

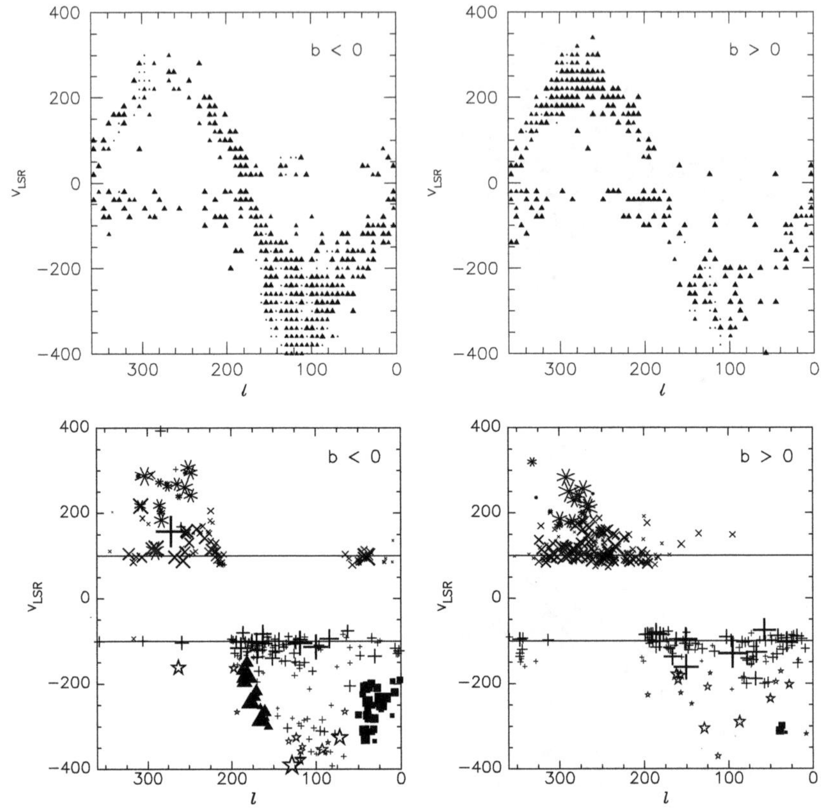

Figure 4. Comparison of the simulated longitude-velocity distribution of the HVCs with the observed situation. Radial velocities are relative to the LSR. *Top*: Simulated kinematic distribution of clouds with H I column densities greater than 3×10^{18} cm^{-2}, plotted separately for $b<0°$ and for $b>0°$. *Bottom*: Longitude-velocity diagram of the observed HVC ensemble, as compiled by WvW91. The symbols are proportional in size to the flux from the individual clouds, and are keyed to the individual complexes defined by WvW91. The MS, OA, and A, C, and M complexes are not included in this Figure (see text). Clouds with LSR velocities $|v_{\rm LSR}|<80$ km s^{-1} are not considered here as HVCs, regardless of their location. Figure from Blitz et al. (1999a).

a group of clouds known to be of tidal origin (Mathewson et al. 1974) and is thus not well represented by our model. Complex C is by far the largest cloud in the HVC ensemble and covers more than 1600 deg^2. Clouds A and M, which are also quite large, have similar velocities and may be related to this large complex. If the A, C, and M complexes are gravitationally bound, they must be tidally unstable and thus quite nearby. These complexes are also very elongated, consistent with tidal shearing. Thus, because of their apparent proximity and tidal shearing, they are also not well represented by our dynamical model and are also excluded from the comparison.

Figure 4 is a comparison of the longitude-velocity plots of the observa-

TABLE 1. Mean derived HVC properties (from Blitz et al. 1999a).

Quantity	Value
H I mass	1.9×10^7 M$_\odot$
Total Neutral Gas Mass	2.7×10^7 M$_\odot$
Total Mass	2.8×10^8 M$_\odot$
Diameter	28 kpc
Distance	1 Mpc
$n_{\rm HI}$	0.7×10^{-4} cm^{-3}

tions and the simulations. Again, the Magellanic Stream and the A, C, and M complexes are removed. The simulations reproduce the sinusoidal envelope of the cloud ensemble, the offset toward negative LSR velocities (due to the motion of the Milky Way toward the Local Group barycenter), and the magnitude of the envelope of the distribution. This agreement in the quantitative aspects of the longitude-velocity distribution is particularly noteworthy because of the absence of free parameters in the model.

3. Implications and speculations

If we accept the model at face value, it implies that the HVCs are formed with the earliest structures in the Universe and are the building blocks from which the Milky Way and M 31 formed. The HVCs that we see today would then be the left over building blocks that have not yet been accreted by either galaxy. If the Local Group is not unique, it suggests that structures similar to the HVCs are responsible for *all* initial galaxy formation, though there would be large differences in how galaxy evolution proceeds, depending on the density of the environment (see below).

If the HVCs are almost as old as the Universe, they must be gravitationally bound and tidally stable. If they have typical distances of 1 Mpc, as the model suggests, then the observed angular sizes and velocity dispersions imply that, to be self-gravitating, about 90% of the matter in the HVCs must be dark; the dark matter may be either baryonic or non-baryonic. Table 1 gives mean derived parameters for the HVCs. For example, the clouds could have a 90% ionization fraction, in which case the emission measure would be $\simeq 10^{-2}$ cm^{-6} pc, an undetectable value at present. If the dark matter is the same as that in the halo of the Milky Way and M 31, the ratio of luminous to dark matter is comparable in the HVCs and the galaxies, just what one would expect if the Milky Way and M 31 were assembled from HVCs. In this case, the HVCs would also have characteristic masses of $\simeq 10^8$ M$_\odot$, similar to the mini-halos postulated by Ikeuchi (1986) and Rees (1986) to be the first structures to form after recombination.

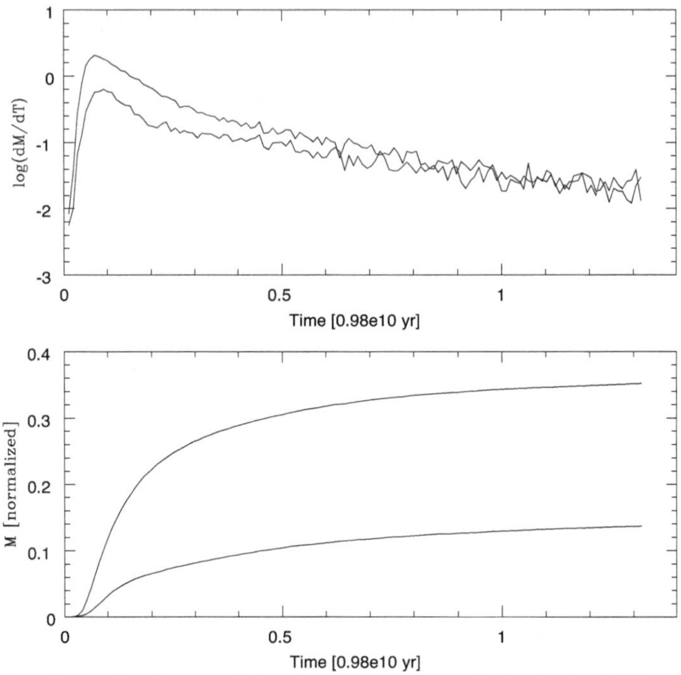

Figure 5. *Top*: Normalized rate of simulated accretion of clouds for M 31 (upper line) and for the Milky Way (lower line). The M 31 accretion rate is typically about twice that of the Milky Way. After about 3×10^9 yr, the accretion rate becomes nearly exponential with an e-folding time of about 5×10^9 yr. At the current epoch, the accretion rate is flattening out, and is equivalent to about 7.5 $M_\odot\,yr^{-1}$ (about 10 times that of the H I alone) for the Milky Way. *Bottom*: Normalized accreted mass for M 31 (upper line) and for the Milky Way (lower line). The plot shows that most of the mass is accreted at early times and that additional mass is being added to both galaxies quite slowly at the current epoch. Figure from Blitz et al. (1999a).

Our dynamical model allows us to calculate a mass-accretion history, which is shown in Fig. 5. The present-day mass-accretion rate is estimated to be 0.8 to 1.2 $M_\odot\,yr^{-1}$, approximately what is needed to fuel the present-day star formation in the Milky Way (e.g. Blitz 1996). The orbits of some of the HVCs are likely to cross in the region between M 31 and the Galaxy, giving rise to collisions between the HVCs. Typical collision velocities can be estimated from Fig. 2 to be $\simeq$200 km s^{-1}, possibly giving rise to an X-ray halo surrounding the Milky Way and M 31. This gas would probably have a temperature of about 10^6 K, and might be detectable. To predict whether or not such an X-ray halo exists, and how large it would be requires adding hydrodynamics to our simulation. The cooling time for the gas would be $>10^{10}$ yr. Collisions between clouds could also be the source of the X-ray halos around poor groups and in denser extragalactic environments.

In our picture, the growth of the Galactic Disk is fueled by the gradual

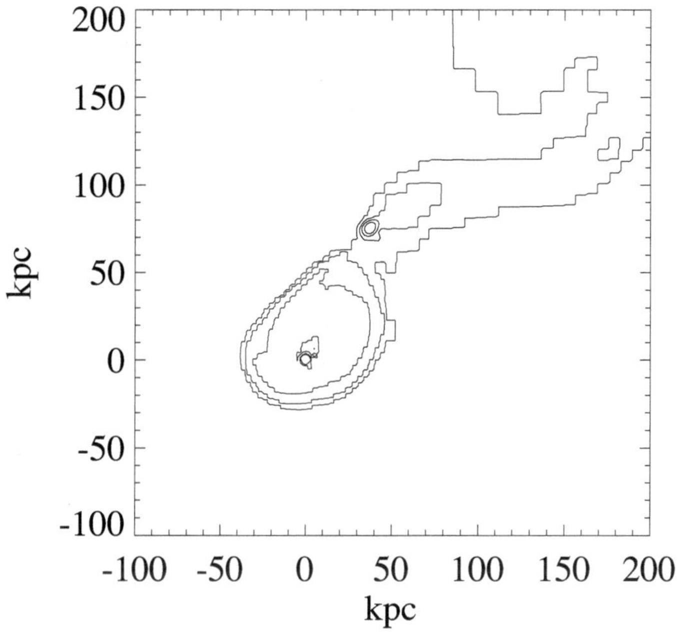

Figure 6. This figure from Kepner (1998) shows the projected hydrogen distribution in a disk galaxy in a numerical simulation by J. Kepner and G. Bryan. In the AMR hydrodynamics simulation of a Λ-dominated CDM universe, they focused their high-resolution mesh on a binary galaxy system with properties similar to the Milky Way and M 31. In their simulation, the galactic disks are built up primarily by the gradual accretion of gas clouds with properties similar to those deduced for the HVCs. In this time-slice, the main disk is accreting a gas cloud with properties similar to complex H. The accreted cloud is being ram-pressure stripped by the hot gaseous halo and is falling toward the disk. Since it has relatively high angular momentum, it eventually settles at the edge of the pre-existing disk.

accretion of HVCs and is consistent with numerical simulations. Figure 6, for example, shows the results of a hydrodynamical simulation in which a cloud similar to complex H is being accreted by a disk galaxy similar to the Milky Way. The gas streamer shown in the simulation is similar to that seen in higher-contrast versions of Fig. 1. Our picture of Milky Way formation is thus more consistent with the episodic accretion model of Searle & Zinn (1978) than it is with the Eggen et al. (1962) model. Episodic evolution, furthermore, would lead to metallicity correlations consistent with trends seen in disk stars (Edvardsson et al. 1993).

Finally, the HVCs might be the z=0 analogues of the Lyα-absorbing systems. If the HVCs are indeed ubiquitous in the Universe, they would correspond in column density to the Lyman limit systems. The frequency distribution of the Lyα absorbers is a power law with a slope of -1.4 over 8 orders of magnitude in column density (Wolfe 1993). The slope of the frequency distribution of column densities in the HVCs has the same value of -1.4.

4. Predictions and comparison with other observations

In our original paper (Blitz et al. 1999a), we made several predictions based on our model.

1. The HVCs should have substantially subsolar metallicities. These are not expected to be zero, since no extragalactic gas has primordial abundances.

2. The HVCs should have low internal pressures, inconsistent with a Galactic origin. If the clouds are self-gravitating, then the internal pressure within the clouds can be given by

$$P/k = 3\pi\alpha G\Sigma^2/20k, \tag{3}$$

where Σ is the gas surface density, α=2 for self-gravitating clouds, and k is Boltzmann's constant (Bertoldi & McKee 1992). For self-gravitating HVCs bound by their H I alone and a surface density of 3 to 30×10^{18} cm^{-2}, the expected mean hydrostatic pressure within a cloud is 0.016 to 1.6 K cm^{-3}. If the cloud is in a dark-matter potential with 10 times the H I mass, as we expect, the internal pressure would be about 10 times higher, thus pressures on the order of 0.1 to 10 K cm^{-3} are expected.

3. There should be analogues of the HVCs in other extragalactic systems.

4. There should be Hα emission associated with the HVCs, at a level at least as great as that detected toward the clouds in the Magellanic Stream, if the HVCs are Galactic in origin. If they are extragalactic, the emission measures should be $\lesssim$0.1 cm^{-6} pc.

Several groups have reported new observations that are consistent with our predictions and appear to contradict the models with a Galactic HVC origin such as the Galactic Fountain model:

1. Wakker et al. (1999) reported a measurement of subsolar metallicity on a line of sight toward Mrk 290 through complex C. They detected S II, a species in the dominant ionization state and which is not depleted onto grains. Wakker et al. (1999) obtained values for both the atomic and ionized gas components, and concluded that the abundance along this line of sight is only 0.094 solar. They concluded that this cloud represents an accretion event of an extragalactic cloud, in agreement with our predictions. Complex C has by far the largest angular size of any HVC, and is probably the nearest. If this cloud is of extragalactic origin, it suggests that the other smaller clouds are also extragalactic.

TABLE 2. Extragalactic HVC analogues

Mass	Diameter	Galaxy	Reference
$M_\odot$	kpc		
1.6×10^8	16	M 101	van der Hulst & Sancisi (1988)
1.2×10^7	5	M 101	van der Hulst & Sancisi (1988)
1×10^8	25	NGC 5668	Schulman et al. (1996)
5×10^7	7	UM 422C	Taylor et al. (1995; 1996)
1.6×10^8 (?)	16 (?)	UM 456B	Taylor et al. (1995; 1996)
1.4×10^8 (?)	8 (?)	F495-IVB	Taylor et al. (1996)
7.9×10^7	38	NGC 628	Kamphuis & Briggs (1992)
9.5×10^7	47	NGC 628	Kamphuis & Briggs (1992)
2.1×10^8	6	NGC 3227	Mundell et al. (1995)
2×10^7	28	Local Group	This paper; Table 1

2. Sembach et al. (1999) observed the ionized edges of HVCs in the direction of Mrk 509 and PKS 2155−304, and detected strong C IV absorption, with little or no C II or Si II. The authors concluded that the clouds are low-density ($n_{\rm H}\simeq10^{-4}{\rm cm}^{-3}$), large (greater than several kiloparsecs) clouds, with $P/k\simeq2$ K cm^{-3}. This pressure is just in the range expected for self-gravitating, dark-matter-confined HVCs; it is four orders of magnitude less than the pressure in the midplane of the Milky Way and two orders of magnitude less than expected in the Galactic Halo (Wolfire et al. 1995). The density is also in good agreement with the derived value given in Table 1.

3. Extragalactic analogues of the HVCs have been seen toward a number of galaxies. A list of such clouds is given in Table 2. These clouds were found serendipitously in the course of mapping other objects. Many are seen in projection against other galaxies, some are seen as distinct objects separated in both position and velocity from the parent galaxy. Such clouds would appear as HVCs if viewed from the target galaxy. Clearly, numerous extragalactic H I clouds have been found with properties similar to those of the HVCs given in Table 1.

Several blind surveys of H I have been undertaken, notably by Zwaan et al. (1996), and more recently by Spitzak & Schneider (1998a, b). In the Zwaan et al. (1996) survey, no extragalactic analogues were found without optical counterparts. Spitzak & Schneider (1998b) found one cloud without an optical counterpart. Neither survey is particularly sensitive to H I masses typical of what we expect from extragalactic HVCs, though a few probably should have been detected in the Zwaan et al. (1996) survey if it is as sensitive as was claimed. It is difficult to predict how many HVC analogues should be detected, since the number associated with a galaxy group or cluster probably depends sensitively on the density of the environment. It is therefore difficult to assess whether the non-detections in the blind searches (except for Spitzak & Schneider 1998b) are significant. A targeted, high-sensitivity survey in the direction of a good Local Group analogue might well decide this issue.

4. Hα has been detected toward clouds A, C, and M (Tufte et al. 1998), as well as toward the Magellanic Stream (Weiner & Williams 1996). There is some question as to whether the emission toward the Magellanic Stream is due to photoionization from the galactic ionizing radiation that leaks out of the plane of the Milky Way, or is due to shock heating as the clouds pass though the diffuse gas in the Galactic Halo. However, regardless of what produces the Hα, HVCs that are of Local Group origin should have lower emission measures than those detected toward either cloud complex. The Hα measurements are a critical test of the Local Group model, and observations toward the very-high-velocity clouds ($|v_{\rm LSR}|>200$ km s^{-1}) should give emission measures no higher than $\sim$0.1 cm^{-6} pc. Measurements are currently underway by the Wisconsin group using their *WHAM* instrument, by a Maryland-Carnegie group and by a group in Australia. If the HVCs are galactic (distances <50 kpc), their Hα surface brightnesses should be at least as large as those already detected. Results should be available within the next year.

The low metallicity, pressure and density detected along several lines of sight support a Local Group origin. The detection of extragalactic H I clouds with properties similar to those inferred if the HVCs are Local Group objects suggests that such clouds do exist in intergalactic regions. The relative paucity of HVC analogues seen in blind H I surveys may simply be a result of insufficient sensitivity or sky coverage, and more sensitive observations perhaps directed toward poor galaxy groups might usefully be undertaken. Hα measurements will provide a critical test for distinguishing between galactic and extragalactic locations for the HVCs.

5. Other possible origins

One possibility other than a Local Group origin for the HVCs is that the clouds are extensive tidal debris from previous passages either of the Magellanic Clouds or other nearby dwarfs. Bland-Hawthorn et al. (1998) have suggested that the so-called "Smith Cloud" is related to tidal streaming associated with the passage of the Sgr dwarf (Ibata et al. 1994). However, an important constraint on tidal models is the crossing time for HVCs, which provides an upper limit to the lifetime of unbound HVCs, and which can be written as follows:

$$t_c = 17.1 \frac{\Omega^{1/2}\, r_{\rm kpc}}{\Delta v}\ \mathrm{Myr}, \tag{4}$$

where $r_{\rm kpc}$ is the distance from the Galactic Center in kpc and Δv is the FWHM of the H I line averaged over the cloud. The mean value of Δv for the HVCs is 30 km s^{-1}, and the median value of Ω is 1.5 deg^2 (Blitz et al. 1999a). Thus the crossing time for a typical HVC is about 1 Myr kpc^{-1}, or about 50 Myr at the distance of the Magellanic Clouds. HVCs at that distance cannot be gravitationally bound (except perhaps for complex C, H, and the Anti-Center complex). Clouds at 50 kpc therefore double in size in a crossing time, which corresponds to a decrease in density of an order of magnitude. Thus HVCs from a tidal origin should not be able to survive for more than 1 or 2 crossing times. The orbital time for the Magellanic Clouds is about 1.5 Gyr, far too long for HVCs to have been the result of a prior passage. The Magellanic Stream itself has a much larger Ω than the typical HVC, and thus has a longer crossing time. Nevertheless, the Magellanic Stream is identified over about 1/4 of the sky, suggesting that those H I clouds survive for as long as 400 Myr. No other dwarf companions are close enough to the Milky Way to have produced the extensive tidal debris that would be necessary to explain the HVCs [NB: Note, however, the recent discovery of the CMa dwarf galaxy].

The Galactic Fountain model postulates that the HVCs are H I clouds which have condensed from gas expelled into the halo by supernovae and stellar winds (Shapiro & Field 1976; Bregman 1980). It has long been known that the Galactic Fountain model cannot produce H I clouds with radial velocities in excess of the circular speed of the Galaxy of about 220 km s^{-1}. Many HVCs have significantly larger velocities. Furthermore, the recent evidence for low HVC pressures and densities, the low metallicities, and the inability of the fountain models to reproduce the observed features seen in Figs. 3 and 4 make the Galactic Fountain untenable for the majority of HVCs. Nevertheless, even the Local Group model seems to produce insufficient numbers of clouds at low LSR velocities (clouds near $v_{\rm LSR}$=100 km s^{-1} – see Fig. 4). Some of these clouds are not fully separated in velocity from the main galactic emission and may therefore yet be part of the normal galactic emission or a Galactic Fountain phenomenon. Just as Hα measurements become a good test of the Local Group origin for the HVCs, it may be that the lower-velocity HVCs will be relatively bright Hα emitters.

References

Babul, A., Rees, M.J. 1992, MNRAS, 255, 346
Barkana, R., Loeb, A. 1999, ApJ, 523, 54

Bertoldi, F., McKee, C.F. 1992, ApJ, 395, 140
Bland-Hawthorn, J., Veilleux, S., Cecil, G.N., Putman, M.E., Gibson, B.K., Maloney, P.R. 1998, MNRAS, 299, 611
Blitz, L. 1996, in IAU Symp. 170, CO: Twenty-Five Years of Millimeter-Wave Spectroscopy, eds. W.B. Latter et al. (Dordrecht: Kluwer), 11
Blitz, L., Spergel, D.N., Teuben, P.J., Hartmann, D., Burton, W.B. 1999a, ApJ, 514, 818
Blitz, L., Spergel, D.N., Teuben, P.J., Hartmann, D., Burton, W.B. 1999b, ASP Conf. Ser., 166, Stromlo workshop on High-Velocity Clouds, eds. B.K. Gibson, M.E. Putnam (San Francisco: ASP), 125
Bothun, G., Impey, C., McCaugh S. 1997, PASP, 109, 745
Braun, R., Burton, W.B. 1999, A&A, 341, 437
Bregman, J.N. 1980, ApJ, 236, 577
Dalcanton, J.J., Spergel, D.N., Gunn, J.E., Schmidt, M., Schneider, D.P. 1997, AJ, 114, 2178
Edvardsson, B., Anderson, J., Gustafsson, B., Lambert, D.L., Nissen, F.E., Tomkin, J. 1993, A&A, 275, 101
Efstathiou, G., Frenk, C.S., White, S.D.M., Davis, M. 1988, MNRAS, 235, 715
Eggen, O.J., Lynden-Bell, D., Sandage, A. 1962, ApJ, 136, 748
Gelb, J.M., Bertschinger, E. 1994, ApJ, 436, 467
Hartmann, D., Burton, W.B. 1997, Atlas of Galactic Neutral Hydrogen (Cambridge: Cambridge University Press)
Ibata, R., Gilmore, G., Irwin, M. 1994, Nature, 370, 19
Ikeuchi, S. 1986, ApSpSci, 118, 509
Kamphuis, J., Briggs, F. 1992, A&A, 253, 335
Kepner, J.V. 1998, Ph.D. thesis, Princeton University
Kepner, J.V., Babul, A., Spergel, D.N. 1997, ApJ, 487, 61
Klypin, A., Kravtsov, A.V., Valenzuela, O., Prada, F. 1999, ApJ, 522, 82
Lockman, F.J. 2003, ApJ, 591, L33
Loveday, J., Tresse, L., Maddox, S., 1999, MNRAS, 310, 281
Mateo, M.L. 1998, ARA&A, 36, 435
Mathewson, D.S., Cleary, M.N., Murray, J.D. 1974, ApJ, 190, 291
Mundell, C.G., Pedlar, A., Axon, D.J., Meaburn, J., Unger, S. 1995, MNRAS, 277, 641
Muriel, H., Valotto, C.A., Lambas, D.G. 1998, ApJ, 506, 540
Oort, J.H. 1964, in IAU Symp. 20, The Galaxy and the Magellanic Clouds, ed. F.J. Kerr (Canberra: Australian Acad. Sci.), 130
Press, W.H., Schechter, P.L. 1974, ApJ, 330, 579
Rees, M.J. 1986, MNRAS, 218, 25P
Schulman, E., Bregman, J.N., Brinks, E., Roberts, M.S. 1996, AJ, 112, 960
Searle, L., Zinn, R. 1978, ApJ, 225, 357
Sembach, K., Savage, B.D., Lu, L., Murphy , E.M. 1999, ApJ, 515, 108
Shapiro, P.R., Field, G.B. 1976, ApJ, 205, 762
Spitzak, J.G., Schneider, S. 1998a, ApJ, 507, L9
Spitzak, J.G., Schneider, S. 1998b, ApJS, 119, 159
Taylor, C.L., Brinks, E., Grashuis, R.M., Skillman, E.D. 1995, ApJS, 99, 427
Taylor, C.L., Brinks, E., Grashuis, R.M., Skillman, E.D. 1996, ApJS, 102, 189
Tufte, S.L., Reynolds, R.J., Haffner, L.M. 1998, ApJ, 504, 773
van der Hulst, J.M., Sancisi, R. 1988, AJ, 95, 1354
Wakker, B.P., van Woerden, H. 1991, A&A, 250, 509
Wakker, B.P., Howk, J.C., Savage, B.D., van Woerden, H., Tufte, S.L., Schwarz, U.J., Benjamin, R., Reynolds, R.J., Peletier, R.F., Kalberla, P.M.W. 1999, Nature, 402, 388
Weiner, B.J., Williams, T.B. 1996, AJ, 111, 1156
Wolfe, A.M. 1993, in Relativistic Astrophysics and Cosmology, eds. C.W. Akerlof, M.A. Srednicki (New York: NY Academy of Science), 281
Wolfire, M.G., McKee, C.F., Hollenbach, D., Tielens, A.G.G.M. 1995, ApJ, 453, 673
Zwaan, M.A., Briggs, F.H., Sprayberry, D., Sorar, E. 1996, ApJ, 490, 173

15. COMPACT, ISOLATED HIGH-VELOCITY CLOUDS

W. BUTLER BURTON
Leiden Observatory, The Netherlands, and
National Radio Astronomy Observatory, Charlottesville, USA
burton@starband.net

ROBERT BRAUN
Netherlands Foundation for Research in Astronomy
Dwingeloo, The Netherlands; rbraun@nfra.nl

AND

VINCENT DE HEIJ
Leiden Observatory, The Netherlands
vhey@cbs.nl

Abstract. We consider the class of compact, isolated, high-velocity H I clouds, CHVCs, which are sharply bounded in angular extent, with no kinematic or spatial connection to other H I features down to a limiting column density of 1.5×10^{18} cm^{-2}. We describe the automated search algorithm developed by de Heij et al. (2002b) and applied by them to the Leiden-Dwingeloo Survey north of $\delta=-28°$ and by Putman et al. (2002) to the *Parkes HIPASS* data south of $\delta=0°$, resulting in an all-sky catalog numbering 246 CHVCs. We argue that these objects are more likely to represent a single phenomenon in a similar evolutionary state than would a sample that included any of the major HVC complexes. Five principal observables are defined for the CHVC population: (1) the spatial deployment of the objects on the sky, (2) the kinematic distribution, (3) the number distribution of observed H I column densities, (4) the number distribution of angular sizes, and (5) the number distribution of line widths. We show that the spatial and kinematic deployments of the ensemble of CHVCs contain various clues regarding their characteristic distance. These clues are not compatible with a location of the ensemble within the Galaxy proper. The deployments resemble in several regards those of the Local Group galaxies.

We describe a model testing the hypothesis that the CHVCs are a Local Group phenomenon, principally populating extensive halos around M 31 and the Milky Way. The agreement of the model with the data is judged by

H. van Woerden et al. (eds.), High-Velocity Clouds, 313–340

extracting the observables from simulations, in a manner consistent with the sensitivities of the observations and explicitly taking account of galactic obscuration. We show that models in which the CHVCs are the H I counterparts of dark-matter halos evolving in the Local Group potential provide a good match to the observables, if account is taken of tidal and ram-pressure disruption, the consequences of obscuration due to Galactic H I, and of differing sensitivities and selection effects pertaining to the surveys. Only a small fraction of the CHVCs which would be concentrated near M 31 would have been seen with *LDS* sensitivities. We describe testable predictions against which the model can be judged, using new data with sensitivity equal to or better than that of the *HIPASS* material. The predictions have largely been verified by recent *GBT* and *WSRT* observations made in the M 31 field by Thilker et al. (2004) and Braun & Thilker (2004), respectively.

A representative sample of CHVCs has been studied with high (sub-arcminute) angular resolution using the *WSRT*, and with high N(H I) sensitivity ($<10^{17}$ cm^{-2}) using the *Arecibo* telescope. The picture that emerges indicates a nested morphology of cold neutral medium (CNM) cores shielded by warm neutral medium (WNM) cocoons, plausibly surrounded by a warm ionized medium (WNM) halo. These observations lead to indirect constraints on the distances to a number of individual CHVCs, ranging from 150 to 850 kpc.

1. Introduction

H I structures found lying in the velocity regime harboring the high-velocity clouds manifest themselves in different forms; considered collectively, these clouds either do not represent a single astronomical phenomenon, or (more likely) do not represent the phenomenon at a single stage in its evolution, with all members seen under the same physical circumstances. The variety of positional and kinematic properties may correspond to differing evolutionary histories. The Magellanic Stream, for example, is identified as associated with the Magellanic Clouds because of positional and kinematic coincidences of part of the Stream with the Clouds. The Stream appears to be constrained to follow the orbit of the Clouds, and thus may lie at distances between some 50 and 150 kpc (Gardiner & Noguchi 1996). There are several other major streams of HVCs, notably complexes M, C, and A, which can also be traced over many tens of degrees and which, like the Magellanic Stream, individually show substantial spatial and kinematic coherence. The distance of only one complex has been bracketed with any certainty: absorption-line measurements (van Woerden et al. 1999; Wakker 2001) toward complex A place it within the range $4<D<10$ kpc. The histories of the Magellanic Stream and of complex A (and probably of the other large complexes as well) have been multifaceted, with exposure to strong radiation fields and tidal distortions by the Milky Way.

Braun & Burton (1999) identified the class of compact, isolated, high-velocity H I clouds, CHVCs, as a subset of the anomalous-velocity gas which might be characteristic of a single class of objects, whose members plausibly originated under common circumstances and share a common subsequent evolutionary history. CHVCs are sharply bounded in angular extent, with no kinematic or spatial connection to other H I features down to a limiting column density of 1.5×10^{18} cm^{-2}. The discussion of the CHVCs hypothesized that the large HVC complexes (other than the Magellanic Stream) were once similar objects, but are now at relatively close range.

We describe in Sect. 2 the automated search algorithm developed by de Heij et al. (2002b) and applied by them to the Leiden-Dwingeloo Survey north of δ=$-28°$ and by Putman et al. (2002) to the *Parkes HIPASS* data south of δ=0°, resulting in an all-sky catalog numbering 246 CHVCs. We argue that these objects represent a more uniform physical situation than would a sample which included any of the major HVC complexes. In Sect. 3 five principal observables are defined for the CHVC population: (1) the spatial deployment of the objects on the sky, (2) the kinematic distribution, (3) the number distribution of observed H I column densities, (4) the number distribution of angular sizes, and (5) the number distribution of line widths. We show that the spatial and kinematic deployments of the ensemble of CHVCs contain various clues regarding their characteristic distance. These clues are not compatible with a location of the ensemble within the Galaxy proper. In Sect. 4, we show that a model in which the CHVCs are the H I counterparts of dark-matter halos evolving in the Local Group potential provides a good match to the observables, but only if the simulation is sampled as if it were being observed with the differing sensitivities and selection effects of the available observations, if explicit account is taken of obscuration by H I in the Galaxy, and if account is taken of tidal- and ram-pressure disruption. These considerations lead to predictions regarding the limitations of the *LDS* for the detection of a population of CHVCs clustered near M 31; these predictions have been borne out by the CHVC analogs found in the outer halo of M 31 in higher-sensitivity data from the *Green Bank* Telescope (*GBT*) and the Westerbork Synthesis Radio Telescope (*WSRT*), reported by Thilker et al. (2004) and Braun & Thilker (2004), respectively. In Sect. 5, we show *WSRT* and *Arecibo* high-resolution imaging of a selection of individual CHVCs, stressing the importance of the synthesis observations to study of the cold cores and of the single-dish observations to study the diffuse halos. The high-resolution imaging supports several indications of substantial distances.

2. An all-sky CHVC catalog extracted from the *LDS* and *HIPASS*

Obtaining an adequate observational foundation for the anomalous-velocity H I has been a persistent and continuing challenge. The Braun & Burton (1999) sample was obtained by visual inspection of the Leiden-Dwingeloo Survey (*LDS*, Hartmann & Burton 1997). The *LDS* was observed with the 25-m *Dwingeloo* telescope, whose FWHM beam subtends 36′, and covered the sky north of $\delta = -30°$ on a $0\overset{\circ}{.}5$ grid. The effective velocity coverage of the *LDS* spans $-450 < v_{\rm LSR} < +400$ km s^{-1}, resolved into channels 1.03 km s^{-1} wide. The nominal 1-σ brightness-temperature sensitivity is 0.07 K.

The Parkes All-Sky Survey (*HIPASS*, Barnes et al. 2001) advanced the observational foundations of HVC work. The *HIPASS* material was Nyquist-sampled at the angular resolution of $15\overset{\prime}{.}5$ afforded by the *Parkes* 64-m telescope, covering the sky south of $\delta = 0°$ and reaching a 1-σ rms brightness-temperature sensitivity of approximately 10 mK, in spectral channels 26 km s^{-1} wide. The *HIPASS* spectra were searched for HVC signatures over the range $-700 < v_{\rm LSR} < +500$ km s^{-1}. The point-source sensitivity, angular resolution, and extent of velocity coverage of the *HIPASS* material are superior to those of the *LDS*, but the velocity resolution of the *LDS* is superior to that of the *HIPASS*. The column density sensitivity to emission filling the telescope beam is approximately the same for both surveys, corresponding to an rms of about 0.5×10^{18} cm^{-2}. We comment below on the inhomogeneities introduced in an all-sky CHVC catalog by the differing observational parameters of the two surveys.

An automated algorithm for extracting anomalous-velocity H I clouds from survey material has been developed by de Heij et al. (2002b), and applied by them to the *LDS* and by Putman et al. (2002) to the *HIPASS*. The algorithm sought objects isolated in position and in velocity, down to N(H I) values below about 1.5×10^{18} cm^{-2}. Initial application of the algorithm resulted in a list of all objects satisfying the search criteria, and thus included not only compact high-velocity clouds, but also structures that are part of HVC complexes, the intermediate-velocity features, and even the gaseous disk of our Galaxy, as well as features which were subsequently eliminated as due to excessive noise, radio interference, the non-square response of the receiver bandpass, or other imperfections in the data.

To avoid possible contamination from emission associated with our Galaxy and the intermediate-velocity complexes, all clouds which did not satisfy a deviation-velocity constraint were removed from the preliminary catalog. Wakker (1991) defined $v_{\rm DEV}$ as the excess velocity of a feature compared to velocities allowed by a simple model of our Galaxy. In order to constrain possible Milky Way contamination more accurately than would be possible with a constant limit on $v_{\rm DEV}$, we modified the definition, allowing the limiting velocity to vary with direction, in order to account for the warp of the Milky Way H I disk, for its flare to increasing thickness with increasing galactocentric distance, and for its lopsidedness in plan view. The gas-layer model was derived from the Voskes & Burton (1999) analysis of combined southern and northern H I survey data. De Heij et al. (2002c) describe how discrimination against CHVCs in the "H I zone of avoidance" (which, unlike the optical counterpart, is velocity-dependent as well as position-dependent) influences the subsequent analysis: the swath of this zone of avoidance is distorted after coordinate transformation from the $v_{\rm LSR}$ to the $v_{\rm GSR}$ or $v_{\rm LGSR}$ frames, corresponding to the Galactic Standard of Rest or the Local Group Standard of Rest, respectively, which are shown by the kinematics to be more relevant to the CHVC ensemble.

The degree of isolation of the clouds extracted using the algorithm was determined from velocity-integrated images covering an area measuring 10° by 10° centered on the position and velocity of the cloud in question. An ellipse was fit to the contour with a value of half the maximum brightness of the cloud to allow tabulation of size and orientation. The degree of isolation was assessed on the basis of the lowest significant contour level of N(H I) commensurate with the data sensitivity, corresponding to about 1.5×10^{18} cm^{-2} over the typical 25 km s^{-1} FWHM line width in both the *LDS* and *HIPASS* surveys. We demanded that this contour be closed, with its greatest radial extent less than the 10° by 10° image size, and that there not be confusion from any nearby extended emission near our cut-off column density. Both the degree of isolation and the total angular size (measured at our cut-off level) will therefore be dependent on data sensitivity. While unfortunate, this is an inevitable limitation of real data. A distinction was made between the cleanest isolated objects, CHVCs, and the somewhat ambiguous candidates, CHVC:s and CHVC?s, which had some degree of confusing emission in their 10°×10° environment. While isolated objects as large as about 8° could have been found by the de Heij et al. (2002b) algorithm, the distribution is strongly peaked with a median angular size of only 1° and a maximum FWHM of 2.°2. This small median angular size motivates the use of the term "compact" in the object designation. It is not ruled out that some of the features which we tabulated as isolated in l, b, v space would be shown under scrutiny of more sensitive data to be embedded in a weaker envelope, or even to be part of a large, but relatively weak, complex or stream. And it is indeed expected that deeper observations will result in the detection of additional CHVCs: the modeling discussed by de Heij et al. (2002c) leads to specific predictions in this regard.

Independent confirmation was required for all of the CHVCs in the *LDS* portion of the all-sky catalog. The *LDS* is not free of a pernicious type of radio interference (RFI), which occasionally produced a broad, weak spectral feature, of the same general form as the signature of a high-velocity cloud, or of a nearby galaxy. This RFI was sometimes of short temporal duration, not long enough that a persistent telltale spoor in a single frequency band would reveal the signal as spurious. Confirmation is all the more important because the isolated objects being sought appear commonly at only a few *LDS* lattice points, or even at only one. Confirmation was not required in the *HIPASS* case, because that survey accumulated independent data in the multi-beam feeds. All of the *LDS* CHVCs were confirmed either from published data or from new observations made with the *WSRT*, operating in a total-power observing mode whereby auto-correlation spectra were obtained from all 14 individual 25-meter antennas, rather than the more usual cross-correlation spectra.

The all-sky catalog of CHVCs includes 67 found in the *LDS* at $\delta>0°$ by de Heij et al. (2002b) and 179 found in the *HIPASS* material at $\delta<0°$ by Putman et al. (2002). The data on CHVCs found by both surveys in the overlap zone $-30°<\delta<0°$ are recorded in the all-sky catalog using the *HIPASS* parameters. The detection statistics in the overlap zone supports discussion of the relative completeness of the two surveys.

3. Principal morphological observables of the ensemble of CHVCs

3.1. OBSERVED SPATIAL DEPLOYMENT OF THE CHVCS

Figure 1 shows the distribution on the sky of the CHVCs in the combined *LDS* and *HIPASS* catalog. The CHVCs show no obvious tendency to cluster in streams or complexes. In particular, there is no preference in the sky distribution for an accumulation near the galactic equator, toward the inner Galaxy, or in association with any of the known HVC complexes or the Magellanic Stream. Unlike the situation pertaining for the principal HVC complexes, which show a substantially higher total H I flux in the northern galactic hemisphere than in the southern, the CHVCs in the all-sky catalog are more numerous in the southern hemisphere than in the northern. The overdensity in the southern hemisphere is primarily due to the higher sensitivity of the *HIPASS* material to small objects compared to that of the *LDS*; but, as we mention briefly below and discuss in more detail in de Heij et al. (2002c), a slight enhancement remains after taking the observational parameters into account. The locations of Local Group galaxies tabulated by Mateo (1998) are plotted in Fig. 1 as filled circles for comparison.

3.2. OBSERVED KINEMATIC DEPLOYMENT OF THE CHVCS

The longitude distribution of the CHVC motions, measured with respect to the GSR frame, is shown in the upper panel of Fig. 2; the latitude distribution of v_{GSR} is shown in the middle panel. Also plotted is the velocity/longitude distribution of the galaxies in the Local Group. The CHVCs are confined within a definite kinematic envelope, narrower in extent than the spectral coverage of the surveys; we stress below that this confinement is not a selection effect.

The CHVC velocities can be considered in different reference frames. The mean velocity for the all-sky ensemble of CHVCs measured in the LSR frame is -45 km s^{-1}; although not clearly centrally peaked, the dispersion of the v_{LSR} kinematics is formally 238 km s^{-1}. In the GSR frame, the mean velocity decreases to -63 km s^{-1}, with a dispersion of 128 km s^{-1}; in the Local Group Standard of Rest frame, the mean velocity is -60 km s^{-1}, with a dispersion of 112 km s^{-1}. The mean velocities of the 27 dwarf galaxies in the Local Group with known radial velocities tabulated by Mateo (1998) are -57, -22, and $+4$ km s^{-1}, in the LSR, GSR, and LGSR reference frames, respectively, with the corresponding values of the dispersion being 196, 104, and 78 km s^{-1}. Although the dispersion of CHVC velocities is formally somewhat less in the LGSR frame than in the GSR one, it will require more study before the significance of this difference can be judged.

3.3. OBSERVED SIZE AND COLUMN DENSITY DISTRIBUTIONS

In addition to the spatial and kinematic distributions, the measured CHVC sizes, line widths, and column densities are also important observables against which interpretations must be judged. The distributions of CHVC sizes and column densities are shown in the lower panels of Fig. 2. Although the only limit on angular size we have imposed is the $10° \times 10°$ dimension of our initial N(H I) image of each candidate, the size distribution is nevertheless strongly peaked with a median value of less than 1° FWHM, and furthermore does not extend beyond 2.°2. Sharply bounded anomalous-velocity H I objects apparently are, indeed, compact. The distribution of velocity FWHM is strongly peaked at about 25 km s^{-1}, although the distribution does extend out to 100 km s^{-1}. Most CHVCs are spatially unresolved in the *LDS* beam, and many are unresolved in the *HIPASS* beam; although only some 20 CHVCs have been imaged at higher angular resolution, the indications are that the broadest velocity widths are contributed by differing systemic velocities of several individual cores, each with a relatively narrow width, constituting a single CHVC.

The characteristic N(H I) shown in Fig. 2 is also of interest. The significance of H I features which are isolated in column density down to a level as low as 1.5×10^{18} cm^{-2} is that this is about an order of magnitude lower than the critical column density identified at the edges of nearby galaxies (Maloney 1993; Corbelli & Salpeter 1993), $\sim 2 \times 10^{19}$ cm^{-2}, where the ion-

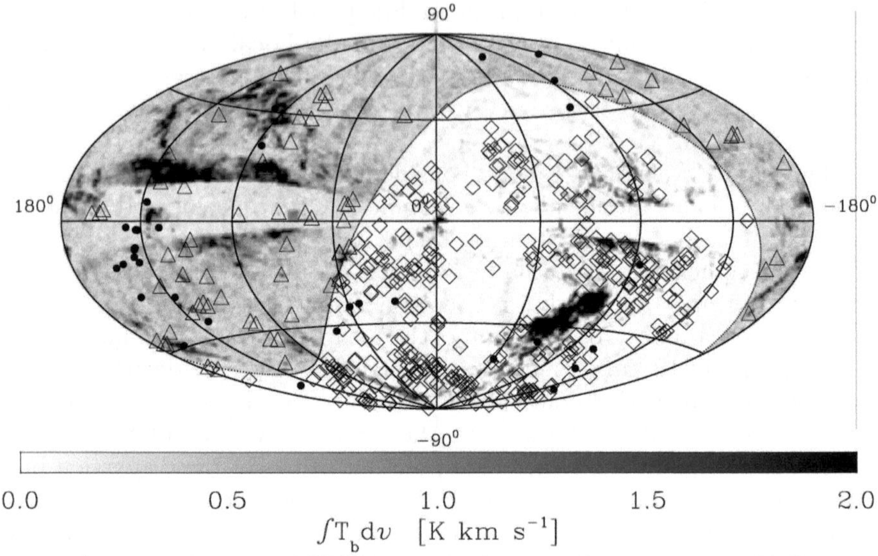

Figure 1. Spatial deployment of CHVCs over the sky. Triangles represent the *LDS* sample cataloged by de Heij et al. (2002b) at $\delta>0°$; diamonds represent the *HIPASS* sample of Putman et al. (2002) at $\delta<0°$. Filled circles correspond to the galaxies comprising the Local Group, as tabulated by Mateo (1998). The background greys show the H I column densities from an integration of observed temperatures over velocities ranging from $v_{\rm LSR}=-450$ km s^{-1} to $+400$ km s^{-1}, but excluding emission with $|v_{\rm DEV}|<70$ km s^{-1}; the generally-grey background in the *LDS* regime reflects the higher noise signature there compared to that of the *HIPASS* regime. The larger number of southern CHVCs, and specifically the apparent demarcation in the source densities at $\delta=0°$, are largely due to the differing responses of the two data sets, as discussed briefly in Sect. 4.2 and accounted for in our modeling.

ized fraction is thought to increase dramatically due to the extragalactic radiation field. These isolated objects will need to provide their own shielding to ionizing radiation. In this case, their small median angular size is consistent with substantial distances, since the partially-ionized H I skin in a power-law ionizing photon field has a typical exponential scale length of 1 kpc (Corbelli & Salpeter 1993).

3.4. COMMENTS ON SAMPLE COMPLETENESS AND HOMOGENEITY

Our CHVC tabulation is probably not incomplete as a consequence of the velocity-range limits of the observations. Although the *LDS* was searched only over the range $-450<v_{\rm LSR}<+350$ km s^{-1}, there are indications that we did not miss many (if any) clouds because of this limited interval. The feature with the most extreme negative velocity found in the *LDS* is CHVC 110.6−07.0−465, discovered by Hulsbosch (1978); the Wakker & van Woerden (1991) tabulation, which relied on survey data covering the

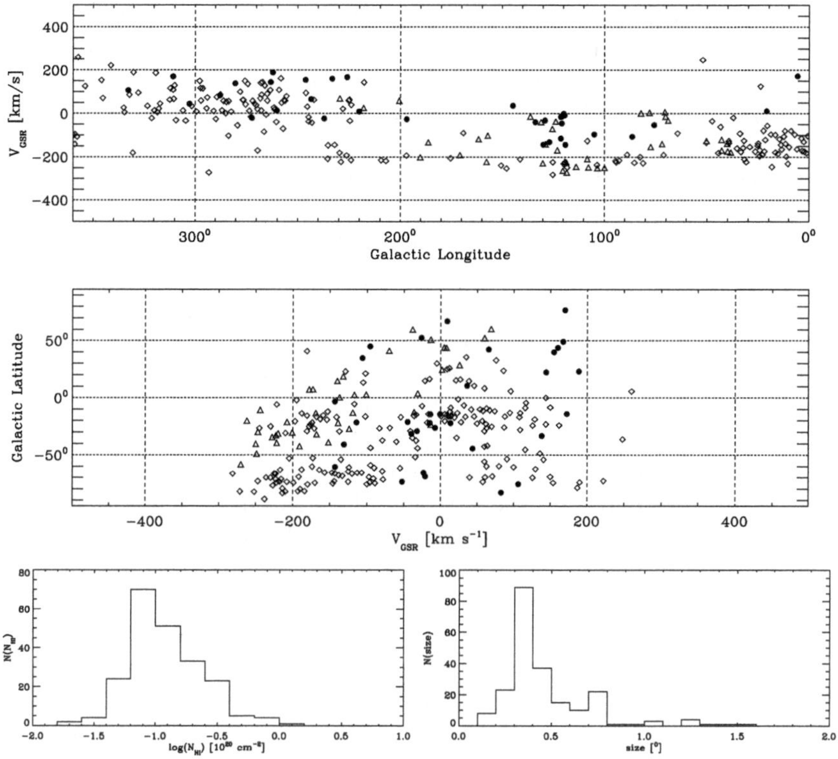

Figure 2. Observed kinematic deployment and size and column density distributions of the CHVC ensemble, from de Heij et al. (2002c). These properties, together with the spatial deployment plotted in Fig. 1, constitute the principal observables of the ensemble. The upper two panels show the variation of $v_{\rm GSR}$ plotted against galactic longitude, and against latitude, respectively. Triangles represent CHVCs identified at $\delta > 0°$ in the *LDS*, and diamonds represent CHVCs found in the *HIPASS*; filled circles show the kinematic distribution of the Local Group galaxies listed by Mateo (1998). The histograms in the bottom row show the observed peak H I column density distribution, on the left, and the observed FWHM angular size distribution of the CHVC population detected in the combined *LDS* and *HIPASS* samples, on the right.

range $-900 < v_{\rm LSR} < +750$ km s^{-1}, found no high-velocity cloud at a more negative velocity. The M 31 CHVC population recently studied by Thilker et al. (2004) using the *GBT* found faint objects out to -515 km s^{-1}; Braun & Thilker (2004), using the *WSRT*, found CHVC 115.4−23.1−503, a discrete feature in the M 31 field at -503 km s^{-1}. The *HIPASS* search by Putman et al. (2002) sought anomalous-velocity emission over the range $-700 < v_{\rm LSR} < +500$ km s^{-1}. Ten of the CHVCs cataloged by Putman et al. have $v_{\rm LSR} < -300$ km s^{-1}, but the most extreme negative velocity is only -353 km s^{-1}, for CHVC 125.1−66.4−353.

In terms of the Local Group deployment variously discussed by Blitz et al. (1999), Braun & Burton (1999), Putman & Moore (2002), and de Heij et al. (2002c), the most extreme negative velocities would be found in the general region of the barycenter of the Local Group, whereas the most extreme positive LSR velocities, which would also be more modest in amplitude than the extreme negative velocities, would be expected in the general region of the Local Group anti-barycenter. Only 7 of the *HIPASS* CHVCs have $v_{\rm LSR}$ greater than +300 km s^{-1}, and only one has a velocity greater than +350 km s^{-1}, namely CHVC 258.2−23.9+359. All of the 7 CHVCs with substantial positive velocities ($v_{\rm LSR}$>+300 km s^{-1}) lie deep in the third longitude quadrant, or in the early fourth, and thus in the general vicinity of the anti-barycenter.

In view of these detection statistics, it seems unlikely that the velocity-range limits either of the *LDS* or of the *HIPASS* have caused a significant number of CHVCs to be missed. Thus, the true velocity extent, as well as the non-zero mean in the LSR frame, of the anomalous-velocity ensemble appear well represented by the extrema of −515 km s^{-1} and +359 km s^{-1}.

It is appropriate to comment on certain selection effects which might cause systematic distortions to the true distribution of CHVCs. A demarcation at δ=0° separates the CHVC catalog entries based on the *LDS* from those based on *HIPASS*. The lower point-source sensitivity of the *LDS* is likely to have discriminated against some CHVCs. To the extent that the objects at positive and negative LSR velocities are not evenly distributed over positive and negative declinations, this may introduce some bias in the kinematic distribution. There is also a mild asymmetric discrimination against detecting CHVCs due to their submersion in the "H I zone of avoidance", i.e. near $v_{\rm LSR}$=0 km s^{-1}, with the velocity discrimination slightly skewed to positive $v_{\rm GSR}$-values in the GSR frame, as discussed by de Heij et al. (2002c).

If the *HIPASS* and *LDS* CHVC catalogs are compared or if they are used together, for example to investigate the all-sky properties of the CHVC ensemble, then due attention should be given to the higher expected detection rate in the southern material, because of its greater sensitivity to small objects and its denser angular sampling. To correct for the higher detection rate in the south, de Heij et al. (2002c) considered the likelihood that a given cloud which is observed in the *HIPASS* data also will be observed in the *LDS*. This issue could be directly confronted, since the two surveys overlap in the declination range $-28°\leq\delta\leq+2°$.

The *HIPASS* 1-σ rms noise figure is 10 mK, for a channel 26 km s^{-1} wide and a FWHM beam of 15′; the corresponding *LDS* rms value is 70 mK, for a channel width of 1.03 km s^{-1} and a FWHM beam of 36′. After smoothing both surveys to the same spectral resolution of 26 km s^{-1}, the 1-σ limiting column density (for emission filling each beam) is 0.47 and 0.64×10^{18} cm^{-2}

in the *HIPASS* and *LDS*, respectively. Thus, while the sensitivity to well-resolved sources is comparable, the point-source sensitivity of *HIPASS* is greater by about a factor of 8. On the other hand, the band-pass calibration of the *HIPASS* data relies on reference spectra, deemed empty of H I emission, that are offset by only a few degrees on the sky. Even with the MINMED5 method of baseline determination employed by Putman et al. (2002), there is significant filtering of extended emission, which complicates the assessment of object isolation down to a low column density limit.

The velocity resolution of the Leiden-Dwingeloo survey is 1.03 $\mathrm{km\,s^{-1}}$ and this resolution should suffice, as all known CHVCs have substantially broader velocity widths. Although the FWHM line width of most of the CHVCs tabulated here is greater than 20 $\mathrm{km\,s^{-1}}$, a few are considerably narrower, and it is not unexpected that some would be missing from the *HIPASS* compilation because they are diluted by the relatively coarse *HIPASS* velocity resolution.

The differences in point-source sensitivity on the one hand and sensitivity to very extended structures on the other, results in a larger number of faint-source detections in *HIPASS*, but also in a different designation for some of the brighter clouds which the surveys have in common. Comparison of the results derived by applying the search algorithm to both surveys allows assessment of the robustness of the selection criteria and of the completeness of the *LDS* catalog. The comparison discussed by de Heij et al. (2002c) shows the influence of survey sensitivity on the number of objects found. Above a peak temperature of 0.45 K, the *LDS* results are as complete as those based on *HIPASS*; between T_{peak}=0.20 K and 0.45 K, the *LDS* results recovered 83% of the clouds found in the *HIPASS* data. The completeness of the *LDS* catalog drops rapidly at lower values of the peak temperature: *HIPASS* clouds fainter than 0.20 K are almost completely absent from the *LDS* catalog. The incompleteness at low peak temperatures will be more important for very compact CHVCs.

3.5. INITIAL CONCLUSIONS FROM THE ENSEMBLE OBSERVABLES

Some general conclusions may be drawn from an initial inspection of the all-sky CHVC catalog. We noted above the roughly even distribution over the sky, the constrained kinematics within a definite envelope, the net infall toward the Galaxy, and the substantially smaller velocity dispersion in the GSR and LGSR frames than in the LSR one.

The entries in the CHVC catalog show spectral FWHM values ranging from 5.9 $\mathrm{km\,s^{-1}}$, characteristic of a very narrow, cold H I feature in the conventional gaseous disk, to 95.4 $\mathrm{km\,s^{-1}}$, characteristic of some external galaxies. Interferometric observations of some of the broad objects have shown that the large width is contributed by individual cores within a common envelope. But the distinction between the H I properties of CHVCs and dwarf galaxies with very weak star formation, or proto-galaxies with no star formation, remains to be made. The broadest object found in the Braun & Burton (1999) search for CHVCs was revealed as a low-surface-brightness spiral galaxy (Burton et al. 1999). Warned by the presence of this interloper, we searched the Digital Sky Survey in the direction of each of the CHVCs listed in the *LDS* portion of the catalog, but found no optical counterparts to the objects listed; a recent deeper search by Simon & Blitz (2002) also found no optical counterparts. The narrowest widths, even without higher-resolution imaging, serve to meaningfully constrain the kinetic temperature of the gas, as well as the line-of-sight component of any rotation or shear in a single object, and the range of kinematics if the object should be an unresolved collection of subunits moving at different velocities.

CHVCs near the galactic equator display the horizontal component of their space motion. The middle panel of Fig. 2 shows that the radial motions at low $|b|$ are as large as those at high latitudes, and furthermore that the CHVC distribution does not avoid the galactic equator. Large horizontal motions are difficult to account for in terms of a Galactic Fountain (Shapiro & Field 1976; Bregman 1980). Burton (1997) has noted that anomalous-velocity clouds do not contaminate the H I terminal-velocity locus in ways which would be expected if they pervaded the Galactic Disk, and that this observation constrains the clouds either to be an uncommon component of the Milky Way Disk, confined to the immediate vicinity of the Sun, or else to be typically at large distances beyond the Milky Way Disk. We note also that the lines of sight in the directions of each of the low-b CHVCs traverse some tens of kpc of the disk before exiting the Milky Way: unless one is prepared to accept these CHVCs as boring through the conventional disk at hypersonic speeds (for which there is no evidence), and atypical in view of the cleanliness of the terminal-velocity locus, then their distances are constrained to be large.

Similarly, CHVCs located near the galactic poles offer unambiguous information on the vertical, z, component of their space motion. The vertical motions are substantial, with recession velocities approximately equal in number and amplitude to approach velocities; the vertical motions are, furthermore, of approximately the same amplitude as the horizontal ones. This situation also is incompatible with the precepts of the fountain model, which predicts negative z-velocities for material returning in a fountain flow. Furthermore, the values of the z-velocities are predicted not to exceed the velocity of free fall, some 200 $\mathrm{km\,s^{-1}}$. In fact, vertical motions of twice the free fall value are observed. It would be interesting to investigate if some of the CHVC objects are even moving with velocities in excess of a plausible value of the Milky Way escape velocity.

The aspects of the spatial and kinematic topology of the class mentioned above are difficult to account for if the CHVCs are viewed as a Milky Way population, in particular if they are viewed as consequences of a Galactic Fountain (see also Blitz 2001). These same aspects would seem to discourage revival of several of the earlier Milky Way mechanisms (reviewed, for example, by Oort 1966) suggested for the general HVC phenomenon, including ejection from the Galactic Nucleus, association with a galactic spiral arm at high latitude, and ejection following a nearby supernova explosion.

4. The CHVC ensemble modeled as a Local Group population, with the model sampled as if observed

The spatial and kinematic distributions of the all-sky CHVC ensemble are consistent with a dynamically cold population at a substantial distance from the Galaxy and with a significant infall velocity with respect to either or both of the Galaxy and the Local Group barycenter. Braun & Burton (2000) suggested that the CHVCs might represent the dark-matter halos predicted by Klypin et al. (1999) and Moore et al. (1999) in the context of the hierarchical-structure paradigm of galactic evolution. These halos would contain no stars, or only a few; most of their visible matter would be in the form of atomic hydrogen. Although many of the halos would already have been accreted by the Galaxy or M 31, a substantial number would still populate the Local Group; most of these would be concentrated around the two dominant Local Group galaxies, although some would be located in the far field. Those passing close to either the Milky Way or M 31 would be ram-pressure stripped of their gas and tidally disrupted by the gravitational field. Near the Milky Way, the tidally-distorted features would correspond to the HVC complexes observed.

4.1. DESCRIPTION OF THE LOCAL GROUP MODEL

Recently several simulations have been performed to test the hypothesis that the CHVCs are the remaining building blocks of the Local Group. Putman & Moore (2002) compared the results of the full n-body simulation described by Moore et al. (2001) with various spatial and kinematic properties of the CHVC distribution, as well as with properties of the more general HVC phenomenon, without regard to object size and degree of isolation. Putman & Moore (2002) were led to reject the Local Group deployment of CHVCs, for reasons which we debate below. Blitz et al. (1999) had earlier performed a restricted two-body analysis of the motion of clouds in the Local Group. In their attempt to model the HVC distributions, Blitz et al. (1999) modeled the dynamics of dark-matter halos in the Local Group, and found support for the Local Group hypothesis when compared qualitatively with the deployment of a sample of anomalous-velocity H I containing most HVCs, but excluding the large complexes and the Magellanic Stream, for which plausible or measured distance constraints are available.

De Heij et al. (2002c) judge the results of their simulations against the properties of the all-sky CHVC sample, viewing the simulated data *as if it were observed* with the *LDS* and *HIPASS* surveys. Simulations were calculated for a range of models, characterized by the following parameters. The spatial distribution of test particles was varied between different degrees of concentration around the Galaxy and M 31. Object positions were chosen in agreement with the assumed spatial density distribution of the ensemble, but otherwise randomly. Velocities were given by the velocity field illustrated in Fig. 3. This field was obtained, as described in de Heij et al. (2002c), from a separate set of simulations of motions in the expanding Universe, in which fixed values were assumed for the Hubble constant and the age of the Universe, and various values for the M/L ratio of galaxies surrounding the Local Group; the velocity field illustrated follows if the M/L ratio of galaxies is adjusted to yield the current relative velocities of M 31 and the Milky Way. The H I mass of a test cloud was randomly set in agreement with a power-law mass distribution between a specified upper mass limit and a lower mass limit. The physical size and line width of each object follow from the choice of the power-law index, β. Once all these parameters were set, we determined the observed peak column density and angular size. Objects in the northern hemisphere were convolved with a beam appropriate to the *LDS*, while those at $\delta<0°$ were convolved with the *HIPASS* beam. Simulated objects were considered detected if the peak brightness temperature exceeded the detection threshold of the relevant survey, i.e. the *LDS* for objects at $\delta>0°$, and *HIPASS* otherwise. Furthermore, in order for a test object to be retained, its deviation velocity was required to exceed 70 $\mathrm{km\,s^{-1}}$ in the LSR frame. In addition, each simulated cloud was judged whether or not it was stable against both tidal disruption and ram-pressure stripping by the Milky Way and M 31; if not, it was removed from further consideration. A cloud was regarded stable against ram-pressure stripping if the gas pressure at its center exceeded the ram pressure, $P_{\rm ram}=n_{\rm halo}V^2$, for a cloud moving with velocity V through a gaseous halo with density $n_{\rm halo}$. A cloud was considered tidally disrupted if the gravitational tidal field of either the Galaxy or M 31 exceeded the self-gravity of the cloud at its location. We continued simulating additional objects following this prescription until the number of detected model clouds was equal to the number of CHVCs in our observed all-sky sample.

A wide range of parameter combinations was simulated, spanning populations tightly concentrated around the Galaxy to essentially uniformly

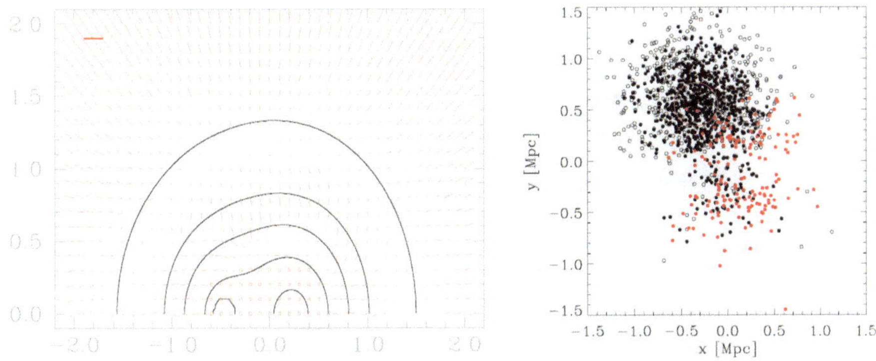

Figure 3. *Left panel:* Velocity field assumed in the Local Group model of CHVC deployment described in the text, from de Heij et al. (2002c). The Milky Way is located at $x=-0.47$ Mpc, $y=0.0$ Mpc, and M 31 at $x=0.23$ Mpc, $y=0.0$ Mpc. The best-fitting distribution of test particles representing the CHVCs is given by two gaussian functions, one centered on the Milky Way and the other on M 31; the contours show schematically the relative density levels of the CHVC test particles. The vectors give the average velocity of test particles in a 10×10 kpc box centered on each grid point (the length of the thick-line vector in the upper left corner corresponds to 200 km s^{-1}). Squares are drawn if the velocity dispersion of the particles within the box exceeds 100 km s^{-1}.
Right panel: Distribution of synthetic CHVCs shown in an (x,y) projection, for the Local Group simulation characterized by the best-fitting parameters described in the last paragraph of Sect. 4.1. The Galaxy is located at the origin; a red circle locates M 31. A symbol marks the location of each of the input CHVCs entering the simulation. But not all of the input clouds can exist in the simulated environment: black symbols indicate those input clouds which are destroyed by tidal and ram-pressure stripping influences of M 31 and the Galaxy. Furthermore, not all surviving input clouds are sufficiently massive to be detected by surveys with the properties of the *LDS* or *HIPASS*: open symbols indicate surviving input clouds which would be too weak to be detected by the *LDS* or by *HIPASS*, respectively, depending on their declination. Red symbols indicate those input clouds which would be perceived as CHVCs if the simulation were sampled as if observed by the *LDS* and *HIPASS*.

distributed throughout the Local Group. The agreement of each simulation with the data was assessed by the simultaneous fit to all of the principal observables noted previously. The best-fitting populations have a maximum H I mass of 10^7 M$_\odot$, a slope of the H I mass distribution in the range -1.7 to -1.8, and a gaussian dispersion for their spatial distributions of between 150 and 200 kpc centered on both the Milky Way and M 31. Given its greater mean distance, only a small fraction of the M 31 sub-population is predicted to have been detected in the *LDS* or earlier, even less sensitive, surveys.

4.2. SIMULATED CHVC DISTRIBUTIONS, PERCEIVED AS IF OBSERVED BY THE *LDS* OR *HIPASS*

An appreciation of the physical appearance of these Local Group models is provided by the right-hand panel of Fig. 3, showing a perpendicular projec-

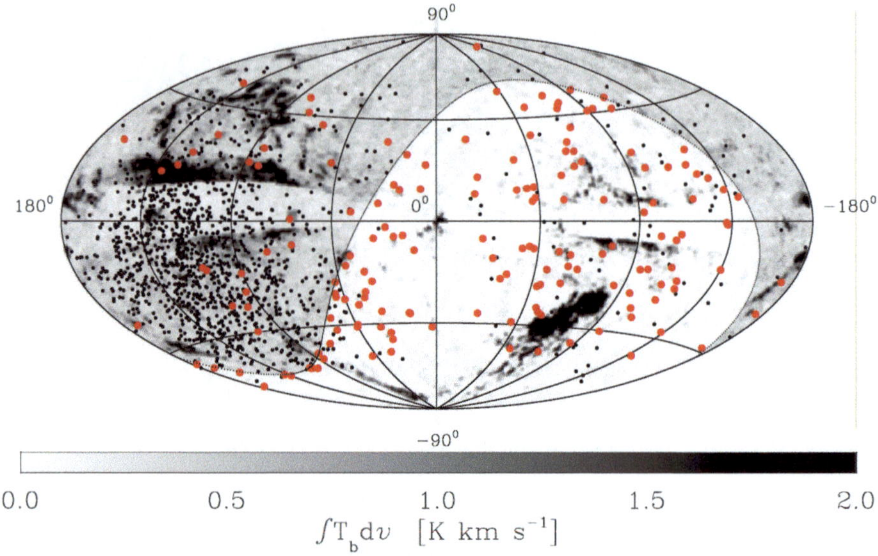

Figure 4. Sky deployment of synthetic CHVCs corresponding to the simulation of a population in the Local Group with the parameters described in the text and by de Heij et al. (2002c). Black dots correspond to input clouds which would not enter a catalog of CHVCs based on the *LDS* or *HIPASS* data, either because the input clouds have been destroyed (tidally, or by ram-pressure stripping) as a consequence of their proximity to M 31 or to the Galaxy, or else because their perceived flux would be below the detection thresholds of the *LDS* or *HIPASS*, depending on their declination. Red symbols indicate clouds which would be detected were the simulation sampled as if by the *LDS* or *HIPASS*. The background grey image, drawn as in Fig. 1, shows the demarcation between the lower-sensitivity *LDS* regime and the higher sensitivity *HIPASS* regime. Many of the input clouds are clustered around M 31 and the region of the barycenter of the Local Group, and lie thus in the northern regime of the sky observed by the *LDS*, where the simulated observations are relatively less sensitive. The magnitude of the overdensity of red dots in the southern hemisphere does not reflect the true distribution of input CHVCs, but the combined effects of the simulated physical environment and the observational parameters conforming to the real situation. The sky deployment of the simulated CHVC population can be compared with the observed situation shown in Fig. 1.

tion of the model population. The (x,y) plane in the figure is the extended galactic plane, with the Galaxy centered at $(x,y)=(0,0)$ and the positive z axis corresponding to positive b. The intrinsic distribution of objects is an elongated cloud encompassing both the Galaxy and M 31, dominated in number by the M 31 concentration. The objects that have at some point in their history approached so closely to either of these galaxies that their H I would not survive the ram-pressure or tidal stripping are indicated by the filled black circles. Cloud disruption was substantially more important in the M 31 concentration than in the Galaxy concentration. The objects that are too faint to have been detected by the *LDS* or *HIPASS* observations, depending on declination, are indicated by open circles. *The bulk of*

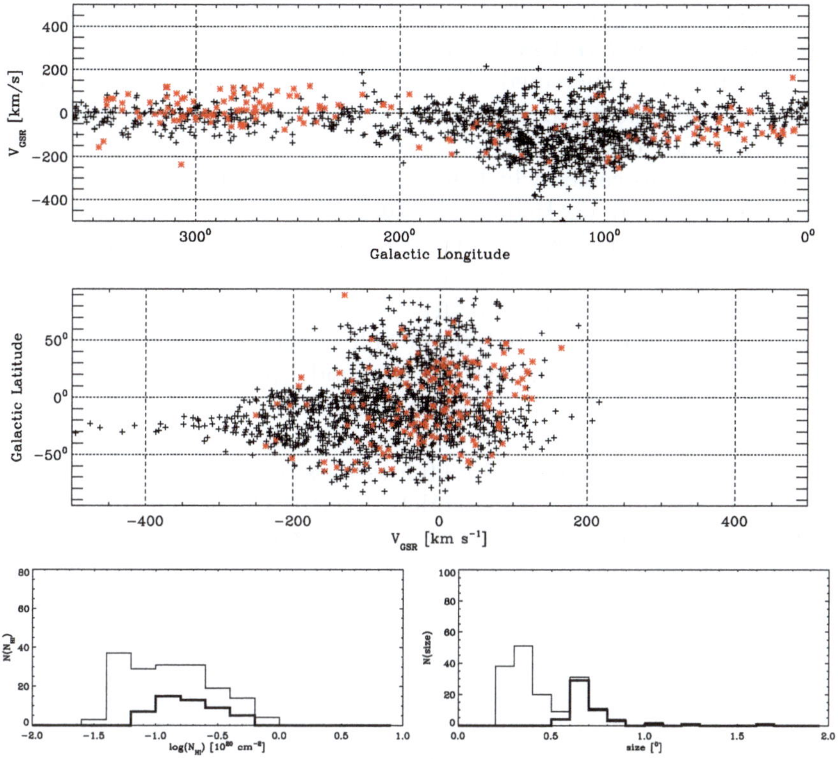

Figure 5. Kinematic deployment and perceived size and column density distributions of the simulated CHVC ensemble in the Local Group, plotted for comparison with the observed distributions shown in Fig. 2. The two upper panels show the variation of $v_{\rm GSR}$ with galactic longitude and latitude, respectively. Black symbols refer to input clouds which do not survive in the Local Group environment or which do not present sufficient flux to be detected by the Leiden-Dwingeloo or Parkes surveys; red symbols indicate clouds which would be detected if the input ensemble were sampled with the observational parameters characterizing the *LDS* and *HIPASS* data. The two panels in the bottom row show the peak N(H I) distribution which is perceived in the simulation, on the left, and the perceived angular size distribution of the simulated CHVC population, on the right. The thick-line histograms indicate the *LDS* contributions to the total detections at $\delta > 0°$.

the M 31 sub-concentration is not detected in our CHVC simulation for two reasons: (1) these objects have a larger average distance than the objects in the Galactic sub-concentration, and (2) the M 31 sub-concentration, around l=120°, b=−20°, is located primarily in the northern celestial hemisphere, where the lower *LDS* sensitivity compromises detection. The objections posed by Putman & Moore (2002) to the Local Group deployment models, based on the failure to observe a clustering of CHVCs near M 31 or the Local Group barycenter, seem accounted for by the crucial role of survey sensitivity in determining what is seen of such Local Group cloud populations. The red symbols in Figs. 4 and 5 indicate objects detectable with the relevant *LDS* or *HIPASS* sensitivities, while the black symbols indicate those that remain undetected. The *LDS*, in particular, is not sufficiently sensitive to have detected the majority of CHVCs at distances of more than a few hundred kpc.

The various processes which influence the perceived distributions are further quantified by de Heij et al. (2002c). About three quarters of the simulated populations were classified as disrupted by ram-pressure or tidal stripping, while 80% of the remaining objects were deemed too faint to detect with the *LDS* (in the north) or *HIPASS* (in the south). Obscuration by Galactic H I eliminated about half of the otherwise detectable objects. The total H I mass involved in the model populations shown here is 4.3×10^9 $M_\odot$. About 75% of this mass had already been consumed by M 31 and the Galaxy via cloud disruption, leaving only 25% still in circulation, although distributed over some 1200 low-mass objects.

If the de Heij et al. (2002c) models describe the actual distribution of CHVCs, then the prediction follows that future deeper surveys will detect large numbers of objects at high negative LSR velocities in the general vicinity (about $60^\circ\times60^\circ$) of M 31. To make this prediction more specific, we have imagined the sensitivity afforded by the current *HIPASS* survey in the south extended to the entire northern hemisphere. A high concentration of about 250 faint, newly detected CHVCs is predicted in the Local Group barycenter direction once *HIPASS* sensitivity is available, for example from the current Jodrell Bank *HIJASS* effort (Kilborn 2002; Lang et al. 2003).

One of the most suggestive attributes of the CHVC population in favor of a Local Group deployment is the modest concentration of objects which are currently detected in the general direction of M 31, near l=120°, b=−20°, i.e. not far from the direction of the barycenter of the Local Group (see Fig. 1). These objects have extreme negative velocities in the GSR reference frame (see Fig. 2). While this is a natural consequence of the Local Group models, it does not follow from models simulating a distribution throughout an extended Galactic Halo, nor is it a consequence of obscuration by Galactic H I.

4.3. DETECTION OF A CHVC POPULATION SURROUNDING M 31

The prediction made by de Heij et al. (2002c) that higher-sensitivity H I data would reveal CHVC analogs surrounding M 31 has been borne out by recent *GBT* and *WSRT* observations of the Andromeda field (Thilker et al. 2004; Braun & Thilker 2004). Figure 6 shows the Thilker et al. (2004) results for the high-velocity H I in the M 31 *GBT* field, with H I emission of M 31 itself masked out. The *GBT* observations reveal some 20 discrete features

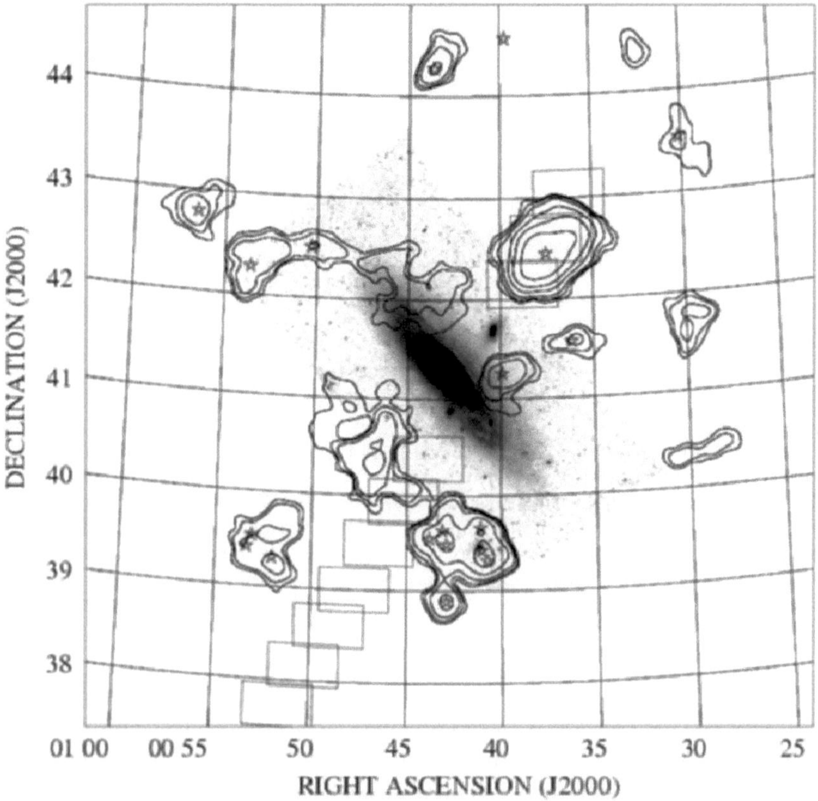

Figure 6. Total H I column density observed in the field of M 31 with the *GBT* by Thilker et al. (2004). H I emission from Andromeda's rotating disk has been masked out; a V-band optical image of the galaxy is superposed. Contours show $N(\mathrm{H\,I})$ at levels of 0.5, 1, 2, 10, and 20×10^{18} cm^{-2}. The star symbols indicate the location of discrete clouds.

in a $7^\circ\times7^\circ$ region around M 31 which are judged to be likely analogs of the CHVC ensemble found clustered around the Milky Way. The *GBT* data show H I masses in the range 10^5 to 10^7 $M_\odot$, with the majority of detections near the limiting flux density. Braun & Thilker (2004) used the *WSRT* array in auto-correlation mode to survey a field extending 60° in R.A. and 30° in declination, centered near the direction of M 31 and including the direction to M 33. The rms column density sensitivity of 4×10^{16} cm^{-2} was an order of magnitude better than that of earlier survey work. As in the *GBT* data, a mix of discrete and diffuse anomalous-velocity gas was detected at velocities consistent with association with M 31. In the large *WSRT* field, some 95 discrete CHVCs were found. The *GBT* and *WSRT* data confirm the extragalactic counterparts of the CHVC ensemble distributed in the distant halo of the Milky Way.

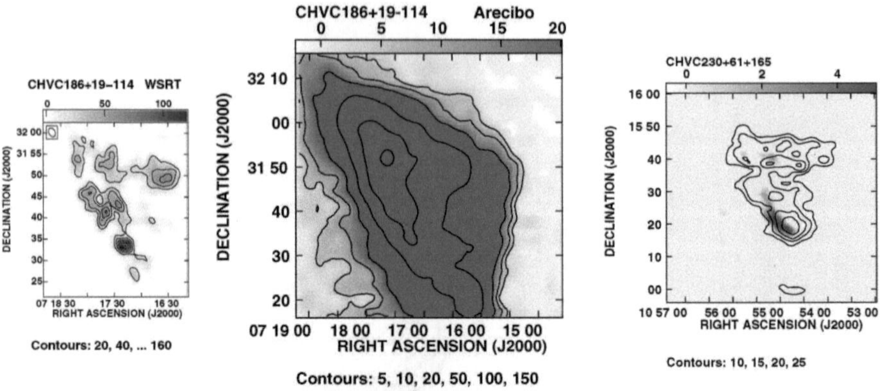

Figure 7. Nested core/halo morphology which characterizes those few CHVCs which have been observed at high resolution by both interferometric and filled-aperture instruments. The two left-hand panels show CHVC 186+19−114 as seen with the *WSRT* and *Arecibo*, respectively; the right-hand one shows an overlay for CHVC 230+61+165, with contours for the *Arecibo* H I data, and the greyscale images for the *WSRT* data. The cores are better revealed in synthesis data than in filled-aperture data because of their small angular scales; the halos, on the other hand, are better revealed in filled-aperture data than in synthesis data, because of their diffuse morphology and large angular scales. Typically only a few tenths of the total flux observed in a single-dish spectrum is recovered in synthesis data.

5. Properties of individual CHVCs imaged at high resolution

The all-sky CHVC catalog is based on survey data of modest angular resolution: most CHVCs are largely unresolved by the 36′ beam of the *Dwingeloo* telescope, undersampled on a 0°.5 lattice, although some details in individual CHVCs are revealed by the 15′ resolution of the Nyquist-sampled *HIPASS* material. Wakker & Schwarz (1991) used the *WSRT* to image two CHVCs; subsequently Braun & Burton (2000) imaged six additional CHVC fields, and de Heij et al. (2002a) six more. The angular and kinematic resolution afforded by the *WSRT* suit it well in important regards to detailed studies of the CHVC class of objects. Diffuse structures extending over more than about 10′ are, however, not adequately imaged by the interferometer, unless precautions are taken to account for the missing short spacings. Burton et al. (2001) observed several *WSRT* CHVC fields with the *Arecibo* telescope, and showed that a combination of high-resolution filled-aperture and synthesis data is crucial to determining the intrinsic properties of the CHVCs. This is especially important for CHVC targets which have sizes comparable to the field of view of most synthesis instruments.

5.1. NESTED CORE/HALO MORPHOLOGY

The CHVCs imaged with the *WSRT* at resolutions of about 1′ show a characteristic morphology, in which one or more low-dispersion (FWHM line widths in the range 2 to 10 $\mathrm{km\,s^{-1}}$) cores are distributed over some tens of arcmin: the larger FWHM seen in the single-dish spectra reflects the differing systemic velocities of the cores, but is dominated by the intrinsically-large velocity dispersion of the diffuse gas comprising the halo. Figure 7 shows combined *WSRT* (from Braun & Burton 2000) and *Arecibo* data (from Burton et al. 2001) for two CHVCs. The left-hand panel shows that CHVC 186+19−114 comprises a half-dozen cores, embedded in a common halo. The unresolved FWHM of 20 $\mathrm{km\,s^{-1}}$ seen in the single-dish spectrum represents the diffuse halo gas as well as modest differences in the systemic velocities of the narrow-dispersion cores. The right-hand panel shows a similar situation for CHVC 230+61+165: several cores, resolved in the *WSRT* data, are embedded in a common diffuse envelope seen in the *Arecibo* map, with a FWHM of 26 $\mathrm{km\,s^{-1}}$, well represented by a single gaussian.

The combined *WSRT*/*Arecibo* data thus indicate a morphology of one or more cores in the cool neutral medium (CNM) phase, surrounded by a diffuse halo of warm neutral medium (WNM). Such a core/halo morphology was found earlier in some extended HVCs, e.g. by Giovanelli et al. (1973) in detailed H I mapping using the *Green Bank* 300-ft telescope. The FWHM line width of the CHVC halo gas is typically 25 $\mathrm{km\,s^{-1}}$, consistent with the expected WNM thermal line width of 10^4 K gas. The FWHM of the cores is consistent with CNM temperatures of 100 to 500 K. Some of the cores show a kinematic gradient consistent with rotation. The halos do not share the kinematic gradients of the cores, however, implying that an external origin of the core gradient, e.g. by tidal shear, is unlikely.

The *Arecibo* data have also shown that the halo edge profiles are exponential, and can be followed to the limiting $N(\mathrm{H\,I})$ sensitivity of about 2×10^{17} $\mathrm{cm^{-2}}$. This exponential form implies that measurements of apparent angular sizes and total flux densities of CHVC halos will depend on the resolution as well as on the sensitivity of the data; larger areal filling factors will be returned by deeper observations (see Lockman et al. 2002). The exponential edges also suggest that the outer envelopes of CHVCs are not tidally truncated, consistent with the objects residing at substantial distances from the Milky Way. The sources studied by Burton et al. (2001) indicate a characteristic central halo column density of $4.1\pm3.2\times10^{19}$ $\mathrm{cm^{-2}}$, and a characteristic exponential scale length of $420''\pm90''$. For plausible values of the thermal pressure at the core/halo interface, these edge profiles support distance estimates which range between 150 and 850 kpc.

5.2. ROTATION OF CORES

Under higher resolution than afforded by the *LDS* or *HIPASS*, the characteristic width of individual CHVC components narrows as objects are

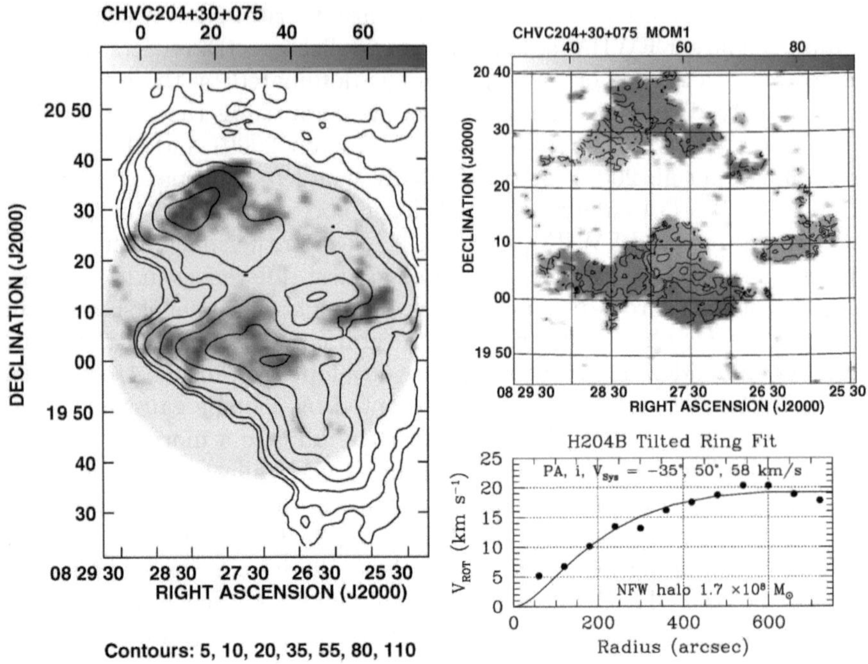

Figure 8. *Left:* Overlay of *WSRT* and *Arecibo* H I data (from Braun & Burton 2000 and Burton et al. 2001, respectively) for CHVC 204+30+075, showing several cores embedded in a common diffuse halo. Seen unresolved in the 36′ beam of the 25-m *Dwingeloo* telescope, this source has a velocity FWHM of 34 $\mathrm{km\,s^{-1}}$. *Upper right:* Intensity-weighted line-of-sight velocity, derived from *WSRT* data at an angular resolution of 1′. Contours of v_{LSR} show systematic kinematic gradients across the two principal components of the CHVC, consistent with rotation; contours are drawn in steps of 5 $\mathrm{km\,s^{-1}}$ from 40 to 85 $\mathrm{km\,s^{-1}}$. *Lower right:* Rotation velocities resulting from the tilted-ring model applied to the kinematic gradient of the lower-declination component of CHVC 204+30+075. The best-fit position, inclination, and systemic velocity are indicated. The line shows the rotation curve corresponding to a Navarro et al. (1997) cold dark matter (CDM) halo of the indicated mass.

resolved into several principal units, moving relative to each other. If the objects with multiple cores are to be stable, distances of order several hundred kpc are required. Some of the compact cores imaged owe their large total width seen in single-dish data to velocity gradients. Several of the CHVC cores resolved with the *WSRT* exhibit kinematic gradients consistent with rotation. Figure 8 shows that the resolved *WSRT* image of CHVC 204+30+075 has two principal components, each of which is elongated and shows, furthermore, a systematic velocity gradient along the major axis. The velocity gradients exhibited by the two principal components of CHVC 204+30+075 can be modeled in terms of circular rotation in a flattened disk system. Figure 8 shows the results of fitting a standard tilted-ring model to one of the cores. The fit shows velocity rising slowly but continuously with radius to an amplitude of some 20 km s^{-1}, and then flattening to a constant value beyond about 500 or 600 arcsec. Estimates of the contained dynamical mass follow from the rotation curve if the distance is assumed, and the total gas mass follows from the integrated H I flux. At an assumed distance of 0.7 Mpc (see Braun & Burton 2000), the two principal clumps of CHVC 204+30+075 have $M_{\rm dyn} = 10^{8.1}$ and $10^{8.3}$ M$_\odot$, and gas masses (including H I and helium of 40% by mass) of $10^{6.5}$ and $10^{6.9}$ M$_\odot$, respectively, for the upper and lower concentrations shown in Fig. 8. The dark-to-visible mass ratios for these concentrations are 36 and 29, respectively. The shape of the modeled rotation curves for both of the CHVC 204+30+075 components resembles that of the standard cold-dark-matter halo presented by Navarro et al. (1997). At the assumed distance of 0.7 Mpc, the Navarro et al. (1997) halos fitted to the two components have masses of $10^{7.8}$ M$_\odot$ (within 9.3 kpc) and $10^{8.2}$ M$_\odot$ (within 12.6 kpc).

A similar analysis has recently been applied by Robishaw et al. (2002) to a compact H I cloud lying near the Local Group dwarf galaxy LGS 3. The cloud and the dwarf are apparently interacting, leading Robishaw et al. (2002) to assign a plausible distance of 0.7 Mpc to the cloud; they then derive a dynamical mass for the rotating high-velocity cloud of 2×10^7 M$_\odot$.

5.3. A VERY COLD CORE

CHVC 125+41−207 (WW 84) is representative of the compact objects in several regards. The *WSRT* image of CHVC 125+41−207 observed by Braun & Burton (2000) and shown in Fig. 9 reveals several cool, quiescent cores embedded in a diffuse, warmer halo. The spectrum plotted in the lower left of the figure was observed toward the brightest of these cores: it has a line width unresolved by the 2 km s^{-1} resolution of the *WSRT* imaging. The velocity channels adjacent to the line peak have intensities below 20% of the maximum value. Such a width is one of the narrowest measured in H I emission – H I line widths are almost always dominated by line blending and mass motions – and constrains both the kinetic temperature and the amount of turbulence within the core. An upper limit to the thermal-broadening FWHM of 2 km s^{-1} corresponds to an upper limit to the kinetic temperature, T_k, of 85 K. As it happens, the physical situation is yet more tightly constrained, because the brightness temperature T_B in this core was observed to be 75 K; thus a lower limit to the opacity τ follows from $T_B = T_k\ (1 - e^{-\tau})$, yielding $\tau \geq 2$. Any broadening which might be due to macroscopic turbulence is less than 1 km s^{-1}. Such quiescence would be most unusual in a Milky Way environment.

A fortuitous property of this CHVC is that it lies near the direction of the Seyfert galaxy Mrk 205, located at $(\alpha, \delta) = (12^{\rm h}21^{\rm m}44\overset{\rm s}{.}5, +75°18'40'')$.

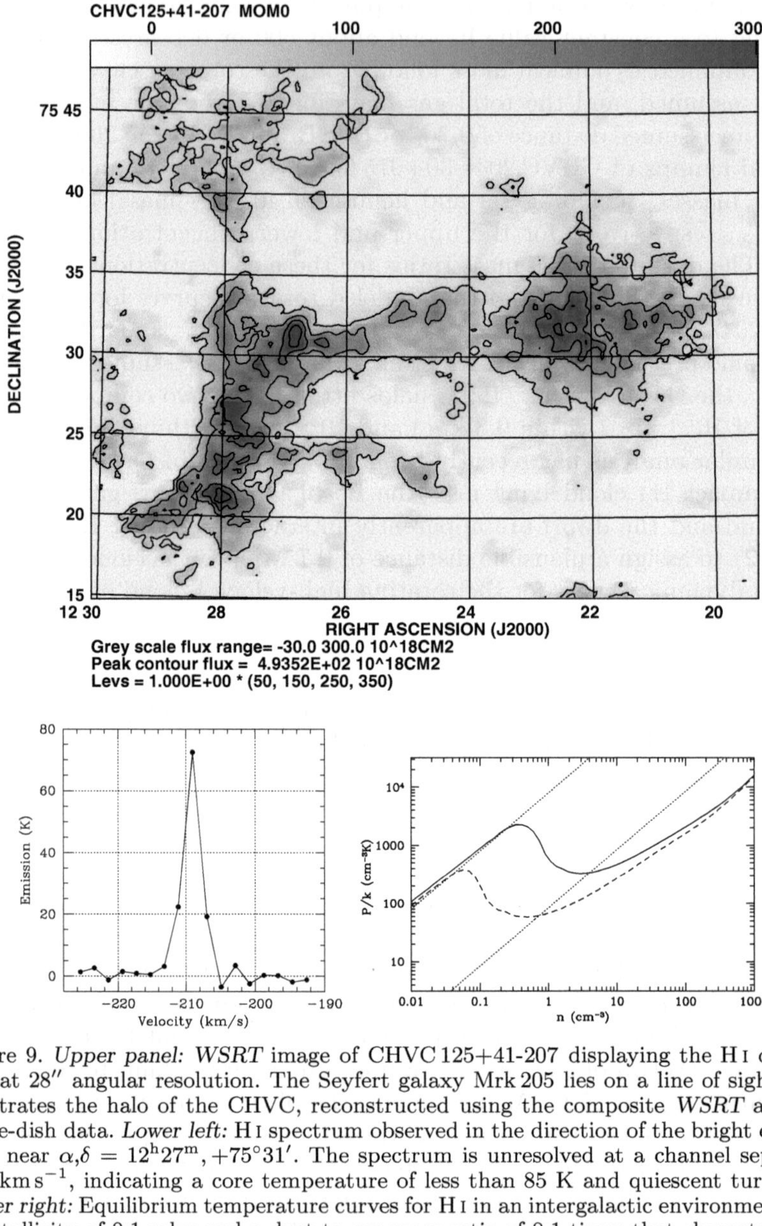

Figure 9. *Upper panel: WSRT* image of CHVC 125+41-207 displaying the H I distribution at 28″ angular resolution. The Seyfert galaxy Mrk 205 lies on a line of sight which penetrates the halo of the CHVC, reconstructed using the composite *WSRT* and *LDS* single-dish data. *Lower left:* H I spectrum observed in the direction of the bright emission knot near $\alpha,\delta = 12^h27^m, +75^\circ31'$. The spectrum is unresolved at a channel separation of 2 $km\,s^{-1}$, indicating a core temperature of less than 85 K and quiescent turbulence. *Lower right:* Equilibrium temperature curves for H I in an intergalactic environment (with a metallicity of 0.1 solar and a dust-to-gas mass ratio of 0.1 times that characteristic of the solar neighborhood), shown for two values of the neutral shielding column density, namely 10^{19} cm^{-2} (solid curve) and 10^{20} cm^{-2} (dashed curve). The 85 K kinetic temperature tightly constrains the volume density; a distance of order 500 kpc then follows from the measured column density and angular size. Figure from Braun & Burton (1999).

Bowen & Blades (1993) measured Mg II absorption toward Mrk 205, which allows determination of a metallicity along this line-of-sight which is substantially subsolar. (In general, the small angular size of CHVCs, and the amount of substructure being revealed at high resolution, will render it difficult to find suitable background sources to support metallicity measurements.) Wolfire et al. (1995) show that a cool H I phase is stable under extragalactic conditions if a sufficiently high column density of shielding gas is present and if the thermal pressure is high. Calculations of equilibrium conditions which would pertain in the Local Group environment, characterized by a metallicity of 10% of the solar value and a dust-to-gas ratio of 10% of that in the solar neighborhood, were communicated to us by Wolfire, Sternberg, Hollenbach, and McKee. The tightly constrained temperature found for CHVC 125+41−207 then allowed Braun & Burton (2000) to estimate the distance to this object. Figure 9 shows, as the solid and dashed curves respectively, two bracketing values of the shielding column density, namely 10^{19} cm^{-2} and 10^{20} cm^{-2}. The diagonal dotted lines in this figure correspond to fixed kinetic temperatures of 8000 and 85 K. The equilibrium volume densities $n_{\rm HI}$ corresponding to the observed value of T_k=85 K lie in the range 0.65 to 3.5 cm^{-3}. Provided with this range of volume densities, and having measured both the column density, N(H I), of the cool core and its angular size θ, the distance to CHVC 125+41−207 follows directly from $D=N(\mathrm{H\,I})/(n_{\rm HI}/\theta)$ (assuming crude spherical symmetry), yielding a value in the range 210 to 1100 kpc.

5.4. CHVCS AS GASEOUS TRACERS OF DARK-MATTER SATELLITES

The CHVC analogs found near M 31 by Thilker et al. (2004) in *GBT* observations and by Braun & Thilker (2004) in *WSRT* data can be assigned plausible distances because of their evident association with M 31, and these distances lead to plausible H I masses. Figure 10 shows the plot made by Thilker et al. (2004) of the observed H I masses versus FWHM line widths for the discrete clouds shown in Fig. 6. The line widths increase systematically with increasing H I mass, in a manner consistent with a dark-matter component with a total-mass to H I mass ratio of 100:1. The detected number of H I clouds of a given line width within the currently-surveyed region is also fully consistent with the expectations for H I counterparts of dark mini-halos in the numerical simulations of Klypin et al. (1999) and Moore et al. (1999) as parameterized by Sternberg et al. (2002).

6. Concluding remarks

The concept of a distinct class of compact, isolated high-velocity clouds has been objectively developed by application of a search algorithm to

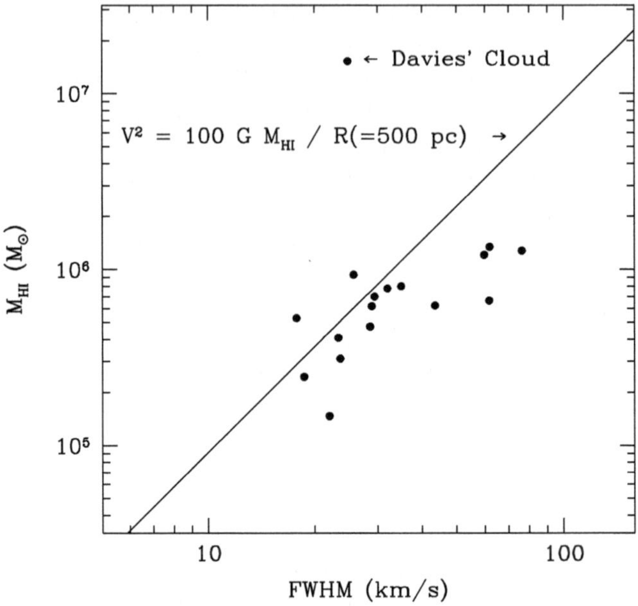

Figure 10. H I mass plotted against FWHM line width for the discrete clouds found by Thilker et al. (2004) in their *GBT* observations of the M 31 field. The line shows the relationship of H I mass and line width expected for gravitationally bound clouds, assuming a ratio of dark matter to H I mass of 100:1 and a characteristic size of 0.5 kpc for each cloud core.

the *LDS* and *HIPASS* datasets, resulting in an all-sky catalog of CHVCs, high-contrast H I features which are at best only marginally resolved with half-degree angular resolution and which are sharply bounded, such that emission at column densities above a few times 10^{18} cm^{-2} is confined to less than about 2° diameter. We identify five principal observables from the all-sky catalog: the spatial and kinematic deployments, and the number distributions of angular sizes, N(H I), and line width. We show that agreement of models with the data must be judged by extracting these same observables from the simulations in a manner consistent with the sensitivities of the observations, and explicitly taking into account the limitations imposed by obscuration due to foreground H I in the Milky Way. General considerations of these observables show compatibility with characteristic distance scales of hundreds of kpc.

Since the inception of high-velocity cloud research, the possibility of an extragalactic deployment of these clouds has been critically considered as a possibility; among others, the discussions by Oort (1966, 1970, 1981), Verschuur (1969, 1975), Giovanelli (1981), Bajaja et al. (1987), Wakker & van Woerden (1997), Blitz et al. (1999), and Braun & Burton (1999) are particularly relevant here. Our simulations of a Local Group population of

CHVCs are treated as if real observations were involved. We have shown that models in which the CHVCs are the H I counterparts of dark-matter halos evolving in the Local Group can provide a good match to the observables, if the simulations are "observed" in accordance with the *LDS* and *HIPASS* parameters. Only a small fraction of the CHVCs which would be concentrated near M 31 would be seen with *LDS* sensitivities. De Heij et al. (2002c) describe testable predictions against which the models may be confronted when new data, e.g. from the *HIJASS* survey, provide about the same velocity coverage, angular resolution, and sensitivity in the north as the *HIPASS* effort has provided in the south. Meanwhile, the predictions have been largely verified by the *GBT* and *WSRT* data in the M 31 fields observed by Thilker et al. (2004) and Braun & Thilker (2004), respectively.

A representative sample of CHVCs has been studied with high angular resolution (sub-arcminute) using the *WSRT* and with high N(H I) sensitivity ($<10^{17}$ cm^{-2}) using the *Arecibo* telescope. The picture that emerges is a nested morphology of CNM cores shielded by WNM cocoons, and plausibly surrounded by a WNM halo. The cores contribute typically about 40% of the H I flux, while covering about 15% of the surface of the CHVC. The imaging also supports three independent forms of distance estimates. A distance for CHVC 125+41−207 follows from assuming rough spherical symmetry and equating the well-constrained volume and column densities of the compact cores. A second type of distance constraint (coupled with a dark-to-visible mass ratio) follows from consideration of the stability of CHVCs having multiple cores with large relative velocities in a common envelope. Similar constraints follow from an interpretation of the large observed velocity gradients in some cores as arising from rotation. A third distance indication follows from the measured scale length of the WNM halos, together with an assumed thermal pressure at the core/halo interface. These observations lead to indirect constraints on the distances, ranging from 150 to 850 kpc. Thus the evidence available indicates that CHVCs have characteristic sizes of about 2 kpc and maximum H I masses of about 10^7 $M_\odot$. The failure to detect stars would imply that CHVCs might be primitive sub-galactic objects dominated by dark-matter halos, plausibly the missing Local Group satellite systems predicted by Klypin et al. (1999) and Moore et al. (1999).

References

Bajaja, E., Morras, R., Pöppel, W.G.L. 1987, Pub. Astr. Inst. Czech. Ac. Sci., 69, 237
Barnes, D.G., Staveley-Smith, L., de Blok, W.J.G., et 36 al. 2001, MNRAS, 322, 486
Blitz, L. 2001, in ASP Conf. Ser. 240, Gas and Galaxy Evolution, eds. J.E. Hibbard, M. Rupen, J.H. van Gorkom (San Francisco: ASP), 469
Blitz, L., Spergel, D.N., Teuben, P.J., Hartmann, D., Burton, W.B. 1999, ApJ, 514, 818
Bowen, D.V., Blades, J.C. 1993, ApJ, 403, L55

Braun, R., Burton, W.B. 1999, A&A, 341, 437
Braun, R., Burton, W.B. 2000, A&A, 354, 853
Braun, R., Thilker, D.A. 2004, A&A, 417, 421
Bregman, J.N. 1980, ApJ, 236, 577
Burton, W.B. 1997, in Proc. 156th WE-Heraeus Seminar Bad Honnef, The Physics of the Galactic Halo, eds. H. Lesch, R.-J. Dettmann, U. Mebold, R. Schlickeiser (Berlin: Akademie Verlag), 15
Burton, W.B., Braun, R., Walterbos, R.A.M., Hoopes, C.G. 1999, AJ, 117, 194
Burton, W.B., Braun, R., Chengalur, J.N. 2001, A&A, 369, 616
Corbelli, E., Salpeter, E.E. 1993, ApJ, 419, 104
de Heij, V., Braun, R., Burton, W.B. 2002a, A&A, 391, 67
de Heij, V., Braun, R., Burton, W.B. 2002b, A&A, 391, 159
de Heij, V., Braun, R., Burton, W.B. 2002c, A&A, 392, 417
Gardiner, L.T., Noguchi, M. 1996, MNRAS, 278, 191
Giovanelli, R. 1981, AJ, 86, 1468
Giovanelli, R., Verschuur, G.L., Cram, T.R. 1973, A&AS, 12, 209
Hartmann, D., Burton, W.B. 1997, Atlas of Galactic Neutral Hydrogen (Cambridge: Cambridge University Press)
Hulsbosch, A.N.M. 1978, A&A, 66, L5
Kilborn, V.A. 2002, in ASP Conf. Ser. 276, Seeing through the Dust, eds. R. Taylor, T. Landecker, T. Willis (San Francisco: ASP), 80
Klypin, A., Kravtsov, A.V., Valenzuela, O., Prada, F. 1999, ApJ, 522, 82
Lang, R.H., Boyce, P.J., Kilborn, V.A., et 8 al. 2003, MNRAS, 342, 738
Lockman, F.J., Murphy, E.M., Petty-Powell, S., Urick, V.J. 2002, ApJS, 140, 331
Maloney, P. 1993, ApJ, 414, 41
Mateo, M.L. 1998, ARA&A, 36, 435
Moore, B., Ghigna, S., Governato, G., Lake, G., Quinn, T., Stadel, J., Tozzi, P. 1999, ApJ, 524, L19
Moore, B., Calcaneo-Roldan, C., Stadel, J., Quinn, T., Lake, G., Ghigna, S., Governato, F. 2001, Phys. Rev. D., 64f3508M
Navarro, J.F., Frenk, C.S., White, S.D.M. 1997, ApJ, 490, 493
Oort, J.H. 1966, BAN, 18, 421
Oort, J.H. 1970, A&A, 7, 381
Oort, J.H. 1981, A&A, 94, 359
Putman, M.E., Moore, B. 2002, in ASP Conf. Ser. 254, Extragalactic Gas at Low Redshift, eds. J. Mulchaey, J. Stocke (San Francisco: ASP), 245
Putman, M.E., de Heij, V., Staveley-Smith, L., et 29 al. 2002, AJ, 123, 873
Robishaw, T., Simon, J.D., Blitz, L. 2002, ApJ, 580, L129
Shapiro, P.R., Field, G.B. 1976, ApJ, 205, 762
Simon, J.D., Blitz, L. 2002, ApJ, 574, 726
Sternberg, A., McKee, C.F., Wolfire, M.G. 2002, ApJS, 143, 419
Thilker, D.A., Braun, R., Walterbos, R.A.M., Corbelli, E., Lockman, F.J., Murphy, E., Maddalena, R. 2004, ApJ, 601, L39
van Woerden, H., Schwarz, U.J., Peletier, R.F., Wakker, B.P., Kalberla, P.M.W. 1999, Nature, 400, 138
Verschuur, G.L. 1969, ApJ, 156, 771
Verschuur, G.L. 1975, ARA&A, 13, 257
Voskes, T., Burton, W.B. 1999, in ASP Conf. Ser. 168, New Perspectives on the Interstellar Medium, eds. A.R. Taylor, T.L. Landecker, G. Joncas (San Francisco: ASP), 375
Wakker, B.P. 1991, A&A, 250, 499
Wakker, B.P. 2001, ApJS, 136, 463
Wakker, B.P., Schwarz, U.J. 1991, A&A, 250, 484
Wakker, B.P., van Woerden, H. 1991, A&A, 250, 509
Wakker, B.P., van Woerden, H. 1997, ARA&A, 35, 217
Wolfire, M.G., Hollenbach, D., McKee, C.F., Bakes, E.L.O. 1995, ApJ, 443, 152

16. THE ORIGIN OF THE HIGH-VELOCITY CLOUDS

JOEL N. BREGMAN
Department of Astronomy, University of Michigan, Ann Arbor, USA; jbregman@astro.lsa.umich.edu

Abstract. There are three leading models for the origin of the high-velocity clouds of neutral hydrogen: galactic fountains, material lost from dwarf galaxies, and the ancient remnants of galaxy formation. The tidal-stripping picture has strong observational support in that the Magellanic Stream is connected to the Magellanic Clouds and such phenomena are seen in other galaxy groups. Also, absorption in at least two sightlines through the Stream shows approximately the metallicity expected for Magellanic Cloud material. The Magellanic Clouds dominate the H I content of the ensemble of smaller galaxies in the Local Group. So, although other Local Group dwarf galaxies may have had their neutral gas removed, it would be a minor component compared to that of the Magellanic Stream. One unsolved issue in this area is whether the presence of other high-velocity gas complexes could be due to gas stripped in previous orbits of the Clouds past the Milky Way.

Galactic fountains are another formation mechanism of intermediate- and high-velocity gas, the main question being the degree of the mass flux involved. This model requires that hot gas, created by supernovae in the disk, escapes into the halo. Theoretical calculations support superbubble breakout, while X-ray, radio, and optical observations reveal the presence of halo gas extending several kiloparsecs above the plane in the Milky Way and in other spiral galaxies. Calculations predict that neutral clouds form from this hot halo gas, both on the way up and on the way down. This mechanism is probably responsible for most of the intermediate-velocity gas and for some of the high-velocity gas (within 10 kpc). We discuss the evidence that clouds have impacted the disk, and the role of cosmic rays populating the halo with hot gas.

In the third model, involving gas that never formed into galaxies, the high-velocity clouds are distributed around the Local Group at distances of up to several hundred kiloparsecs. The total H I mass in this model is comparable to if not greater than the gaseous mass within galaxies, so the

H. van Woerden et al. (eds.), High-Velocity Clouds, 341–369

clouds could be of cosmological importance. A primary prediction of this model, their large distances, has not been confirmed, but efforts examining this issue are continuing.

1. Introduction

The H I high-velocity clouds (HVCs) were discovered in 1963 as part of an exploratory program. Subsequent observational studies at 21 cm quantified the basic nature of this phenomenon and it became clear that there were significant implications for the Galaxy, as well as more general issues involving galactic evolution and the intergalactic medium (Oort 1970). In an effort to understand this phenomenon, much energy has been spent in developing models for the origin of the HVCs, along with predictions for the velocity and spatial distributions of the clouds. There are three general classes of models: a Galactic model, where clouds are produced by events within the Galaxy; a dwarf-galaxy model, where the interaction of dwarfs with the Galaxy produces HVCs; and a Local Group model, where the HVCs represent material that fell into potential wells but failed to form normal galaxies. As will be argued, at least two of these mechanisms must be occurring, and they each have important implications.

For Galactic models, such as galactic fountains, HVCs are formed in the halo due to the flow or ejection of material from the Galactic Disk. The vitality of this type of flow is determined by the rate of supernovae, the frequency with which they form in associations, the magnetic field in the gas, and the topology of the interstellar medium. Consequently, a clear understanding of this phenomenon would be revealing for a number of important issues in the interstellar medium. This mechanism naturally produces a Galactic Halo of gas, and understanding the halo gas is essential for cosmic-ray issues, since cosmic rays spend 90% of their time in the Galactic Halo.

The implications for the external models of HVCs are also very exciting. The model that involves dwarf disruption leads to the inflow of gas onto the Galaxy, possibly at a rate consistent with that necessary for the models of galactic stellar evolution. In the past decade, it has been recognized that the disruption of the stellar component of dwarf galaxies contributes significantly to the stellar content of the Galactic Bulge (and possibly other regions as well). It seems only natural that the neutral gas from dwarfs would be an important component as well. Finally, a recent and especially exciting model posits that the HVCs are part of the galaxy-formation process. In this picture, a significant fraction of gas failed to form into normal optically-bright galaxies, and instead it falls into shallow potential wells of dark matter. If this model is correct, there is more neutral gas floating around the Local Group than in galaxies. Therefore, the HVCs may be tracers of three very important processes, and sorting out which are the most likely is a worthy goal. In this chapter, we will develop the theoretical concepts for each of the models, although with particular emphasis on galactic fountains since the other models are discussed in detail in other chapters.

2. Galactic models for the HVCs

2.1. OVERVIEW

In a paper that began the development of the field, Spitzer (1956) suggested that if there were hot gas in the Galaxy, it would have a natural scale height much larger than the disk and thereby form a gaseous Galactic Halo. Spitzer's motivation for suggesting a hot halo was the need to pressure-confine cool gas clouds that had been discovered to lie in the halo of the Galaxy (Münch 1952; Münch & Zirin 1961). Following this work, the field lay dormant for nearly two decades until soft X-ray surveys by several groups led to the understanding that there is a considerable amount of 10^6 K gas in the Galaxy (Bunner et al. 1969, 1971; Henry et al. 1971; Garmire & Riegler 1972; review by McCammon & Sanders 1990). This discovery of the hot component led to the multi-phase model of the interstellar medium (ISM), in which a considerable fraction of the volume of the ISM was hot, and the primary source of energy was supernovae (McKee & Ostriker 1977; Cox 1979, 1981). In these models, supernovae were assumed to occur randomly in space and when many overlapped, a tunnel network of hot gas would be created (Smith 1977). The tunnels would connect to the halo and were the conduits by which hot gas flowed upward from the disk. This basic picture was changed substantially owing to a variety of observational and theoretical studies of superbubbles during the 1980s (Weaver et al. 1977; Tomisaka et al. 1981; Tomisaka & Ikeuchi 1986; Mac Low & McCray 1988; Mac Low et al. 1989). These superbubbles are similar to ordinary supernova remnants, but larger than can be created by a single supernova. They are produced by multiple supernovae, implying the presence of many young stars, which are seen within many bubbles in the form of OB associations. A typical superbubble is produced by approximately 20 to 100 supernovae, which is similar to the number of supernova progenitors expected in an association. If all of the Type II supernovae occur in associations, only 10 to 25% of supernovae (Type Ia) occur at random locations in space, outside of associations (van den Bergh & Tammann 1991). This understanding of clustered supernovae was brought into theoretical models for the structure of hot gas in the Galaxy (Heiles 1990; Cioffi & Shull 1991), and it became recognized that the primary flow of hot gas into the halo was more likely to occur through superbubble breakout than through a tunnel network. Even when supernovae are not strongly clustered, superbubbles still are eventually produced, although more slowly than when supernovae occur in associations (Cioffi & Shull 1991; Rosen & Bregman 1995). The flow of hot gas into the halo and the production of high-velocity H I became viewed as closely related to the evolution of these superbubble structures.

Parallel to the improvement in our understanding of multi-component models for the disk ISM, there were a series of investigations that examined the consequences of gas escaping to the halo. These studies generally sidestepped the issue of how the gas flowed into the halo, assuming either a flow rate of hot gas or an equilibrium state in the halo. Oort (1970) suggested that "superexplosions" within the disk ejected material into the halo, but the concept was not well developed. Shapiro & Field (1976) first discussed the fate of hot gas that flowed away from the disk and suffered radiative losses, which became known as the "Galactic Fountain" model. In this model, the hot buoyant gas rises into the halo, radiatively cools, and returns to the disk as cooled material. There are a few possible fates for hot gas flowing from the disk (Chevalier & Oegerle 1979; Habe et al. 1981; Kahn 1981). If the gas temperature is sufficiently high, so that it exceeds the escape temperature, it will flow away from the Galaxy as a Galactic Wind. If the gas is below the escape temperature, it remains bound to the Galaxy, and the nature of the flow depends upon the cooling timescale and the flow time (equivalently, the sound crossing time for gas in hydrostatic equilibrium). If the cooling time is long compared to the time required for gas to flow to one density scale height, the gas distribution will approach hydrostatic equilibrium before cooling becomes important. In this case, the thickness of the gaseous halo is simply given by the equations of hydrostatic equilibrium and is approximately $h = 6.5\ (T/10^6\ \mathrm{K})\ (g/10^{-8}\mathrm{cm\ s^{-2}})^{-1}$ kpc, where g is the gravitational force perpendicular to the plane and T is the temperature of the hot halo gas. The production of cool gas occurs in a quiescent medium and the velocity of the cooled gas is initially downward (having lost buoyancy as it cooled) and accelerates downward onto the disk. The other case is when the cooling time is shorter than the flow time. This leads to a gaseous halo whose height is less than the hydrostatic scale height and is given by the product of the sound speed and the cooling time. Also, since the gas cools as it rises, cooled gas is produced in the upward flow and is therefore born with positive velocity away from the disk, before stopping and falling back downward to the disk. In this limit, the cooled gas will have both positive and negative velocities.

It is rather easy to determine that in the present epoch, galactic fountains are more likely than galactic winds from spiral galaxies, and this was elegantly discussed by Kahn (1981, 1992), who compared the temperature of a mature supernova remnant or superbubble and showed it to be significantly lower than the escape temperature (modifications by cosmic rays are discussed below). Since it appeared that a Galactic Fountain was likely, provided that gas could escape into the Halo, a primary question became whether the superbubbles would break out sufficiently often to adequately fuel such a flow.

2.2. THE NECESSITY FOR SUPERBUBBLE BREAKOUT

Although it is clear that superbubbles will break out of the disk if magnetic fields are unimportant, there has been some controversy about whether breakout will occur when magnetic fields are included. Perhaps the most realistic treatment is the recent work by Tomisaka (1998 and references therein), in which he considers a gaseous disk with a magnetic field that is independent of height and one where it scales with the square root of the density, which is probably the most appropriate case. For a constant magnetic field of B=5 μG, the magnetic field has a substantial pressure (P_{mag}/k=7200 K cm^{-3}), and this can prevent superbubbles from breaking out into the halo. However, for the more realistic case where $B \propto \rho^{1/2}$, normalized so that B=5 μG at z=0, superbubble breakout occurs and the evolution is similar in time with the B=0 case. For frozen-in field lines, one expects the $B \propto \rho^{1/2}$ to be appropriate, so we conclude that, although magnetic fields alter the shape of superbubbles (longer along the field direction in the disk), they do not hinder significantly superbubble breakout.

In addition to the theoretical calculations on the subject of breakout, one can appeal directly to the observations, since it is possible to search for superbubbles extending into galaxy halos as well as gas that has escaped into the halo. In the Galaxy, there are several observations indicating that hot gas has escaped into the halo. Perhaps the most direct evidence is the presence of X-ray-emitting gas surrounding the disk. It was difficult to isolate this component prior to *ROSAT*, but the sensitive all-sky survey at several energies, along with shadowing observations of Galactic X-ray emission by neutral disk clouds (e.g. Burrows & Mendenhall 1991), has led to the identification of the hot halo. Two separate groups obtain rather similar results (Pietz et al. 1998; Snowden et al. 1998), finding that the X-ray halo has a luminosity of approximately 4×10^{39} erg s^{-1} and that it is a flattened structure. This latter result is not particularly constraining and it implies that the thickness of the X-ray emission is probably a few kpc. The observations indicate that the X-ray emission brightens toward the bulge region, indicating that the hot halo is probably denser in the inner part of the Galaxy and in the bulge. The temperature of the X-ray-emitting gas, 1 to 2×10^6 K, is well below the escape temperature, so it is gravitationally bound to the Galaxy.

There also is evidence that this hot gas in the halo has cooled, as provided by the presence of the high-ionization UV resonance lines. Some of the UV resonance lines might be produced by photoionization of gas near 10^4 K, but there is good evidence that the high-ionization lines, such as those of Si IV, C IV, and N V (e.g. Savage & Sembach 1994) are due to gas that is cooling. For the Si IV and C IV absorption lines, there was the surprising result that the ratio of the column densities of these two species was a fairly constant value. So, even if there were adequate ionizing photons, it was difficult to understand why the ionization state of the gas was nearly identical in different parts of the halo. An extremely appealing theory, suggested by Benjamin & Shapiro (1993; see also Benjamin 1994), resolves this problem. They argued that in a Galactic Fountain, gas cooling from 10^6 K would be opaque to its own radiation, causing it to "self-ionize", so that its ionization state is determined by the cooling process itself. Furthermore, they showed that the absorption strength of N V was naturally reproduced also. Emission from the C IV $\lambda\lambda$1548, 1550 lines and the O III] λ1663 line was detected by Martin & Bowyer (1990), who showed that they are consistent with being produced from a cooling Galactic Fountain with a mass-flux rate of 6 to 25 $M_\odot\,yr^{-1}$ (see also Ch. 11). Along these lines, the Far Ultraviolet Spectroscopic Explorer (*FUSE*) has detected emission from the O VI $\lambda\lambda$1031, 1037 doublet (Shelton et al. 2001). Subsequent observations have shown that the contribution from the Local Bubble is small and that all observed emission lies beyond a cold filament that is 230 pc away (Shelton 2003). The emission is consistent with originating entirely from the halo, and when combined with O VI absorption-line studies (Savage et al. 2003), electron densities may be inferred (0.005 cm^{-3}) that are similar to those inferred from pulsar dispersion measures. If the gas is near the collisional ionization temperature for O VI, the gas pressure is near 3500 K cm^{-3}, which is comparable to independent estimates for the thick disk (Shelton et al. 2001). The mass cooling rate inferred from these observations is similar to that inferred by Martin & Bowyer (1990).

In external galaxies, there is good evidence for hot gaseous halos and subsequent galactic fountains, which can be studied most easily in edge-on galaxies. Several edge-on galaxies have been detected in Hα emission (review by Rand 1998a) and three exhibit X-ray-emitting halos (NGC 891, NGC 4631, and NGC 4565), with the most pronounced case being NGC 891, which has been studied extensively (Rand 1997, 1998b).

The fairly-isolated edge-on galaxy NGC 891 ($D{\sim}10$ Mpc) is approximately the same spectral type and luminosity as the Milky Way (van der Kruit & Searle 1981), although with a slightly elevated star formation rate as judged by its far-IR luminosity (Rice et al. 1988). It is known to have a prominent and fairly smooth halo from optical emission-line and radio-continuum observations, and *ROSAT* PSPC and HRI observations revealed it to have a substantial X-ray halo with a surface brightness scale height of 100″ (FWHM; density scale height of 5 kpc; Bregman & Pildis 1994; Bregman & Houck 1997). In addition, it has a prominent optical halo (Rand et al. 1990; Keppel et al. 1991; Rand 1996, 1997), a radio-continuum halo similar in size to the X-ray and Hα halo (Allen et al. 1978; Sukumar & Allen 1991), and H I extending into the halo as well (Rupen 1991; Swaters 1996). There is even good evidence for superbubbles and chimneys that have broken into the halo, as seen in the beautiful optical extinction image of Howk & Savage (1997). Along the disk of NGC 891, the X-ray emission declines similarly to the traditional star formation measures of CO (1-0) and the submillimeter continuum (Scoville et al. 1993; García-Burillo & Guélin 1995; Guélin et al. 1993; Bregman & Houck 1997), suggesting that the escape of material into the halo is related to star formation, which is a result independently determined by comparing the far-infrared and Hα halo properties of other edge-on galaxies (Rand 1998a). It is difficult to deduce the temperature from the PSPC data, because disk emission becomes confused with emission from the halo, but the best analysis reveals a mean temperature for the gaseous component of 3.5×10^6 K. For a filling factor of unity, the characteristic pressure is $P/k{\sim}10^5$ K cm^{-3}, comparable to that of the molecular gas in the disk. The bolometric X-ray luminosity of the gas is 3.9×10^{39} erg s^{-1}, only 0.4% of the energy input by supernovae into the disk. The total hot gas mass is 4.0×10^7 M$_\odot$ and the hot gas cooling rate is 0.087 M$_\odot$ yr^{-1}, below the predictions of galactic fountain theory (Shapiro & Field 1976; Bregman 1980).

This value for the X-ray luminosity is similar to that seen in the edge-on galaxies NGC 4631 and NGC 4565, where L_X=3 to 10×10^{39} erg s^{-1} (Wang et al. 1995; Volger et al. 1995, 1996; Wang et al. 2001) and for the Milky Way, where L_X=7×10^{39} erg s^{-1} (Pietz et al. 1998). These observed X-ray luminosities are less than those produced by a fountain of 10 M$_\odot$ yr^{-1}, which, for T=10^6 K, would be L_X=1.3×10^{41} erg s^{-1}. This might seem to point to a rather weak galactic fountain in these galaxies, but it is probably an effect due to most of the energy being emitted at wavelengths that are not accessible. That is, most of the energy radiated by supernovae comes out at temperatures that are typically 0.5 to 1×10^6 K, which is easily absorbed by the neutral and warm ionized gas in the Galaxy (largely by He I and He II). We have several good examples of gas at these temperatures, such as our Local Bubble, a modest superbubble that has been reheated recently by a supernova (e.g. Cox & Reynolds 1987; McCammon & Sanders 1990). This problem of absorption of soft X-ray emission is manifest in the fact that the diffuse emission from face-on spiral galaxies (${\sim}10^{40}$ erg s^{-1}) is also about two orders of magnitude lower than the energy input from supernovae ($\sim 10^{42}$ erg s^{-1}; e.g. Fabbiano 1989). For an SNR in the disk with only 2×10^{20} cm^{-2} of neutral gas above it, the optical depth to 0.1 keV photons (T=10^6 K) is 11, so the photons are trapped. Once these photons are absorbed by the gas, eventually becoming Lyman continuum photons, they have a short mean free path and can be absorbed by dust, which reradiates the energy in the far infrared. The observed far-IR luminosities of spiral galaxies are significantly greater than 10^{42} erg s^{-1}, so it is difficult to isolate this component. Rather, one must use other tracers of cooling gas, such as emission from the O VI line, which occurs as gas passes through the 3×10^6 K range. In addition to the detection of O VI from the Milky Way, it has been detected at two locations in the halo of NGC 4631 by *FUSE* (Murphy et al. 2001; Otte et al. 2003), and subsequent observations should provide us with an independent determination of the cooling rate (preliminary results place the rate above 1 $M_\odot$ yr^{-1}). To summarize, the soft X-ray observations of the Milky Way and external galaxies place a lower limit to the mass-flux rate of a Galactic Fountain of 0.1 $M_\odot$ yr^{-1}, but the UV line emission from the Milky Way indicates that the mass flux is $\sim$10 $M_\odot$ yr^{-1}, which is what one would expect if some fraction ($\sim$10%) of the energy input from supernovae went into expelling gas from the plane and driving a fountain.

2.3. GALACTIC FOUNTAIN DYNAMICS

The dynamics of a Galactic Fountain depend upon the temperature that the gas has at breakout. One important characteristic temperature is the value that it has when the vertical isothermal scale height is comparable to the radial length scale of the disk. For a constant gravitational field, the hydrostatic scale height would be $h = 6.5\ (T/10^6\ \mathrm{K})\ (g/10^{-8}\ \mathrm{cm\ s^{-2}})^{-1}$ kpc, and the radial scale length of the disk is comparable at 10^6 K. For bound gas with temperatures comparable to or greater than 10^6 K, the gas flows upward, and if its radiative cooling time is longer than the flow time, it finds that it is not in radial equilibrium with the gravitational field for two reasons. First, the radial component of the gravitational field is lower at $z{\sim}6$ kpc than in the midplane, so a test particle would move radially outward until it found an equilibrium point. Secondly, there may be a radial pressure gradient in the halo that could drive an outward flow. Such a pressure gradient would naturally arise if the rate that gas flows into the halo is proportional to the star formation rate, which is greater interior to the solar circle.

The nature of the vertical velocity of the cooled gas depends upon its initial temperature, for a constant pressure. Gas with a flow time shorter than the radiative cooling time will have risen to the hydrostatic scale height, at which point the flow velocities have decreased; this is typical of gas whose initial temperature is $>10^6$ K. For such flows, cooled gas is formed with only a modest upward or downward velocity, but subsequently, if the gas flows ballistically, its downward (and inward) velocities increase, so there is a preference to observe material falling downward onto the plane. Since these are the clouds that obtain the greatest velocities, the model predicts that clouds with the largest downward vertical velocities and radial velocities (toward the central part of the Galaxy) formed from the hottest material in the halo, and that there should be very few clouds moving in the opposite direction (upward from the disk or outward in the Galaxy). Unfortunately, this prediction is not unique. It is similar to that expected from material falling into the Galaxy from an external source, such as a nearby dwarf galaxy. However, measurements of metallicity and distance should help to distinguish between these two possibilities.

If the gas has a temperature of 5×10^5 K or less, it cools in the upward flow, so there can be comparable amounts of cool gas rising and falling (Houck & Bregman 1990; Kahn 1994). The velocity of the gas when it cools is typically the sound speed of the original gas that began the flow, or about 50 to 100 $\mathrm{km\,s^{-1}}$. Also, the height achieved by the gas is lower, typically 0.5 to 2 kpc. Both of these characteristics are typical of the intermediate-velocity clouds of H I, and it is likely that this low-temperature component of the Galactic Fountain is responsible for producing this material. Numerical hydrodynamic simulations (Rosen & Bregman 1995) show that the upward- or downward-moving gas occurs in large structures, comparable to the size of a superbubble (0.2 to 0.5 kpc). Consequently, from a random location in the disk, the view out of the plane can be dominated by such large structures, so that the observer sees primarily either infalling or outflowing gas. This may be the case from the location of the Sun, where we see mainly inflowing intermediate-velocity gas in the northern hemisphere (Danly 1989). It is this material that has a scale height similar to the Reynolds layer, a characteristic that has been seen in various numerical hydrodynamic simulations (Rosen & Bregman 1995; de Avillez 2000). The free-fall time from 1 kpc is about 25 Myr, so the cycle time for this cool fountain is 50 Myr.

In the Galactic Fountain model, the velocities of the fastest cooled gas are dependent upon the model chosen. The reason for this has to do with the way in which the cooled gas interacts with its surroundings once it has formed. The assumption that the clouds move ballistically following their formation (e.g. Bregman 1980; Kahn 1981) leads to the maximum velocities that clouds can achieve. However, depending upon the effective viscosity between the cloud and its surroundings, hydrodynamic drag may reduce the velocity of the infalling clouds, especially when they reach the denser part of the halo near the disk (Benjamin & Danly 1997). Numerical hydrodynamic calculations are not yet sufficiently detailed to address these issues, as they would need resolution adequate to resolve hydrodynamic instabilities and they must include effects that damp instabilities, such as magnetic fields. However, there are some exciting developments in the area of detailed three-dimensional modeling, such as those of de Avillez (2000).

Many interesting insights into the behavior of galactic fountains can be obtained by assuming a smooth and steady flow of gas into the halo, and this is the approach that has been followed by most of the analytic and numerical calculations until the last several years. With the advent of large computers, it has become possible to simulate the hydrodynamics of the interstellar medium and its interactions with supernovae on scales of kpc. By making such calculations, it is no longer necessary to distinguish between disk and halo or to assume quiescent conditions within which superbubbles grow. These calculations determine the vertical flow-rate of hot gas as well as the distribution of temperatures and the mass-flux rate onto the disk.

One approach is that of Rosen et al. (1993; also Rosen & Bregman 1995) in which the calculations employ two-dimensional hydrodynamics where separate gaseous and stellar fluids interact through star formation, stellar mass loss, supernovae, and stellar wind heating. These calculations lead to an interstellar medium with low-density regions that are warm (10^4 to 10^5 K) or hot (10^5 to 10^6 K), surrounded by cool gas ($<3\times10^3$ K), although the relative proportions of these different components depend upon the supernova rate. It is easy to identify regions in which multiple supernovae have occurred, but these "superbubbles" are usually quite irregular in shape. A range of supernova heating rates was used, and the model that best reproduced the scale heights of the cold, warm, and hot phases led to a supernova rate that is close to that observed in the Galaxy. The resulting filling factors in the disk are typically 50% for the hot phase, 25% to 30% for the warm gas, and 20% for the cold gas, although most of the mass of the gas is in the cold phase. The structure of the ISM depends upon the heating rate, so at lower values for the heating rate there were few dramatic breakouts of hot gas into the halo. The disk appeared to be "better behaved", and the volume filling factor of the hot gas was much less. For higher heating rates, nearly all the gas is hot, with cold clouds literally appearing as islands in a sea of hot gas.

The simulation shown in Fig. 1 (Plewa & Rosen; private communication) is similar to the models calculated by Rosen & Bregman (1995), but with greater resolution and accuracy. This case represents the model that we believe is most similar to the properties of the Galaxy near the solar circle. Since the gas tries to approach pressure equilibrium, regions of high gas density correspond to regions of low temperature. Although it is not

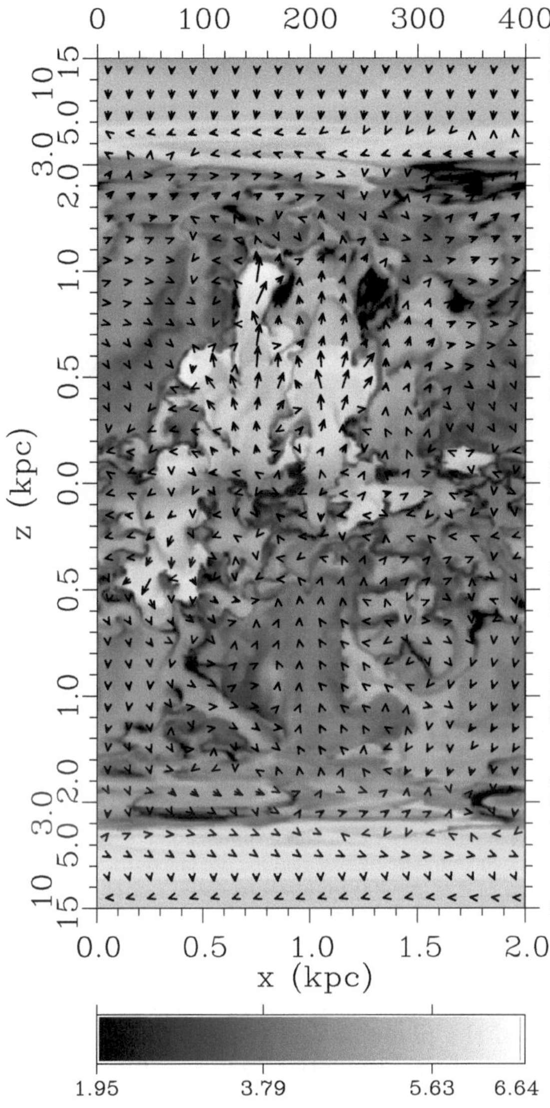

Figure 1. Temperature and velocity distribution in a 2-D simulation of the galactic ISM by Plewa & Rosen (private communication). The midplane runs horizontally across the middle of the simulation, which has an extent of 2 kpc in the plane and ±15 kpc out of the plane. The gas tries to approach pressure equilibrium, so, at constant z, regions of low temperature are the densest, and the hottest regions are low-density bubbles. There are examples of bubbles confined to the disk (x, z=1.7 kpc, 0.1 kpc), superbubble or chimney breakout (x, z=0.75 kpc, 1 kpc), as well as organized return flow to the disk (x, z=0.9 kpc, −1 kpc).

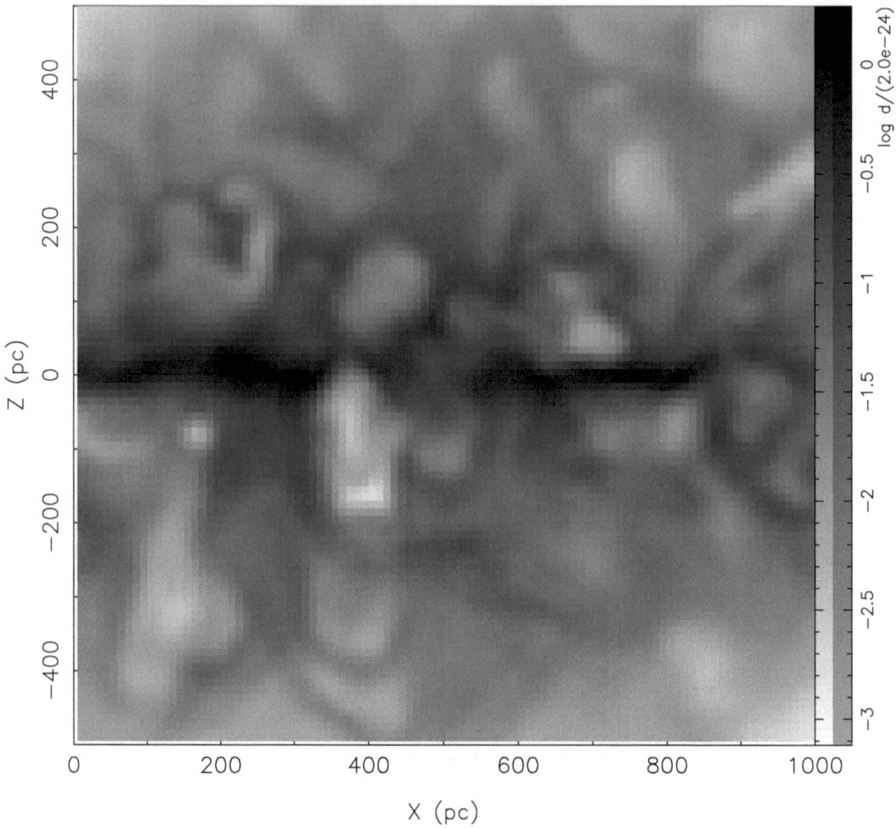

Figure 2. Gas density for a slice through a three-dimensional simulation (at y=30 pc) in which there is an OB association; the lighter regions are the hot low-density bubbles. Two chimneys are developing, one at x=400 pc (z=0 to −400 pc) and another at x=150 pc (z=0 to −400 pc). Figure from de Avillez (2000).

immediately apparent from this figure, most of the cold gas lies within a few hundred parsecs of the plane, although cold clouds are found several kiloparsecs away from the plane as well (e.g. x, z=0.5 kpc, −2.5 kpc). At this particular moment of the simulation, which is typical, breakout events are occurring at a few locations (e.g. x, z=0.3 kpc, −0.5 kpc; x, z=0.8 kpc, 1.2 kpc), producing elongated plumes or "chimneys" that can rise to several kpc. One surprise of the calculations was that the cooling and the infall also occur on similarly well-defined structures. This leads to regions of infalling cool gas (e.g. 0.9 kpc$<x<$1.2 kpc, −2 kpc$<z<$−0.4 kpc; wider at some locations), which to an observer in the midplane would cover 40° on the sky at its widest location (z=−0.8 kpc). Hydrodynamically, this is like a type of convective flow where the natural length scales are those of superbubble breakout and of the scale height of the gas when it cools. For the observer,

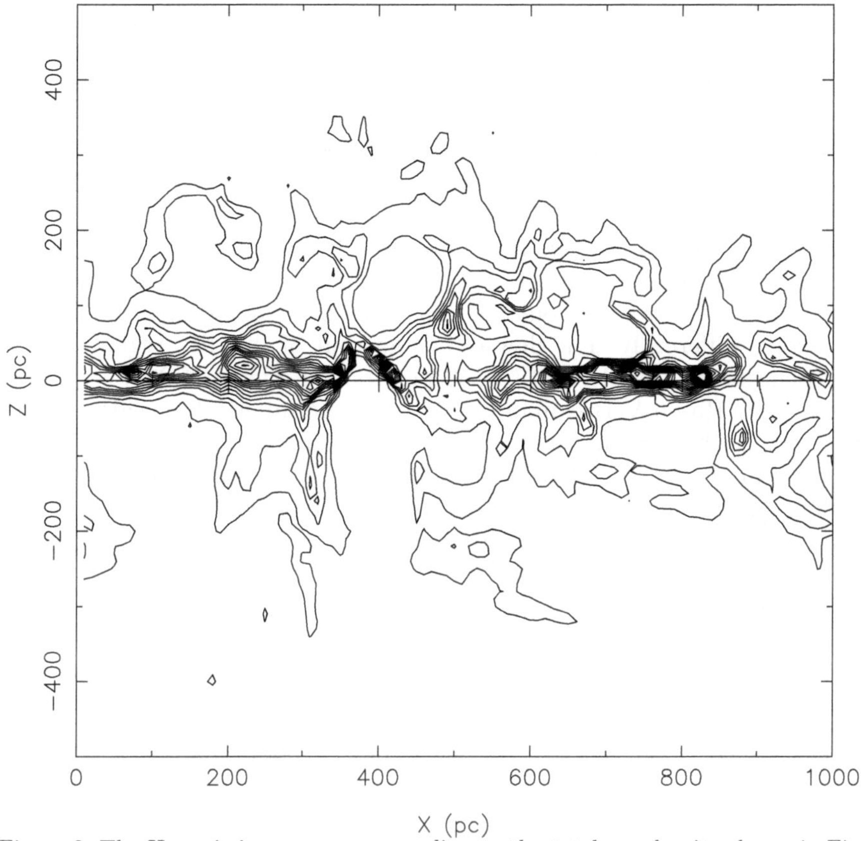

Figure 3. The H I emission map corresponding to the total gas density shown in Fig. 2, with 20 equally-spaced contours varying between 10^{18} cm^{-2} and 10^{20} cm^{-2}. There are a variety of irregularly-shaped closed bubbles, broken bubbles, and one feature that would be considered an H I chimney by observers, at x=400 pc (z=0 to −400 pc). Figure from de Avillez (2000).

this naturally predicts structures that are a significant fraction of a radian on the sky, generally with velocities in the range 50 to 100 km s^{-1}. These calculations have a number of limitations: they are two-dimensional rather than three-dimensional, magnetic fields are not included, and numerical viscosity probably artificially reduces the infall velocity of the clouds.

Another effort of global modeling has been developed by de Avillez and collaborators (de Avillez 2000; de Avillez & MacLow 2001, 2002; de Avillez & Berry 2001), and it holds promise for producing yet more detailed results. These calculations are three-dimensional in nature, which is possible because rather than using uniform-size cells, they use the adaptive-mesh technology that adds cells only where more detailed calculations are warranted. In this model, there is a single fluid (no stellar fluid), so supernovae are assumed to occur both in clustered and non-clustered fashion. The computational grid has dimensions of 1 kpc$\times$1 kpc in the plane, and extends to $\pm$4 kpc in z. Qualitatively, the structure of the ISM is similar to that of Rosen & Bregman (1995) in which hot low-density regions are surrounded or bounded by cool higher-density regions (Figs. 2, 3). However, the filling factor of the hot gas ($>10^5$ K) is only about 10%, significantly less than the value calculated by Rosen & Bregman (1995). One reason for the difference may be that the energy input rate from supernovae used by de Avillez (2000) is about a factor of two lower than that of Rosen & Bregman (1995), and another reason could be that two- and three-dimensional blast waves evolve differently. The flow of hot gas out of the disk region occurs through channels of width 100 to 150 pc, about a factor of 2 to 3 smaller than found by Rosen & Bregman (1995) or in the superbubble calculations of Tomisaka (1998), which may merely reflect the number of supernovae that occur in their associations. Additional semi-analytic calculations for the filling factor have been performed by Ferriere (1995), who finds a hot gas filling factor of 30% in the midplane at the solar circle, rising to 90% at a height of 2 kpc, although in a more detailed treatment (Ferriere 1998) she finds a filling factor in the plane at the solar circle to be closer to 20%, with a decrease as a function of height. Her calculation also uses a lower supernova heating rate than Rosen & Bregman (1995), and this is responsible for most of the difference between the estimated filling factors.

An area in which these three-dimensional calculations excel is their ability to see important small-scale structures, such as the breakup of superbubble shells and the flow of clouds returning to the disk. Clouds are identified flowing toward the disk at up to 90 km s^{-1}, indicating that their technique has made significant improvements in dealing with artificial viscosity. There is the hope that such calculations will eventually be able to predict cloud-size distributions, which would be very valuable for comparison with observations.

2.4. THE ROLE OF COSMIC RAYS AND THE POSSIBILITY OF A GALACTIC WIND

The presence of cosmic rays in the halos of galaxies is well-established, as they are seen directly in external systems by the non-thermal radio emission from cosmic-ray electrons (e.g. Sukumar & Allen 1991). In the Milky Way, cosmic rays are detected directly, as is their synchrotron emission. In the Milky Way, the abundance of cosmic-ray secondaries, such as ^{2}H, ^{3}He, Li, Be, and B requires that most of the cosmic rays have traversed a mean column of $\rho ct{\sim}5$ g cm^{-2}, where c is the speed of light, t is the timescale, and ρ is the density of interstellar gas. The timescale is set by the abundances of the radioactive secondaries, ^{10}Be, ^{26}Al, and ^{36}Cl, which leads to a mean time $>2\times10^7$ yr and implies that the mean density of the ISM experienced by the cosmic rays is $<5\times10^{-25}$ g cm^{-3}, which is a factor of about four times smaller than the known mean density of interstellar gas (atomic plus molecular content). This implies a containment volume several times larger than the disk and leads to the conclusion that cosmic rays spend most of their time in the Galactic Halo.

The structure and properties of the cosmic-ray halo are a matter of ongoing effort, but much progress has been made, particularly with efforts that take into account the wide range of available data, such as by Strong & Moskalenko (1998, 2000, and references therein). They consider two classes of models, one in which cosmic-ray transport occurs through diffusion plus convection, and the other in which it occurs through diffusion plus reacceleration of the cosmic rays. They find that the diffusion-plus-convection model does not provide a good fit to all their data (especially the low-energy data), and the presence of the convection term leads to a significant *ad hoc* break in the diffusion coefficient, which seems unnatural. The diffusion-plus-reacceleration model is more successful in fitting the data and leads to limits on the height of the propagation region, z_h: 4 kpc$<z_h<$12 kpc. This propagation height is generally larger than that found by previous authors, who found $z_h\leq$4 kpc (Webber et al. 1992; Bloemen et al. 1993; Lukasiak et al. 1994), but Strong & Moskalenko (1998) argue that the additional constraints that they considered lead to more accurate values of z_h. If they are correct, then the propagation height of the cosmic rays is very similar to the height expected from galactic fountains, which would provide a consistent picture between the two interstellar components.

In the context of halo gas, one might expect cosmic rays to contribute in two ways, pressure support and the outflow of gas. Pressure support has been considered for static halos (e.g. Chevalier & Oegerle 1979), but radiative cooling causes gas to lose its buoyancy and such material cannot be supported by cosmic rays since it would be Rayleigh-Taylor unstable. A more realistic model is to consider cosmic-ray pressure in a dynamical situation, such as the galactic fountain picture, and Kahn (1994) has developed this line of investigation. He assumed a continuous flow into the halo and a cosmic-ray pressure that was typical of the Galaxy, and found that the place where the gas became supersonic and the nature of the flow near the disk does not change greatly, but after the gas cools cosmic-ray pressure will become important in driving the flow. This solution showed somewhat higher velocities when cosmic rays were included, but since the cosmic rays were not the dominant form of energy (when averaged over the fountain cycle), the general nature of flow was not changed.

It may also be possible for cosmic rays to drive a Galactic Wind from the Milky Way, and this has been the subject of several investigations. If one estimates the energy going into cosmic-ray production, based upon the cosmic-ray energy density (about 1 eV cm^{-3}) and the cosmic-ray timescale, $\sim 10^{40.5-41}$ erg s^{-1} is required. Strictly from these energy considerations, a Galactic Wind of ~0.3 to 1 $M_\odot$ yr^{-1} would occur, so this is an upper limit to the mass flux of a wind. However, a Galactic Wind can have a profound effect on the gas content of the disk or bulge as well as on the angular momentum loss. There are several constraints on the possible properties of a present-day wind. One can set a limit on the mass-loss rate of a wind so that the cold gas in the disk is not removed on a timescale short compared to a Hubble time. The neutral hydrogen content of the Galaxy is about 5×10^9 $M_\odot$, so we infer that mass-loss rates ≥ 0.5 $M_\odot$ yr^{-1} are too great.

In his discussion of the effects of cosmic rays, Kahn (1994) considered the possibility that galactic winds occurred from the top of a galactic fountain. He shows that only a small fraction of the gas is likely to attain a high velocity due to the effect of the cosmic rays. Averaged over a long period, Kahn estimates that ~2% of the gas could rise upward in the fountain and be lost from the Galaxy as a wind (about 0.15 $M_\odot$ yr^{-1}).

A different set of models for galactic winds have been developed by groups mainly concerned with cosmic-ray propagation and evolution. First, we should point out that the term "cosmic-ray winds" is sometimes misleading in these models, as it can be used to mean that advection is present in addition to diffusion. However, it also is used to describe true galactic winds that are driven by the energy stored in cosmic rays, as originally discussed by Ipavich (1975) and more recently by Breitschwerdt et al. (1993) and by Zirakashvili et al. (1996). The latter consider a cosmic-ray driven wind that is coupled to the plasma by resonant excitation of small-scale magnetic field fluctuations. In their solution, the gas in the lower halo has a density of 10^{-3} cm^{-3} (z=1 kpc) and a velocity of 20 to 30 km s^{-1}. This leads to a mass-loss rate that would deplete the disk gas inside the solar circle in only 3 Gyr, so it is unlikely to be applicable today. However, there is some flexibility as to the mass loading and other details of the model, and it is possible to obtain winds with mass-loss rates <1 $M_\odot$ yr^{-1} (Breitschwerdt et al. 1991). One of the shortcomings of these calculations is that although the proposed energy sources (self-excited hydromagnetic waves) are sensible and likely, there is no direct evidence for their existence in this environment. Also, although these models are self-consistent, their predictions for the convective transport of cosmic rays, which is different from the convection examined by Strong & Moskalenko (1998), has not been evaluated extensively. Such detailed evaluations, along the lines of Strong & Moskalenko (1998), are likely to lead to interesting constraints on the model. These concerns aside, it certainly is possible for cosmic rays to drive a wind, provided that the mass-loss rate is small (<0.5 $M_\odot$ yr^{-1}).

3. Gas streams from dwarf galaxies

Models for the Magellanic Stream involve either ram-pressure stripping or tidal stripping as the gas removal mechanism. Once the gas is no longer gravitationally bound to the Large and Small Magellanic Clouds (LMC and SMC), it will be interacting with the gaseous environment of the Local Group. Little is known about the Local Group medium, but we can infer the likely intergalactic gas conditions from studies of other galaxy groups. The gas in galaxy groups will become thermalized over time and will assume the temperature corresponding to the depth of the potential well. Since the one-dimensional velocity dispersion of groups is typically 200 to 400 $\mathrm{km\,s^{-1}}$, the gas temperature will be 3 to 10×10^6 K (0.3 to 1 keV) and it will emit in the X-ray band. Groups can be difficult to identify with certainty, due to foreground and background contamination, so much of the effort has gone into the Hickson Compact Groups or large, nearby groups where contamination is less of a problem. Such systems have been studied extensively, for example by Ponman et al. (1996), Pildis et al. (1995), and Mulchaey et al. (1996), and although these analyses lead to slightly different results, some general properties of the gaseous environment can be inferred. The gas densities in the centers of the groups are typically 10^{-4} to 10^{-3} $\mathrm{cm^{-3}}$ (although not all groups are detected), the temperatures are about 10^7 K, and the core radii are approximately 20 to 150 kpc. This leads to a central gas pressure of $P/k{\sim}10^3$ to 10^4 $\mathrm{K\,cm^{-3}}$, and the pressure drops primarily as the density decreases.

With these densities, one can estimate if ram-pressure stripping will be effective in removing gas from galaxies such as the LMC or SMC. To do so, one must have the gas density and other parameters in the vicinity of the dwarf galaxies, and the radial distribution of density is often described in groups with a beta model, which has the form

$$n(r) = \frac{n_o}{(1+(r/r_c)^2)^{3\beta/2}}$$

where r_c is the core radius and n_o is the ambient gas density at r=0. Assuming a central gas density of 10^{-4} $\mathrm{cm^{-3}}$, a typical value of β between $\frac{1}{2}$ and $\frac{2}{3}$, and a core radius of 30 kpc, the density at 60 kpc from the Galaxy (the approximate distance of the LMC and SMC) is $\sim10^{-4.5}$ $\mathrm{cm^{-3}}$. The estimated space velocity of the LMC and SMC is 275 $\mathrm{km\,s^{-1}}$ relative to the Galaxy, which we assume to be at rest with respect to the dilute group medium. The resulting ram pressure is $nm_pv^2 = 4\times10^{-14}$ $\mathrm{g\,cm^{-1}\,s^{-2}}$. In competition with the pressure force is the maximum gravitational force that a galaxy can exert to retain its material (e.g. Gunn & Gott 1972). For the LMC and SMC, a Plummer-type potential was used by Gardiner & Noguchi (1996; recent revisions to the potential by Gardiner (1999) do not change this result significantly), and for a radius of 3 kpc and an H I column density of 10^{20} cm^{-2}, the gravitational force is $\sim 2\times10^{-12}$ g cm^{-1} s^{-2} for the LMC, and 2×10^{-13} g cm^{-1} s^{-2} for the SMC. Ram-pressure stripping is insufficient to strip gas from the LMC by almost two orders of magnitude, and this is the source of the argument that it is unimportant in removing much gas from that galaxy. However, for the SMC, the ram-pressure force is only a factor of five less than the gravitational force, so given the uncertainty in the value of the ram pressure, it may be an important factor, such as in the outermost parts of the SMC where the gravitational force and the gas column density are lower. Also, ram-pressure stripping is likely to be important when less massive dwarfs pass near the Galaxy, since the dwarfs within 200 kpc of the Galaxy are $\sim$30 times less massive than the SMC.

The prevailing model for the origin of the Magellanic Stream is by a tidal gravitational interaction with the Milky Way, as the Magellanic Clouds make a close passage (Gardiner & Noguchi (1996) and references therein). Such tidal interactions lead to both a leading and a trailing feature. The tail is much more prominent, and it is this feature that is matched to the H I observations that form most of the Magellanic Stream (Mathewson & Ford 1984). In addition, there should also be a weaker Leading Arm, and until recently, there was little convincing evidence for this phenomenon. However, the new H I Parkes All-Sky Survey (*HIPASS*), which is more sensitive and complete than previous surveys, has revealed a number of new features, including a likely detection of the Leading Arm (counter-stream; Putman et al. 1998; Ch. 5). Although the agreement between models and theory is not perfect, it is very encouraging. The identification of both the leading and trailing features with H I structures is extremely strong support for the tidal model.

There are still some problems with this model, but they are likely to be overcome by improving the realism of the models. For example, tidal models would remove both gas and stars from the Magellanic Clouds, but few stars are seen in the Stream (not to be confused with the Bridge between the LMC and SMC, where stars are observed). It would seem that this tidal encounter was unreasonably selective in the creation of a tidal tail, but this concern is valid only if the gas and stars are cospatial. In many galaxies, the H I disk is considerably larger than the stellar disk (Roberts & Haynes 1994), and if that was the situation for the LMC and SMC, H I would have been stripped away but with very few associated stars. Other discrepancies between theory and observation are reduced when the consequences of drag are included (Gardiner 1999). The drag would be between the ambient cluster medium and the Stream (and the Leading Arm), and from the calculations given above, it is expected that drag and ram-pressure confinement would become important once the gas expanded and reached densities lower than seen in the Magellanic Clouds. Improvements occur in the location of the Leading Arm feature and in the velocities along the Stream. Finally, the compressed heads of several of the gas complexes in the Magellanic Stream are further evidence for ram-pressure effects due to an ambient medium.

The Magellanic Stream is the most apparent example of the removal of gas from companion galaxies, material now falling onto the Milky Way. There is another example in that the interaction of the Sagittarius Dwarf with the Galaxy may have led to the Smith Cloud (Bland-Hawthorn et al. 1998) (also known as complex GCP). One might expect gas removal from dwarfs to be a common phenomenon, because dwarfs are plentiful in the Local Group, with 13 lying within 400 kpc of the Galaxy and seven within 100 kpc (Mateo 1998). The masses of these dwarfs are significantly smaller than those of the LMC and SMC, making ram-pressure stripping likely to occur when the dwarfs pass through regions where the ambient Local Group density exceeds $\sim 10^{-5}$ cm^{-3}. Both ram-pressure stripping and tidal stripping naturally produce long streams, which are typical of some of the HVC complexes. Also, these dwarfs easily obtain the radial velocities that are typical of the HVCs, including the very-high-velocity clouds. Furthermore, this scenario is consistent with the strength of the metal lines, although with two exceptions (Wakker et al. 1999; Richter et al. 2001), they are poor indicators of the metallicity of the gas, which would be an important diagnostic of the phenomenon. On the negative side of the balance, only a few dwarfs are detected in H I, and there are not yet theoretical models evaluating whether the phenomenon occurs with sufficient frequency to explain the observations.

One of the concerns is the state of the gas once it has been stripped from a galaxy. The gaseous material will expand until it is in pressure equilibrium with its surroundings, with contributions from thermal pressure, magnetic pressure, and ram pressure. The pressure of the hot ambient medium is most likely dominated by the thermal component, but the pressure of the stripped gas is likely to be a combination of magnetic and thermal pressures. If the gas expands adiabatically after it has been stripped out, the thermal pressure decreases as R^{-5}, where R is the cloud size, while the magnetic pressure depends on cloud size as $B^2 \propto R^{-4}$ for the frozen-in field condition (BR^2 is a constant). As the cloud expands, the ratio of the magnetic to thermal pressure increases linearly with cloud size, so that the magnetic pressure becomes dominant. However, heating of the gas by ambient photons could raise its temperature, violating the adiabatic assumption and increasing the thermal pressure. In the case of the Magellanic Stream, if we assume that the Stream is as deep as it is wide, the mean density in the Stream is $\sim 10^{-2}$ cm^{-3} and the thermal pressure is $\sim 1\,(T_{\rm gas}/100\ {\rm K})\ {\rm K\,cm^{-3}}$.

If this is an accurate measure of the ambient group pressure, then the ambient density is extremely low, $\sim 10^{-6}$ to 10^{-7} cm^{-3}. With a magnetic field in the Stream gas of only 0.6 μG, the magnetic pressure in the cloud would be 10^2 K cm^{-3}, implying an ambient density of $\sim 10^{-4}$ to 10^{-5} cm^{-3}, which is along the lines of the densities seen in other galaxy groups.

4. The Local Group debris model

The most recent model for the origin of the HVCs is that of Blitz et al. (1999; Ch. 14), who argue that these clouds are not part of the Milky Way or of the dwarf galaxies, but represent the material that failed to form galaxies. This material is in gravitationally-bound systems, where most of the gravity is provided by the dark matter. These bound units are small compared to the size of galaxies, originally with potential wells whose velocity dispersion is characterized by tens of km s^{-1}, deep enough to accumulate cooled baryons. These "mini-halos" would collect into filaments, the nearby ones falling into the Local Group. If these mini-halos interact with the existing medium in the Local Group, there would be an accretion shock, although the (unknown) nature of the Local Group medium at large distances will determine the strength of the shock and its ability to remove the gas from these dark-matter potential wells. As this theory is still in an early stage, the simulations were intentionally simplistic (Blitz et al. 1999). They calculate the dynamics through an N-body simulation, which leaves out the potentially important contributions of compressional gas heating and photoionization from stars and AGNs over the history of the Universe. It is assumed that the gas that settles into these potential wells is neutral today and always had a temperature below the escape temperature of its local potential well. Other cosmological simulations include the effects of photoionization and gas dynamical processes, yet they do not have the fine resolution to follow the evolution of these proposed mini-halos. Calculations that include these processes and make an effort to calculate the ionization state of the gas (e.g. Hellsten et al. 1998) find that for collapsed regions that would be typical of galaxy groups the gas temperature is $\sim 10^6$ K; higher-resolution calculations should address the fate of these mini-halos.

A dramatic consequence of this theory is that the mean distance of the HVCs is $\sim$1 Mpc, compared to $\sim$10 kpc for the Galactic Fountain model and $\sim$100 kpc for gas stripped from dwarf galaxies. This has the implication that the total mass of H I involved is $\sim 10^{11}$ $M_{\odot}$, which means that there is an order of magnitude more gas outside of galaxies than gravitationally bound to galaxies. This model makes some predictions that can be tested, the clearest prediction being the distance to the clouds. It is already known that several of the cloud complexes are within about 10 kpc.

Complexes A and M have direct distance determinations that place them closer than 10 kpc (van Woerden et al. 1999; Ch. 10), while complex C is probably at a similar distance, based on the H I fingers connecting it with low-velocity gas, the X-ray enhancement that is indicative of an interaction with galactic material, and its Hα emissivity (Tufte et al. 1998; Putman et al. 2003; Ch. 8). However, Blitz et al. (1999) argue that these are the nearby manifestations of their mini-halos and not typical of most HVCs.

Alternatively, it should be possible to detect this large amount of gas in other nearby groups of galaxies, since it is expected to be a universal phenomenon. There have been a series of observations of galaxy groups as well as "blind" H I surveys (Zwaan et al. 1997; Kraan-Korteweg et al. 1998). The clouds are expected to be about $10^{7.5}$ $M_\odot$, so Zwaan & Briggs (2000) surveyed the halos of about 300 cataloged galaxies and the environments of 14 groups with the *Arecibo* telescope, reaching a sensitivity of 10^7 $M_\odot$, but they failed to find H I that was not associated with optical galaxies. Subsequently, Zwaan (2001) pressed down to lower mass limits (7×10^6 $M_\odot$) for five groups similar in richness to the Local Group, but all detections were associated with optical galaxies. Most recently, de Blok et al. (2002) used the Australia Telescope Compact Array (*ATCA*) to conduct a sparse survey of the Centaurus A and Sculptor galaxy groups, with limiting sensitivities of 3×10^6 $M_\odot$. They detected previously-known group members, but no new galaxies or H I clouds. Their failure to uncover a population of H I satellites in these groups and clusters is in conflict with the Local Group model suggested by Blitz.

5. The impact of HVCs onto the disk

There are some HVCs that are within a few kiloparsecs of the disk and have a negative velocity, so it seems inevitable that the gas will eventually crash into the disk. Many of the HVCs are long complexes, so one might wonder whether some part of the stream has already struck the disk, and exploration of this issue has been an active research topic in recent years. The type of interaction that one can expect for gas falling onto a disk depends upon its degree of homogeneity. Steady accretion of a uniform and volume-filling gas will be decelerated by a standoff shock surrounding the galaxy, after which the energy is radiated away and the accreted gas gently settles onto the disk (Cox & Smith 1976). However, accretion of discrete clouds that are much denser than the hot halo is more likely to be characterized as ballistic motion modified by drag forces. The impact of such HVCs on the disk should have a noticeable effect, based on energetic arguments. A typical HVC has a radius of $\sim1^\circ$, a column density of about 3×10^{19} cm^{-2}, a distance of 5 kpc, and a velocity of 100 km s^{-1}, which gives it a mass of about 10^4 $M_\odot$, and an energy of $\sim 10^{51}$ erg, comparable to a SNR. Some clouds are more massive and many come in complexes, so the energy deposited in the disk can reach up to the energy associated with some superbubbles. However, since the column density of the disk is typically an order of magnitude greater than that of the clouds, momentum transfer will decelerate the HVC by an order of magnitude, so that the disk material near the impact site and the HVC will have a final velocity of about 10 $\mathrm{km\,s^{-1}}$. In this picture, the HVC will not penetrate the disk.

Theoretical interaction models for the impact of gas onto the disk have been developed in which a cloud falls ballistically onto a uniform disk, and some simulations include a magnetic field (Tenorio-Tagle et al. 1987; Rand & Stone 1996; Santillán et al. 1999). The impact leads to a shock at the interface between the cloud and the disk as the cloud drives into the disk and creates a cavity. For an impact of ~ 100 $\mathrm{km\,s^{-1}}$, the characteristic temperature is near 10^6 K, so soft X-rays are emitted. The addition of magnetic fields leads to additional new effects (Santillán et al. 1999), especially for field lines parallel to the disk (Fig. 4). The impact of the cloud compresses the field lines, which increases the field pressure and tension, effects that help prevent the clouds from penetrating far into the disk. Also, the impact induces a complex pattern of MHD waves that carry away the energy of the disturbance, spreading it out over large distances. These simulations assume that the medium and the magnetic field are uniform, but the more natural situation is that the disk is highly structured, so material falling into low-density regions falls deeper into the disk than infalling clouds that strike dense material. In some cases they largely pass through the disk (Rosen & Bregman 1995; de Avillez 2000). Although the latter two simulations provide a more realistic model for the disk, they lack magnetic fields, and the numerical viscosity can affect cloud motions.

A soft X-ray survey of the sky might detect these impact events, but identifying them uniquely as the impacts of HVCs can be difficult because they can be confused with supernovae that have occurred in the upper disk and broken out. One difference between a supernova breakout and an HVC impact is that the former would have H I ejecta moving away from the disk owing to the shredding of the upward moving neutral gas shell, although Santillán et al. (1999) show that, owing to the tension stored in magnetic fields, some material can be ejected following an impact.

Another signature of an HVC impact that is less ambiguous is the presence of an HVC at the location of a soft X-ray enhancement. This would occur if one were fortunate enough to see the HVC before the shock wave moved through it entirely, or if one of several of the clouds in a complex has interacted with the disk. The time for a shock wave to move through a typical HVC is about 10^6 yr, for a typical cloud size of 100 pc (1° at

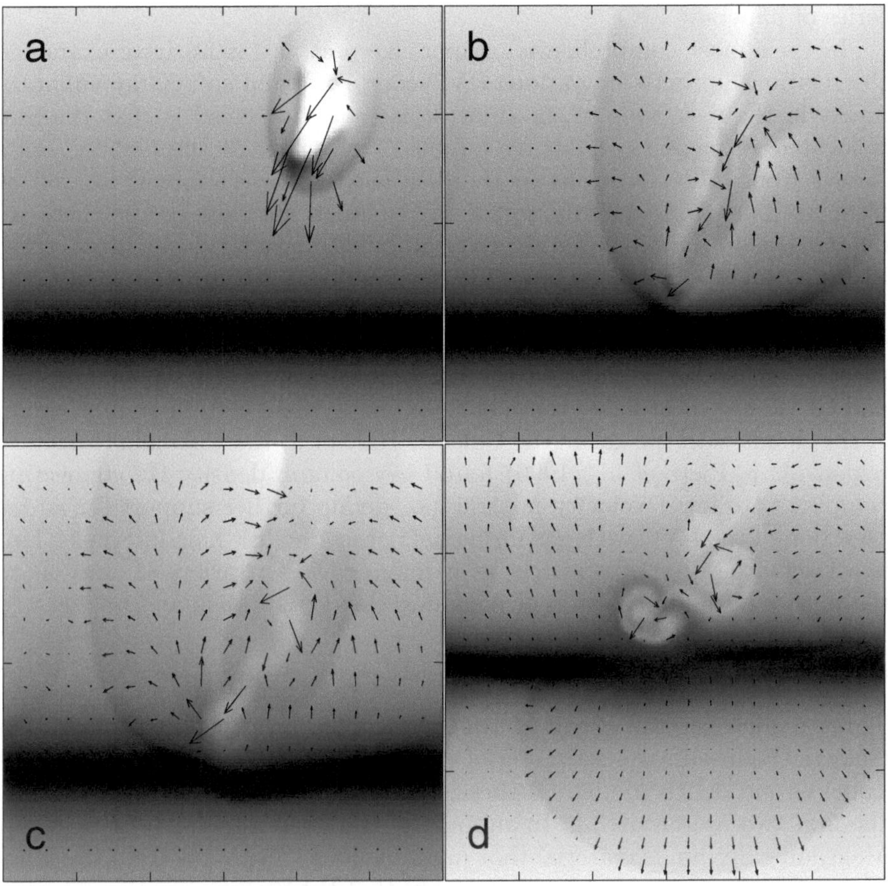

Figure 4. The oblique collision of an HVC with the Galactic Disk, including a magnetic field parallel to the z-axis. The midplane is at z=0 kpc and the distance between tick marks is 500 pc for both the x axis (the in-plane coordinate) and the vertical, z, axis. The sequence shows the density (gray log scale) and the velocity field at 3.2, 9.5, 19, and 41.2 Myr, for which the maximum velocity values are 173, 80, 56, and 38 km s^{-1}. Figure from Santillán et al. (1999).

5 kpc), and since the free-fall time from 5 kpc is about 6×10^7 yr, there would have to be 100 HVCs with this average height in order to "catch" one of the HVCs hitting the disk. Since Wakker & van Woerden (1991) have cataloged nearly 600 HVCs, it is feasible for about 10 to be seen during the interaction phase with the disk. One difficulty is that, if most of the velocity of the HVC is directed downward, the radial velocity seen from our location might not distinguish it as an HVC. An easier and more likely association would be between one of the long streams and a soft X-ray enhancement, and that is the approach that Herbstmeier et al. (1995) have taken in a

series of papers.

In their most comprehensive paper, Kerp et al. (1999) have examined the relationship between various HVC complexes and soft X-ray enhancements in the disk. One of the difficulties that they faced is that the Sun is located in an X-ray-emitting bubble of plasma and the Galaxy emits X-rays from other locations in the disk (SNRs, superbubbles) as well as from the halo. Consequently, soft X-ray emission fills the sky and there are fluctuations in that emission. Kerp et al. (1999) have tried to address these issues by incorporating a model for the soft X-ray background due to the Local Bubble and the Halo (Pietz et al. 1998). This X-ray background is smooth enough that some fluctuations appear to stand out. In particular, there is excess X-ray emission toward regions near HVC complexes C, D, and GCN. The most likely cause of this enhancement is from an interaction of the clouds with the gas in the Galactic Halo or at the disk-halo transition region. Also, Pietz et al. (1996) found low column density H I connecting complex C to lower-velocity neutral gas, giving further support for an interaction of the HVC with gas in the halo or at the disk-halo interface. This is a fascinating line of research, as it may provide a variety of clues as to the nature of the interaction of incoming clouds with disk and halo gas.

6. Discussion and summary

Three models for the origin of the HVCs have been presented here: the Galactic Fountain, the stripping of material from dwarf galaxies, and the Local Group debris model. Of these, there is strong evidence that at least the first two are occurring. In support of the dwarf-stripping picture, the Magellanic Stream is material that unambiguously has come from the Magellanic Clouds. Originally, it was thought that the Stream was the only material torn from the Magellanic Clouds, but recent work suggests that there is a Leading Arm of gas on approximately the other side of the Magellanic Clouds. There are two open questions regarding this model that need to be developed further: stripping of gas from other dwarfs near the Milky Way, and material torn from the Magellanic Clouds on previous passages. These interactions, largely gravitational, possibly augmented by ram-pressure effects with the hot halo of the Galaxy, naturally produce long streams, a feature of other HVC complexes. Also, these dwarfs easily obtain the radial velocities that are typical of the HVCs, including the very-high-velocity clouds (LSR velocities exceeding 200 $\mathrm{km\,s^{-1}}$). Furthermore, this scenario is consistent with the strength of the metal lines, although with two exceptions (Wakker et al. 1999; Richter et al. 2001), these are poor indicators of the metallicity of the gas, which would be an important diagnostic of the phenomenon.

In order of importance of their contributions to the HVCs, the Local Group dwarfs probably contribute less detectable gas than material from the Magellanic Clouds, especially if one considers previous passages of the Magellanic Clouds as well. First, the sum of the dynamical masses of all the dwarfs within 200 kpc of the Milky Way is 3×10^8 $M_\odot$ (after Mateo 1998), which is a factor of three less than the mass of the SMC and a factor of 60 less than the LMC. Even if half the mass of the dwarfs had originally been H I (3×10^8 $M_\odot$), it is still several times less than the mass of H I found in the LMC, SMC, and the Stream together ($\sim10^9$ $M_\odot$). Secondly, the LMC and SMC are closer to the Milky Way than the other dwarfs (except Sagittarius), so at these smaller distances, lower-mass clouds are more easily detected. Nearer clouds have a greater chance to be beam-filling, which also helps in their detection and study. In an effort to understand the importance of the Magellanic Clouds in producing HVCs, it is necessary to extend tidal-stripping models to include previous passages of the Clouds past the Galaxy, and we hope that theorists will investigate this issue in the near future.

In external galaxies, there are a variety of observations that support the presence of galactic fountains: extraplanar gas is seen in Hα emission and is related to the star formation rate; X-ray halos are observed; and IVCs and HVCs are seen projected on the disk of external galaxies. The question is not really whether galactic fountains occur, but what is the mass-flux rate and for which gaseous features are they responsible? The mass-flux issue could be settled if we knew the energy being deposited into the halo, or alternatively, the cooling rate of hot extraplanar gas. As discussed above, most of the energy is probably emitted in the extreme ultraviolet region that is heavily absorbed by galactic neutral gas, so high-ionization diffuse UV line emission will be the best diagnostic. The single such UV emission-line study supports a mass flux rate $\sim$10 $M_\odot$ yr^{-1} (Martin & Bowyer 1990), but additional observations would certainly be welcome.

On the issue of the HVCs that may be explained by galactic fountains, a comparison between the model and the data was carried out by Wakker (1990). The most challenging observations to reproduce were the Anti-Center clouds, where it was difficult to achieve a velocity sufficiently high to agree with the observations. Another aspect of the observations that is troubling is the extremely large coherent structures in certain complexes. For example, complex A is a radian in length, and since its distance is constrained, it has a size of $\sim$5 kpc, which is larger than the coherent structures produced in calculations, although modern simulations on a galactic scale (10 kpc) have not been conducted. Another feature that galactic fountains are unlikely to produce is the very-high-velocity clouds (VHVCs; clouds with LSR velocities exceeding 200 km s^{-1}), which have velocities near the escape velocity of the system. The only way to produce these clouds through the galactic fountain mechanism would be to have a halo hot enough that it extended to rather large radii (30 to 100 kpc), so that the infall velocity would be large, but the temperature measurements of the hot halo from X-ray observations rule out such a scenario (which would require temperatures in the range 3 to 10×10^6 K).

Observations that appear to be explained by galactic fountains are the intermediate-velocity clouds, where the models are able to reproduce the scale heights and column densities of the cold, warm, and hot components of the interstellar medium (Rosen & Bregman 1995). In this model, the size of coherent structures and the north-south asymmetry in the intermediate-velocity clouds are local effects being dominated by a small number of (former) superbubbles. These are reproduced, along with the scale heights and column densities. Also, some of the high-velocity clouds are likely to be produced by galactic fountains (up to about 200 $\mathrm{km\,s^{-1}}$), but probably not the largest angular structures.

The third mechanism for the origin of the HVCs, the Local Group debris model (Blitz et al. 1999), is appealing since it would account for the gas not used in galaxy formation. Also, in the bottom-up model of structure formation, one expects a large number of small dark-matter halos (e.g. Miralda-Escude & Rees 1993). Detailed models predict hundreds of dwarfs in the Local Group (Klypin et al. 1999; Moore et al. 1999), and if these are the HVCs, it removes the large discrepancy between these theoretical predictions and the number of known dwarf galaxies. However, there are theoretical models that remove the problem of having many dwarf-sized objects simply by adding photoionization to the models of formation (e.g. Somerville 2002). The critical test of this model will be the distance determinations of more clouds, although for clouds at a distance of $\sim$1 Mpc, this is a challenging task. Direct distance determinations are few, although van Woerden et al. (1999) were able to constrain the distance to complex A between 4 and 10 kpc (z=2.5 to 7 kpc), and there are a few upper limits to other clouds of several kpc (Ch. 10). Another approach to obtaining distances is through their Hα emission, which is caused by UV galactic light that is absorbed by the clouds. This emission is detected in nearly all clouds, and its brightness places them within 100 kpc of the Milky Way and probably closer (Tufte et al. 1998, 2002; Maloney & Putman 2003; Putman et al. 2003; Weiner 2003).

Another prediction of the Local Group model is that, if it is a universal model, such clouds should be present in other groups and clusters. Zwaan and coworkers have searched for these clouds in nearby groups, such as Fornax, Centaurus, and Sculptor, as well as poorer groups more similar to the Local Group (Zwaan et al. 1997, 2001; Zwaan 2001; de Blok et al.

2002). In all cases, they fail to find the H I clouds. In addition, Charlton et al. (2000) point out that the Local Group model would produce too many strong quasar absorption lines. At the moment, the weight of evidence is moving against the Local Group model, but further work will no doubt be required to reach a strong conclusion along these lines.

References

Allen, R.J., Baldwin, J.E., Sancisi, R. 1978, A&A, 62, 397
Benjamin, R.A. 1994, Ph.D. thesis, University of Texas at Austin
Benjamin, R.A., Shapiro, P.R. 1993, in The Evolution of Galaxies and Their Environments, 338
Benjamin, R.A., Danly, L. 1997, ApJ, 481, 764
Bland-Hawthorn, J., Veilleux, S., Cecil, G.N., Putman, M.E., Gibson, B.K., Maloney, P.R. 1998, MNRAS, 299, 611
Blitz, L., Spergel, D.N., Teuben, P.J., Hartmann, D., Burton, W.B. 1999, ApJ, 514, 818
Bloemen, H., Dogiel, V.A., Dorman, V.I., Ptuskin, V.S. 1993, A&A, 267, 372
Bregman, J.N. 1980, ApJ, 236, 577
Bregman, J.N., Pildis, R.A. 1994, ApJ, 420, 570
Bregman, J.N., Houck, J. 1997, ApJ, 485, 159
Breitschwerdt, D., McKenzie, J.F., Völk, H.J. 1991, A&A, 245, 79
Breitschwerdt, D., McKenzie, J.F., Völk, H.J. 1993, A&A, 269, 54
Bunner, A.N., Coleman, P.L., Kraushaar, W.L., McCammon, D., Pavlmieri, T.M., Shilepsky, A., Ulmer, M.P. 1969, Nature, 223, 1222
Bunner, A.N., Coleman, P.L., Kraushaar, W.L., McCammon, D. 1971, ApJ, 167, L3
Burrows, D.N., Mendenhall, J.A. 1991, Nature, 351, 629
Charlton, J.C., Churchill, C.W., Rigby, J.R. 2000, ApJ, 544, 702
Chevalier, R.A., Oegerle, W.R. 1979, ApJ, 227, 398
Cioffi, D.F., Shull, J.M. 1991, ApJ, 367, 96
Cox, D.P. 1979, ApJ, 234, 863
Cox, D.P. 1981, ApJ, 245, 534
Cox, D.P., Reynolds, R.J. 1987, ARA&A, 25, 303
Cox, D.P., Smith, B.W. 1976, ApJ, 203, 361
Danly, L. 1989, ApJ, 342, 785
de Avillez, M.A. 2000, MNRAS, 315, 479
de Avillez, M.A., Mac Low, M.-M. 2001, ApJ, 551, 57
de Avillez, M.A., Berry, D.L. 2001, MNRAS, 328, 708
de Avillez, M.A., Mac Low, M.-M. 2002, ApJ, 581, 1047
de Blok, W.J.G., Zwaan, M.A., Dijkstra, M., Briggs, F.H., Freeman, K.C. 2002, A&A, 382, 43
Fabbiano, G. 1989, ARA&A, 27, 87
Ferriere, K. 1995, ApJ, 441, 281
Ferriere, K. 1998, ApJ, 503, 700
Garcá-Burillo, S., Guélin, M. 1995, A&A, 299, 657
Gardiner, L.T. 1999, in ASP Conf. Ser. 166, Stromlo Workshop on High-Velocity Clouds, eds. B.K. Gibson, M.E. Putman (San Francisco: ASP), 292
Gardiner, L.T., Noguchi, M. 1996, MNRAS, 278, 191
Garmire, G., Riegler, G.R. 1972, A&A, 21, 131
Guélin, M, Zylka, R., Mezger, P.G., Haslam, C.G.T., Kreysa, E., Lemke, R., Sievers, A.W. 1993, A&A, 279, L37
Gunn, J.E., Gott, J.R. 1972, ApJ, 176, 1
Habe, A., Ikeuchi, S., Tanaka, Y.D. 1981, PASJ, 33, 23
Heiles, C. 1990, ApJ, 354, 483

Hellsten, U., Gneden, N.Y., Miralda-Escude, J. 1998, ApJ, 509, 56
Henry, R.C., Fritz, G., Meekins, J.F., Chubb, T., Friedman, H. 1971, ApJ, 163, L73
Herbstmeier, U., Mebold, U., Snowden, S.L., Hartmann, D., Burton, W.B., Moritz, P., Kalberla, P.M.W., Egger, R. 1995, A&A, 298, 606
Houck, J.R., Bregman, J.N. 1990, ApJ, 352, 506
Howk, J.C., Savage, B.D. 1997, AJ, 114, 2463
Ipavich, F. 1975, ApJ, 196, 107
Kahn, F.D. 1981, in Investigating the Universe: Papers presented to Zdenek Kopal on the occasion of his retirement, September 1981 (Dordrecht, Reidel), 1
Kahn, F.D. 1992, Irish Astronomical Journal, 20, 278
Kahn, F.D. 1994, ApSpSci, 216, 325
Keppel, J.W., Dettmar, R.-J., Gallagher III, J.S., Roberts, M.S. 1991, ApJ, 374, 507
Kerp, J., Burton, W.B., Egger, R., Freyberg, M.J., Hartmann, D., Kalberla, P.M.W., Mebold, U., Pietz, J. 1999, A&A, 342, 213
Klypin, A., Kravtsov, A.V., Valenzuela, O., Prada, F. 1999, ApJ, 522, 82
Kraan-Korteweg, R.C., van Driel, W., Briggs, F., Binggeli, B., Mostefaoui, T.I. 1998, A&AS, 135, 255
Lukasiak, A., Ferrando, P., McDonald, F.B., Webber, W.R. 1994, ApJ, 423, 426
Mac Low, M.-M., McCray, R. 1988, ApJ, 324, 776
Mac Low, M.-M., McCray, R., Norman, M.L. 1989, ApJ, 337, 141
Maloney, P.R., Putman, M.E. 2003, ApJ, 589, 270
Martin, C., Bowyer, S. 1990, ApJ, 350, 242
Mateo, M.L. 1998, ARA&A, 36, 435
Mathewson, D.S., Ford, V.L. 1984, in IAU Symp. 108, Structure and Evolution of the Magellanic Clouds, eds. S. van den Bergh, K.S. de Boer (Dordrecht: Reidel), 125
McCammon, D., Sanders, W.T. 1990, ARA&A, 28, 657
McKee, C.F., Ostriker, J.P. 1977, ApJ, 218, 148
Miralda-Escude, J., Rees, M. 1993, MNRAS, 260, 61
Moore, G., Ghigna, S., Governato, F., Lake, G., Quinn, T., Stadel, J., Tozzi, P. 1999, ApJ, 524, L19
Mulchaey, J.S., Davis, D.S., Mushotzky, R.F., Burstein, D. 1996, ApJ, 456, 80
Münch, G. 1952, PASP, 64, 312
Münch, G., Zirin, H. 1961, ApJ, 133, 11
Murphy, E.M., Otte, B., Howk, J.C., Wang, Q.D., Oegerle, W.R., Sembach, K.R., and the *FUSE* Science Team 2001, BAAS, 199, 9710
Oort, J.H. 1970, A&A, 7, 381
Otte, B., Murphy, E.M., Howk, J.C., Wang, Q.D., Oegerle, W.R., Sembach, K.R. 2003, ApJ, 591, 821
Pietz, J., Kerp, J., Kalberla, P.M.W., Mebold, U., Burton, W.B., Hartmann, D. 1996, A&A, 308, L37
Pietz, J., Kerp, J., Kalberla, P.M.W., Burton, W.B., Hartmann, D., Mebold, U. 1998, A&A, 332, 55
Pildis, R.A., Bregman, J.N., Evrard, A.E. 1995, ApJ, 443, 514
Ponman, T.J., Bourner, P.D.J., Ebeling, H., Bohringer, H. 1996, MNRAS, 283, 690
Putman, M.E., Gibson, B.K, Staveley-Smith, L., et 23 al. 1998, Nature, 394, 752
Putman, M.E., Bland-Hawthorn, J., Veilleux, S., Gibson, B.K., Freeman, K.C., Maloney, P.R. 2003, ApJ, 597, 948
Rand, R.J. 1996, ApJ, 462, 712
Rand, R.J. 1997, ApJ, 474, 129
Rand, R.J. 1998a, PASAu, 15, 106
Rand, R.J. 1998b, ApJ, 501, 137
Rand, R.J., Stone, J.M. 1996, AJ, 111, 190
Rand, R.J., Kulkarni, S.R., Hester, J.J. 1990, ApJ, 352, L1
Rice, W., Lonsdale, C.J., Soifer, B.T., Neugebaur, G., Koplan, E.L., Lloyd, L.A., de Jong, T., Habing, H.J. 1988, ApJS, 68, 91.

Richter, P., Sembach, K.R., Wakker, B.P., Savage, B.D., Tripp, T.M., Murphy, E.M., Kalberla, P.M.W., Jenkins, E.B. 2001, ApJ, 559, 318
Roberts, M.S., Haynes, M.P. 1994, ARA&A, 32, 115
Rosen, A., Bregman, J.N. 1995, ApJ, 440, 634
Rosen, A., Bregman, J.N., Norman, M.L. 1993, ApJ, 413, 137
Rupen, M. 1991, AJ, 102, 48
Santillán, A., Franco, J., Martos, M. 1999, ApJ, 515, 657
Savage, B.D., Sembach, K.R. 1994, ApJ, 434, 145
Savage, B.D., Sembach, K.R., Wakker, B.P., Richter, P., Meade, M., Jenkins, E.B., Shull, J.M., Moos, H.W., Sonneborn, G. 2003, ApJS, 146, 125
Scoville, N.Z., Thakkar, D., Carlstrom, J.E., Sargent, A.E. 1993, ApJ, 404, L59
Shapiro, P.R., Field, G.B. 1976, ApJ, 205, 762
Shelton, R.L. 2003, ApJ, 589, 261
Shelton, R.L., Kruk, J.W., Murphy, E.M., et 22 al. 2001, ApJ, 560, 730
Smith, B.W. 1977, ApJ, 211, 404
Snowden, S.L., Egger, R., Finkbeiner, D.P., Freyberg, M.J., Plucinsky, P.P. 1998, ApJ, 493, 715
Somerville, R.S. 2002, ApJ, 572, L23
Spitzer, L. 1956, ApJ, 124, 20
Strong, A.W., Moskalenko, I.V. 1998, ApJ, 509, 212
Strong, A.W., Moskalenko, I.V. 2000, in Topics in Cosmic Ray Astrophysics (Horizons in World Physics, 230), ed. M.A. DuVernois (New York: Nova Scientific), 81
Sukumar, S., Allen, R.J. 1991, ApJ, 382, 100
Swaters, R. 1996, Ph.D. thesis, Rijks Universiteit Groningen
Tenorio-Tagle, G., Franco, J., Bodenheimer, P., Rozyczka, M. 1987, A&A, 179, 219
Tomisaka, K. 1998, MNRAS, 298, 797
Tomisaka, K., Ikeuchi, S. 1986, PASJ, 38, 697
Tomisaka, K., Habe, A., Ikeuchi, S. 1981, ApSpSci, 78, 273
Tufte, S.L., Reynolds, R.J., Haffner, L.M. 1998, ApJ, 504, 773
Tufte, S.L., Wilson, J.D., Madsen, G.J., Haffner, L.M., Reynolds, R.J. 2002, ApJ, 572, L153
van den Bergh, S., Tammann, G.A. 1991, ARA&A, 29, 363
van der Kruit, P.C., Searle, L. 1981, A&A, 95, 116
van Woerden, H., Schwarz, U.J., Peletier, R.F., Wakker, B.P., Kalberla, P.M.W. 1999, Nature, 400, 138
Volger, A., Pietsch, W., Kahabka, P. 1995, AdSpR, 16, 139
Volger, A., Pietsch, W., Kahabka, P. 1996, A&A, 305, 71
Wakker, B.P. 1990, Ph.D. thesis, Rijks Universiteit Groningen
Wakker, B.P., van Woerden, H. 1991, A&A, 250, 509
Wakker, B.P., Howk, J.C., Savage, B.D., van Woerden, H., Tufte, S.L., Schwarz, U.J., Benjamin, R., Reynolds, R.J., Peletier, R.F., Kalberla, P.M.W. 1999, Nature, 402, 388
Wang, Q.D., Walterbos, R.A.M., Steakley, M.F., Norman, C.A., Braun, R. 1995, ApJ, 439, 176
Wang, Q.D., Immler, S., Walterbos, R., Lauroesch, J.T., Breitschwerdt, D. 2001, ApJ, 555, L99
Weaver, R., Castor, J., McCray, R., Shapiro, P., Moore, R. 1977, ApJ, 218, 377
Webber, W.R., Lee, M.A., Gupta, M. 1992, ApJ, 390, 96
Weiner, B.J. 2003, in ASSL Conf. Proc. 281, The IGM/Galaxy Connection, eds. J.L. Rosenberg, M.E. Putman (Dordrecht: Kluwer), 163
Zirakashvili, V.N., Breitschwerdt, D., Ptuskin, V.S., Völk, H.J. 1996, A&A, 311, 113
Zwaan, M.A. 2001, MNRAS, 325, 1142
Zwaan, M.A., Briggs, F.H. 2000, ApJ, 530, L61
Zwaan, M.A., Briggs, F.H., Sprayberry, D., Sorar, E. 1997, ApJ, 490, 173

17. UNSOLVED MYSTERIES OF HIGH-VELOCITY CLOUDS

ROBERT A. BENJAMIN
Department of Physics, University of Wisconsin-Whitewater, USA; benjamir@uww.edu

Abstract. This contribution explores three "unsolved" mysteries of the high-velocity clouds: what surrounds HVCs, what makes up HVCs, and are HVCs important? Most models of HVCs assume, for convenience, that they exist in an isothermal, static Galactic Halo, but this chapter shows that there is considerable uncertainty in the density, temperature, ionization, and velocity structure of the gaseous environs of the Galaxy. Most treatments of HVCs concentrate on the neutral gas, but the ionized gas and possible dark matter could be the dominant component by mass. There are even hints that some HVCs could be completely ionized. Magnetic fields, dust, and molecular gas also play a relatively unconstrained role in shaping HVC structure. Depending on the total mass reservoir represented by HVCs, they could play an important role in galactic chemical evolution, and may provide our sole way of studying the low-mass dark-matter potential wells expected in ΛCDM models of galaxy formation. The more we learn about HVCs, the more mysterious and tantalizing they become.

1. Introduction

For four decades, the two principal unsolved mysteries of HVCs have been: how far away are they and where do they come from? The status of these discussions is addressed by several contributions to this volume. It is now clear that there is not a single answer. Clouds like complex M lie only a few kiloparsecs from the disk, have solar metallicity, and appear to be galactic in origin. Clouds like the Magellanic Stream are $\sim$50 kpc away, are dynamically associated with nearby galaxies, and have a metallicity similar to their parent galaxy. Clouds like complex A and C lie at intermediate distances, but with uncertain origins. And finally, some clouds have an origin that is still uncertain, particularly the compact high-velocity clouds. These questions rightly deserve all the scrutiny that the astronomical community can bring to bear, since answering them may lead to fundamental insights

H. van Woerden et al. (eds.), High-Velocity Clouds, 371–389

on issues of galaxy formation and evolution.

For this chapter, I would like to highlight different, but related questions. First, what surrounds high-velocity clouds? Second, is neutral hydrogen all there is to high-velocity clouds? And this contribution ends with possibly the most significant unsolved mystery of high-velocity clouds: are high-velocity clouds important?

2. What surrounds HVCs?

UV and optical absorption-line work indicates that some HVCs lie within a few kiloparsecs of the Galactic Disk (Wakker 2001). On the other hand, some clouds might lie up to a megaparsec from the Galaxy. At these distances, HVCs provide one of the few probes of the gaseous Galactic Halo and of the gas between the galaxies of the Local Group. Here I discuss what we know, and more importantly, what we don't know, about the temperature, density, and velocity of the gas in the halo of the Milky Way and the gaseous medium of the Local Group.

2.1. EXISTENCE OF THE GASEOUS GALACTIC HALO

Even before HVCs were identified as a class, it was known that there were neutral clouds well above (up to $\sim$1 kpc) the mid-plane of the Galaxy (Münch & Zirin 1961). The existence of these clouds led Lyman Spitzer to suggest that the Galaxy is surrounded by a hot gaseous corona, with $T \geq 10^6$ K (Spitzer 1956). Without an external medium to provide pressure confinement, Spitzer reasoned, neutral clouds would dissipate in a few sound-crossing times. Most researchers take a low-density confining medium for granted now, but it was not always so. Consider the following statement regarding clouds whose random motions take them out of the galactic plane (Oort 1954): *"It will take these clouds about 50 million years before they return to the denser gas layer near the plane. But in about half this time the clouds will practically disintegrate, and therefore return to the galactic layer as a more or less continuous stream instead of as discrete clouds."*

Using estimates for the galactic gravitational potential, gas hotter than $\sim 10^6$ K could support itself up to scaleheights greater than 1 kpc by thermal pressure alone (see Fig. 1). By analogy with the solar corona, Spitzer suggested that this was evidence for the existence of a Galactic Corona. He further suggested that one way to confirm such a halo would be the detection of the UV resonance absorption lines of O VI and other highly-ionized atoms. The recent discovery of O VI in HVCs (Sembach et al. 2000, 2003) would appear to confirm this prediction. These observations force us to think much harder about the gaseous surroundings of the Galaxy. What surrounds HVCs is an integral part of what they are. Without an exter-

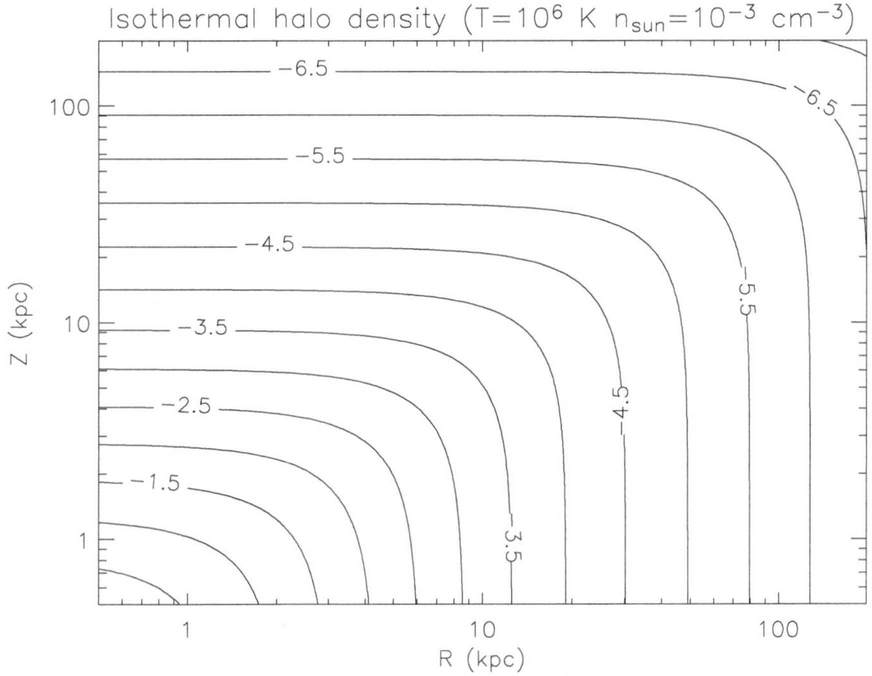

Figure 1. Hydrogen particle density as a function of position for an isothermal halo with T=10^6 K, using the galactic potential model 2 of Dehnen & Binney (1998). The density at the position of the Sun is assumed to be $n_H(R_\odot)$=10^{-3} cm^{-3}. Although the physics determining the density structure of the halo is likely to be considerably more complex than such a simple model, isothermal or adiabatic halos constrained by a few observational limits are used to predict the density of the gas surrounding high-velocity clouds.

nal medium they would not exist, and the interaction with the external medium almost certainly affects their structure and evolution. Given that an extended gaseous halo for the Galaxy exists, what are its fundamental parameters?

2.2. DENSITY STRUCTURE OF THE GASEOUS GALACTIC HALO

Measuring the density of a (nearly) invisible confining medium at the boundaries of intergalactic space is a tough problem, but there have been several attempts to do this using HVCs. Here I describe the methods, their advantages and disadvantages, and then note some new avenues that provide independent information on the density and temperature of this gas.

2.2.1. *Evidence for and implications of confined HVCs*

Early attempts to study the overall sky distribution of HVCs led to the realization that many of the "clouds" are not single, isolated clouds, but rather form "complexes", with coherent velocities found over large sections of the sky. Given their large angular sizes, studies of HVC complexes are more advanced than studies of isolated clouds, since there are more background sources available for metallicity measurements and distance constraints. One can also profitably study the velocity gradients and dispersions of these complexes for clues to their origin. Ever-higher angular resolution 21-cm observations have indicated significant substructure in these complexes. Observations of the Magellanic Stream (Putman et al. 2002; Stanimirovic et al. 2002), for example, show that this well-defined and well-modeled complex breaks into numerous elongated clumps, some of which are quite cometary in appearance.

At the same time, significant attention has also been brought to bear on the sub-population of "compact" high-velocity clouds (CHVCs). These are small, isolated clouds, and it has been suggested that it is these clouds that are the most likely to be scattered throughout the Local Group (rather than being immersed in the gaseous Galactic Halo). It is not clear they really are a distinctly separate class of clouds (Blitz et al., Ch. 14), but they have attracted much interest lately. Initial observations indicated that the clouds were deserving of the name "compact" HVCs, but subsequent deeper observations indicate that even these isolated clouds show elongated structures that suggest an interaction with an ambient medium.

Finally, both HVCs and CHVCs occasionally show evidence of "core-halo" structures, with narrow-linewidth H I cores and wider-linewidth H I envelopes (Giovanelli & Haynes 1976; Schwarz & Oort 1981; Wakker & Schwarz 1991; Braun & Burton 2000). This indicates a sufficiently-high bounding pressure to allow a two-phase medium structure, either due to a high ambient halo pressure (which could be due to proximity to the Galaxy) or ram pressure associated with the interaction with ambient gas.

Both morphology and the phase structure of HVCs therefore demonstrate the effects of this ambient medium. Comparison of these observations with models for interactions can thus be used to constrain the density of the halo gas at the position of the cloud. Model assumptions in the literature include the following:

Thermal pressure balance. Dating back to Spitzer (1956), the assumption here is that the internal thermal pressure of the cloud balances the external thermal pressure of the confining medium:

$$n_{halo} \cong n_{\rm HVC} \frac{T_{\rm HVC}}{T_{\rm halo}}.$$

This approximation has been used by numerous authors, most recently by Espresate et al. (2002) and Stanimirovic et al. (2002). The key assumptions that one must make are the conversion of H I line width to $T_{\rm HVC}$ and the assumed value for the temperature of the halo gas, usually taken to be isothermal with a temperature of $T_{\rm halo} \cong 10^6$ K. One must also know the distance of the cloud in order to convert the observed column density to $n_{\rm HVC}$. With all of these caveats, typical numbers derived by this method are $n_{\rm HVC}=10^{-4}$ to 10^{-3} cm^{-3}. Of course, probably the most important caveat in this approximation is the assumption that the thermal pressure is the dominant source of pressure. This is discussed in the next section.

Ram pressure-thermal pressure balance. This approximation relaxes one of the principal assumptions above, and takes into account the fact that the ram pressure associated with an HVC's motion can be a significant contribution to the total pressure. In the case where ram pressure dominates,

$$n_{\rm halo} \propto n_{\rm HVC} \frac{T_{\rm HVC}}{v_{\rm HVC}^2}.$$

This approach was used, for example, by Weiner & Williams (1996) to interpret their Magellanic Stream observations. The resulting density estimates are not unlike those for the first method, for the simple reason that the assumed temperature of the halo gas turns out to be approximately equivalent to the cloud velocities in the galactic gravitational potential. The same caveats about non-thermal pressures mentioned above also apply here. A modification of this approach was suggested by Benjamin & Danly (1996), who balanced deceleration by ram pressure with gravitational acceleration. Both approaches require independently determined cloud distances to determine halo density as a function of position.

Evaporation limit. If neutral HVCs are immersed in a hot ionized halo, they are prone to evaporation. The higher the halo density, the higher the evaporative mass flux. Using an approach similar to Cowie & McKee (1976), Murali (2000) estimates that the halo density must be less than $n_{\rm halo}=10^{-5}$ to 10^{-7} cm^{-3}, or else HVCs will evaporate in less than 500 Myr. The sensitivity of cloud evaporation to the magnetic configuration makes this density estimate highly uncertain, but the principle that the evaporative processes provide limits on the ambient density is an important and underappreciated one.

2.2.2. *Direct measurements of gaseous halo density?*

The recent detection of zero-redshift X-ray absorption due to O VII and O VIII ions in *CHANDRA* and *XMM-NEWTON* spectra of several bright extragalactic sources has opened up the possibility of *directly* measuring the density of the gaseous halo gas as well as of the gas filling the Local Group. These observations, summarized nicely by Sembach (2004), typically yield column densities of these ions in the range 10^{15} to a few $\times 10^{16}$ cm^{-2}. The path length occupied by an O^{+6} ion, for example, can be written as

$$L = (6.6\ \mathrm{kpc})\ f_{occ}^{-1} \left(\frac{A(O)}{4.9 \times 10^{-4}}\right)^{-1} \left(\frac{N(O^{+6})}{10^{16}\ \mathrm{cm}^{-2}}\right) \left(\frac{n_e}{10^{-3}\ \mathrm{cm}^{-3}}\right)^{-1},$$

where the column density of highly-ionized oxygen $N(\mathrm{O}^{+6})$ and halo electron density (n_e) have been normalized to their estimated values, the oxygen abundance $A(\mathrm{O})$ comes from Allende-Prieto et al. (2001), and the f_{occ} term is a (linear) filling factor. Photoionization models which simultaneously predict the observed column densities of O VII and O VIII ions *and* O VI, detected in the ultraviolet, predict densities of $n_e \sim 10^{-6}$ cm^{-3}, yielding surprising large, but not impossible, pathlengths (Nicastro et al. 2001). However, this assumes that all three ions are completely intermixed. If the O VI arises in an interface zone between highly ionized halo gas and a colder HVC, this extremely low density is no longer required.

An additional constraint on the density of this gas comes from soft X-ray detections of O VII and O VIII emission lines (McCammon et al. 2002) and *ROSAT* 3/4 keV band observations which are (presumably) dominated by these emission lines. If the density of the O VII and O VIII bearing gas were significantly higher than $n_e = 10^{-3}$ cm^{-3}, these emission constraints would be violated (assuming the gas is in ionization equilibrium). However, the fact that the highest observed column density of O VII is toward 3C 273, a direction that passes through a 3/4 keV X-ray enhancement due to the North Polar Spur, hints that this ionized gas is not completely independent of the Galaxy (Fang et al. 2003).

By constraining the occupation length in the above equation, one can more confidently constrain the density of the ambient medium surrounding HVCs. X-ray absorption studies of a set of sources outside the Galactic Disk, but within the Local Group, should help to clarify the distribution of this gas. The recent detection of O VII absorption (log N(O VII)=16.2 to 16.7) toward the low-mass X-ray binary 4U 1820−303 (Futamoto et al. 2004) in the globular cluster NGC 6624 is an important first step. Since this globular cluster is only 7.6±0.4 kpc distant (Kuulkers et al. 2003), and the column density of O VII is similar to that on the extragalactic lines of sight, it raises the possibility that nearly *all* of the detected O VII and O VIII could lie within the galactic environs. Assuming a solar metallicity, anywhere from 10% to 50% of the ionized gas indicated by the dispersion measure of pulsars in this globular cluster (Gomez et al. 2001) could arise in hot gas. The detection of radio pulsars in Local Group galaxies would also constrain the density of this hot ionized gas. The asymptotic dispersion measure DM $\sin|b|$ for galactic pulsars is approximately 22 cm^{-3} pc. For an intergalactic density of 10^{-4} cm^{-3}, a pulsar in the LMC should have a 20% higher value; a pulsar in Leo I would have a 100% higher value.

2.3. TEMPERATURE STRUCTURE OF THE GASEOUS GALACTIC HALO

The consensus assumption regarding the gaseous envelope of the Galaxy is that it is coronal and usually isothermal (e.g. Wolfire et al. 1995; Ferrara & Field 1997), so I would like to start by considering an alternative point of view. Although I feel that it is certainly a reasonable assumption and is consistent with several models, we lack compelling proof as to the temperature of gas several kiloparsecs away from the plane of the Galaxy.

Arguments for a warm ($T\cong10^4$ K) gaseous halo. First, it is interesting to note that the original motivation for the Spitzer hot halo has disappeared! We now know that from a hundred parsecs up to a few kiloparsecs, the warm ($T\cong10^4$ K) ionized medium is the dominant component of the ISM, with pressure support from kinematic, magnetic, and cosmic-ray pressures (Reynolds 1989; Boulares & Cox 1990), rather than thermal pressure. This layer is analogous to the solar chromosphere, since it is at a similar temperature, characterized by optical line emission, and is sufficient to provide the pressure confinement of clouds up to a few kiloparsecs. What happens above this point? Several models have suggested that some combination of cosmic-ray pressure, magnetic buoyancy, and centrifugal forces could drive a Galactic Wind (Chevalier & Fransson 1984; Breitschwerdt et al. 1991; Parker 1992). Such a wind would have a low initial gas temperature, and adiabatic expansion would drive the temperature even lower. Most of these models do not worry about additional heating sources, for example heating by Alfven waves, magnetic reconnection, or external photoionization from the meta-galactic radiation field.

Even if the gas near the disk was primarily coronal, the effects of adiabatic expansion would serve to lower the gas temperature of the outflowing gas. For adiabatic gas with γ=5/3, $T\propto\rho^{1/3}$. Assuming a constant-velocity spherical outflow, $\rho\propto r^{-2}$, indicates that $T(r)=T(r_0)(r/r_0)^{-2/3}$. If we assume the gas temperature is 10^6 K at a reference point of r_0=2 kpc, then by the time this gas reaches 50 kpc (the distance of the Magellanic Stream) the temperature has dropped to 116,000 K. This is without including the effects of radiative cooling, which can be severe if the gas is at a high-enough density as it reaches $T\sim10^5$ K, the peak of cooling efficiency (Sutherland & Dopita 1994). Breitschwerdt & Schmutzler (1999) have calculated the structure of such an outflow, and confirm this magnitude drop in gas temperature for an initially coronal wind. Again, this situation is analogous to the Sun, where the solar wind electron temperature drops from $T=1.2\times10^6$ K (r_o=1.175 $R_\odot$) in the corona to $T=3\times10^5$ K when it reaches the Earth's orbit (r=215 $R_\odot$). The alert reader will note that this drop in temperature is not as much as the adiabatic approximation predicts: T=37,000 K. This is the effect of additional heating sources, an effect that also seems likely to be present in any Galactic Wind.

Given the low density of halo gas and significant sources of non-thermal and even thermal pressures, it seems likely that gas would be accelerated into a wind. If potential heating sources are found to be negligible, this wind would adiabatically expand, cool, and the bulk of the gaseous Galactic Halo could be at low temperatures. Like the Solar Wind, the wind would terminate when its ram pressure balances the pressure of the inter-Local Group medium, a highly uncertain parameter.

Arguments for a hot ($T \cong 10^6$ K) gaseous halo. Since Spitzer's hypothesis of a hot Galactic Corona, many theoretical mechanisms have been suggested for heating the diffuse gas well above the Galaxy. Mechanical heating of the disk ISM by supernovae and stellar winds could result in over-pressured hot gas which flows out of the disk in buoyant hot bubbles, "galactic fountains" (Shapiro & Field 1976; Bregman 1980), or galactic winds (Bregman 1978). Gas may also be heated *in situ* by upward traveling waves that steepen into shocks in the low-density galactic atmosphere (Walters & Cox 2001). Reconnection of magnetic fields could also heat the low-density gas (Raymond 1992; Parker 1993; Zimmer et al. 1997). Finally, hot intergalactic gas may settle onto the Galaxy in a cooling flow (Sembach et al. 2000).

There are three direct observational lines of evidence purporting to show a hot halo. First, analyses of broadband X-ray emission maps from the Wisconsin sounding-rocket surveys (McCammon & Sanders 1990), *HEAO* (Garmire et al. 1992), and *ROSAT* (Pietz et al. 1998; Kuntz & Snowden 2000; Almy et al. 2000) all suggest the presence of an emission component with an angular distribution that is consistent with expectations for gas filling the gravitational potential of the Galaxy. X-ray shadowing studies indicate that at least some of this emission originates more than a few hundred parsecs above the plane (Burrows & Mendenhall 1991; Snowden et al. 1994). Second, there are many recent detections of highly-ionized gas in absorption along many sight lines, including high-velocity O VI and C IV ultraviolet (Sembach et al. 1999, 2003; Fox et al. 2004) and O VII and O VIII X-ray absorption lines (Nicastro et al. 2001; Fang et al. 2003). Third, there is the detection of UV (Martin & Bowyer 1990; Shelton 1999; Dixon et al. 2001) and X-ray (Sanders et al. 2001; McCammon et al. 2002) emission lines, which appear to be brightest at high latitude. Finally, the fact that nearby edge-on spiral galaxies show extended X-ray (Bregman & Pildis 1994) and O VI emission (Otte et al. 2003) is indirect evidence for the existence of a hot halo.

Because the X-ray and most UV emission measurements have little or no velocity resolution (which would allow us to spatially locate the gas), and because they are weighted by n_e^2, they are principally evidence of hot gas in the nearby halo (within a few kiloparsecs of the Sun), where the density is expected to be higher. Absorption-line data provide much more significant constraints, since they are clearly associated with HVCs, some of which, like the Magellanic Stream, are at known distances. In a thorough comparison of available models with high-quality data for complex C, Fox et al. (2004) argue that the data are most consistent with a conductive interface between hot halo and cool HVC gas. Is it possible that a model of a warm, $T \cong 10^4$ K, halo could also be consistent with observations? In such a model, the O VII and O VIII absorption would need to be explained by frozen-in ionization from an adiabatically expanding outflow. The absorption associated with HVCs would come either from shocks surrounding the clouds, or from compression and recombination or charge exchange on the face of the cloud. It would be valuable to see if such a model could adequately explain all of the available observations.

2.4. VELOCITY STRUCTURE OF THE GASEOUS GALACTIC HALO

Given that the density and temperature of the gas surrounding HVCs is so poorly known, it is no surprise that the velocity structure of gas more than a few kiloparsecs from the galactic plane is also *terra incognita*. Possibilities include a global outflow or wind from the full Galactic Disk (Zirakashvili et al. 1996), a bipolar outflow from only the central regions of the Galaxy (Bland-Hawthorn & Cohen 2003), a large, extended rotating halo (Collins et al. 2003), a global accretion flow from Local Group gas (Oort 1970), or a mixture of all of the above phenomena. In the absence of any compelling evidence, most models have assumed a static (and isothermal) halo (Ferrara & Field 1994; Wolfire et al. 1995). Both models and observations of outflows from other galaxies suggest that galactic outflows could achieve speeds up to $\sim$1000 km s^{-1}, twice the escape speed from the Galaxy. Such flows, if present, could dwarf the infall or rotational velocity that the gas might otherwise have.

Studies of cometary or elongated HVCs can constrain these possibilities, with the principal uncertainty being the difficulty in untangling the distance uncertainties from the independent motion of both HVC and inter-HVC medium. Studies of cometary morphologies and velocity bridges (Pietz et al. 1996; Brüns et al. 2000), for example, show lots of evidence for interactions, but no clear patterns. Perhaps the most hopeful possibility lies in the detailed study of the Magellanic Stream, which spans over 100 degrees across the sky, and has a reasonably constrained model for both cloud distances and orbital velocities (Putman et al. 2002). Studies of HVCs near the galactic equator, such as complex H (Lockman 2003), could also be used to constrain the velocity of the extended gaseous Galactic Disk.

2.5. THE UNSOLVED MYSTERIES

To summarize this section, studies of HVCs indicate that the gas density in the halo of the Galaxy and the Local Group is probably somewhere between 10^{-7} and 10^{-3} cm^{-3}; the temperature of this gas lies between 10^4 and 10^7 K; it is not known whether this gas is or is not in ionization equilibrium; and the gas is either static, infalling, rotating, flowing outward at $\sim$1000 $\mathrm{km\,s^{-1}}$, or all of the above. Yet the physical state of this gas is undeniably important in shaping the structure and evolution of HVCs. It is for this reason that I suggest it as one of the unsolved mysteries of HVCs.

3. HVCs and H I: is that all there is?

For decades, the study of high-velocity clouds had been the exclusive domain of the 21-cm radio astronomer. The development of new detectors over a wide range of wavelengths is now making it possible to address whether the neutral hydrogen represents the bulk of the clouds or just the tip of the iceberg. The following section discusses all of the possible ingredients that might go into constructing a high-velocity cloud.

3.1. HVCS AND MOLECULAR GAS

Studies of H I line widths in HVCs indicate that at least some of these clouds have warm, $T\sim5000$ K, neutral envelopes and cold, $T\sim100$ K, cores. A surprise was the detection of molecular hydrogen in some HVCs (Richter et al. 1999). Comparing the data to models of molecular hydrogen formation and destruction allows one to estimate the volume density of molecular gas. This comparison indicates that the region occupied by molecules is tiny, typically only a pathlength of $\sim$100 astronomical units (Richter et al. 2003). Although the fraction of HVCs made up of molecules is negligible, the very presence of molecular gas provides information on the pressure within the cloud as well as the nature of time-dependent molecule formation.

3.2. HVCS AND DUST

A true mystery of HVCs lies in the non-detection of dust in these clouds. Searches with *IRAS* for infrared emission from these clouds have yielded only upper limits (Wakker & Boulanger 1986), indicating either a lack of dust or a lack of a source to heat the dust to detectable emission levels. Abundance depletion patterns hint that for some HVCs there may yet be dust to be found. Since the same processes that inject metals into these clouds ought to inject dust as well, it would be remarkable if no evidence of dust were to come to light, and would indicate something important about the ability of dust to survive in intergalactic space.

3.3. HVCS AND IONIZED GAS

Studies of Hα emission from HVCs show that ionized gas is frequently, and arguably always, present (Tufte et al. 1998, 2002; Putman et al. 2003). It had been suggested that the brightness of Hα emission from HVCs could be used as distance indicator; clouds near the Galaxy would glow brighter than distant clouds, which would intercept a geometrically-diluted ionizing flux (Bland-Hawthorn & Maloney 1999, 2002; Maloney & Bland-Hawthorn 1999). The data are more ambiguous. The Hα emission appears to be patchy across cloud complexes, and the emission from the Magellanic Stream is much brighter than expected. A complete Hα map of a few HVCs might clarify the matter. Until these discrepancies are understood, the use of Hα emission as a surrogate for distance should be viewed with some skepticism.

Is this ionized gas a substantial fraction of the cloud mass? For directions for which both emission and absorption observations are available, the ionized pathlength is estimated to be only ~25% (Wakker et al. 1999), but these measurements are limited to regions where there is a significant neutral column density. Recent models of the phase structure of HVCs indicate that the ionized gas could dominate the total baryonic mass (Sternberg et al. 2002; Maloney & Putman 2003; Gnat & Sternberg 2004). If the clouds are at Local Group type distances, $D{\sim}0.4$ to 1 Mpc, they have sufficiently low density that a huge column density is necessary to shield the neutral HVC core, yielding $M_{\rm ionized}/M_{\rm neutral}{\gtrsim}100$! This number goes down considerably if the clouds are circumgalactic, $D{\lesssim}300$ kpc. Depending on the bounding pressure of the ambient medium and the possible presence of dark matter, models predict that the ionized gas mass could be anywhere from 0.2 to 10 times the neutral mass. Thus, estimates of the total gas mass of HVCs based on neutral H I only could be serious underestimates of the total mass.

The Hα observations constrain the presence of ionized envelopes around HVCs, but ultraviolet absorption-line studies hint at the presence of purely-ionized HVCs. Studies of *FUSE* O VI absorption in the vicinity of HVCs indicate the presence of absorption by highly-ionized gas at similarly high velocities, but separated by several degrees from the neutral clouds (Wakker et al. 2003; Sembach et al. 2003). Does this indicate large ionized extents for HVCs and HVC complexes? Or does it indicate the existence of completely separate, fully-ionized HVCs? Or are these just very low column density neutral clouds such as those traced by Na I and Ca II optical absorption lines, summarized in Wakker & van Woerden (1997). Velocity-resolved emission-line observations in Hα and other transitions could potentially uncover a new population of objects and an additional gas mass reservoir for galactic evolution.

3.4. HVCS AND MAGNETIC FIELDS

Most researchers are happy to neglect the role of magnetic fields in affecting the structure and nature of HVCs. This is easy to justify, given the lack of detection of magnetic fields in HVCs, save the 21-cm Zeeman measurement indicating a $\sim$10 μG field in a core of complex A (Kazés et al. 1991). But of course, absence of evidence is not evidence of absence. There are several reasons why it is important to search for evidence of magnetic fields in HVCs. First, HVCs with an extragalactic origin contain information about the magnetic-field processing in an environment very unlike the gas associated with the Galaxy, and could lead to insight into the generation of magnetic fields in the early Universe. Second, as this gas is accreted, it adds magnetic flux as well as mass and metals to the Galaxy, a fact that should be taken into account in models of galactic magnetic field generation (Zweibel 2004). Third, magnetic fields can supply an important source of pressure that would need to be taken into account in future models of HVC structure and interaction. Fourth, magnetic fields could supply a tension that could act to stabilize HVCs against hydrodynamic instabilities, which could otherwise act to destroy the clouds on short timescales (Benjamin 1999; Gregori et al. 1999, 2000; Konz et al. 2002). Fifth, magnetic reconnection on the face of HVCs could produce local and detectable heating (Zimmer et al. 1997). Finally, HVCs are potentially sites of long-lasting shocks which, like supernova shocks, can serve as sites for particle acceleration in the gaseous halo (Hedrick & Cox 1977).

3.5. HVCS AND STARS

Dwarf galaxies can have H I masses from as low as 2.6×10^4 $M_\odot$ (Sculptor) to $\sim10^8$ $M_\odot$ (IC 10) (Mateo 1998). If some of the observed HVCs were at distances of several hundred kpc, they would be approaching the upper end of this range. It is therefore reasonable to ask if there is evidence of stellar content in these objects. Several studies have used the positions of HVCs to target searches for stars (Simon & Blitz 2002; Davis et al. 2002; Willman et al. 2002). These searches have yielded only upper limits down to a surface brightnesses of $\mu_{V,0}$=26.7 mag arcsec^{-2}.

A related question is whether a dynamical or morphological connection can be made between stellar streams and HVC complexes. Searches for a stellar counterpart to the gaseous Magellanic Stream, for example, have yielded only upper limits. As our knowledge of stellar features in the Galactic Halo improves, it is useful to ask if any HVCs match up with these structures. Putman et al. (2004) argue for a connection between the Anti-Center complex of HVCs and the Sagittarius Dwarf galaxy. This identification could be corroborated by a comparison of the metallicity of this complex with the metallicities of the stars in the Sagittarius Dwarf galaxy.

3.6. HVCS AND DARK MATTER

The suggestion by Blitz et al. (1999) that at least some fraction of the observed HVCs are spread through the Local Group, as opposed to within a few hundred kiloparsecs of the Galaxy, has proponents (Braun & Burton 1999; Nicastro et al. 2003) and problems (Maloney & Putman 2003; Putman et al. 2003; Sternberg et al. 2002; Wakker 2004). However, one novel element of this work seems likely to remain a permanent feature of the debate on the nature of HVCs, namely, do HVCs trace dark matter? Models of galaxy formation in the Cold Dark Matter paradigm generically predict a large number of satellite "galaxies" around our Galaxy, nearly 2000 more than are observed (Klypin et al. 1999; Moore et al. 1999). Could these objects be traced by the "condensation" of H I within their potential wells?

This scenario has several attractive features. First, the increased central pressure in the clouds would help to explain the existence of core-halo structure for clouds in regions of extremely low external pressure (Wolfire et al. 1995; Sternberg et al. 2002). Second, it has been shown that the potential well could also stabilize clouds against hydro-magnetic instabilities (Quilis & Moore 2001). Finally, the presence of dark-matter potential wells could provide a way to *create* neutral HVCs – only when the bounding pressure associated with the gaseous halo becomes high enough will clouds start to self-shield themselves. However, there is a balance between a high-enough pressure to create neutral HVCs and such a high ram pressure that the gas is separated from the potential well.

4. Are high-velocity clouds important...

Are high-velocity clouds important? The widespread interest in HVCs right now is due to their mystery and their *potential* for being important. Importance is in the eye of the beholder, of course. If your interest is the inner workings of the Galaxy, then that subset of HVCs that is part of a galactic circulation is of great interest for understanding the flows of energy and mass within the Galaxy. But to the community, it seems that the pressing issue of the day is working toward an understanding of the assembly and evolution of galaxies. In this domain, it seems that there are two major areas in which HVCs might be extremely important: as a major source of (low-metallicity) gas for galactic star formation and as a tracer of dark-matter potential wells in intergalactic space.

4.1. ...FOR GALACTIC CHEMICAL EVOLUTION?

Galaxies are in the business of converting gas into stars. To the extent that HVCs turn out to be extragalactic, the gas associated with these clouds provides the Galaxy with additional fuel. But is this a significant amount of fuel? Does it leave an observable imprint on the Galaxy?

Models of chemical evolution of the Galaxy have long indicated that the Galaxy is not a "closed box". In particular, there is the long-standing "G-dwarf" problem, in which low-mass stars have a much narrower metallicity distribution than would be expected if they were forming out of an ISM of constantly increasing metallicity. There are a wide range of chemical-evolution models with different time-dependent inflow rates that could explain this fact (c.f. Pagel 1997). For these models, a reasonable declining inflow rate with a current mass influx of approximately $\dot{M}$=1 $M_\odot\,yr^{-1}$ of gas with ~0.1 $Z_\odot$ would be sufficient to explain the stellar abundance distributions. Note that even this modest mass-flux rate, maintained for several Gyrs, could provide the same amount of mass as the Galaxy's current H I disk, $M_{HI,Galaxy}$=3.5 to 7.0×10^9 $M_\odot$ (Dame 1993). This approximate value can be considered the dividing line between "important" and "unimportant".

What sources of gas mass reach this level? To contribute a total mass of about 5×10^9 $M_\odot$ requires the equivalent of either an SMC every gigayear, or five smaller irregulars like Leo A or Sextans A every gigayear. The only gas-rich dwarf galaxies currently in sight, the LMC and SMC, have an estimated orbital decay time of ~5 Gyr due to dynamical friction (Binney & Merrifield 2000), and the next closest gas-rich dwarf (NGC 6822 with M(H I)~10^8 $M_\odot$) is 490 kpc distant. If dwarf galaxies have supplied the desired mass-flux rate in the past, the Galaxy is about to go through a long dry spell. HVCs might provide a more steady flux of mass, provided there are enough of them with high-enough mass. The current mass census of HVC complexes is provided in Wakker (2001). The Magellanic Stream dominates, with a mass of $M_{HI,MS}$~2×10^8 $M_\odot$. The orbital decay time of this gas has not been reliably calculated, since it depends sensitively on drag forces (which requires knowing the density and velocity of the halo gas), but it is probably safe to say that it is less than the 5 Gyr for the Magellanic system as a whole. Complex C has an estimated mass of M_C~5×10^6 (D/10 kpc) $M_\odot$ and a metallicity of Z~0.14 $Z_\odot$ (see Fox et al. (2004), who scaled the previous determinations to the same solar reference). These clouds, taken together with the other large complexes, make it appear that for the next gigayear or so HVCs could possibly supply the desired level of mass flux.

But are HVCs an inexhaustible supply? The answer to this will probably come from understanding the nature of the compact high-velocity clouds. Although there is no direct evidence ruling out the hypothesis that these small-angular-size clouds lie at distances as large as ~0.5 Mpc (Blitz et al. 1999; Braun & Burton 1999), models of their internal structure (Sternberg et al. 2002; Maloney & Putman 2003) and their sky distribution (de Heij et al. 2002) suggest that they are more likely distributed within ~200 kpc of the Galaxy, yielding a total mass of $M_{\rm CHVC}(\rm H\,I) \sim 3\times10^8$ $M_\odot$, approximately the mass of a single dwarf galaxy. Such calculations could be serious underestimates, however, if the ionized mass of such clouds is as large as some models predict. They also neglect entirely the possibility of fully-ionized high-velocity clouds. A survey for such clouds could really turn up a large amount of additional mass.

4.2. ...FOR UNDERSTANDING DARK MATTER?

Perhaps one of the most influential ideas concerning high-velocity clouds to arise in the past several years is the possibility that at least some HVCs owe their existence to the presence of dark-matter "halos" (Blitz et al. 1999). These halos arise as part of hierarchical structure formation (White & Rees 1978). Different models of the mass distribution of an individual dark-matter halo are given by Navarro et al. (1995) and Burkert (1995), and expectations for the spatial and mass distribution of such halos in the Local Group are given by Klypin et al. (1999) and Moore et al. (1999). These models predict ~2000 dark-matter halos with dark-matter masses in the range of $M_{\rm halo}(\rm DM)=0.5$ to 10×10^9 $M_\odot$ (Putman & Moore 2002). If such halos contain a small percentage of this mass in the form of neutral H I, they would be easily detectable. The stakes for this question are high. If HVCs are a manifestation of dark-matter halos, they provide the *only* observational evidence concerning the physical nature of such halos.

This realization has sparked several attempts to develop models of the structures of individual clouds (Sternberg et al. 2002; Maloney & Putman 2003; Gnat & Sternberg 2004), of cloud populations (de Heij et al. 2002; Putman & Moore 2002), and of cloud dynamics (Quilis & Moore 2001), all of which involve dark matter. These models have then been compared with observations of HVCs (principally the CHVCs), with instructive results. Models in which clouds are beyond about ~200 kpc tend to have low bounding pressure and low central densities, requiring a large column density of ionized gas to shield the neutral cores from the extragalactic radiation field. Thus, the more distant the cloud, the larger its ionization fraction. However, no model has successfully produced the observed Hα fluxes or the O VI column densities (Putman et al. 2003; Sembach et al. 2003), indicating that the models are incomplete, perhaps seriously so.

Will we ever know for sure that HVCs contain dark matter? Despite some claims of evidence for dark matter in individual clouds (Robishaw et al. 2002), models of individual clouds both with and without dark matter have a sufficiently large number of free parameters (e.g. nature of the bounding medium, magnetic fields, distance uncertainties), that one can not rule out either option. Studies of the population as a whole might be more fruitful, since ΛCDM models seem to have fairly inflexible predictions for the mass and spatial distribution of dark-matter halos. Ultimately, the decision may rest on plausibility. Which is easier: to create and sustain the observed HVCs with dark-matter halos or without?

Additional evidence that could be incorporated is deep H I searches in other galaxy groups for clouds analogous to HVCs (Lo & Sargent 1979; Kraan-Korteweg et al. 1999; Zwaan & Briggs 2000; de Blok et al. 2002; Pisano et al. 2004). A recent deep search around M 31 suggests this galaxy has a halo of circumgalactic HVCs, not unlike the expectation from the ΛCDM models (Thilker et al. 2004). Sensitive searches in other groups could help to confirm or refute the expectations for the distribution of these clouds.

5. Concluding thoughts

Progress on HVCs has been extremely rapid in the last several years. New wavebands have opened up, yielding new insights. New issues have been raised, and new models have been developed to try to synthesize our understanding of these enigmatic objects. But all of these recent discoveries have served to highlight just how incomplete our knowledge of HVCs really is. My hope is that, when it comes time to write a similar article ten years from now, we will have the distances to all of the HVC complexes (together with their metallicities), reliable detection and characterization of the gas surrounding high-velocity clouds, and a more mature understanding of the role of dark matter in HVCs. There will always be unsolved mysteries associated with HVCs, but if we could make headway on these three fronts, we will be making real progress.

Acknowledgements

The author would like to acknowledge influential conversations with Don Cox, Bart Wakker, Ellen Zweibel, and Mary Putman, as well as the financial support provided by NASA ATP grant 5-12128.

References

Allende-Prieto, C, Barklem, P.S., Asplund, M., Ruiz Cobo, B. 2001, ApJ, 558, 830
Almy, R.C., McCammon, D., Digel, S.W., Bronfman, L., May, J. 2000, ApJ, 545, 290
Benjamin, R.A. 1999, in ASP Conf. Ser. 166, Stromlo Workshop on High-Velocity Clouds, eds. B.K. Gibson, M.E. Putman (San Francisco: ASP), 147
Benjamin, R.A., Danly, L. 1997, ApJ, 481, 764
Binney, J., Merrifield, M. 2000, Galactic Astronomy (Princeton: Princeton University

Press)
Bland-Hawthorn, J., Maloney, P.R. 1999, ApJ, 510, L33; erratum in ApJ, 550, 231L
Bland-Hawthorn, J., Maloney, P.R. 2002, in ASP Conf. Ser. 254, Extragalactic Gas at Low Redshift, eds. J.S. Mulchaey, J. Stocke (San Francisco: ASP), 267
Bland-Hawthorn, J., Cohen, M. 2003, ApJ, 582, 246
Blitz, L., Spergel, D., Teuben, P., Hartmann, D., Burton, W.B. 1999, ApJ, 514, 818
Boulares, A., Cox,, D.P. 1990, ApJ, 365, 544
Braun, R., Burton, W.B. 1999, A&A, 351, 437
Braun, R., Burton, W.B. 2000, A&A, 354, 853
Bregman, J.N. 1978, ApJ, 224, 768
Bregman, J.N. 1980, ApJ, 236, 577
Bregman, J.N., Pildis, R.A. 1994, ApJ, 420, 570
Breitschwerdt, D., Schmutzler, T. 1999, A&A, 347, 650
Breitschwerdt, D., McKenzie, J.F., Völk, H.J. 1991, A&A, 245, 79
Brüns, C., Kerp, J., Kalberla, P.M.W., Mebold, U. 2000, A&A, 357, 120
Burkert, A. 1995, ApJ, 447, L25
Burrows, D.N., Mendenhall, J.A. 1991, Nature, 351, 629
Chevalier, R.A., Fransson, C. 1984, ApJ, 279, 43
Collins, J.A., Shull, J.M., Giroux, M.L. 2003, ApJ, 585, 336
Cowie, L.L., McKee, C.F. 1976, ApJ, 209, 105
Dame, T. 1993, in AIP Conf. Ser. 278, Back to the Galaxy, eds. S.S. Holt, F. Verter, 267
Davis, J., Sabatini, S., Davies, L., Linder, S., Roberts, S., Smith, R., Evans, Rh. 2002, MNRAS, 336, 155
de Blok, W.J.G., Zwaan, M.A., Dijkstra, M., Briggs, F.H., Freeman, K.C. 2002, A&A, 382, 43
de Heij, V., Braun, R., Burton, W.B. 2002, A&A, 392, 417
Dehnen, W., Binney, J. 1998, MNRAS, 294, 429
Dixon, W.V.D., Sallmen, S., Hurwitz, M., Lieu, R. 2001, ApJ, 552, L69
Espresate, J., Cantó, J., Franco, J. 2002, ApJ, 575, 194
Fang, T., Sembach, K.R., Canizares, C.R. 2003, ApJ, 586, 49
Ferrara, A., Field, G.B. 1994, ApJ, 423, 665
Fox, A.J., Savage, B.D., Wakker, B.P., Richter, P., Sembach, K.R., Tripp, T.M. 2004, ApJ, 602, 738
Futamoto, K., Mitsuda, K., Takei, Y., Fujimoto, R., Yamasaki, N. 2004, ApJ, 605, 793
Garmire, G.P., Nousek, J.A., Apparao, K.M.V., Burrows, D.N., Fink, R.L., Kraft, R.P. 1992, ApJ, 399, 693
Giovanelli, R., Haynes, M.P. 1976, MNRAS, 177, 525
Gnat O., Sternberg, A. 2004, ApJ, in press
Gomez, G.C., Benjamin, R.A., Cox, D.P. 2001, AJ, 122, 908
Gregori, G., Miniati, F., Ryu, D., Jones, T.W. 1999, ApJ, 527, 113
Gregori, G., Miniati, F., Ryu, D., Jones, T.W. 2000, ApJ, 543, 775
Hedrick, D., Cox, D.P. 1977, ApJ, 215, 208
Kazès, I., Troland, T.H., Crutcher, R.M. 1991, A&A, 245, L17
Klypin, A., Kravtsov, A.V., Valenzuela, O., Prada, F. 1999, ApJ, 522, 82
Konz, C., Brüns, C., Birk, G.T. 2002, A&A, 391, 713
Kraan-Korteweg, R.C., van Driel, W., Briggs, F., Bingelli, B., Mostefaoui, T.I. 1999, A&AS, 135, 255
Kuntz, K.D., Snowden, S.L. 2000, ApJ, 543, 195
Kuulkers, E., den Hartog, P.R., in't Zand, J.J.M., Verbunt, F.W.M., Harris, W.E., Cocchi, M. 2003, A&A, 399, 663
Lo, K.Y. & Sargent, W.L.W. 1979, ApJ, 227, 756
Lockman, F.J. 2003, ApJ, 591, L33
Maloney, P.R., Bland-Hawthorn, J. 1999, ApJ, 522, L81
Maloney, P.R., Putman, M.E. 2003, ApJ, 589, 270
Martin, C., Bowyer, S. 1990, ApJ, 350, 242

Mateo, M.L. 1998, ARA&A, 36,435
Münch, G., Zirin, H. 1961, ApJ, 133, 11
Murali, C. 2000, ApJ, 529, 81
McCammon, D., Sanders W.T. 1990, ARA&A, 28, 657
McCammon, D., Almy, R., Apodaca, E., et 17 al. 2002, ApJ, 576, 188
Moore, G., Ghigna, S., Governato, F., Lake, G., Quinn, T., Stadel, J., Tozzi, P. 1999, ApJ, 524, L19
Navarro, J.F., Frenk, C.S., White, S.D.M. 1995, MNRAS, 275, 720
Nicastro, F., Zezas, A., Drake, J., Elvis, M., Fiore, F., Fruscione, A., Marengo, M., Mathur, S., Bianchi, S. 2001, ApJ, 573, 157
Nicastro, F., Zezas, A., Elvis, M., Mathur, S., Fiore, F., Cecchi-Pestellini, C., Burke, D., Drake, J., Casella, P. 2003, Nature, 421, 719
Oort, J.H. 1954, BAN, 12, 177
Oort, J.H. 1970, A&A, 7, 381
Otte, B., Murphy, E.M., Howk, J.C., Wang, Q.D., Oegerle, W.R., Sembach, K.R. 2003, ApJ, 591, 821
Pagel, B.E.J. 1997, Nucleosynthesis and Chemical Evolution of Galaxies (Cambridge: Cambridge University Press), p237
Parker, E.N. 1992, ApJ, 401, 137
Parker, E.N. 1993, ApJ, 414, 389
Pietz, J., Kerp, J., Kalberla, P.M.W., Mebold, U., Burton, W.B., Hartmann, D. 1996, A&A, 308, L37
Pietz, J., Kerp, J., Kalberla, P.M.W., Burton, W.B., Hartmann, D., Mebold, U. 1998, A&A, 332, 55
Pisano, D.J., Wakker, B.P., Wilcots, E.M., Fabian, D. 2004, AJ, 127, 199
Putman, M.E., de Heij, V., Staveley-Smith, L., et 29 al. 2002, AJ, 123, 873
Putman, M.E., Bland-Hawthorn, J., Veilleux, S., Gibson, B.K., Freeman, K.C., Maloney, P.R. 2003, ApJ, 597, 948
Putman, M.E., Thom, C., Gibson, B.K., Staveley-Smith, L. 2004, ApJ, 603, L77
Quilis, V., Moore, B. 2001, ApJ, 555, L95
Raymond, J.C. 1992, ApJ, 384, 502
Reynolds, R.J. 1989, ApJ, 339, L29
Richter, P., de Boer, K.S., Widmann, H., Kappelmann, N., Gringel, W., Grewing, M., Barnstedt, J. 1999, Nature, 402, 386
Richter, P., Wakker, B.P., Savage, B.D., Sembach, K.R. 2003, ApJ, 586, 230
Robishaw, T., Simon, J.D., Blitz, L. 2002, ApJ, 580, L129
Sanders, W.T., Edgar, R.J., Kraushaar, W.L., McCammon, D., Morgenthaler, J.P. 2001, ApJ, 554, 694
Schwarz, U.J., Oort, J.H. 1981, A&A, 101, 305
Sembach, K.R. 2004, in The Local Group as an Astrophysical Laboratory: Proceedings of May 2003 STScI Symposium, in press (astro-ph/0311089)
Sembach, K.R., Savage, B.D., Lu, L., Murphy, E.M. 1999, ApJ, 515, 108
Sembach, K.R., Savage, B.D., Shull, J.M., et 13 al. 2000, ApJ, 538, L31
Sembach, K.R., Wakker, B.P., Savage, B.D., Richter, P., Meade, M., Shull, J.M., Jenkins, E.B., Sonneborn, G., Moos, H.W. 2003, ApJS, 146, 165
Shapiro, P.R., Field, G.B. 1976, ApJ, 205, 762
Shelton, R. 1999, ApJ, 521, 217
Simon, J.D., Blitz, L. 2002, ApJ, 574, 726
Snowden, S.L., Hasinger, G., Jahoda, K., Lockman, F.J., McCammon, D., Sanders, W.T. 1994, ApJ, 430, 601
Spitzer, L. 1956, ApJ, 124, 20
Stanimirovic, S., Dickey, J.M., Krco, M., Brooks, A.M. 2002, ApJ, 576, 773
Sternberg, A., McKee, C.F., Wolfire, M.G. 2002, ApJS, 143, 419
Thilker, D.A., Braun, R., Walterbos, R.A.M., Corbelli, E., Lockman, F.J., Murphy, E., Maddalena, R. 2004, ApJ, 601, L39

Tufte, S.L., Reynolds, R.J., Haffner, L.M. 1998, ApJ, 504, 77
Tufte, S.L., Wilson, J.D., Madsen, G.J., Haffner, L.M., Reynolds, R.J. 2002, ApJ, 572, L153
Wakker, B.P. 2001, ApJS, 136, 463
Wakker, B.P. 2004, ApJ, submitted
Wakker, B.P., Boulanger, F. 1986, A&A, 170, 84
Wakker, B.P., Schwarz, U.J. 1991, A&A, 250, 484
Wakker, B.P., Howk, J.C., Savage, B.D., van Woerden, H., Tufte, S.L., Schwarz, U.J., Benjamin, R., Reynolds, R.J., Peletier, R.F., Kalberla, P.M.W. 1999, Nature, 402, 388
Wakker, B.P., Savage, B.D., Sembach, K.R., et 18 al. 2003, ApJS, 146, 1
Wakker, B.P., van Woerden, H. 1997, ARA&A, 35, 217
Walters, M.A., Cox, D.P. 2001, ApJ, 549, 353
Weiner, B.J., Williams, T.B. 1996, AJ, 111, 1156
White, S.D.M., Rees, M.J. 1978, MNRAS, 183, 341
Willman, B., Dalcanton, J., Ivezic, Z., Schneider, D.P., York, D.G. 2002, AJ, 124, 2600
Wolfire, M.G., Hollenbach, D., McKee, C.F., Bakes, E.L.O. 1995, ApJ, 443, 152
Zimmer, F., Lesch, H., Birk, G.T. 1997, A&A, 320, 746
Zirakashvili, V.N., Breitschwerdt, D., Ptuskin, V.S., Völk, H.J. 1996, A&A, 311, 113
Zwaan, M.A., Briggs, R.H. 2000, ApJ, 530, L61
Zweibel, E. 2004, J. Fluid Mech., submitted

INDEX

Astrophysics and Space Science Library

Volume 302: ***Stellar Collapse,*** edited by Chris L. Fryer
Hardbound, ISBN 1-4020-1992-0, April 2004

Volume 301: ***Multiwavelength Cosmology,*** edited by Manolis Plionis
Hardbound, ISBN 1-4020-1971-8, March 2004

Volume 300: ***Scientific Detectors for Astronomy,*** edited by Paola Amico, James W. Beletic, Jenna E. Beletic
Hardbound, ISBN 1-4020-1788-X, February 2004

Volume 299: ***Open Issues in Local Star Fomation,*** edited by Jacques Lépine, Jane Gregorio-Hetem
Hardbound, ISBN 1-4020-1755-3, December 2003

Volume 298: ***Stellar Astrophysics - A Tribute to Helmut A. Abt,*** edited by K.S. Cheng, Kam Ching Leung, T.P. Li
Hardbound, ISBN 1-4020-1683-2, November 2003

Volume 297: ***Radiation Hazard in Space,*** by Leonty I. Miroshnichenko
Hardbound, ISBN 1-4020-1538-0, September 2003

Volume 296: ***Organizations and Strategies in Astronomy, volume 4,*** edited by André Heck
Hardbound, ISBN 1-4020-1526-7, October 2003

Volume 295: ***Integrable Problems of Celestial Mechanics in Spaces of Constant Curvature,*** by T.G. Vozmischeva
Hardbound, ISBN 1-4020-1521-6, October 2003

Volume 294: ***An Introduction to Plasma Astrophysics and Magnetohydrodynamics,*** by Marcel Goossens
Hardbound, ISBN 1-4020-1429-5, August 2003
Paperback, ISBN 1-4020-1433-3, August 2003

Volume 293: ***Physics of the Solar System,*** by Bruno Bertotti, Paolo Farinella, David Vokrouhlický
Hardbound, ISBN 1-4020-1428-7, August 2003
Paperback, ISBN 1-4020-1509-7, August 2003

Volume 292: ***Whatever Shines Should Be Observed,*** by Susan M.P. McKenna-Lawlor
Hardbound, ISBN 1-4020-1424-4, September 2003

Volume 291: ***Dynamical Systems and Cosmology,*** by Alan Coley
Hardbound, ISBN 1-4020-1403-1, November 2003

Volume 290: ***Astronomy Communication,*** edited by André Heck, Claus Madsen
Hardbound, ISBN 1-4020-1345-0, July 2003

Volume 287/8/9: ***The Future of Small Telescopes in the New Millennium,*** edited by Terry D. Oswalt
Hardbound Set only of 3 volumes, ISBN 1-4020-0951-8, July 2003

Volume 286: ***Searching the Heavens and the Earth: The History of Jesuit Observatories,*** by Agustín Udías
Hardbound, ISBN 1-4020-1189-X, October 2003

Volume 285: ***Information Handling in Astronomy - Historical Vistas,*** edited by André Heck
Hardbound, ISBN 1-4020-1178-4, March 2003

Volume 284: ***Light Pollution: The Global View,*** edited by Hugo E. Schwarz
Hardbound, ISBN 1-4020-1174-1, April 2003

Volume 283: ***Mass-Losing Pulsating Stars and Their Circumstellar Matter,*** edited by Y. Nakada, M. Honma, M. Seki
Hardbound, ISBN 1-4020-1162-8, March 2003

Volume 282: ***Radio Recombination Lines,*** by M.A. Gordon, R.L. Sorochenko
Hardbound, ISBN 1-4020-1016-8, November 2002

Volume 281: ***The IGM/Galaxy Connection,*** edited by Jessica L. Rosenberg, Mary E. Putman
Hardbound, ISBN 1-4020-1289-6, April 2003

Volume 280: ***Organizations and Strategies in Astronomy III,*** edited by André Heck
Hardbound, ISBN 1-4020-0812-0, September 2002

Volume 279: ***Plasma Astrophysics , Second Edition,*** by Arnold O. Benz
Hardbound, ISBN 1-4020-0695-0, July 2002

Volume 278: ***Exploring the Secrets of the Aurora,*** by Syun-Ichi Akasofu
Hardbound, ISBN 1-4020-0685-3, August 2002

Volume 277: ***The Sun and Space Weather,*** by Arnold Hanslmeier
Hardbound, ISBN 1-4020-0684-5, July 2002

Volume 276: ***Modern Theoretical and Observational Cosmology,*** edited by Manolis Plionis, Spiros Cotsakis
Hardbound, ISBN 1-4020-0808-2, September 2002

Volume 275: ***History of Oriental Astronomy,*** edited by S.M. Razaullah Ansari
Hardbound, ISBN 1-4020-0657-8, December 2002

Volume 274: ***New Quests in Stellar Astrophysics: The Link Between Stars and Cosmology,*** edited by Miguel Chávez, Alessandro Bressan, Alberto Buzzoni,Divakara Mayya
Hardbound, ISBN 1-4020-0644-6, June 2002

Volume 273: ***Lunar Gravimetry,*** by Rune Floberghagen
Hardbound, ISBN 1-4020-0544-X, May 2002

Volume 272:***Merging Processes in Galaxy Clusters,*** edited by L. Feretti, I.M. Gioia, G. Giovannini
Hardbound, ISBN 1-4020-0531-8, May 2002

Volume 271: ***Astronomy-inspired Atomic and Molecular Physics,*** by A.R.P. Rau
Hardbound, ISBN 1-4020-0467-2, March 2002

Volume 270: ***Dayside and Polar Cap Aurora,*** by Per Even Sandholt, Herbert C. Carlson, Alv Egeland
Hardbound, ISBN 1-4020-0447-8, July 2002

Volume 269: ***Mechanics of Turbulence of Multicomponent Gases,*** by Mikhail Ya. Marov, Aleksander V. Kolesnichenko
Hardbound, ISBN 1-4020-0103-7, December 2001

Volume 268: ***Multielement System Design in Astronomy and Radio Science,*** by Lazarus E. Kopilovich, Leonid G. Sodin
Hardbound, ISBN 1-4020-0069-3, November 2001

Volume 267: ***The Nature of Unidentified Galactic High-Energy Gamma-Ray Sources,*** edited by Alberto Carramiñana, Olaf Reimer, David J. Thompson
Hardbound, ISBN 1-4020-0010-3, October 2001

Volume 266: ***Organizations and Strategies in Astronomy II,*** edited by André Heck
Hardbound, ISBN 0-7923-7172-0, October 2001

Volume 265: ***Post-AGB Objects as a Phase of Stellar Evolution,*** edited by R. Szczerba, S.K. Górny
Hardbound, ISBN 0-7923-7145-3, July 2001

Volume 264: ***The Influence of Binaries on Stellar Population Studies,*** edited by Dany Vanbeveren
Hardbound, ISBN 0-7923-7104-6, July 2001

Volume 262: ***Whistler Phenomena - Short Impulse Propagation,*** by Csaba Ferencz, Orsolya E. Ferencz, Dániel Hamar, János Lichtenberger
Hardbound, ISBN 0-7923-6995-5, June 2001

Volume 261: ***Collisional Processes in the Solar System,*** edited by Mikhail Ya. Marov, Hans Rickman
Hardbound, ISBN 0-7923-6946-7, May 2001

Volume 260: ***Solar Cosmic Rays,*** by Leonty I. Miroshnichenko
Hardbound, ISBN 0-7923-6928-9, May 2001

Volume 259: ***The Dynamic Sun,*** edited by Arnold Hanslmeier, Mauro Messerotti, Astrid Veronig
Hardbound, ISBN 0-7923-6915-7, May 2001

Volume 258: ***Electrohydrodynamics in Dusty and Dirty Plasmas- Gravito-Electrodynamics and EHD,*** by Hiroshi Kikuchi
Hardbound, ISBN 0-7923-6822-3, June 2001

Volume 257: ***Stellar Pulsation - Nonlinear Studies,*** edited by Mine Takeuti, Dimitar D. Sasselov
Hardbound, ISBN 0-7923-6818-5, March 2001

Volume 256: ***Organizations and Strategies in Astronomy,*** edited by André Heck
Hardbound, ISBN 0-7923-6671-9, November 2000

Volume 255: ***The Evolution of the Milky Way- Stars versus Clusters,*** edited by Francesca Matteucci, Franco Giovannelli
Hardbound, ISBN 0-7923-6679-4, January 2001

Volume 254: ***Stellar Astrophysics***, edited by K.S. Cheng, Hoi Fung Chau, Kwing Lam Chan, Kam Ching Leung
Hardbound, ISBN 0-7923-6659-X, November 2000

Volume 253: ***The Chemical Evolution of the Galaxy***, by Francesca Matteucci
Paperback, ISBN 1-4020-1652-2, October 2003
Hardbound, ISBN 0-7923-6552-6, June 2001

Volume 252: ***Optical Detectors for Astronomy II***, edited by Paola Amico, James W. Beletic
Hardbound, ISBN 0-7923-6536-4, December 2000

Volume 251: ***Cosmic Plasma Physics***, by Boris V. Somov
Hardbound, ISBN 0-7923-6512-7, September 2000

Volume 250: ***Information Handling in Astronomy***, edited by André Heck
Hardbound, ISBN 0-7923-6494-5, October 2000

Volume 249: ***The Neutral Upper Atmosphere***, by S.N. Ghosh
Hardbound, ISBN 0-7923-6434-1, July 2002

Volume 247: ***Large Scale Structure Formation***, edited by Reza Mansouri, Robert Brandenberger
Hardbound, ISBN 0-7923-6411-2, August 2000

Volume 246: ***The Legacy of J.C. Kapteyn***, edited by Piet C. van der Kruit, Klaas van Berkel
Paperback, ISBN 1-4020-0374-9, November 2001
Hardbound, ISBN 0-7923-6393-0, August 2000

Volume 245: ***Waves in Dusty Space Plasmas,*** by Frank Verheest
Paperback, ISBN 1-4020-0373-0, November 2001
Hardbound, ISBN 0-7923-6232-2, April 2000

Volume 244: ***The Universe***, edited by Naresh Dadhich, Ajit Kembhavi
Hardbound, ISBN 0-7923-6210-1, August 2000

Volume 243: ***Solar Polarization***, edited by K.N. Nagendra, Jan Olof Stenflo
Hardbound, ISBN 0-7923-5814-7, July 1999

Volume 242: ***Cosmic Perspectives in Space Physics***, by Sukumar Biswas
Hardbound, ISBN 0-7923-5813-9, June 2000

Volume 241: ***Millimeter-Wave Astronomy: Molecular Chemistry & Physics in Space,*** edited by W.F. Wall, Alberto Carramiñana, Luis Carrasco, P.F. Goldsmith
Hardbound, ISBN 0-7923-5581-4, May 1999

Volume 240: ***Numerical Astrophysics,*** edited by Shoken M. Miyama, Kohji Tomisaka,Tomoyuki Hanawa
Hardbound, ISBN 0-7923-5566-0, March 1999

Volume 239: ***Motions in the Solar Atmosphere,*** edited by Arnold Hanslmeier, Mauro Messerotti
Hardbound, ISBN 0-7923-5507-5, February 1999

Volume 238: ***Substorms-4,*** edited by S. Kokubun, Y. Kamide
Hardbound, ISBN 0-7923-5465-6, March 1999

Volume 237: ***Post-Hipparcos Cosmic Candles,*** edited by André Heck, Filippina Caputo
Hardbound, ISBN 0-7923-5348-X, December 1998

Volume 236: ***Laboratory Astrophysics and Space Research,*** edited by P. Ehrenfreund, C. Krafft, H. Kochan, Valerio Pirronello
Hardbound, ISBN 0-7923-5338-2, December 1998

Missing volume numbers have not yet been published.
For further information about this book series we refer you to the following web site:
http://www.wkap.nl/prod/s/ASSL

To contact the Publishing Editor for new book proposals:
Dr. Harry (J.J.) Blom: harry.blom@wkap.nl

MIX
Papier aus verantwortungsvollen Quellen
Paper from responsible sources
FSC® C105338
FSC
www.fsc.org

If you have any concerns about our products,
you can contact us on
ProductSafety@springernature.com

In case Publisher is established outside the EU,
the EU authorized representative is:
Springer Nature Customer Service Center GmbH
Europaplatz 3, 69115 Heidelberg, Germany

Printed by Libri Plureos GmbH
in Hamburg, Germany